Software Engineering for Robotics

Ana Cavalcanti • Brijesh Dongol • Rob Hierons •
Jon Timmis • Jim Woodcock

Editors

Software Engineering for Robotics

 Springer

Editors
Ana Cavalcanti (iD)
Department of Computer Science
University of York
York, UK

Brijesh Dongol (iD)
University of Surrey
Guildford, UK

Rob Hierons (iD)
University of Sheffield
Sheffield, UK

Jon Timmis (iD)
University of Sunderland
Sunderland, UK

Jim Woodcock (iD)
Department of Computer Science
University of York
York, UK

ISBN 978-3-030-66496-1   ISBN 978-3-030-66494-7   (eBook)
https://doi.org/10.1007/978-3-030-66494-7

This Springer imprint is published by the registered company Springer Nature Switzerland AG.
The registered company address is: Gewerbestrasse 11, 6330 Cham, Switzerland

Foreword

As a Professor of Robotics and Autonomous Systems, I have been carrying out research into advanced robotics for over 30 years, half of that as Deputy Director of the Bristol Robotics Laboratory, which is one of the most comprehensive robot research laboratories in the UK. During that period, I have been lucky enough to have the pleasure of getting to know many of this book's editors and chapter co-authors, and, indeed, I have worked closely with some of them. During that period also, all of those involved in writing or coordinating the book's contents have become eminent figures in their own right, in recognition of their individual achievements in advancing the field of Software Engineering for Robotics. As a result, I am pleased and honoured that I have been asked to write this foreword.

Robotics and the sectors of software engineering associated with it are critically linked. There is a growing public perception that robots will soon emerge from the confines of "factory shop-floor"-style repetitive motion sequences to work with us and amongst us, thus improving both our domestic and professional lives. However, for this aspiration to succeed, it is essential that critical aspects of software engineering and advanced robotics develop together in close synchrony. On one hand, robotics provides a physically observable manifestation that enables software systems to interact with human beings and other artefacts in an embedded and instantiated manner. On the other hand, meeting the challenges presented by the many requirements arising from the typically complex nature of those interactions is fully reliant on the success of the software engineering design efforts that are described, demonstrated and proposed in this book.

Taken together, the chapter authors of this book have achieved a tour de force that ranges freely across the wide scope of the field of Software Engineering for Robotics and the exceptional advances currently being made in it. This rich content has been linked together deftly by its editors.

If a reader seeks to discover all that is important in this field, and understand the pivotal issues that it faces, then this book should be one of that reader's first ports of call. By absorbing its content, the reader will come to understand many of the most fundamental, and often interlinked, challenges that must be overcome in order to bring the promised "robot revolution" to successful fruition, including the society-shifting benefits that could be realised by it, as well as the crucial dangers and pitfalls that must be circumvented in doing so.

Professor of Robotics and Autonomous Systems   Tony (Anthony) Pipe
Bristol Robotics Laboratory
University of the West of England
Bristol, UK

Preface

The origin of this book is a 2-day event, entitled RoboSoft, that took place in November 2019 in London. Organised with the generous support of the Royal Academy of Engineering and the University of York, UK, RoboSoft brought together more than 100 scientists, engineers and practitioners from all over the world, representing 70 institutions to discuss the state of the art and practice on Software Engineering for Robotics.

The lively discussions that followed each of the talks and the panel session provide evidence for the importance of this area and for the exciting challenges that lie ahead. The chapters of this book are co-authored by a RoboSoft speaker, and provide a snapshot of the diversity of the work that is being carried out in this area, of the research and insights available and of future possibilities. The list of these speakers, with a short biography, follows.

The topics covered range from programming languages and environments, via approaches for design and verification, to issues of ethics, accident investigation and regulation. In terms of techniques, there are results on product lines, mission specification, component-based development, simulation, testing and proof. Applications range from manufacturing to service robots, to autonomous vehicles, and even to robots that evolve in the real world.

The views put forward by the various authors are also diverse. Some cover approaches that are successful in other areas and can be translated for use in robotics. Some authors state that there is nothing special about Software Engineering for Robotics, while the starting point for others are the peculiarities of this domain. Some take the view that there is no time for toy examples; others use such examples to drive foundational research in robotics. Some focus on the decision aspects of an autonomous robot, with the view that dealing with functional aspects is a standard task. Others understand that models for decision components are formal in nature, so the challenge is verification of functional components. So, the reader, has a breadth of perspectives and arguments, and the opportunity to make their own mind up.

The structure of the book follows that adopted in organising the RoboSoft presentations, to cover the topics of algorithms, modelling, testing, ethics and formal verification. A final chapter summarises the discussions on ethics and regulation, including the material based on the panel at the end of RoboSoft.

The intended readership of this book includes researchers and practitioners with all levels of experience interested in working in the area of robotics, and software engineering more generally. The chapters are all written to be accessible to a wide readership and are self-contained, including explanations of the core concepts that they cover to aid the reader. The chapters also provide links with the material in other chapters to give the book a cohesive and comprehensive account of the state of the art and practice. Finally, the chapters include a discussion of directions for further work.

We are grateful for the contribution of all speakers and attendees of RoboSoft and of all co-authors of the RoboSoft speakers. We also would like to thank the administrators that worked on behalf of the Royal Academy of Engineering and of the University of York to make RoboSoft and, therefore, this book possible. The enthusiasm of all these colleagues has given us all extra encouragement to produce this book as a snapshot of Software Engineering for Robotics as an emergent field of study.

We look forward to working together, as a community, to contribute to this fascinating multidisciplinary area. It will no doubt have a significant and positive impact on the quality of life of our society. We envisage advancements to economy, manufacturing, transportation, health, entertainment, safety and much more. Our hope is that, in 5 or 10 years, we will look back and decide that it is definitely time for a second edition.

York, UK   Ana Cavalcanti
Guildford, UK   Brijesh Dongol
Sheffield, UK   Rob Hierons
Sunderland, UK   Jon Timmis
York, UK   Jim Woodcock
October 2020

Contents

Editors and Contributors

We list here the speakers and organisers of the RoboSoft event. We acknowledge and appreciate also the contribution of all co-authors of the chapters.

Ron Bell (*Engineering Safety Consultants*) is one of the most influential figureheads in the field, having been awarded an OBE. In 1998, he was appointed as one of the five UK members of the binational Channel Tunnel Safety Authority, which is a post he held for 13 years. He chairs one of the two IEC working groups responsible for IEC 61508 (the international standard dealing with safety-critical systems), a post that he has held since 1987. In 2005, he received the IEC 1906 Award for his work on functional safety and IEC 61508. He held a 3-year appointment (2015–2018) as a Royal Academy of Engineering Visiting Professor at Liverpool John Moores University.

Davide Brugali (*University of Bergamo*) graduated in Electronic Engineering at Politecnico di Milano in 1994; he received a PhD in Computer Science from Politecnico di Torino in 1998. Since 2011, he is Associate Professor at the University of Bergamo. He has been Visiting Researcher at the CMU Robotics Institute in 1997 and Visiting Professor at NASA Jet Propulsion Laboratory in 2006. From 2000 to 2020, he has been co-chair of the IEEE RAS Technical Committee on "Software Engineering for Robotics and Automation". He was editor and co-author of a Springer STAR book on *Software Engineering for Experimental Robotic* (2006). He is the main author of the book *Software Development: Case Studies in Java* published by Addison-Wesley in 2005.

Ana Cavalcanti (*University of York*) is Professor of Software Verification and Royal Academy of Engineering Chair in Emerging Technologies working on Software Engineering for Robotics: modelling, validation, simulation and testing. She currently leads the RoboStar research group at the University of York. She held a Royal Society Wolfson Research Merit Award and a Royal Society Industry Fellowship to work with QinetiQ in avionics. She has chaired the programme committee of various well-established international conferences, is on the editorial board of four international journals and is chair of the board of the Formal Methods Europe Association. She is, and has been, Principal Investigator on several large research grants. Her current research is on the theory and practice of verification and testing for robotics.

Brijesh Dongol (*University of Surrey*) is a Senior Lecturer. His research is on the formal techniques and verification methods for concurrent and real-time systems. This includes concurrent objects, transactional memory and associated correctness conditions; weak memory models; algebraic techniques; and hybrid systems. He completed his PhD in 2009 from the University of Queensland. He was a postdoctoral researcher at the University of Sheffield and lecturer at Brunel University London before moving to Surrey. He leads several projects funded by the EPSRC, research institutions and industrial partners. He is a member of the Formal Methods Teaching Committee.

Kerstin Eder (*University of Bristol*) is Professor of Computer Science and leads the Trustworthy Systems Laboratory at the University of Bristol, as well as the Verification and Validation for Safety in Robots research theme at the Bristol Robotics Laboratory. Her research is focused on specification, verification and analysis techniques to verify or explore a system's behaviour in terms of functional correctness, safety, performance and energy efficiency. Kerstin has gained extensive expertise in verifying complex microelectronic designs at leading semiconductor design and EDA companies. She seeks novel combinations of formal methods with state-of-the-art simulation and test-based approaches to achieve solutions that make a difference in practice. She holds a PhD in Computational Logic, an MSc in Artificial Intelligence and an MEng in Informatics. In 2007, she was awarded a Royal Academy of Engineering "Excellence in Engineering" prize.

Gusz Eiben (*Vrije Universiteit Amsterdam*) is Professor of Artificial Intelligence at the Vrije Universiteit Amsterdam, where he leads the Computational Intelligence Group, and Visiting Professor at the University of York, UK. His research lies in the fields of Artificial Intelligence, Artificial Life, and Adaptive Collective Systems. His approach to Artificial Intelligence is based on Evolutionary Computing. Over more than 30 years, he has worked on the theoretical foundations and applications in health, finance and traffic management, built a system to evolve Mondriaan- and Escher-style art and exhibited it in the Haags Gemeentemuseum, researched how artificial societies can emerge in the computer through evolution and learning, invented and tested reproduction mechanisms that use more than two parents and studied how evolutionary processes can be (self-)calibrated. Lately he has been active in Evolutionary Robotics.

Michael Fisher (*University of Manchester*) holds a Royal Academy of Engineering Chair in Emerging Technologies on the theme of "Responsible Autonomous Systems". His research particularly concerns formal verification and autonomous systems, as well as safety and ethics in autonomous robotics. He serves on the editorial boards of both the *Journal of Applied Logic* and the *Annals of Mathematics and Artificial Intelligence* and is a corner editor for the *Journal of Logic and Computation*. He is a Fellow of both the BCS and the IET, is a member of the British Standards Institute's AMT/10 "Robotics" standards committee and is a member of the IEEE's P7009 "Fail-Safe Design of Autonomous System" standards committee. He is involved in a range of EPSRC projects across robotics for hazardous environments and sensor network analysis and is co-chair of the IEEE's international Technical Committee on the Verification of Autonomous Systems.

Arnaud Gotlieb (*Simula Research Laboratory*) is Chief Research Scientist and Research Professor at Simula Research Laboratory in Norway. His research interests are on the application of Artificial Intelligence to the validation of software-intensive systems, cyber-physical systems including industrial robotics, and autonomous systems. Arnaud has co-authored more than 120 publications on Artificial Intelligence and Software Engineering and has developed several tools for testing critical software systems. He was the scientific coordinator of the French ANR-CAVERN project (2008–2011) for Inria and led the Research-Based Innovation Center Certus dedicated to Software Validation and Verification (2011–2019) at Simula. He was recently awarded with the prestigious RCN FRINATEK grant for the T-LARGO project on testing learning robots (2018–2022). He leads the industrial pilot experiments of the H2020 AI4EU Project (2019–2022).

Ibrahim Habli (*University of York*) is a Senior Lecturer. His research interests are in the design and assurance of safety-critical systems, with a particular focus on intelligent systems (for instance, autonomous and connected driving) and digital health (for instance, e-Prescribing and self-management apps). In 2015, he was awarded a Royal Academy of Engineering Industrial Fellowship through which he collaborated with the NHS on evidence-based means for assuring the safety of digital health systems. His research is empirical and industry-informed, with collaborative projects with organisations such as Rolls-Royce, NASA, Jaguar Land Rover and NHS Digital. Ibrahim is an academic lead on the Assuring Autonomy International Programme, a £12 million initiative funded by the Lloyd's Register Foundation and the University of York. He has been a member of several safety standardisation committees (e.g. DO178C, MISRA and BSI).

Rob Hierons (*University of Sheffield*) received a BA in Mathematics (Trinity College, Cambridge) and a PhD in Computer Science (Brunel University). He then joined the Department of Mathematical and Computing Sciences at Goldsmiths College, University of London, before returning to Brunel University in 2000. He was promoted to full professor in 2003 and joined the University of Sheffield in 2018. His research concerns the automated generation of efficient, systematic test suites on the basis of program code, models or specifications. He is joint Editor-in-Chief of the journal of *Software Testing, Verification, and Reliability (STVR)* and is a member of the editorial boards of *The Computer Journal and Formal Aspects of Computing.*

Félix Ingrand (*LAAS/CNRS*) is a tenured researcher at CNRS. After his PhD from the University of Grenoble (1987), he spent 4 years at SRI International (Menlo Park, CA) where he worked on procedural reasoning. He joined the Robotics and Artificial Intelligence Group at CNRS/LAAS in 1991. His work deals with architecture for autonomous systems with an emphasis on the decisional aspect. He has been invited to NASA Ames Research Center to work on various robotics platforms (K9 and Gromit) to study the use of the LAAS Architecture and tools on those platforms, and conduct research on the development of a temporal planner and execution control system based on the IDEA/Europa planner. Recently, Félix has worked on extending the LAAS Architecture to support formal validation, verification and correct controller synthesis.

Mark Lawford (*McMaster University*) is Chair of McMaster University's Department of Computing and Software and Director of the McMaster Centre for Software Certification. He has a BSc (1989) in Engineering Mathematics from Queen's University, Kingston, where he received the University Medal in Engineering Mathematics. His MASc (1992) and PhD (1997) are from the University of Toronto. He worked at Ontario Hydro as a real-time software verification consultant on the Darlington Nuclear Generating Station Shutdown Systems Redesign project, receiving the Ontario Hydro New Technology Award for Automation of Systematic Design Verification of Safety Critical Software in 1999. Since 2012, he has been involved in automotive software research, and in 2014, he was a co-recipient of a Chrysler Innovation Award. He serves on the steering committee of the Software Certification Consortium (SCC). He is a licensed Professional Engineer in Ontario and a senior member of the IEEE.

Pippa Moore (*Civil Aviation Authority*) worked for GEC-Marconi Avionics prior to joining the UK CAA. While with this organisation, she worked on the development of a range of safety-critical flight control computers in both the military and civil fields. She has been a Design Surveyor with the CAA since 1996, specialising in airborne software, airborne electronic hardware and safety assessment. In that time, Pippa has worked as a CAA, JAA and EASA systems specialist on civil aircraft certification projects such as the Boeing 737 and 767, A330/340 and A380. She has also worked on numerous engine certification and validation projects. Additionally, Pippa has worked with the regulatory authority teams for several UAS programmes and undertaken research on aviation safety topics that have directly changed aircraft safety regulations. She has spent the last 3 years as the technical lead for the CAA's Cyber Oversight Programme.

Patrizio Pelliccione (*University of L'Aquila*) is Full Professor at Gran Sasso Science Institute (GSSI) and Associate Professor at the Department of Computer Science and Engineering at Chalmers University of Technology and University of Gothenburg. He received his PhD in 2005 at the University of L'Aquila, and since 2014, he is Docent in Software Engineering, a title given by the University of Gothenburg. His research topics are in software architectures modelling and verification, autonomous systems and formal methods. He has co-authored more than 120 publications in journals and international conferences and workshops. He has been on the programme committees for several top conferences, is a reviewer for top journals and has chaired the programme committee of several international conferences. He is very active in European and national projects. He is the PI for the Co4Robots H2020 EU project for the University of Gothenburg. In his research activity, he has pursued extensive and wide collaboration with industry.

Zeyn Saigol (*Connected Places Catapult*) is Principal Technologist at Connected Places Catapult, specialising in verification, validation and regulatory approval of autonomous vehicles (AVs). He is the technical lead for the MUSICC project, which is a key pillar of the UK Department for Transport's contribution to multinational AV certification. Zeyn is also an interface architect on the VeriCAV project, which is creating a smart simulation testing framework, and is a member of ISO and ASAM committees working on standards for AV testing. His background includes a BSc in Physics from the University of Bristol, master's degrees from Imperial College and the University of Edinburgh and a PhD in AI from the University of Birmingham. His research interests covered planning under uncertainty, knowledge representation, mapping, path planning and machine learning. He has worked with a variety of autonomous systems, including wheeled, flying and marine robots, in roles spanning academia and industry.

Christian Schlegel (*Technische Hochschule Ulm*) is Head of the Service Robotics Research Group, Professor for Real-Time and Autonomous Systems in the Computer Science Department since 2004 and co-opted member of the Faculty of Engineering, Computer Science and Psychology of the University of Ulm. He is the technical lead of the EU H2020 RobMoSys project and the elected coordinator of the euRobotics Topic Group on Software Engineering, System Integration, System Engineering. Christian is co-founder and Associate Editor of the open access journal JOSER—*Journal of Software Engineering for Robotics*—and co-organiser of the series of International Workshop on Domain-Specific Languages and Models for Robotics Systems (DSLRob). He was awarded a Diploma and PhD in Computer Science in 1993 and 2004, respectively.

Rob Skilton (*Remote Applications in Challenging Environments—RACE*) is Head of Research at RACE, a UK centre for Remote Applications in Challenging Environments, where he leads a team specialising in control systems, autonomy and perception for robotic operation and inspection in hazardous environments. Robert graduated with an MSc in Cybernetics in 2011 and is currently studying for a PhD in Autonomous Robotics and Machine Learning at the Surrey Technology for Autonomous Systems and Robotics (STAR) Lab. Robert is a Chartered Engineer, brings experience in developing robotic systems for hazardous environments and has developed numerous robotic and software platforms for use in nuclear and other extreme environments. He has experience from a wide range of roles in industrial engineering and R&D projects including in telerobotics and is currently leading various activities including the Robotics and AI in Nuclear (RAIN) work on the teleoperation of industrial robots.

Jon Timmis (*University of Sunderland*) is Professor of Intelligent and Adaptive Systems and Deputy Vice-Chancellor. He is Visiting Professor at the University of York. Jon has worked for over 20 years in the area of biologically-inspired systems and computational biology. His research cuts across many areas, but the majority of his work revolves around immunology, either developing computational models of immune function (computational immunology), or fault-tolerance achieved via bio-inspired engineering with a focus on the immune system and evolutionary processes. Jon has worked extensively on swarm robotic systems and adaptive autonomous robotic systems. Jon is a previous recipient of a Royal Society Wolfson Research Merit Award and a Royal Academy of Engineering Enterprise Fellowship. Jon co-founded Simomics Ltd in 2014 to commercialise his research.

Alan Winfield (*University of the West of England*) is Professor of Robot Ethics, Visiting Professor at the University of York and Associate Fellow of the Cambridge Centre for the Future of Intelligence. He co-founded and led APD Communications Ltd until taking up appointment at the University of the West of England (UWE), Bristol, in 1992. Alan co-founded the Bristol Robotics Laboratory, where his research is focused on the science, engineering and ethics of cognitive robotics. He is passionate about communicating research and ideas in science, engineering and technology; he led the UK-wide public engagement project "Walking with Robots" and was awarded the 2010 Royal Academy of Engineering Rooke medal for public promotion of engineering. Alan sits on the executive of the IEEE Standards Association Global Initiative on Ethics of Autonomous and Intelligent Systems and chairs Working Group P7001, drafting a new IEEE standard on Transparency of Autonomous Systems. He is a member of the World Economic Forum Global AI Council.

Jim Woodcock (*University of York*) is Professor of Software Engineering, a Fellow of the Royal Academy of Engineering and an award-winning researcher. He is also Professor of Digital Twins at Aarhus University. He has 40 years' experience in formal methods. His research interests are in unifying theories of programming (UTP), robotic digital twins and industrial applications. Formerly, he worked on applying the Z notation to the IBM CICS project, helping to gain a Queen's Award for Technological Achievement in 1992. He created the theory and practical verification for NatWest Bank's Mondex smart-card system, the first commercial product to achieve ITSEC Level E6 (Common Criteria EAL 7). For the last decade, he has researched the theory and practice of cyber-physical systems and robotics. He led the team that developed extensive UTP theories and the Isabelle/UTP theorem prover. He is Editor-in-Chief of the prestigious Springer journal *Formal Aspects of Computing*.

Chapter 1
Software Product Line Engineering for Robotics

Davide Brugali

Abstract The cost of creating new robotics products is significantly related to the complexity of developing robotic software applications that are flexible enough to easily accommodate frequently changing requirements. In various application domains, software product line (SPL) development has proven to be the most effective approach to achieving software flexibility and to face this kind of challenges.

This chapter reviews the fundamental concepts in SPL from a robotics perspective and defines guidelines for the adoption of software product line engineering in robotics. In particular, it discusses the concepts of software variability, domain analysis and modeling, reference architectures, and system configuration.

1 Introduction

Robots are versatile machines that are increasingly being used not only to perform dirty, dangerous, and dull tasks in manufacturing industries but also to achieve societal objectives, such as enhancing safety in transportation, reducing the use of pesticide in agriculture, and improving efficacy in the fight against crime and civilian protection. Compared to the manufacturing work cell, a public road, a corn field, and a crime scene are open-ended environments, which require autonomous robots to be equipped with advanced cognitive capabilities, such as perception, planning, monitoring, coordination, and control, in order to cope with unexpected situations reliably and safely.

Even a simple robotic application, like moving a wheeled robot from place A to place B in an indoor environment, requires several capabilities, such as (1) *sensing* the environment in order to avoid unexpected obstacles (i.e. moving people), (2) *planning* a path from A to B taking into account several constraints (e.g. energy consumption), (3) *controlling* the actuators in order to execute the computed path

D. Brugali (✉)
University of Bergamo, Bergamo, Italy
e-mail: brugali@unibg.it

© Springer Nature Switzerland AG 2021
A. Cavalcanti et al. (eds.), *Software Engineering for Robotics*,
https://doi.org/10.1007/978-3-030-66494-7_1

correctly (i.e. with a given accuracy), and (4) *reasoning* about alternative courses of actions (e.g. waiting for a passage to get clear or plan a different path).

Sensing, planning, controlling, and reasoning are human-like capabilities that can be artificially replicated in an autonomous robot as software systems, which implement data structures and algorithms devised on a large spectrum of theories, from mechanics and control theory to ethology and cognitive sciences. Software plays a key role in the development of robotic systems as it is the medium to embody intelligence in the machine.

In this scenario, the cost of creating new robotics products is significantly related to the complexity of developing robotic software applications that are flexible enough to easily accommodate frequently changing requirements: more advanced tasks requiring high degrees of autonomy, in highly dynamic environments demanding robust and dependable robot control systems, and in collaboration with unskilled users for whom safety, availability, and performance are primary concerns.

In various application domains, software product line (SPL) development has proven to be the most effective approach to achieving software flexibility and to face these kinds of challenges.

An SPL is a set of software-intensive systems that share a common set of features for satisfying a particular market segment's needs. Building software systems according to the product line approach is economical and efficient [16]. Most of the work is about integration, customization, and configuration of common software assets (e.g. architectural models, software components) instead of creation (i.e. software implementation and refactoring).

Few other papers dealing with software product lines for robotics can be found in the literature.

Jørgensen and Joosen [34] propose an approach for developing SPL for industrial robot controllers that is based on the separation between functional and non functional requirements. Components that implement essential functional behavior are implemented as reusable building blocks of a component framework, while components that are primarily involved with non functional requirements belong to the periphery part of the framework.

Jung et al. [35] apply Feature-Oriented Programming (FOP) to automatically generate code from a declarative specification of robot software controllers. As discussed in Sect. 5, FOP is one of the techniques typically used for managing the variability of software product lines.

Weiss et al. [56] conducted a survey on software control systems for ABB industrial manipulators and analyzed the opportunities to define an SPL in order to reduce maintenance costs and shorten time to market. The survey was based on stakeholders' interviews and manual and automated code inspection.

Kang et al. [37] reengineered legacy home service robot applications into product line assets through feature modeling and analysis and designed a new architecture and components for robot applications.

Thompson and Heimdahl [53] introduced the concept of n-dimensional and hierarchical product family and exemplified it by modeling a family of mobile robots, a family of flight guidance systems, and a family of cardiac pacemakers.

Abd Halim et al. [3] present an approach and modeling language for mapping multiple models of software systems at requirements level to an architectural model and exemplified their approach to the development of a software product line of autonomous mobile robots.

Baumgartl et al. [5] propose to use domain-specific languages to easy robot programming for experts in specific application domains (e.g. pick and place in industrial automation) without adequate skills in robotics and software development. Their approach exploits code generators to automatically integrate reusable modules of an SPL according to user specifications. A similar approach is proposed by Dhouib et al. [20].

This chapter reviews the fundamental concepts in SPL from a robotics perspective and defines guidelines for the adoption of software product line engineering in robotics.

It is structured as follows. Section 2 illustrates the concept of *flexibility* in the context of different disciplines and gives an interpretation of flexibility related to robotic systems. Section 3 presents the challenges and approaches for developing software product lines in the robotics domain. It introduces the two main development phases of an SPL, i.e. *domain engineering* and *application engineering*. Section 4 clarifies the concept of *domain* and provides guidelines to exploit in robotics two well-known software domain analysis approaches, namely, *Feature-Oriented Domain Analysis* [36] and *Stability-Oriented Domain Analysis* [22]. It also illustrates the concept of reference architecture, i.e. the main outcome of domain engineering. Section 5 illustrates the application engineering phase through a case study and presents a software tool for modeling robotic software product lines and configuration of robotic systems. Finally, Sect. 6 draws relevant conclusions.

2 System and Software Flexibility

Flexibility is a concept that has different meanings in different disciplines. Intrinsic to the notion of flexibility is the ability or potential to change and adapt to a range of states. The common ground on which all disciplines agree is that flexibility is needed in order to cope with uncertainty and change, and that it implies an ease of modification and an absence of irreversible or rigid commitments [50].

Flexibility in Decision Theory Mandelbaum and Buzacott [40] define the concept of flexibility in the context of decision theory. They consider a two-period decision problem and a set of decision choices in each period. The definition and measure of flexibility is thus the number of options still open in the second period after a decision has been made in the first period. In other words, flexibility is the number of remaining alternatives after a first commitment is made: more remaining choices reflect increased flexibility.

Flexibility in Manufacturing Systems Manufacturing is another area that is strongly associated with issues of flexibility [51]. Flexibility is an attribute of a manufacturing system that is capable of reconfiguring manufacturing resources in order to produce efficiently different products in different volumes.

Flexibility in Engineering Design In [49] flexibility is defined as a property of a system that allows it to respond to changes in its initial objectives and requirements. In this context, flexibility implies that the system has been designed with certain characteristics (e.g. additional parameters, a particular architecture) that may not be necessary or justified in the context of optimizing the system for the immediate mission (or set of requirements) for which it is being fielded [1].

Flexibility in Software Engineering The IEEE Standard Glossary of Software Engineering Terminology defines flexibility as the *ease with which a system or component can be modified for use in applications or environments other than those for which it was specifically designed*. The term environment refers to the complete range of elements in an installation that interact with the software system. This includes the computer and network platform, the controlled robotic hardware, and the middleware infrastructure. More specifically, flexibility is concerned with the interoperability among independently developed components (e.g. components interfacing heterogeneous robotic devices), the reusability of individual components in different application contexts (e.g. a motion planner for static or dynamic environments), and the component system reconfigurability at runtime (e.g. adaptable robot behaviors).

More specifically, flexibility of software artifacts depends upon the type of the artifact and the evolution of the artifact's requirements. Flexibility is related to the type of software artifact in the sense that properties belonging to the source code, the architecture, and the technology used can improve or degrade it. Flexibility depends upon the type of change request. In case that it is possible to foresee change requests that are likely to be implemented in the future, it would be much easier to develop the program with an architecture that might support those changes.

Flexibility in Autonomous Robots Autonomous robots are intelligent machines capable of performing tasks in the world by themselves, without explicit human control [33]. They can differ significantly for the purpose for which they are designed (what they are for), the mechanical form (how they are structured), and their capabilities (what they are able to do). It is this diversity in form and function that requires a model for flexibility and efficiency beyond those developed in other application domains.

Robot system development is a multi-phase decision problem. At each phase, specific technologies must be selected, which may lock in the subsequent development phases. Typically, the design of the robot embodiment comes first as it strongly depends on the user requirements related to the operational environment and to the task to be performed. For example, a rover for floor cleaning and an autonomous harvesting machine significantly differ in terms of robotic devices, i.e. sensors and actuators.

In a subsequent phase, the development of the software control system is constrained by the robot embodiment, which thus reduces the number of options for implementing typical robot capabilities, such as motion control, self-localization, and obstacle detection.

Professional service robots (e.g. humanoid and non-humanoid robots for retail, restaurants, healthcare, and other customer-facing applications) are typically equipped with redundant hardware and software resources in order to increase their robustness and dependability. For example, a common hardware configuration includes an RGB camera and a laser rangefinder for obstacle detection. More economic configurations replace the laser rangefinder with infrared or ultrasonic sensors. The same sensors can be used for self-localization in combination with GPS sensors, inertial measurement units (IMU), and wheel encoders. Similarly, the robot control system implements multiple versions of common functionality (e.g. localization, obstacle avoidance, mapping, etc.) in order to exploit available resources at best.

In this context, flexibility means the ease with which these resources can be composed and the software control system reconfigured for different operational conditions. For example, the GPS sensor does not provide reliable measures inside a building, the RGB camera requires adequate illumination, and different localization techniques are adequate for indoor and outdoor environments.

Very often, open-source software libraries implementing the most common robotic functionality are not easily reusable because they have been designed to meet the quality, functional, and technical requirements of very specific application scenarios. Typically, they are tied to specific robot hardware and the assumptions and constraints about tasks and operational environments are hidden and hard coded in the software implementation.

Robotic application reengineering frequently occurs in three circumstances:

- When the application migrates to a different robot platform or scales up from a single-robot to a multi-robot system. Robotics is a discipline that grounds its roots into mechanics, electronics, computer science, and cognitive sciences. In order to build complex systems, Robotics integrates the most advanced results from a variety of disciplines: mechanics, electronics, control theory, computer science, and cognitive sciences. Thus, robotic control systems are highly dependent on the evolution of the underlying technologies.
- When the same robotic system is used in different environments. Robotics is a research field that pursues the ambitious goal of developing fully autonomous robots that operate in open-ended everyday environments, such as hospitals, private homes, and public roads. Although the key capabilities of an autonomous robot (perception, navigation, manipulation) are common to most applications, their implementations often require tailoring to the characteristics of the operational environment (indoor/outdoor, mostly static/highly dynamic, etc.).
- When the same robotic equipment is used to perform different tasks. As an example, a robot for autonomous logistics requires different configurations of

Table 1.1 Development costs of the three EFFIROB prototypes.

	Container transport	Floor cleaning	Care utensil
Total prototype cost	2,663,094.93 €	5,983,490.86 €	8,115,678.09 €
Of which software development	2,078,207.95 €	4,796,908.35 €	6,428,873.08 €

the navigation parameters (e.g. maximum speed, maximum acceleration, type of trajectory) depending on the type of load (e.g. fragile or perishable).

The EFFIROB project [30] has conducted a study with the aim of analysing the technical and economic feasibility of new service robotics applications and has developed an approach to estimate the effort required to implement new service robots.

Table 1.1 reports the *total cost* of each prototype, which includes material cost, administration cost, software development cost, and system designing cost. The table shows that the cost of *software development* is around the 80% of the total cost. These data indicate that the impact of software reuse in the software development process is negligible, despite the similarities in robotic capabilities of the three prototypes, such as perception, mapping, navigation, and planning.

The EFFIROB study also indicates that cost estimates in early product phases are subject to high uncertainty, mainly due to the lack of detailed knowledge of the final product in terms of environment and mission parameters. As a consequence, the majority of the costs are incurred during production, although up to 90% of the life-cycle costs are determined by decisions taken in planning and development phases as depicted in Fig. 1.1 and discussed by Ehrlenspiel et al. [21].

This situation is typical for single system development, where the requirements for a single scenario are considered. Software variability is increasingly reduced every time a design decision is taken from the analysis of existing technologies and user needs down to software and system deployment. During development, the number of potential systems decreases until finally at runtime there is exactly one system. Subsequently, any change to system requirements demands, potentially extensive, editing of the source code [55].

In various application domains, software product line (SPL) development has proven to be the most effective approach to face these kinds of challenges. The concept of software product line (SPL) is based upon an appealing idea: instead of considering the various examples of service robots individually, the co-development of a family of similar systems is planned from the beginning. Variability is a planned or anticipated change, not a change due to errors and maintenance.

As discussed in Sect. 4, the sources of variability need to be identified right at the beginning of the robot system development process in a phase called domain engineering. Typically, the sources of variability fall into three categories related to the robot embodiment (e.g. the application might require a single robot or a fleet of mobile robots), the robot environment (e.g. the application requires the robot to navigate both indoor and outdoor), and the robot task (e.g. the application consists in heterogeneous tasks, such as guiding a patient or following a nurse in a hospital).

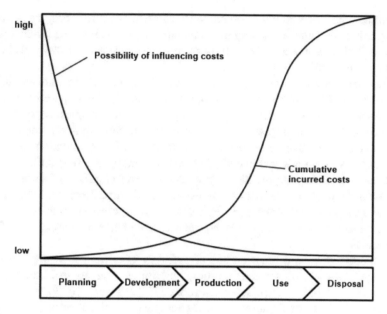

Fig. 1.1 Influence on costs vs. incurrence of costs over product life cycle. © Springer-Verlag Berlin Heidelberg 2007. Reprinted with permission.

Before addressing robotic domain engineering in details, the next section illustrates the state-of-the-art approaches to SPL engineering and analyzes their applicability to robotics.

3   Robotic Software Product Lines

Typically, an SPL is a strategic investment for organizations that want to achieve customer value through large commercial diversity of their products with a minimum of technical diversity at minimal cost. This business strategy is called *software mass customization* [38] and can be adopted according to the following three models.

The *proactive* approach consists of designing a complete SPL from scratch to support the full scope of products. This is possible when the organization knows precisely the requirements of its products and can define their common software infrastructure and overall architecture. In robotics, the proactive approach can be reasonably applied to very specific domains, such as robotic planetary exploration, where big organizations like NASA or the European Space Agency (ESA) develop all the software for their robot control systems in-house, because of the uniqueness of their equipment.

The *extractive* approach consists of reengineering a pre existing set of systems that have a significant amount of similarities in order to eliminate duplicated code and enhance interoperability of common functionalities. In robotics, the *extractive* approach has been applied in very mature domains, such as Flexible Manufacturing Systems (FMS), where a strong commitment of robotics and industry stakeholders can be ensured [56].

The *reactive* approach consists in revising the design of the software architecture of an existing SPL in order to accommodate the unpredictable requirements of new products and to exploit well rapidly evolving robotic technologies. In robotics, the reactive approach is appropriate to the development of SPL for autonomous service robots that require advanced capabilities, such as mobile manipulation, to perform complex tasks (e.g. logistics) in everyday environments (e.g. a hospital).

Nevertheless, adopting the SPL approach for the development of service robots is particularly challenging, because these advanced capabilities are implemented reusing open-source libraries typically resulting from research projects, which are not coordinated by key stakeholders, such as robot manufacturers or software foundations.

As a consequence, a huge corpus of software systems, which implement the entire spectrum of robot functionalities, algorithms, and control paradigms, is potentially available as open-source libraries [7]. Unfortunately, their reuse even in slightly different application scenarios requires significant effort, because the assumptions about the computational and robotic hardware, the software infras-tructure, and the robot operational environments are hidden and hard coded in the software implementation.

In this context, we propose a new approach to the development of SPLs for autonomous service robots that is a mix of the three approaches discussed above.

It is *proactive* as it is applied to the development of SPLs for functional sub-domains (i.e. motion planning, robust navigation, 3D perception). By focusing not on an entire robot application, but only on specific subsystems, experts in specific functionalities can define stable software architectures that promote the development of reusable components. It is *extractive* as it promotes the development of core assets for the SPL (i.e. reusable components) through the refactoring and harmonization of existing open-source libraries [9]. It is *reactive* as it supports the development of higher-level SPLs for specific application domains (e.g. hospital logistics) as composition of SPLs for functional sub-domains. This means that reusable components and subsystems are composed on the fly according to ad hoc software architectures that meet the frequently changing requirements of robotic applications.

The approach does not assume that the developed assets (e.g. reusable compo-nents, stable architectures) and the product derivation are under the control of a single developing organization as it is typical for SPLs. It envisages three main stakeholders:

- The community of researchers, who keep implementing new algorithms for common robot functionalities as open-source libraries, increasing the variability of robotic control systems

- Specialized software houses, who design SPLs for specific robotic sub-domains (e.g. robot navigation) and SPLs for robot application domains (e.g. logistics)
- System integrators, who develop robot applications with increasingly challenging requirements by customizing functional subsystems and applications to be deployed on specific robotic systems

SPL development is split in two phases: domain engineering and application engineering.

Domain engineering identifies parts of the software system that are stable and parts that remain variable and are not fully defined during design or even implementation. This facilitates the development of different versions of a software system. The outcome of the domain engineering phase is a repository of common software assets (e.g. requirements specification, architectural and design models, software components) that share many (structural, behavioral, etc.) commonalities and together address a particular domain.

Each new application is built in the application engineering phase by configuring the SPL, i.e. by selecting the variants (e.g. functionalities, software resources) that meet specific application requirements.

The next two sections illustrate these two phases in the context of software development for robotics.

4   Domain Engineering

Domain engineering is a set of activities aimed at developing reusable artifacts and flexible software systems within a domain. The term domain is used to denote or group a set of systems (e.g. mobile robots, humanoid robots) or functional areas within systems (e.g. motion planning, 3D mapping, etc.), or applications (hospital logistics, professional cleaning services, etc.) that exhibit similar characteristics or fulfill similar requirements.

Developing flexible software systems requires insight into the future and an understanding of how their requirements are likely to evolve. Thus, the fundamental task of the domain engineering process consists in analysing existing systems in a given domain (e.g. various types of mobile robots, such as wheeled, legged, flying), identifying the relevant concepts in the domain (e.g. robot kinematics), describing their key characteristics (e.g. degrees of freedom, motion constraints, etc.), and identifying similarities, differences, and recurrent properties that affect the design of software control systems.

In software engineering, this task is called *domain analysis and modeling*. Czarnecki et al. [19] define a domain model as *an explicit representation of the common and the variable properties of the systems in a domain, the semantics of the properties and domain concepts, and the dependencies between the variable properties.*

The rest of this section illustrates the application in robotics of two software domain analysis and modeling approaches, namely, *Feature-Oriented Domain Analysis* [36] and *Stability-Oriented Domain Analysis* [23], and of the concept of reference architecture, i.e. the main outcome of domain engineering.

Feature-Oriented Domain Analysis is a method for the identification of prominent or distinctive features of software systems in a domain and for the graphical representation of commonalities among them.

Stability-Oriented Domain Analysis is a complementary approach to FODA that provides guidelines for the identification of stable features of software systems, i.e. features that are not affected by the evolution of application requirements or system technologies.

A reference architecture [43] is a software architecture that encompasses the knowledge about how to design concrete architectures of systems of a given application domain. As such it is used as a foundation for the design of the software architecture of an SPL.

4.1   Feature-Oriented Domain Analysis

A fundamental tenet of domain engineering is that substantial reuse of knowledge, experience, and software artifacts can be achieved by performing some sort of commonality and variability analysis to identify those aspects that are common to all applications within the domain, and those aspects that distinguish one application from another [18].

A system configuration is an arrangement of components and associated options and settings that completely implements a software product. Variants may exclude others (e.g. the selection of a component implementing an indoor navigation algorithm excludes the choice of components providing GPS-based localization services), or one option may make the integration of a second one a necessity (e.g. a component implementing a visual odometry algorithm depends on a component that supplies images of the environment). Hence, only a subset of all combinations are admissible configurations.

In order to model and symbolically represent a domain variability, its variants and the constraints between them, a formalism called feature models was introduced in 1990 in the context of the Feature-Oriented Domain Analysis (FODA) approach [36].

Figures 1.2 and 1.3 show the feature models (FM) that capture the variability in robot perception capabilities and robot manipulation capabilities of a RoboCup@Work system as described in a recent paper by Brugali et al. [15]. In particular, Fig. 1.2 shows that the robot has the capabilities to detect obstacles and recognize containers of different colors, service areas with and without cavities, and different types of objects. Figure 1.3 shows that a mobile manipulation robot can exploit alternative approaches for local planning (e.g. DWA and VFH) and for motion control.

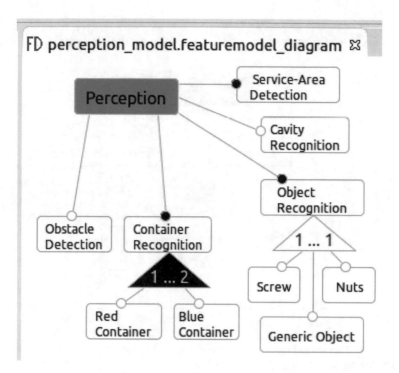

Fig. 1.2 FM for robot perception.

Feature models are organized as a tree and the root feature, also called **concept**, defines the application family. Parent features are connected to children features by means of edges, which represent containment relationships.

Features can be discerned by two main categories: **mandatory** and **optional**. *Mandatory* features have to be present in all the applications of the product line (commonalities). They are graphically depicted by means of a black circle on the top, like *Global Planner* in Fig. 1.3. *Optional* features by contrast can be present but are not mandatory (variation points). They are depicted by means of a white circle on the top, like *Cavity Recognition* in Fig. 1.2.

In these feature models, each feature corresponds to a variation point (e.g. the Local Planner functionality) or a concrete variant (e.g. the Dynamic Window Approach (DWA) algorithm for local planning). A black circle on a child node (e.g. Global Planner) indicates that the feature is mandatory, while a white circle indicates that the feature is optional. White triangles indicate that the child features are mutually exclusive, while black triangles indicate the cardinality of the OR containment association. For example, the perception system can be configured with algorithms that can recognize only one specific type of object (e.g. Screw, Nut) or a generic object. Similarly, it can be configured to recognize only one or both types of containers (Red and Blue).

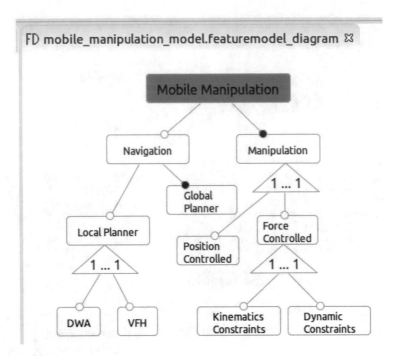

Fig. 1.3 FM for mobile manipulation.

Feature models also define two kinds of constraints between the features: **requires** and **excludes**. These constraints allow the definition of a subset of valid configurations. The *requires* constraint means that if a feature A is selected, then a feature B also has to be selected. The *excludes* constraint on the contrary means that if a feature A is selected, then a feature B cannot be selected. Constraints are defined in a textual way and are not graphically represented in a feature diagram.

4.2 Stability-Oriented Domain Analysis

A complementary approach to feature-oriented domain analysis exploits the concept of software stability [23], which can be defined as a software system's resilience to changes in the original requirements specification. In order to enhance reuse, the software development process should focus on those aspects of the domain that will remain stable over time.

If the entities in the domain have not changed, and have not changed their relationships to each other, for a long time, we can assume that the relationships between the entities will continue to hold through future changes.

Therefore, the logical structure of every concrete software system should match the invariant structure of the domain in order to ensure a stable core design and, thus, stable software artifacts [24]. Those changes that are introduced to the software project will then be in the periphery, since the core was based on something that has remained and will remain stable. Since whatever changes must be made to the software in the future will be in this "periphery", it will only be these small external modules that will have to be (re)engineered.

In order to support stability analysis, the concept of an Enduring Business Theme (EBT) was first introduced in [17]. An EBT is an aspect of a domain that is unlikely to change, since it is part of the essence of the domain. After the appropriate endur- ing themes of the business are understood and catalogued, a software framework can be constructed that enables the development of business objects, which offer application-specific functionality and support the corresponding EBT. Business objects have stable interfaces and relationships among each other but are internally implemented on top of more transient components called industrial objects.

Fayad et al. [23] sketched some heuristics for guiding the identification of EBTs. In particular, one of the criteria is their tangibility. "If an object in a model represents a concrete entity, then it is most likely an industrial object". Tangibility is considered one of the driving factors for the instability of industrial objects. The reason is that physical entities, such as a piece of machinery, are highly sensitive to technological evolutions.

A milestone paper of Rodney Brooks [8] identifies a set of properties (we say EBTs) of every robotic system, among which three are of interest for our discussion: *Embodiment*, *Situatedness*, and *Intelligence*. We exemplify these properties with reference to a family of commercial products by PAL Robotics (http://pal-robotics. com/) that we have studied in detail.

Embodiment Robot embodiment refers to the mechanical structure with sensors and actuators that allows the robot to experience the world directly. The robot receives stimuli from the external world and executes actions that cause changes in the world state. Simulated robots may be situated in a virtual environment, but they are certainly not embodied. Figure 1.4 shows an example of how to manage hardware variability. All the products share a common mobile platform and can be customized with a variety of hardware devices.

Fixed Shelves   Safety Box   Roller Conveyor   TIAGo Base   Standard Boxes Stacking   Lifter

Fig. 1.4 A family of mobile platforms by PAL Robotics. ©PAL Robotics 2021. All rights reserved.

Fig. 1.5 Mostly static. ©PAL Robotics 2021. All rights reserved.

Situatedness Robot situatedness refers to existence in a complex, dynamic, and unstructured environment that strongly affects the robot behavior. For example, the environment is a museum full of people where a mobile robot guides tourists and illustrates masterworks or a manufacturing work cell where an industrial manipulator handles workpieces. Situatedness implies that the robot behavior is influenced by the characteristics of the operational environment. For example, a GPS sensor cannot be used inside a building, and, outdoor, it can provide only a rough estimate of the robot pose. A stereoscopic vision system can provide accurate 3D information about the surrounding environment, but is highly sensitive to environmental lighting conditions. A laser rangefinder is also highly accurate, but cannot detect transparent surfaces, such as a sliding glass doors. Thus, the robot engineer should take into account these characteristics during the development of a robot control system in order to make its behavior robust to unexpected environmental changes. Figures 1.5, 1.6, and 1.7 show three environments with different characteristics that affect the robot behavior. A robot for warehouse inventory (Fig. 1.5) operates in an environment that is supposed to be mostly static, i.e. there are not moving obstacles except the human operator. In this scenario, the robot can exploit a geometric map of the environment and a laser rangefinder for precise self-localization and path planning. A robot that is expected to interact

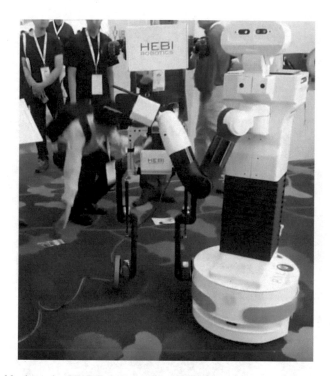

Fig. 1.6 Highly dynamic.

socially with humans (Fig. 1.6) should be equipped with additional sensors for detecting and recognizing people and efficient algorithms for path planning and obstacle avoidance in a highly dynamic environment. Outdoor navigation (Fig. 1.7) can benefit from GPS data for self-localization but requires an advanced vision system for place recognition.

Intelligence Robot intelligence refers to the ability to express adequate and useful behaviors while interacting with the dynamic environment. Intelligence is perceived as "what humans do, pretty much all the time" [8], and mobility is considered at the basis of every ordinary human behavior. Figures 1.8, 1.9, and 1.10 show three tasks that require specific robot capabilities.

Logistic tasks (Fig. 1.8) simply require navigation capability in indoor environments: navigation, manipulation, and human-robot interaction. Housekeeping (Fig. 1.9) in addition requires manipulation capabilities. Social interaction (Fig. 1.10) mostly requires advanced human-robot interaction capabilities in addition to navigation and manipulation capabilities.

These three EBTs are guidelines that help the robot engineer to identify the stable characteristics of a robotic system. Let us consider the case of mobile robots more in detail.

Fig. 1.7 Outdoor. ©PAL Robotics 2021. All rights reserved.

Fig. 1.8 Logistic. ©PAL Robotics 2021. All rights reserved.

Fig. 1.9 Housekeeping. ©PAL Robotics 2021. All rights reserved.

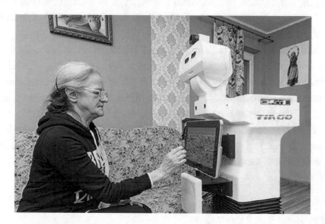

Fig. 1.10 Social interaction. ©PAL Robotics 2021. All rights reserved.

Robot mobility refers to both the robot and its constituent components that move one with respect to the others (*Embodiment*). For example, a humanoid robot is a complex structure comprising many limbs that change their relative positions while the robot is walking. It also applies to parts or devices of the robot that move with respect to the environment. For example, the robot changes the orientation of the head to track a moving object (e.g. a human being). Embodiment helps to answer two questions: Which part of the robot does move? How much does it move?

Robot mobility is a form of robot-environment interaction (*Situatedness*). It refers to the ability of changing its own position concerning the environment and thus of being in different places on different times. Situatedness helps to answer two questions: With respect to what does the robot move? When and where does it move?

Robot mobility is explained in terms of robot behaviors and tasks (*Intelligence*). These concepts help to answer two questions: How does the robot move? Why does the robot move?

In our approach, these three EBTs also help the robotic engineer in defining the feature models of a robotic software product line, i.e. for representing the functional variation points of a robot control system, the variants, and the constraints among them.

Figure 1.11 depicts the feature model of the application requirements for the RoboCup@Work scenario [15]. In this competition, mobile manipulation robots are expected to perform a wide range of manipulation, assembly, and logistic tasks in factory-like environments composed of service areas. Each service area represents a region of the factory having a specific purpose for a particular task, for example, areas to pick objects, to insert objects into object-specific cavities, to place objects into containers, to operate machines, and so forth. Those service areas differ also in terms of height, width, and so forth. Additionally, some environments include static obstacles, whereas others are free of obstacles or even include dynamic obstacles such as other robots and human workers.

The feature model is structured around three main dimensions of variability, namely, the type of task that the robot should perform (*EBT:Intelligence*), the environment characteristics (*EBT:Situatendess*), and the available equipment (*EBT:Embodiment*). For example, the perception system could be a depth sensor (e.g. a Kinect) or a stereo camera (e.g. a BumbleBee sensor), and the rover kinematics could be omnidirectional or differential drive. The task could be a simple pick and place, an object placement on a plain surface, or a precision placement inside a hole. The environment consists of a variety of objects and containers.

4.3   Reference Architectures

The control application of an autonomous robot is typically designed as a (logically) distributed component-based system (see [13] for a survey). A real-world robot control system is composed of tens of software components. For each component providing a robot functionality, tens of different implementations may be available. The initial release of the Robot Operating System (ROS) [46] in year 2010 already contained hundreds of open-source packages (collections of nodes) stored in 15 repositories around the world.

System integration is a crucial phase in the robot application development process [31]. It requires to select, integrate, and fine-tune the functionalities of the autonomous robots according to the environment conditions (often completely beyond the control of the system integrator), the task to be performed (often continuously for long periods without human intervention), and the available resources (subject to failures and malfunctioning).

Component integration requires a common system architecture that defines not only how components interact by means of communication and synchronization

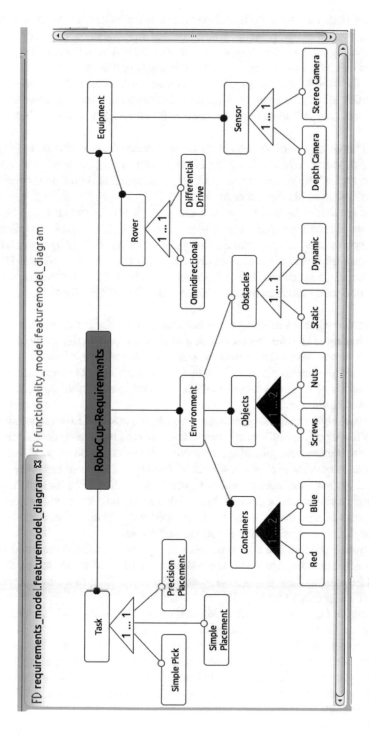

Fig. 1.11 Requirements of the RoboCup@Work scenario.

mechanisms (e.g. the message-based communication infrastructure of the ROS framework), but also the principles and policies that must be enforced by the set of interacting components. These include the allowed variations in the system architecture that, when exercised, become individual applications.

This is the role of the software product line architecture. It specifies both commonalities and differences of a family of component-based systems and the constraints on how software components can be assembled to derive individual products.

Thus, the architecture of an SPL plays the role of reference architecture [43] for a family of products. While refactoring effort can never be completely avoided, it should be a goal of any software product line development to define an architecture that will be mostly stable for future products [54].

In other embedded systems domains, one can observe a strong move toward the definition of reference architectures (e.g. AUTOSAR [26]). In some application domains, such as telecommunications, factory automation, and enterprise information systems, large companies or international committees have defined well-known reference models, such as the ISO-OSI model, the USA-NBS Reference Model for Computer Integrated Manufacturing [41], and AUTOSAR [26] for the automotive domain.

In robotics, the goal of defining reference architectures for autonomous robots is more challenging [11]. The reason can be found in the peculiarity of the robotics domain: robots can have many purposes, many forms, and many functions. While the operational environments of automotive or avionics control systems are well defined in terms of required functionalities, an autonomous robot operates in an open-ended environment.

Despite this high variability in robotic systems, we acknowledge the significant stability in the set of capabilities of autonomous robots [10]: navigation, dexterous manipulation, perception, planning, and control are common capabilities that represent the essence of the robotics domain. We argue that these capabilities can be provided by functional subsystems designed as individual SPL, for which stable software architectures can be defined. In this context, stable means that a significant variety of robot control systems can be developed by configuring (i.e. resolving the robotic variability) and integrating functional subsystems.

The software product lines of functional subsystems can be reused and composed as black-box components [44]. Systems made of subsystems can be further composed in order to design more complex systems resulting in an n-dimensional and hierarchical product family [53].

For example, Fig. 1.12 visualizes the architectural variability of the perception system for the RoboCup@Work scenario. It is graphically represented using the Robot Perception Specification Language (RPSL) [32], a domain-specific language (DSL), which enables domain experts to express the architectural variability of robot perception systems and to model multi-stage perception systems by composing sensing and processing components in a perception graph which yields a directed acyclic graph (DAG).

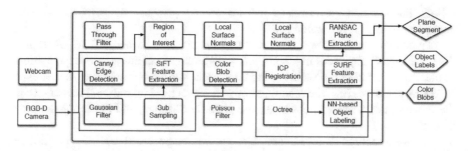

Fig. 1.12 Perception graph for the RoboCup@Work scenario.

Here, sensing components on the left-hand side represent sensors that are typically available on mobile robots such as cameras, depth cameras, and laser scanners. Those sensors produce data that are processed by many and diverse processing components. Those processing components encapsulate perception-related functions (e.g. filters, feature descriptors, and so forth) that produce and consume data in a flow-oriented manner. After performing several processing steps, some output is provided as shown on the right-hand side of Fig. 1.12.

For example, the container recognition feature, which is part of the feature model depicted in Fig. 1.2, is implemented by the perception flow composed of three modules connected by red arrows in Fig. 1.12: *RGB-D Camera, Color Blob Detection, Color Blobs*. An alternative implementation may include a *Gaussian Filter* module that filters images produced by the *RGB-D Camera* module before they are processed by the *Color Blob Detection* module. The *Gaussian Filter* module improves the quality of images when the camera or the object is moving, but reduces the performance of the container recognition functionality.

More, in general, alternative implementations of a robot functionality may substantially differ for non functional properties (NFP), such as performance, robustness, and resource demand. The next section surveys existing approaches for dealing with non functional requirements in robot control systems.

Modeling and Analysis of Non functional Requirements The tight relationship between non functional properties and software architectures has been investigated in several software-intensive domains [4]. For example, dependability is the most important NFP in service-oriented systems [6]. Security and availability are important in telecommunication systems [48].

Software systems for autonomous robots have several NFPs of interest in common with other types of cyber-physical systems (e.g. medical monitoring systems, automatic pilot avionics), such as *Reliability, Availability, Maintainability,* and *Safety* (RAMS properties). Other NFPs are domain-specific. For example, the performance of a robot navigation system is defined by Steinfeld et al. [52] in terms of (i) deviation from planned route, (ii) percentage of obstacles that were successfully avoided, and (iii) percentage of covered area. Similarly, for social robots, the relevant performance metrics are (i) persuasiveness, (ii) trust, (iii) engagement, and (iv) compliance.

In previous work [12], we have presented an extension to the UML MARTE profile called *ARM::Safety* for modeling safety requirements of autonomous navigation. In particular, it makes it possible to model the relationship between the properties of hardware resources (e.g. the scanning rate of a sensor), the operational scenario (e.g. the dynamics of a moving obstacle), the obstacle distance, and the robot stopping distance.

The Architecture Analysis and Design Language (AADL) [25] is another modeling language that makes it possible to annotate architectural models with non functional properties. AADL is supported by the ASSERT Set of Tools for Engineering (TASTE) [45], an open-source development environment dedicated to embedded, real-time systems, which was created under the initiative of the European Space Agency. The TASTE environment is used in the context of the EROCOS project [42] to develop and test an open-source robotics framework to support the development of space robotics applications with space-grade RAMS properties.

The RoQME project [47] aims to develop a domain-specific language to specify NFPs of autonomous robots and a set of tools for their runtime monitoring. In particular, the *effectiveness* of a robotic salesman is defined as the degree to which the autonomous robot is successful in producing the following desired results: (i) acceptance among customers, (ii) attracting the interest of the customers, (iii) engaging customers in the service, (iv) achieving a high level of interaction with the customers, and (v) high ratio of customers served to customers detected.

Gobillot et al. [29] present an extension of the Mauve DSL, which supports both architectural design of a robot control system and the analysis of its temporal properties. As such, the Mauve DSL resembles a simplified version of UML MARTE profile for schedulability analysis. In addition, the Mauve DSL makes it possible to specify behavioral properties of the control system using temporal logic.

5 Application Engineering

The previous section has illustrated an approach to exploit well-known domain engineering techniques for the development of robotic software product lines. According to this approach, domain engineering produces three main outputs: a model of the variability in application requirements (e.g. the feature model in Fig. 1.11), a model of the variability in robot capabilities (e.g. the feature model in Fig. 1.3), and a reference architecture for each functional subsystem (e.g. the perception graph in Fig. 1.12).

These models represent the input of the application engineering phase that produces a model of the configured control system according to the requirements of the specific application at hand.

Each new application is built by configuring the SPL, i.e. by selecting the variants (e.g. functionalities, software resources) that meet specific application requirements. For this purpose, model-driven engineering (MDE) environments simplify system configuration by providing domain-specific languages to model robot variability and model-to-model transformations to resolve the variability in the software control system. A tutorial and survey on model-driven software engineering approaches in robotics can be found in [11].

In particular, the HyperFlex [28] toolchain is an MDE environment for the development and configuration of software product lines for robotics. It provides domain-specific languages and graphical editors for the definition of (i) feature models representing variability in application requirements and variability in robot capabilities, (ii) architectural models of component systems and subsystems implementing robotic functionalities, and (iii) resolution models for linking feature models and architectural models and generating configuration files of the deployed control system.

Resolution models make it possible to link the feature model representing the application requirements and the feature model representing the robot capabilities. For example, the system designer can specify that the feature Precision Placement (see Fig. 1.11) in the Requirements feature model is linked to the feature Dynamic Constraints (see Fig. 1.3) in the Capabilities feature model. Similarly, the feature Static obstacles is linked to feature DWA local planner.

The selection of desired features in the Requirements feature model triggers model-to-model transformations that automatically select the corresponding features in the Capabilities feature model. In turn, this selection triggers a model-to-model transformation that configures the software architecture. Several types of transformations can be defined, such as removing a component, changing the properties of a component, and changing the connections between components. HyperFlex supports the configuration of component-based systems based on the ROS framework.

At deployment time, a system engineer configures the robot control system by simply selecting the relevant features in the Requirements feature model, and HyperFlex generates the configuration file, i.e. the ROS launch file.

As an example, Fig. 1.13 represents the architecture of the *Marker Locator* functional system for detecting visual markers, which is made of two elemental components. The *Marker Detector* receives as input an RGB image and computes the 3D position of the marker. A property called *Pattern type* is used for configuring the detector with the specific type of marker. The *Deblurring Filter* processes the input image for noise reduction. This component is **optional** and is needed only when the images are acquired by a moving camera.

Components define provided and required interfaces, which can be connected to exchange messages according to the topic-based publisher/subscriber paradigm (e.g. the topic *3DPose* in Fig. 1.13).

Figure 1.14 shows the Feature Diagram (left) representing a configuration of the Marker Locator subsystem and the corresponding ROS launch file (right) that is used to start the two ROS nodes of Fig. 1.13. It should be noted that the feature *Moving*

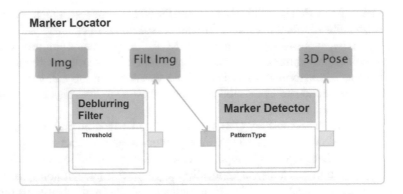

Fig. 1.13 The architectural model of the Marker Locator SPL.

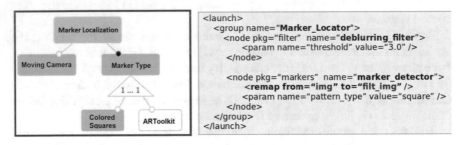

Fig. 1.14 The ROS launch file.

Camera is selected (marked in green). As a consequence, the node *Deblurring Filter* is activated and the node *Marker Detector* subscribes to the topic *Filtered Messages* (see the *remap* tag in Fig. 1.14 right).

If the feature *Moving Camera* is not selected (marked in white), the *remap* tag is removed from the ROS launch file as well as the *node* tag that activates the *deblurring_filter* node.

Most variation points are bound at deployment time, but there may still be variation points that are bound only at runtime. This flexibility enables the dynamic reconfiguration of the component system according to the execution context [14].

A different approach to robotics variability management is discussed by Lotz et al. [39], where the SmartSoft [2] developers propose to model functional and non functional properties for adapting the robot behavior at runtime. In particular the approach addresses two orthogonal levels of variability by means of two domain-specific languages: (a) the variability related to the operations required for completing a certain task and (b) the variability associated to the quality of service.

The first level of variability is managed by configuring and activating the software components and coordinating the robot actions according to the state of the robot and the state of the environment (e.g. the order of the actions needed for cleaning a table changes according to the objects that have to be removed). The second level of variability is managed by configuring certain functional properties (e.g. maximum velocity), based on the value of some context variables (e.g. battery level), in order to optimize non functional properties (e.g. power consumption). These two variability levels are more related to the execution of a specific task (in the paper, the example is a robot delivering coffee), while in our approach we model the variability of functional systems and the variability of the family of applications resulting from the composition of functional systems.

6   Conclusions

This chapter has presented the fundamental concepts related to software product line (SPL) engineering and has shown how this software development technique can be applied to the design and configuration of robot control systems.

SPL engineering is a paradigm shift in robot system engineering, from single system development for specific application requirements to the development of a family of related systems that cover the requirements of an application domain.

The benefit of SPL engineering is the reduction of development costs and time to market of new robotic systems that is achieved thanks to the focus on the design of software artifacts (components and architectures) that are reusable across applications. The shortcoming is the initial investment in the domain engineering phase that may require significant effort and advanced skills in robotics, software engineering, and system engineering.

A recent study by Garcia et al. [27] has analyzed the current practice in software development for robotics from both the academic and industrial perspectives. The study shows that practitioners hardly reuse existing software components and rather develop them from scratch. The most common reason (77.97%) is technical problems (e.g. component's granularity does not fit, missing functionality, incompatible interfaces). These technical problems would be solved by adopting the SPL engineering approach, which promotes the definition of reference architectures for common robot capabilities. Unfortunately, the same study reveals that only 13% of the respondents use SPL engineering and that 32% of the total amount of respondents did not even know about software product line engineering (SPLE).

A wider adoption of SPL engineering in robotics requires initiatives, such as those illustrated in Chap. 3, which favor the synergies between the research community and key industrial players in the robotics domain with the support of international funding agencies.

A significant challenge related to the adoption of SPL engineering in robotics is the verification and validation of a family of related systems. While models, techniques, and architectures exist that support formal validation and verification of individual autonomous robotic system (see Chaps. 8, 9, 11, and 13), their extension to SPL requires them to face the additional problem of combinatorial explosion of alternative configurations for the same system. New approaches are needed for the verification and validation of an SPL as a whole.

A possible approach for addressing this challenge is presented in Chap. 4, which is based on the use of Constraint Programming.

References

1. G.T. Mark, Incorporating flexibility into system design: a novel framework and illustrated developments. Thesis (Masters of Science). Department of Aeronautics and Astronautics, Massachusetts Institute of Technology (2005). https://dspace.mit.edu/handle/1721.1/30363? show=full
2. The Smartsoft Project, (2013). http://smart-robotics.sourceforge.net/
3. S.A. Halim, N.A. Jawawi Dayang, I. Noraini, D. Safaai, An approach for representing domain requirements and domain architecture in software product line, in *Software Product Line - Advanced Topic* (InTech, 2012), pp. 23–42
4. D. Ameller, C. Ayala, J. Cabot, X. Franch, Non-functional requirements in architectural decision making. IEEE Software **30**(2), 61–67 (2013)
5. J. Baumgartl, T. Buchmann, D. Henrich, B. Westfechtel, Towards easy robot programming - using dsls, code generators and software product lines, in *ICSOFT 2013 - Proceedings of the 8th International Joint Conference on Software Technologies, Reykjavík, Iceland, 29–31 July, 2013*, ed. by J. Cordeiro, D.A. Marca, M. van Sinderen (SciTePress, 2013), pp. 548–554
6. H. Becha, D. Amyot, Consumer-centric non-functional properties of soa-based services, in *Proceedings of the 6th International Workshop on Principles of Engineering Service-Oriented and Cloud Systems*, PESOS 2014 (ACM, New York, NY, USA, 2014), pp. 18–27
7. J. Boren, S. Cousins, Exponential growth of ROS [ROS topics]. IEEE Robot. Automat. Mag. **18**, 19–20 (2011)
8. R.A. Brooks, Intelligence without representation. Artificial Intelligence **47**(1), 139–159 (1991)
9. D. Brugali, W. Nowak, L. Gherardi, A. Zakharov, E. Prassler, Component-based refactoring of motion planning libraries, in *2010 IEEE/RSJ International Conference on Intelligent Robots and Systems (IROS)*, pp. 4042–4049, oct. 2010
10. D. Brugali, Stable analysis patterns for robot mobility, in *Software Engineering for Experimental Robotics*, vol. 30 of *Springer Tracts in Advanced Robotics*, ed. by D. Brugali (Springer, Berlin, Heidelberg, 2007), pp. 9–30
11. D. Brugali, Model-driven software engineering in robotics: Models are designed to use the relevant things, thereby reducing the complexity and cost in the field of robotics. IEEE Robot. Automat. Mag. **22**(3), 155–166 (2015)
12. D. Brugali, Modeling and analysis of safety requirements in robot navigation with an extension of uml marte, in *Proceedings of the 2018 IEEE International Conference on Real-Time Computing and Robotics*, RCAR'18 (IEEE, 2018), pp. 439–444
13. D. Brugali, A. Brooks, A. Cowley, C. Côté, A.C. Domínguez-Brito, D. Létourneau, F. Michaud, C. Schlegel, *Trends in Component-Based Robotics* (Springer, Berlin, Heidelberg, 2007), pp. 135–142
14. D. Brugali, R. Capilla, M. Hinchey, Dynamic variability meets robotics. IEEE Computer **48**(12), 94–97 (2015)
15. D. Brugali, N. Hochgeschwender, Software product line engineering for robotic perception systems. Int. J. Semantic Comput. **12**(01), 89–107 (2018)
16. P.C. Clements, L. Northrop, *Software Product Lines: Practices and Patterns*. SEI Series in Software Engineering (Addison-Wesley, 2001)
17. M. Cline, M. Girou, Enduring business themes. Commun. ACM **43**, 101–106 (2000)
18. J. Coplien, D. Hoffman, D. Weiss, Commonality and variability in software engineering. IEEE Software **15**(6), 37–45 (1998)

19. K. Czarnecki, U.W. Eisenecker, *Generative Programming: Methods, Tools, and Applications* (ACM Press/Addison-Wesley Publishing, New York, NY, USA, 2000)
20. S. Dhouib, S. Kchir, S. Stinckwich, T. Ziadi, M. Ziane, Robotml, a domain-specific language to design, simulate and deploy robotic applications, in *Simulation, Modeling, and Programming for Autonomous Robots - Third International Conference, SIMPAR 2012, Tsukuba, Japan, November 5–8, 2012. Proceedings*, vol. 7628 of *Lecture Notes in Computer Science* ed. by I. Noda, N. Ando, D. Brugali, J.J. Kuffner (Springer, 2012), pp. 149–160
21. K. Ehrlenspiel, A. Kiewert, U. Lindemann, *Cost-Efficient Design* (Springer, 2007)
22. M. Fayad, Accomplishing software stability. Commun. ACM **45**(1), 111–115 (2002)
23. M. Fayad, A. Altman, An introduction to software stability. Commun. ACM **44**, 95–98 (2001)
24. M.E. Fayad, D.S. Hamu, D. Brugali, Enterprise frameworks characteristics, criteria, and challenges. Commun. ACM **43**(10), 39–46 (2000)
25. P. Feiler, D. Gluch, J. Hudak, The architecture analysis and design language (aadl): An introduction. Technical Report CMU/SEI-2006-TN-011, Software Engineering Institute, Carnegie Mellon University, Pittsburgh, PA, 2006
26. S. Fürst, J. Mössinger, S. Bunzel, T. Weber, F. Kirschke-Biller, P. Heitkämper, G. Kinkelin, K. Nishikawa, K. Lange, Autosar—a worldwide standard is on the road, in *14th International VDI Congress Electronic Systems for Vehicles, Baden-Baden*, vol. 62, 2009
27. S. Garcia, D. Struber, D. Brugali, T. Berger, P. Pelliccione, Robotics software engineering: A perspective from the service robotics domain, in *ESEC/FSE 2020 - Proceedings of the 2020 ACM Joint European Software Engineering Conference and Symposium on the Foundations of Software Engineering, held virtually, 8–13 November, 2020*, 2020
28. L. Gherardi, D. Brugali, Modeling and reusing robotic software architectures: the HyperFlex toolchain, in *IEEE International Conference on Robotics and Automation (ICRA 2014)*, Hong Kong, China, May 31–June 5 2014 (IEEE, 2014)
29. N. Gobillot, C. Lesire, D. Doose, A modeling framework for software architecture specification and validation, in *Simulation, Modeling, and Programming for Autonomous Robots*, ed. by D. Brugali, J.F. Broenink, T. Kroeger, B.A. MacDonald (Springer International Publishing, Cham, 2014), pp. 303–314
30. M. Hägele, N. Blümlein, O. Kleine, Profitability Analysis of New Service Robot Applications and Their Relevance for Robotics (EFFIROB) (2010). https://www.aal.fraunhofer.de/en/projekte/effirob.html
31. N. Hochgeschwender, L. Gherardi, A. Shakhirmardanov, G.K. Kraetzschmar, D. Brugali, H. Bruyninckx, A model-based approach to software deployment in robotics, in *IEEE/RSJ International Conference on Intelligent Robots and Systems IROS*, pp. 3907–3914, 2013
32. N. Hochgeschwender, S. Schneider, H. Voos, G.K. Kraetzschmar, Declarative specification of robot perception architectures, in *SIMPAR*, 2014
33. F. Iida, Autonomous robots: From biological inspiration to implementation and control. george a. bekey. Artificial Life **13**, 419–421 (2007)
34. B. Jørgensen, W. Joosen, Coping with variability in product-line architectures using component technology, in *Technology of Object-Oriented Languages, Systems and Architectures*, vol. 732 of *The Kluwer International Series in Engineering and Computer Science*, ed. by T. D'Hondt (Springer US, 2003), pp. 208–219
35. E. Jung, C. Kapoor, D. Batory, Automatic code generation for actuator interfacing from a declarative specification, in *2005 IEEE/RSJ International Conference on Intelligent Robots and Systems, 2005. (IROS 2005)*, pp. 2839–2844, 2005
36. K.C. Kang, Feature-oriented domain analysis (FODA) feasibility study. Technical report, DTIC Document, 1990
37. K.C. Kang, M. Kim, J. Lee, B. Kim, Feature-oriented re-engineering of legacy systems into product line assets: a case study, in *Proceedings of the 9th International Conference on Software Product Lines*, SPLC'05 (Springer, Berlin, Heidelberg, 2005), pp. 45–56
38. C.W. Krueger, Easing the transition to software mass customization, in *Revised Papers from the 4th International Workshop on Software Product-Family Engineering*, PFE '01 (Springer, London, UK, 2002), pp. 282–293

39. A. Lotz, J.F. Inglés-Romero, C. Vicente-Chicote, C. Schlegel, Managing run-time variability in robotics software by modeling functional and non-functional behavior, in *Enterprise, Business-Process and Information Systems Modeling* (Springer, 2013), pp. 441–455

40. M. Mandelbaum, J. Buzacott, Flexibility and decision making. Eur. J. Oper. Res. **44**(1), 17–27 (1990)

41. C. McLean, M. Mitchell, E. Barkmeyer, A computer architecture for small-batch manufacturing: Industry will profit from a system that defines the functions of its component-manufacturing modules and standardizes their interfaces. IEEE Spectrum **20**(5), 59–64 (1983)

42. M. Muñoz, G. Montano, M. Wirkus, K.J. Höflinger, D. Silveira, N. Tsiogkas, J. Hugues, H. Bruyninckx, I. Dragomir, A. Muhammad. Esrocos: A robotic operating system for space and terrestrial applications, in *ASTRA 2017*, 2017

43. E.Y. Nakagawa, P.O. Antonino, M. Becker, Reference architecture and product line architecture: A subtle but critical difference, in *Proceedings of the 5th European Conference on Software Architecture*, ECSA'11 (Springer, Berlin, Heidelberg, 2011), pp. 207–211

44. R.C. van Ommering, J. Bosch, Widening the scope of software product lines - from variation to composition, in *Proceedings of the Second International Conference on Software Product Lines*, SPLC 2 (Springer, London, UK, 2002), pp. 328–347

45. M. Perrotin, E. Conquet, J. Delange, A. Schiele, T. Tsiodras, Taste: A real-time software engineering tool-chain overview, status, and future, in *SDL 2011: Integrating System and Software Modeling*, ed. by I. Ober, I. Ober (Springer, Berlin, Heidelberg, 2012), pp. 26–37

46. M. Quigley, B. Gerkey, K. Conley, J. Faust, T. Foote, J. Leibs, E. Berger, R. Wheeler, A. Ng, Ros: an open-source robot operating system, in *ICRA Workshop on Open Source Software*, vol. 3, 2009

47. J.F.I. Romero, How well does my robot work? 2018

48. M. Saadatmand, A. Cicchetti, M. Sjödin, Uml-based modeling of non-functional requirements in telecommunication systems, in *The Sixth International Conference on Software Engineering Advances (ICSEA 2011)*, October 2011

49. J.H. Saleh, D.E. Hastings, D.J. Newman, Flexibility in system design and implications for aerospace systems. Acta Astronautica **53**(12), 927–944 (2003)

50. J.H. Saleh, G. Mark, N.C. Jordan, Flexibility: a multi-disciplinary literature review and a research agenda for designing flexible engineering systems. J. Eng. Des. **20**(3), 307–323 (2009)

51. A.K. Sethi, S.P. Sethi, Flexibility in manufacturing: A survey. Int. J. Flexible Manuf. Syst. **2**(4), 289–328 (1990)

52. A. Steinfeld, T. Fong, D. Kaber, M. Lewis, J. Scholtz, A. Schultz, M. Goodrich, Common metrics for human-robot interaction, in *Proceedings of the 1st ACM SIGCHI/SIGART Conference on Human-robot Interaction*, HRI '06 (ACM, New York, NY, USA, 2006), pp. 33–40

53. J.M. Thompson, M.P.E. Heimdahl, Structuring product family requirements for n-dimensional and hierarchical product lines. Requirements Engineering **8**(1), 42–54 (2003)

54. C. Tischer, B. Boss, A. Müller, A. Thums, R. Acharya, K. Schmid, Developing long-term stable product line architectures, in *Proceedings of the 16th International Software Product Line Conference (SPLCA12)* (vol. 1) (ACM, 2012), pp. 86–95

55. J. van Gurp, J. Bosch, M. Svahnberg, On the notion of variability in software product lines, in *Proceedings Working IEEE/IFIP Conference on Software Architecture*, pp. 45–54, 2001

56. R. Weiss, J. Doppelhamer, H. Koziolek, Bottom-up software product line design: A case study emphasizing the need for stakeholder commitment, in *Proc. 5th SEI Architecture Technology User Network Conference (SATURN'09)*, 2009

Chapter 2
Towards Autonomous Robot Evolution

Agoston E. Eiben, Emma Hart, Jon Timmis, Andy M. Tyrrell, and Alan
F. Winfield

Abstract We outline a perspective on the future of evolutionary robotics and
discuss a long-term vision regarding robots that evolve in the real world. We
argue that such systems offer significant potential for advancing both science and
engineering. For science, evolving robots can be used to investigate fundamental
issues about evolution and the emergence of embodied intelligence. For engineering,
artificial evolution can be used as a tool that produces good designs in difficult
applications in complex unstructured environments with (partially) unknown and
possibly changing conditions. This implies a new paradigm, second-order software
engineering, where instead of directly developing a system for a given application,
we develop an evolutionary system that will develop the target system for us.
Importantly, this also holds for the hardware; with a complete evolutionary robot
system, both the software and the hardware are evolved. In this chapter, we discuss
the long-term vision, elaborate on the main challenges, and present the initial results
of an ongoing research project concerned with the first tangible implementation of
such a robot system.

A. E. Eiben (✉)
Vrije Universiteit Amsterdam, Amsterdam, The Netherlands
e-mail: a.e.eiben@vu.nl

E. Hart
Napier University Edinburgh, Edinburgh, UK
e-mail: E.Hart@napier.ac.uk

J. Timmis
School of Computer Science, University of Sunderland, Sunderland, UK
e-mail: jon.timmis@sunderland.ac.uk

A. M. Tyrrell
Department of Electronic Engineering, University of York, York, UK
e-mail: andy.tyrrell@york.ac.uk

A. F. Winfield
Bristol Robotics Laboratory, UWE Bristol, Bristol, UK
e-mail: alan.winfield@brl.ac.uk

© The Author(s) 2021
A. Cavalcanti et al. (eds.), *Software Engineering for Robotics*,
https://doi.org/10.1007/978-3-030-66494-7_2

29

1   The Grand Vision of Robot Evolution

The long-term vision regarding robot evolution foresees systems of robots that reproduce and undergo evolution in the real world. The interest in such systems is based on perceiving evolution as a generic mechanism that has driven the emergence of life on Earth and created a vast diversity of life forms adapted to all kinds of environmental conditions. Thus, the rationale behind the grand vision can be stated as follows.

> As natural evolution has produced successful life forms for practically all possible environmental niches on Earth, it is plausible that artificial evolution can also produce specialised robots for various environments and tasks.

Robot evolution offers advantages in different areas depending on where the emphasis lies, on robots or on evolution. To put it simply, we can distinguish two perspectives: engineering and science.

As for engineering, artificial evolution can deliver good robot designs for a specific application. In this approach, evolution is (ab)used as an optimiser that is halted when a satisfactory solution is found. Real evolution, however, is not about optimisation but about adaptation that never stops. This potential can be unlocked through open-ended artificial evolution where a robot population continually changes and adapts to possibly unforeseen and changing conditions. In such systems, robots are improving in two dimensions: they are becoming fit for the environment and fit for purpose (as defined by their users).

Robots that evolve in an open-ended 'hands-free' fashion are also very interesting from a scientific perspective. Specifically, they can be regarded as hardware models of natural evolution that are programmable, tuneable, and observable. Hence, they can be employed as a novel research instrument to investigate fundamental issues about evolution and the emergence of embodied intelligence.

1.1   *Researching Evolution Through Robots*

Evolution is a natural phenomenon that has driven the emergence of life on Earth. Studying evolution is traditionally a subject within biological science. However, the invention of the computer and the ability to create artificial (digital) worlds opened the possibility of establishing artificial evolution, that is, computational processes that mimic Darwinian principles. The fundamental insight behind this is the observation that the reproduction-selection cycle that is pivotal to natural evolution is analogous to the generate-and-test loop of search algorithms. This was first postulated a long time ago [7, 42], and by the end of the twentieth century, evolutionary computing (EC) has become a vibrant research area with many applications [21]. Over the past half-century, various evolutionary algorithms

(EAs) have been developed under different names including genetic algorithms, genetic programming, and evolution strategies. They have proven their power on hard problems without analytical models, with non-linear relations among variables and complex objective functions with multiple local optima.

Evolutionary computing mimics natural evolution, but it takes place in a virtual space, whereas natural evolution happens in the real world. The advantage of evolutionary computing systems is that they are programmable, configurable, and observable. However, computer models and simulators are based on abstractions and approximations and may miss crucial aspects of the real world. This leads to the so-called reality gap, the phenomenon that the performance of a solution evolved in simulation decreases once it is transferred to a real robot. This gap can be significant, and observations and conclusions about system behaviour may be mere artefacts [28]. Natural evolutionary systems are quite the opposite. They are certainly real but hardly programmable, configurable, and observable. The combination of the two offering the best of both worlds is called the *Evolution of Things*. The notion and the term was introduced by Eiben et al. in [20] and further discussed in [15] and [22].

The key idea behind the Evolution of Things concept is to have a programmable evolutionary system that works with physical artefacts (see Fig. 2.1). These artefacts can be passive, e.g. sunglasses or airplane wings, or active, e.g. animate things: robots for short. Robots that are able to reproduce and evolve in the real world can be perceived as a physically embodied model of natural evolution [34] and used as a research instrument to study evolutionary phenomena. Fundamental questions that can be investigated include the evolution of (embodied) intelligence, the interplay between the body and the brain, and the impact of the environment on the evolved organisms. To date, such issues can be observed and studied in wetware (living systems) and software (computing systems); the Evolution of Things will realise them in hardware (robotic systems).

Using real robots instead of simulations is interesting, because this guarantees that the observed effects are real and not just artefacts of the simulator. Research with robots also offers advantages with respect to living organisms, because robots

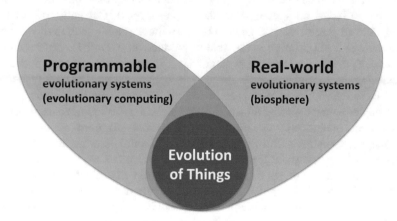

Fig. 2.1 Evolution of Things: the best of both worlds, after [15].

are easily observable (e.g. internal processes and communication can be logged) and controllable. This allows systematic studies under strictly regulated conditions and many repetitions for solid statistics.

1.2 Supervised Robot Evolution: Breeding Farms

Evolutionary algorithms have been successful in solving various design problems, typically regarding inanimate artefacts with a functional or aesthetic value [2, 3, 27]. The potential for designing robots is evident; robots for a certain application can be developed through iterated selection and reproduction cycles until they satisfy the users' criteria.

Obviously, designing robots for structured environments with known and predictable conditions can be done by classic engineering. However, complex unstructured environments with (partially) unknown and possibly changing conditions represent a completely different challenge. Think, for instance, of robots for environmental monitoring in rain forests, exploration of ocean floors, or terraforming on other planets. In such cases, it is hard to determine the optimal morphologies and the control systems driving them. For example, should a robot that operates in the jungle have wheels, legs, or both? What is the optimal arrangement of its sensors? Should that robot be small to manoeuver through narrow openings or should it be big and heavy to trample down obstacles? The number of options considering materials, components, shapes, and sizes is huge, the link between a specific robot makeup and its task performance is poorly understood, and theoretical models and proven design heuristics are lacking.

In such cases, a good robot design can be evolved on a 'robot breeding farm'. This means a human-controlled and human-operated facility that has a mock-up environment resembling the real application conditions and an evolutionary engine that implements the two main components of any evolutionary process: reproduction and selection.

The first one can be realised by a (re)production facility that constructs the robots (phenotypes) based on specification sheets (genotypes). Currently, in 2020, this is a significant engineering challenge because the available technology for rapid prototyping and 3D printing is not able to produce all essential robot ingredients. For instance, printing servo motors and CPUs is beyond reach at the moment.

Realising the second component, selection, is much easier. The quality of a robot can be measured by its performance in the test environment, and parent selection can be implemented by using this quality as fitness. Importantly, the users can steer and accelerate evolution by selecting/deselecting robots for reproduction as they see fit. Doing so, users act akin to farmers who breed animals or plants; this explains the breeding farm metaphor. If needed, the user can also influence (re)production by the robotic equivalent of genetic modifications or directly injecting hand-designed robots into the evolving population. The evolutionary design process on a breeding farm stops after obtaining a good robot. This robot is then put forward as the solution and several copies of it can be produced to be deployed in the real application environment.

1.3   Open-Ended Robot Evolution: Out in the Wild

Robot breeding farms represent one particular approach to artificial evolution. In this approach, evolution is employed as a particular design method, and if the conditions or requirements change, the whole process needs to be repeated.

Enabling robot populations to adapt to previously unknown or changing conditions is essential in applications, where the environment is inaccessible, dangerous, or too costly for humans to approach and work in. Examples include remote areas on Earth, deep seas, and other planets. An evolutionary engine operating autonomously on the spot can mitigate this problem. The first component of this system is a (re)production facility that can make use of local resources and construct a large variety of robots. The second is a twofold selection drive, such that robots become fit for the environment and fit for purpose. Environmental selection (for viability) is for free, as robots with a poor feature set will not be able to operate adequately. Sexual selection, in turn, can be pre-programmed such that robots have a 'basic instinct' to choose mating partners with a high task performance (utility). The evolving robot population will then become increasingly adapted to the given environment and simultaneously become better and better in performing the task(s) predefined by the users. Importantly, over consecutive generations, the bodies and brains of the robot population adjust when the conditions change. In other words, evolution allows the robot population to optimise itself while on the job.

Such an evolving robot colony is very different from a breeding farm because the evolutionary system must be able to operate for extended periods of time without direct human oversight. In contrast to the breeding farm scenario, the evolutionary process never stops, and there is no such thing as *the* final solution in the form of any given robot makeup. In this respect, open-ended robot evolution is closer to biological evolution, while a breeding farm is more a (r)evolutionary approach to designing and optimising robots.

2   Evolutionary Robotics

The discipline that is concerned with robot evolution is evolutionary robotics (ER), a research area that applies EAs to design and optimise the bodies (morphology, hardware), the brains (controller, software), or both for simulated or real autonomous robots [5, 13, 14, 23, 37, 38, 43]. Over the last 20 years, the ER field has addressed the evolution of robot controllers in fixed morphologies with considerable success, but evolving the bodies has received much less attention. This is somewhat understandable, given the difficulty of implementing such systems in the real world, i.e. the lack of technologies for automated (re)production of robots. However, advances in robotics, 3D printing, and automated assembly mean it is now timely to work on physical robot systems with evolvable morphologies [45].

Systems within Artificial Life have addressed the evolution of morphologies (and control systems) of virtual creatures in simulated worlds, e.g. in the pioneering works by Sims, Bongard, and Pfeifer [4, 39, 41] and several subsequent studies

[1, 9]. This was brought closer to the physical instantiation by jointly evolving robotic shapes and control in a simulator and 3D printing the evolved shapes [33]. However, evolution took place in simulation and only the final product was materialised; furthermore, the robot had no sensors. The self-reproducing machines of Zykov et al. were modular robots that were designed or evolved to be able to make exact clones of themselves without variation, and they did not undergo evolution in the real world [46, 47]. More recent work has evolved morphologies composed of novel, soft materials, but once again only the final morphologies were constructed, and in this case they were confined to operating within a pressurised chamber rather than operating in a real-world environment [26]. A related sub-field of evolutionary design has concerned itself with constructible objects, but here again evolution generally has taken place in software, with only the end result being constructed. A few projects employed fitness evaluations on the hardware itself, but these systems produced inert objects with no controllers that passively underwent evolution [32, 40].

A separate category of related work contains biologically motivated studies with evolution implemented in populations of robots [24, 44]. Using real hardware is extremely challenging, and, to the best of our knowledge, there has been only one successful project, that of Long et al. investigating the evolution of vertebrae through swimming robot fish [10, 34]. The project faced huge practical problems, for instance, manual construction of new generations took weeks, which severely limited the number of experiments and the number of generations per experiment.

One of the most relevant related works with respect to the autonomous robot evolution concept is the study of Brodbeck et al. which investigated the morphological evolution of physical robots through model-free phenotype development [8]. Being model-free means that the system does not employ simulations—all robots are physically constructed. As noted by the authors, this avoids the reality gap but raises two new problems: the birth problem and the speed problem—identified previously as challenge 2 and challenge 4 in [20]. The system demonstrates a solution to the birth problem in real hardware based on modular robot morphologies. Two types of cubic modules (active and passive) form the raw materials, and robot bodies are constructed by stacking and gluing a handful of such modules. The robots do not have an on-board controller, they are driven by an external PC, and their task is to locomote. Also the EA runs on the external PC with populations of size 10 and fitness defined by the travelled distance in a given time interval. Robot genomes encode the bodies implicitly by specifying the sequence of operations to build them by a robotic arm, dubbed the mother robot. The construction of new robots ('birth process') is hands-free in some of the reported experiments but requires human assistance in some others.

Finally, the RoboGen system and the Robot Baby Project represent a significant stepping stone towards autonomous robot evolution [1, 30]. The RoboGen system features modular robots as phenotypes, a corresponding space of genotypes that specify the morphology and the controller of a robot, and a simulator that can simulate the behaviour of one single robot in a given environment [1]. Evolution is implemented by a classic evolutionary algorithm that maintains a population of genotypes; executes selection, crossover, and mutation; and calls the simulator for each fitness evaluation. The system is applied to evolve robots in simulation, and physical counterparts of the simulated robots can be easily constructed by 3D printing and manually assembling their components.

RoboGen was not meant to and has never been used to physically create each robot during an evolutionary process, but it inspired the robot design of the Robot Baby Project [30]. This project is a proof-of-concept study to demonstrate robot reproduction, and ultimately evolution, in a real-world habitat populated by robots. A key feature is that robots coexist and (inter)act in the same physical space where they can 'live and work', 'meet and mate', thus producing offspring and—over the long run—many generations of evolving robots. This feature is unique, since in other related works a traditional EA performs evolution, where robots are manifested one by one and evaluated in isolation during the fitness evaluation steps. The Robot Baby Project was based on the Triangle of Life system model (discussed in Sect. 4) and implemented all system components in the simplest possible form, cutting several corners. The experiment started with an initial population of two robots and ran a complete life cycle resulting in a new robot, parented by the first two. Even though the individual steps were simplified to the greatest extent possible, the whole system validated the underlying concepts, illuminated the generic workflow for the creation of more complex incarnations, and identified the most important open issues (Fig. 2.2).

Fig. 2.2 The 'first family' produced by the Robot Baby Project. The blue 'spider' and the green 'gecko' in the background are the parents, the robot in front of them is their offspring. See [30] for details.

3　Evolvable Robot Hardware: Challenges and Directions

Consider now a long-term vision for evolvable robot hardware—freed from the limitations of current materials and fabrication technologies. We see two possible futures: one in which a *universal robot manufacturing cell* evolves and fabricates robots to order and the second in which *smart proto-robot matter* evolves itself into more complex, interesting (and possibly useful) forms. The first of these visions falls squarely within the engineering paradigm; it implies supervised evolution against a specification and requires a (possibly compact and self-contained) manufacturing infrastructure. The second vision is decidedly bio-inspired, even bio-mimetic, and falls within the Artificial Life paradigm: specifications might be loose or non-existent and hence evolution open-ended rather than optimising towards some fitness. Here the infrastructure would consist of some environment in which the robot matter 'lives', recombining or replicating and collectively adapting in response to its environment; in this vision, robots would, in some sense, physically self-reproduce.

3.1　An Engineering Approach

3.1.1　What Would We Evolve?

A robot is a complex interconnected system of subsystems, and any strategy for evolving a robot's body must be based upon the physical characteristics of those subsystems. It is helpful to categorise a robot's physical subsystems according to their function and then consider the implication of each of these for evolving robot bodies. Table 2.1 identifies seven functions, each of which would normally be present in a mobile robot able to both sense its environment and act on that environment.

Table 2.1 Functions normally present in mobile robots to sense and act on the environment

Function	Examples of the physical instantiation of the function
Sensing	Cameras, microphones, accelerometers, IR sensors, touch sensors, location sensors (GPS)
Signalling	LEDs, loudspeakers, Wi-Fi or Bluetooth
Actuation (for moving or for manipulating)	Motors driving wheels, legs, or propulsion for lift (rotors); motors driving actuators for arms and grippers
Energy	A source of electrical power for the robot's electronics and motors
Control	Micro-controllers, or equivalent, to provide robot control
Physical structure, skeleton, or chassis	Metal and/or plastic structure which determines and maintains the organisation of the robot's functional subsystems above
Interconnections	The connections (wiring) between sensing, signalling, actuation, energy, and control subsystems

Consider the physical components required for each of the seven functions outlined above. In principle any or all of these could be evolved. In practice, however, we would not (and perhaps could not) evolve the actual subsystems for some of these functions: sensing, signalling, energy (i.e. the batteries), or control (i.e. the controller hardware), for example. An exception to this general rule would be when we are interested in evolving a particular sensing modality, in which case all other physical aspects would be fixed; an example can be found in [36], in which the auditory system for a Khepera mobile robot was evolved, modelled on Cricket audition.

For sensing and signalling elements, we would normally predetermine[1] and fix the actual sensing and signalling devices and their subsystems. The number, type, and especially the position of sensing and signalling elements are however a different matter. It would be of great interest to allow evolution to discover, from the space of sensing and signalling subsystems, which ones, how many, and where to position these in order to optimise the robot's overall functionality. For the energy and controller hardware, the major decisions must also be predetermined; the power source might be singular or distributed, for instance; but from an evolutionary perspective, providing the energy and controller elements are out of the way—mounted perhaps towards the centre of the physical chassis, then their precise position is of no real interest.

Consider now the actuation elements. Although the motors, or equivalent devices, will need to be predetermined, the physical structures they drive are likely to be a particular focus for evolution. Take mobility, for example; much depends on the kind of motility we are seeking. For a simple wheeled robot, for instance, we might want to evolve the size of the wheels, the number of drive wheels, and their position on the robot's body. For example, Lund [35] describes the evolution of a Lego robot with 3 different wheel types and 25 possible wheel positions. But for walking, flying, or swimming robots with much more complex actuated structures (legs, wings, or fins), or snake-like robots in which the whole of a multi-segmented body is actuated, then it is likely that there will be many parameters we would want evolution to explore and optimise. Similarly, if we want to evolve structures for gripping or manipulating objects, then the number, size, and physical arrangement of the articulated parts would need to be evolvable parameters.

Consider next the robot's physical skeleton or chassis. This passive structure defines the shape of the robot. Although the function of the robot may appear to be primarily determined by its sensing, actuation, and control (software), this is not so: having the correct physical structure and hence physical separation and organisation of sensors and actuators is critical. This 3D structure provides the robot's body-centric frame of reference, within which the position and orientation of the rest of the robot's subsystems are defined. A robot's physical chassis might,

---

[1] With an appropriate level of care. A poor choice of sensing devices, for instance, could have a major negative impact on the evolved robot. It would be hard to evolve, for instance, a wall-climbing robot without an effective sensor for detecting up and down.

at its simplest, be a single stiff component—like the one-piece stiff plastic moulded chassis of the e-puck educational mobile robot—or it may be a complex skeleton with a large number of individual structural elements flexibly connected like the compliant anthropomimetic humanoid ECCE robot [12].[2] In either case, it is likely that we would want evolution to explore the space of possible morphologies for the robot's physical structure. However, the way we would parameterise the robot's physical structure would be different in each case. A relatively simple stiff single-piece chassis can be described with a voxel map, for instance. A more complex skeleton-like structure would be better parameterised in the same way as actuated structures, as a dataset specifying the number, shape, and physical arrangement of its component parts.

Finally, consider the interconnect—the wiring between the electronic subsystems. It is safe to assume a 'standard' wiring model in which (a) every sensing, signalling, and actuation subsystem is wired to the controller subsystem and (b) the controller is wired to the energy source (unless power is distributed among the electronic subsystems). There would seem to be no need for this interconnect to be evolved.

In summary the key elements of the robot we would—in the general case—seek to evolve are:

- The number, type, and position of sensing, signalling, and actuation subsystems
- For the actuation subsystems, the number, shape, and physical arrangement of the articulated parts
- The 3D shape of the robot's physical structure or chassis or, for more complex skeleton like structures, the number, shape, and physical arrangement of the parts of the skeleton

3.1.2   How Would We Physically Build the Evolved Robot's Body?

Consider now the question of how we would physically instantiate the evolved robot body. We are presented with a number of options, at least in principle. These are:

(1) Hand-crafted method   As specified by the genotype, fashion the evolved shapes and structures (i.e. chassis) by hand from raw materials, then hand-assemble these structures together with the predetermined subsystems (i.e. sensors, motors, controller, energy, and wiring), using the genotype to determine the position and orientation of these subsystems.

(2) Hand-constructed method   As specified by the genotype, hand-construct the evolved shapes and structures from pre-manufactured parts (i.e. Lego), then hand-assemble these structures together with the predetermined subsystems, using the genotype to determine the position and orientation of these subsystems.

---

[2]Neither the e-puck or ECCE robot was evolved.

(3) **Semi-automated method** Automate part of the process, by CNC machining or 3D printing the evolved shapes and structures, according to their specification in the genotype, then hand-assemble these structures together with the predetermined subsystems, using the genotype to determine the position and orientation of these subsystems.

(4) **Automated construction method** Fully automate method 2 with a machine that can undertake both the construction of evolved shapes and structures (from, i.e. Lego or equivalent) while also selecting and correctly placing the predetermined subsystems.

(5) **Automated manufacture method** Fully automate method 3 by integrating the process of (say) 3D printing the evolved shapes and structures, with a system for selecting and correctly placing the predetermined subsystems. This method matches the vision of the universal robot manufacturing cell suggested at the start of Sect. 3.

To the best of our knowledge, method 5 has not, to date, been reported in the evolutionary robotics literature. There are several interesting examples of evolved then hand-constructed robots using method 2. For instance, [35] describes the co-evolution of a Lego robot body and its controller in which the evolved robot is physically constructed and tested. Here evolution explores a robot body space with 3 different wheel types, 25 possible wheel positions, and 11 sensor positions. Lund observes that although the body search space is small, with 825 possible solutions, the search space is actually much larger when taking into account the co-evolved controller parameters. Method 3 has been demonstrated in the Golem project [33], although in that project the artificial creatures evolved and physically realised contained no sensing or signalling systems, nor controller or energy, so they were not self-contained autonomous robots. Recently, method 4 has been used in [8], where the automated construction method stacked and glued a handful of modules on each other to form a robot. Also these robots do not have sensing or signalling systems, nor an on-board controller; they are driven by an external PC and their task is to locomote.

3.2 A Bio-inspired (Modular or Multi-cellular) Approach

Biological systems are inherently multi-cellular. It could be argued that modular robotic systems are a natural analogue for multi-cellular systems: one can abstract the fact that a single robotic unit in a modular system could be thought of as a cell, and a modular system a multi-cellular collective. When considering the evolution of these systems, much of what we have discussed in Sect. 3.1 stands, but we have a further consideration to make, and that is now we are not concerned with a single robot acting on its own but a number of robots that have joined together to make a larger single unit that can perform tasks that a single unit cannot.

Fig. 2.3 A SYMBRION organism, comprised of seven modules. Image by W Liu.

Such a modular system may well be able to share resources, such as energy; modules will have a common communication bus with which they can share data, for instance, the SYMBRION robotic platform described by Kernbach et al. in [31] consists of three different types of robots that can join together into a single unit—as shown in Fig. 2.3. Within such a modular system, a review by Eiben et al. [19] is a good example of starting to view the modular robotic collective as a single entity: an 'organism'. Eiben et al. discuss a variety of mechanisms that might be used to evolve not only single robot systems but also modular (and swarm) robotic systems and provide an evolutionary framework to evolve the controllers for robots within that collective, but taking into account the fact that it is a modular collective: so in effect, evolving at the level of a single modular collective. They demonstrate the on-board evolutionary approach using a number of robots, but not when functioning as a single, modular unit.

When considering evolving a controller at the organism level, we are not only concerned with the performance of an individual—we must now consider the performance of the organism with respect to some goal (which may be something as simple as surviving in the environment) and also the performance with respect to the environment: for instance, by implicitly deriving fitness from the environment, rather than having an explicit fitness measure, as shown in the work by Bredech et al. [6] where the environment is taken as the model to drive the performance of the collective.

The work by Zykov et al. [47] demonstrates probably the closest to an evolving modular robotic system on the Molecube platform. In this work, the authors make use of genetic algorithms to evolve various morphologies of the modular system and then, given those morphologies, evolve suitable controllers. This was done to provide a suitable gradient for the evolutionary algorithm to permit the (ultimate) discovery of self-replicating modular systems.

However, as yet, there is still no work that has fully demonstrated the evolution of both structure and function of a modular robotic system, in which evolution is fully embodied in the modular units themselves.

4   The Autonomous Robot Evolution Project

The autonomous robot evolution (ARE) project[3] is concerned with developing the first real robot system with evolvable morphologies. It exceeds its precursors, the works of Long [34], Brodbeck et al. [8], and the Robot Baby Project [30], in several important aspects. The key features that distinguish the ARE system are as follows:

1. Complex, truly autonomous robots with an on-board controller, several sensors, and various actuators undergo *open-ended evolution* in the *real world*.
2. The (re)production mechanism is capable of both manufacturing and assembling a *diverse* range of physical robots autonomously, without human interaction.
3. *Asynchronous* evolution in that robots from different generations coexist in the real world, since producing a new robot does not imply the removal of an existing robot.
4. *Evolution* at a population level is combined with *individual learning* mechanisms which enable a robot to improve its inherited controller during its lifetime in a real environment.
5. Robots are not evolved for a toy application but for a *challenging* task: for example, the exploration of a nuclear cave. This requires that robots develop several morphology-dependent skills, for instance, (targeted) locomotion and object manipulation.

A particularly interesting aspect of the ARE system is the manner the controller software for the robots is developed. Currently, the common approach is to have the software for robots developed by human programmers through the usual development cycle. A system where robots evolve and learn represents a significantly different method, where the controller software of the robots is algorithmically produced.

Specifically, in an evolutionary robot system, both the hardware and the software of a robot, that is, both its bodyplan and its controller, are encoded by its genotype. This genotype is the artificial equivalent of the DNA in living organisms, and it forms the specification sheet that determines the phenotype, the actual robot. By

---

[3] See https://www.york.ac.uk/robot-lab/are/.

the very nature of an evolutionary system, these genotypes undergo crossover and mutation, resulting in new genotypes that encode new robots including new pieces of controller software. Over consecutive generations, selection (and learning) will increase the level of fitness as defined by the given application, thus increasing the functionality of the underlying software.

This implies a new paradigm, second-order software engineering, where instead of directly developing a system for a given application, we develop an evolutionary system that will develop the target system for us. Note that this also holds for the hardware; with a full-blown evolutionary robot system, both the software and the hardware can be evolved.

4.1   Overall System Architecture

A framework for 'evolving robots in real time and real space' has been suggested in Eiben et al. [18]. This framework, named the Triangle of Life, lays out a generic system architecture. A tangible implementation of it is envisaged by the notion of an EvoSphere as introduced and discussed in [17]. An EvoSphere represents a generic design template for an evolutionary robot habitat and forms the basis of the physical environment in the ARE project.

The Triangle of Life consists of three stages: morphogenesis, infancy, and mature life, as illustrated in Fig. 2.4. Consequently, an EvoSphere consists of three components, the Robot Fabricator, the Training Facility, and the Arena. The Robot Fabricator is where new robots are created (morphogenesis). This will be discussed in the next section in detail.

The Training Facility provides a suitable environment for individuals to learn during infancy, providing feedback, perhaps via a computer vision system or a human user, so individual robots can learn to control their—possibly unique—body to acquire basic skills (locomotion, object manipulation) and to perform some sim-

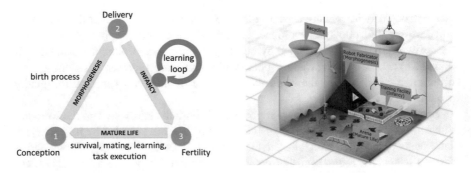

Fig. 2.4 Left: Generic system architecture for robot evolution conceptualised by the Triangle of Life; cf. [18]. Right: Illustration of an EvoSphere after [17]. It consists of three main components belonging to the three edges of the triangle, plus a recycling facility and equipment for observation.

ple tasks. Let us note that the learning methods in the Infancy stage/Training Facility can be of any type, e.g. evolutionary (neuro-evolution, genetic programming, CMA-ES) or non-evolutionary (reinforcement learning, simulated annealing, Bayesian optimisation). In the first case, we get a system with two evolutionary loops: the outer loop that forms the Triangle of Life that is evolving bodies and brains, and the inner loop 'under the hood' of the learning method that is improving the brain in the given body of a newborn robot.

The Training Facility increases the chances of success in the Arena and plays an important role: it prevents reproduction of poorly performing robots and saves resources. It also enables refinement of controllers learned in a simulated environment that do not transfer properly due to the reality gap. If a robot successfully acquires the required set of skills, it is declared a fertile adult and can start its mature life. To this end, it enters the Arena, which represents the world where the robots must survive and perform user-defined tasks and may be selected for reproduction. The selection mechanism can be innate in the robots (by choosing mating partners) or executed by an overseer, which can be algorithmic or a human 'breeder'.

Here it is important to note that different types of robot evolutionary systems can be distinguished depending on the human involvement in selection and reproduction. The involvement of humans can be a technical necessity, for instance, if automated construction is infeasible or if adopting a breeding farm approach. Table 2.2 shows the four possible combinations and the corresponding types of evolutionary robot systems.

The ARE project has started with a system of Type 3 by building a Robot Fabricator that can autonomously produce new robots [25]. For a Type 2 implementation, the information needed by the selection mechanism(s) needs to be obtained and processed automatically. For instance, an overhead camera system can observe the robots' behaviour and play the role of the matchmaker. Alternatively, robots can monitor their own performance and query each other before engaging in reproduction. A Type 4 system, like an autonomously evolving robot colony working on terraforming on another planet, is more than a simple addition of Types 2 and 3. For instance, it requires that 'newborn' robots are moved from the Robot Fabricator to the Arena and activated and observed there as they go about the given task automatically, without humans in the loop.

An essential feature of the EvoSphere concept and the ARE system is centralised, externalised reproduction. For reasons of ethics and safety, we reject distributed reproduction systems, e.g. self-replicators or the robotic equivalents of cell division, eggs, or pregnancy, and deliberately choose for one single facility that can produce

Table 2.2 Different types of robot evolutionary systems depending on the human involvement

	Selection with human assistance (breeding)	Selection without human assistance
Reproduction with human assistance	Type 1	Type 2
Reproduction without human assistance	Type 3	Type 4

new robots. This facility, the Robot Fabricator, serves as an emergency switch that can stop evolution if the users deem it necessary.

A general problem with evolving robots in the real world is the speed of evolution. To produce enough generations for significant progress can take much time and resource. In principle, this problem can be mitigated by a hybrid evolutionary system, where physical robot evolution is combined with virtual robot evolution in a simulator. This hybridisation combines the advantages of the real and virtual worlds. Physical evolution is accelerated by the virtual component that can find good robot features with less time and fewer resources than physical evolution, while simulated evolution benefits from the influx of genes that are tested favourably in the real world. In an advanced system, the physical trials can help improve the accuracy of the simulator as well, thus reducing the reality gap.

In the ARE system, deep integration of virtual and physical robot evolution is possible. In essence, two concurrently running implementations of the Triangle of Life are foreseen, one in a virtual environment and one in the physical environment. A key feature behind the integrated system is using the same genetic representation in both worlds. This enables cross-breeding so a new robot in either environment could have physical or virtual parents, or a combination of both. Additionally, it enables copying robots between environments simply by transferring a genotype from one environment to another and applying the appropriate morphogenesis protocol to create the corresponding phenotype (a virtual or a physical robot). The integration of the virtual and the physical subsystems requires a specific system component, the Ecosystem Manager, which optimises the working of the hybrid evolutionary system, maximising task performance.

4.2   The ARE Robot Fabricator

The ARE project development roughly follows the robotic life cycle represented by the Triangle of Life. The first priority is to establish a way to physically (re)produce robots. Our approach follows method 5 in Sect. 3.1.2 in that we have designed an automated robot fabricator, which we call RoboFab. The robot's genome encodes for both its hardware and software, that is, both its bodyplan and controller. The part of the genome that encodes the bodyplan determines both the 3D shape of the robot's skeleton and the number, type, and disposition of 'organs' on the skeleton.

We define organs as 'active components which individual robots can use to perform their task(s). It is expected that each robot will have a single "brain" organ, which contains the electronic control hardware and a battery to provide the robot with power. Other organs provide sensing and actuation, with their type and location being specified in the genome written by the evolutionary algorithm' [25]. The use of hand-designed organs in this way does not diminish the biological plausibility of ARE. Complex organisms are assemblages of pre-evolved components. Thus, although crude by comparison to their biological counterparts, our organs represent a pragmatic engineering compromise which does reflect the modularity evident in biological evolution, in materio [11].

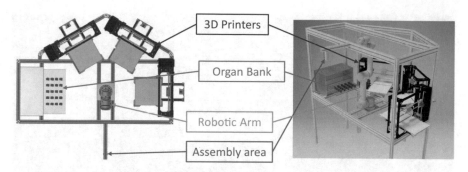

Fig. 2.5 CAD concept drawings of the ARE Robot Fabricator design, from [25]. © MIT Press 2019. All rights reserved. Reprinted with permission.

As shown in Fig. 2.5, an ARE RoboFab has four major subsystems: (i) up to three 3D printers, (ii) an organ bank, (iii) an assembly fixture, and (iv) a centrally positioned robot arm (multi-axis manipulator). The purpose of each of these subsystems is outlined as follows:

(i) The 3D printers are used to print the evolved robot's skeleton, which might be a single part or several. With more than one 3D printer, we can speed up the process by 3D printing skeletons for several different evolved robots in parallel, or—for robots with multi-part skeletons—each part can be printed in parallel.

(ii) The organ bank contains a set of prefabricated organs, organised so that the robot arm can pick organs ready for placing within the part-built robot.

(iii) The assembly fixture is designed to hold (and if necessary rotate) the robot's core skeleton while organs are placed and wired up.

(iv) The robot arm is the engine of RoboFab. Fitted with a special gripper, the robot arm is responsible for assembling the complete robot.

A RoboFab equipped with one 3D printer is shown in Fig. 2.6. The fabrication and assembly sequence has six stages listed below:

1. RoboFab receives the required coordinates of the organs and one or more mesh files of the shape of the skeleton.
2. The skeleton is 3D printed.
3. The robot arm fetches the core 'brain' organ from the organ bank and clips it into the skeleton on the print bed.
4. The robot arm then lifts the core organ and skeleton assemblage off the print bed and attaches it to the assembly fixture.
5. The robot arm then picks and places the required organs from the organ bank, clipping them into place on the skeleton.
6. Finally the robot arm wires each organ to the core organ to complete the robot.

We have demonstrated the successful automated fabrication of several robot bodyplans (steps 1–6 of the sequence above), and one example robot is shown in Fig. 2.7. For a more detailed description of the ARE RoboFab, see [25].

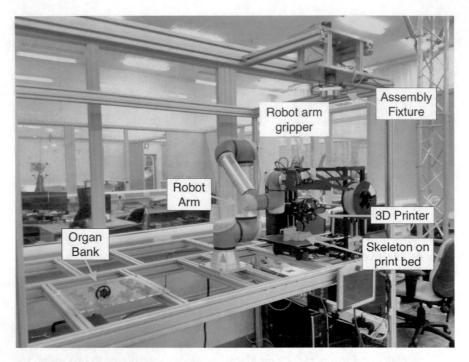

Fig. 2.6 An ARE RoboFab.

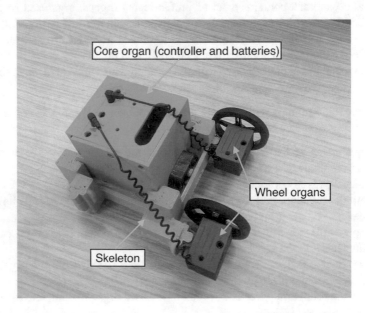

Fig. 2.7 A complete robot, fabricated, assembled, and wired by the RoboFab. This evolved robot has a total of three organs: the core 'brain' organ and two wheel organs.

5   Concluding Remarks

Autonomous robot evolution is a long-term prospect with great potential advantages for science and engineering. Conceptually, evolving robot systems belong to the more generic category of embodied artificial evolution as discussed in [20] and must indeed meet the challenges outlined therein.

1. **Body types**: An appropriate design space of all possible robot makeups needs to be specified. This defines the search space evolution will traverse searching for solutions with high values of (application-dependent) fitness.
2. **How to start**: The implementation of a robot (re)production system that delivers physical robot offspring with some form of inheritance is a critical prerequisite.
3. **How to stop**: To prevent the 'Jurassic Park problem', the system must have a kill switch to stop evolution if necessary.
4. **Evolvability and rate of evolution**: To be useful, an evolving robot system must exhibit a high degree of evolvability and a high rate of evolution. In practice, the system must make good progress in real time and have short reproduction cycles and large improvements per generation.
5. **Process control and methodology**: Evolving robot systems combine open-ended and directed evolution on the fly. An evolution manager—be it human, algorithmic, or combined—should be able to perform online process monitoring and steering in line with the given application objectives and user preferences.
6. **Body-mind co-evolution and lifetime learning**: Randomised reproduction of robot bodies and brains can lead to a mismatch between the body and the brain of newborn robots. This implies the need for a lifetime learning process to align the inherited brain with the inherited body and increase fitness and task performance quickly [16].

Arguably, challenges 1 and 2 form the first priority for any real implementation. In terms of the Triangle of Life model, they are positioned on the first side of the triangle. Thus, the initial steps on the road towards autonomous robot evolution must deal with the Robot Fabricator.

Therefore, the ARE project started by creating the ARE RoboFab. These efforts uncovered previously unknown issues regarding manufacturability and viability. In short, the problem is that a genotype, that is, the specification sheet for a new robot, can be formally correct but infeasible in different ways. Here we can distinguish structural and behavioural infeasibility. For instance, the genotype can describe a robot that is not constructable because its body parts overlap or is constructable in principle, but cannot be manufactured by the given machinery, the 3D printer, and the automated assembler. Another form of infeasibility occurs when a robot can be manufactured, but its bodyplan is lacking essential components for reasonable functioning, for instance, if the robot has no sensors or no actuators. Then the robot is manufacturable, but not viable. These issues are strongly rooted in the morphogenesis of physical robots and have, therefore, not been encountered before.

Ongoing research is concerned with discovering the essence of these issues and understanding how they affect the evolutionary process.

As stated by challenge 6 above, a morphologically evolving robot system needs lifetime learning. Such a component sits on the second side of the Triangle of Life and forms a logical second phase in the development. A fundamental issue here is that the learning methods must work on arbitrary robot morphologies that can possibly be produced by evolution, regardless of the size, the shape, and the arrangement of body parts. Furthermore, the newborn robots in a realistic system need to learn multiple skills. This is largely uncharted territory since most of the existing work in this area concerns learning methods for a given, fixed robot morphology and only one or two tasks. An additional challenge is the need for extreme sample efficiency, as the learning algorithm should work in a few trials and with little prior information. A possibly helpful trick here is to implement a Lamarckian combination of evolution and learning. This means that individually learned skills are coded into the genome and thus become inheritable. This requires more research, but the first results are promising [29]. Last but not least, evolution and learning are capable of delivering working controller software and perform what we call second-order software engineering. However, this software is only subject to functional testing and does not undergo the usual quality testing, debugging, and verification protocols. Clearly, further thought is required to understand how testing and improving the quality of software could be integrated into an autonomous evolutionary system in the future.

Ultimately, the ambition of the ARE project is to deliver the first-ever integrated system that can produce generations of robots in a real-world setting. On the long term, the aim is to produce a foundation for the future of a type of robotic design and manufacture that can change our thinking of robot production and open up possibilities for deploying robots in remote, inaccessible, and challenging environments.

Acknowledgements The ARE project is supported by EPSRC under the following grants: EP/R03561X, EP/R035679, and EP/R035733. A.E. Eiben also receives funding from the ARE@VU project supported by the Vrije Universiteit Amsterdam. The authors are indebted to the research assistants M. Angus, E. Buchanan, M. De Carlo, L. Le Goff, M. Hale, W. Li, and R. Woolley.

References

1. J. Auerbach, D. Aydin, A. Maesani, P. Kornatowski, T. Cieslewski, G. Heitz, P. Fernando, I. Loshchilov, L. Daler, D. Floreano, RoboGen: robot generation through artificial evolution, in *Artificial Life 14: Proceedings of the Fourteenth International Conference on the Synthesis and Simulation of Living Systems*, ed. by H. Sayama, J. Rieffel, S. Risi, R. Doursat, H. Lipson (The MIT Press, jul 2014), pp. 136–137
2. P. Bentley, D. Corne, *Creative Evolutionary Systems* (Morgan Kaufmann, 2002)
3. P. Bentley (ed.), *Evolutionary Design by Computers* (Morgan Kaufmann, 1999)

4. J. Bongard, Morphological change in machines accelerates the evolution of robust behavior. Proc. Natl. Acad. Sci. **108**(4), 1234–1239 (2011)
5. J. Bongard, Evolutionary robotics. Commun. ACM **56**(8), 74–85 (2013)
6. N. Bredeche, J.M. Montanier, W. Liu, A.F.T. Winfield, Environment-driven distributed evolutionary adaptation in a population of autonomous robotic agents. Math. Comput. Modell. Dyn. Syst. **18**(1), 101–129 (2012)
7. H.J. Bremermann, M. Rogson, S. Salaff, in *Global properties of evolution processes*, ed. by H.H. Pattee, E.A. Edlsack, L. Fein, A.B. Callahan. Natural Automata and Useful Simulations (Spartan Books, Washington DC, 1966), pp. 3–41
8. L. Brodbeck, S. Hauser, F. Iida, Morphological evolution of physical robots through model-free phenotype development. PloS one **10**(6), e0128444 (2015)
9. N. Cheney, R. MacCurdy, J. Clune, H. Lipson, Unshackling evolution: Evolving soft robots with multiple materials and a powerful generative encoding, in *Proceedings of the 15th Annual Conference on Genetic and Evolutionary Computation*, ed. by C. Blum (ACM, 2013), pp. 167–174
10. A Cho, The accidental roboticist. Science **346**(6206), 192–194 (2014)
11. J. Clune, J.B. Mouret, H. Lipson, The evolutionary origins of modularity. Proc. Roy. Soc. B **280**(1755), 20122863 (2013)
12. A. Diamond, R. Knight, D. Devereux, O. Holland, Anthropomimetic robots: Concept, construction and modelling. Int. J. Adv. Robot. Syst. **9**, 1–14 (2012)
13. S. Doncieux, N. Bredeche, J.-B. Mouret, (eds.), *New Horizons in evolutionary robotics*, vol. 341 of *Studies in Computational Intelligence* (Springer, 2011)
14. S. Doncieux, N. Bredeche, J.-B. Mouret, A.E. Eiben, Evolutionary robotics: what, why, and where to. Frontiers Robot. AI **2**, 4 (2015)
15. A.E. Eiben, In Vivo Veritas: towards the evolution of things, in *Proc. of PPSN XIII*, ed. by T. Bartz-Beielstein, J. Branke, B. Filipič, J. Smith, LNCS 8672 (Springer, 2014), pp. 24–39
16. A.E. Eiben, E. Hart, If it evolves it needs to learn, in *Proceedings of the 2020 Genetic and Evolutionary Computation Conference Companion*, GECCO '20 (Association for Computing Machinery, New York, NY, USA, 2020), pp. 1383–1384
17. A.E. Eiben, EvoSphere: The world of robot evolution, in *Proc. of the Theory and Practice of Natural Computing 2015*, ed. by A.-H. Dediu, L. Magdalena, C. Martín-Vide. LNCS 9477 (Springer, 2015), pp. 3–19
18. A.E. Eiben, N. Bredeche, M. Hoogendoorn, J. Stradner, J Timmis, A.M. Tyrrell, A. Winfield, The triangle of life: Evolving robots in real-time and real-space, in *Proc. of the 12th European Conference on the Synthesis and Simulation of Living Systems (ECAL 2013)*, ed. by P. Lio, O. Miglino, G. Nicosia, S. Nolfi, M. Pavone (MIT Press, 2013), pp. 1056–1063
19. A.E. Eiben, E. Haasdijk, N. Bredeche, *Embodied, On-Line, On-Board Evolution for Autonomous Robotics* (Springer, 2010), pp. 387–388
20. A.E. Eiben, S. Kernbach, E. Haasdijk, Embodied artificial evolution – artificial evolutionary systems in the 21st century. Evolutionary Intelligence **5**(4), 261–272 (2012)
21. A.E. Eiben, J.E. Smith. *Introduction to Evolutionary Computing* (Springer, Berlin, Heidelberg, 2003)
22. A.E Eiben, J.E. Smith, From evolutionary computation to the evolution of things. Nature **521**(7553), 476 (2015)
23. D. Floreano, P. Husbands, S. Nolfi, Evolutionary robotics, in *Handbook of Robotics* (1st edn.), ed. by O. Siciliano, B. Khatib. (Springer, 2008), pp. 1423–1451
24. D. Floreano, L. Keller, Evolution of adaptive behaviour in robots by means of darwinian selection. PLoS Biology **8**(1), e1000292 (2010)
25. M.F. Hale, E. Buchanan, A.F. Winfield, J. Timmis, E. Hart, A.E. Eiben, M. Angus, F. Veenstra, W. Li, R. Woolley, et al., The are robot fabricator: How to (re)produce robots that can evolve in the real world, in *The 2019 Conference on Artificial Life* (MIT Press, 2019), pp. 95–102
26. J. Hiller, H. Lipson, Automatic design and manufacture of soft robots. IEEE Trans. Robot. **28**(2), 457–466 (2012)

27. G.S. Hornby, J.D. Lohn, D.S. Linden, Computer-automated evolution of an x-band antenna for NASA's space technology 5 mission. Evolutionary Computation 19(1), 1–23 (2011)
28. N. Jakobi, P. Husbands, I. Harvey, Noise and the reality gap: The use of simulation in evolutionary robotics, in *European Conference on Artificial Life* (Springer, 1995), pp. 704–720
29. M. Jelisavcic, K. Glette, E. Haasdijk, A.E. Eiben, Lamarckian evolution of simulated modular robots. Frontiers Robot. AI **6**, 9 (2019)
30. M. Jelisavcic, M. De Carlo, E. Hupkes, P. Eustratiadis, J. Orlowski, E. Haasdijk, J.E. Auerbach, A.E. Eiben, Real-world evolution of robot morphologies: A proof of concept. Artificial Life **23**(2), 206–235 (2017)
31. S. Kernbach, E. Meister, O. Scholz, R. Humza, J. Liedke, L. Ricotti, J. Jemai, J. Havlik, W. Liu, Evolutionary robotics: The next-generation-platform for on-line and on-board artificial evolution, in *Proc. of the IEEE Congress on Evolutionary Computation* (IEEE, 2009), pp. 18–21
32. T. Kuehn, J. Rieffel, Automatically designing and printing 3-D objects with EvoFab 0.2, in *Proc. of the European Conference on the Synthesis and Simulation of Living Systems (ECAL 2012)*, ed. by M.A. Bedau (MIT Press, 2012), pp. 372–378
33. H. Lipson, J.B. Pollack, Automatic design and manufacture of robotic lifeforms. Nature **406**, 974–978 (2000)
34. J. Long, *Darwin's Devices: What Evolving Robots Can Teach Us About the History of Life and the Future of Technology* (Basic Books, 2012)
35. H.H. Lund, *Morpho-Functional Machines: The New Species*, chapter Co-evolving control and morphology with LEGO Robots (Springer, 2003)
36. H.H. Lund, J. Hallam, W. Lee, Evolving robot morphology, in *IEEE International Conference on Evolutionary Computation*, pp. 197–202, 1997
37. S. Nolfi, J. Bongard, P. Husbands, D. Floreano, Evolutionary robotics, in *Handbook of Robotics* (2nd edn.), ed. by O. Siciliano, B. Khatib (Springer, 2016), pp. 2035–2068
38. S. Nolfi, D. Floreano, *Evolutionary Robotics: The Biology, Intelligence, and Technology of Self-organizing Machines* (MIT press, 2000)
39. R. Pfeifer, J. Bongard, *How the Body Shapes the Way We Think: A New View of Intelligence* (MIT press, 2007)
40. J. Rieffel, D. Sayles, EvoFab: A fully embodied evolutionary fabricator, in *Proceedings of the International Conference on Evolvable Systems (ICES 2010)*, ed. by G. Tempesti, A.M. Tyrrell, J.F. Miller (Springer, 2010), pp. 372–380
41. K. Sims, Evolving 3D morphology and behavior by competition. Artificial Life **1**(4), 353–372 (1994)
42. A.M. Turing, Intelligent machinery, national physical laboratory report, 1948, in *Mechanical Intelligence: Collected Works of A.M. Turing*, ed. by D.C. Ince (North-Holland, Amsterdam, 1948), pp. 107–128
43. P.A. Vargas, E.A. Di Paolo, I. Harvey, P. Husbands, *The Horizons of Evolutionary Robotics* (The MIT Press, 2014)
44. M. Waibel, D. Floreano, L. Keller, A quantitative test of Hamilton's rule for the evolution of altruism. PLOS Biology **9**(5), e1000615 (2011)
45. A.F.T. Winfield, J. Timmis, Evolvable robot hardware, in *Evolvable Hardware*, ed. by M.A. Trefzer, A.M. Tyrrell (Springer, 2015), pp. 331–348
46. V. Zykov, E. Mytilinaios, B. Adams, H. Lipson, Self-reproducing machines. Nature **435**(7039), 163–164 (2005)
47. V. Zykov, E. Mytilinaios, M. Desnoyer, H. Lipson, Evolved and designed self-reproducing modular robotics. IEEE Trans. Robot. **23**(2), 308–319 (2007)

Chapter 3
Composition, Separation of Roles and Model-Driven Approaches as Enabler of a Robotics Software Ecosystem

Christian Schlegel, Alex Lotz, Matthias Lutz, and Dennis Stampfer

Abstract Successful engineering principles for building software systems rely on the separation of concerns for mastering complexity. However, just working on different concerns of a system in a collaborative way is not good enough for economically feasible tailored solutions. A successful approach for this is the *composition* of complex systems out of commodity building blocks. These come *as is* and can be represented as blocks with ports via data sheets. Data sheets are models and allow a proper selection and configuration as well as the prediction of the behavior of a building block in a specific context. This chapter explains how model-driven approaches can be used to support separation of roles and composition for robotics software systems. The models, open-source tools, open-source robotics software components and fully deployable robotics software systems shape a robotics software ecosystem.

1 Aims and Challenges of Software Engineering for Robotics

Many definitions refer to a *robot* as a machine that is able to perform a variety of tasks and that is reprogrammable in order to become multifunctional. Thus, as soon as a machine reaches some level of flexibility and versatility in its applicability, we call it a *robot*. In this way, the most advanced robots are called *service robots*. A service robot shall not only be able to fulfill a multitude of different tasks. It is expected to do this successfully and with some robustness even under varying circumstances, as it is quite normal in open-ended environments and when a robot needs to share its workspace with others [1].

C. Schlegel (✉) · A. Lotz · M. Lutz · D. Stampfer
Technische Hochschule Ulm, Ulm, Germany
e-mail: christian.schlegel@thu.de; alex.lotz@thu.de; matthias.lutz@thu.de; dennis.stampfer@thu.de
http://www.servicerobotik-ulm.de

© The Author(s) 2021
A. Cavalcanti et al. (eds.), *Software Engineering for Robotics*,
https://doi.org/10.1007/978-3-030-66494-7_3

53

The aspect of reprogramming is deeply rooted in the notion of a robot as is the challenge of making its flexibility exploitable. In the ideal case, that can be done with as little effort as possible and by the user itself. Of course, there is the full spectrum from simple robots (automated devices) to autonomous systems (making decisions on their own and being able to revise decisions), but software is an indispensable tool for implementing and operating all of them.

The question now is whether Software Engineering for Robotics is different from software engineering for other domains or whether it is different with respect to software engineering in general [2]. The short answer is *no*, there is nothing special at all about Software Engineering for Robotics.

At first glance, that statement seems to be contradictory to the daily experience: it often ends up in a costly IT project or software project when integrating a robot into existing infrastructure or when tailoring a robot to a new use case. Studies identified software as the road block (or at least as the bottleneck) in bringing robots to new application domains [3, 4]. Indeed, Software Engineering for Robotics needs to be mastered in a way that is economically feasible for the application domains so that the flexibility of robots can be exploited.

Figure 3.1 outlines a perspective for a business ecosystem for robotics software systems. Our aim is to introduce into the domain of robotics software the mechanisms and structures of a business ecosystem to let robotics exploit all the benefits of value networks. In this section, we first give an introduction into business

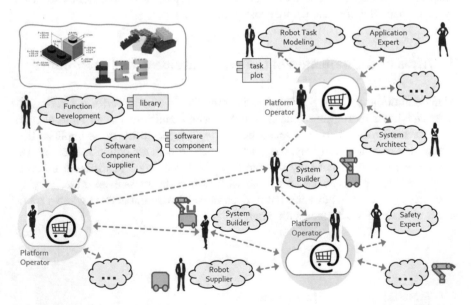

Fig. 3.1 Software engineering is about mastering interfaces between building blocks, competencies and responsibilities. Properly identified and agreed interfaces are the enabler for *composition* and for *separation of roles* and thus for a robotics software business ecosystem that comes with marketplaces for composable assets. Building complex software systems for robotics can become as easy as building with Lego bricks.

ecosystems and into software technologies used in robotics. The subsequent sections then outline how to arrange software technologies used in robotics to shape a robotics software business ecosystem.

1.1 Carving Out the Specifics of Software for Robotic Systems

Unfortunately, markets become more and more volatile, products have shorter market life time, and the ongoing trend to customization is another driver that further increases the number of product variants [5]. Quantities of the same product are not high enough anymore to ensure the return of investment of highly specialized and most advanced single-purpose machines via a single product.

Thus, standardized automation solutions able to handle a single scenario are not sufficient anymore. An expensive machine either must cope with a product mix (flexibility), or the effort for adjusting it to the next scenario must be low enough (adaptability). Here is where users expect that robots are such kinds of machines: (1) flexible and adaptable and (2) matching the high demands in fulfilling complex manufacturing steps or whatever other tasks, even in workspaces shared with others or even when robust task fulfillment requires at least a minimum level of autonomy.

Right now, one can basically go for the offered families of standardized robots that come with a reasonable price tag. However, there are many scenarios that one cannot address with them. They are limited in flexibility and adaptability—very often so limited that one faces problems in properly addressing a specific scenario. On the other hand, custom-made robots are also not yet an option as they are still far from being economically feasible under the constraints of an application domain. Although they can be made to perfectly fit requirements, this is then priced accordingly, and such a robot nowadays still does not fulfill the level of flexibility and adaptability which one expects from it to get on the safe side for a return of investment [6].

To summarize, the changeability and flexibility of robotic systems often is not yet economically exploitable to face the pressure of lot-size one. Robots are in many cases not yet a tool that users can adjust themselves to match their daily changing needs. In robotics, costs of changes are not in proper relation to the similarity of an existing solution (Fig. 3.2). Unfortunately, expectations in many application domains where robotics can contribute cannot yet be fulfilled due to high efforts and costs.

The challenge is nothing less than coming up with a different way of building robotic systems. Otherwise, the pull from application domains for robotic solutions helping the application domains to reach their next level of automation cannot be fulfilled. Robots need to make progress in fulfilling the promise of being flexible and adaptable machines. Indeed, software and software tools play a major role in implementing functionalities of robots, in enabling robots to perform more advanced tasks and in empowering users to command robots to their benefits. Thus, one cannot help but admit that the current practice of how we build and manage robotics software systems needs to change in order to address the above-outlined challenges.

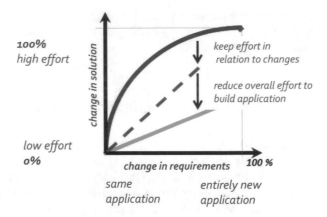

Fig. 3.2 In robotics, costs of changes are not in proper relation to the similarity of an already existing solution. This needs to be resolved in order to enable the widespread use of robotics.

A different way of building robotic systems should tackle questions like: (1) Can we reason about robotic systems before we build them? (2) Can we answer what-if questions and can we find *adequate* solutions? (3) Can we put together systems out of configurable as-is building blocks? (4) Can we bring effort and costs in relation to the similarity to an already existing application? (5) Can we trace the fulfillment of safety requirements? (6) Can we explain what the system does? (7) Can we generate enough trust into the systems—and how and by what means?

Of course, the solutions and approaches are not specific to robotics but to complex systems in general. In some sense, the domain of advanced service robotics is a moving target. As soon as a challenge in advanced service robotics is solved, that tends to be called automation subsequently. Thus, advanced service robotics just always pools all the cutting-edge challenges. That is why the focus of this chapter is not on all the positive examples of software systems for robotics, which made the implementation and deployment of all the successfully working robotic systems possible. The focus is on the challenge of Software Engineering for Robotics: advances here allow for opening up new markets for robotic systems and allow for entering new application domains by robots.

1.2 The Power of Ecosystems and the Power of Separation of Roles

Ecosystems are dynamic and co-evolving communities of diverse actors who create and capture new value through both collaboration and competition [7]. A distinctive characteristic of many ecosystems is that they form to achieve something that lies beyond the effective scope and capabilities of any individual actor (or even group of broadly similar actors).

Ecosystems show benefits in coping with the ever-increasing complexity of many processes, products and services and the decrease of lot sizes [8]. More and more, services or products must be tailored to very specific requirements in order to make them a fit. This fit can best be achieved by individually selecting building blocks and putting them together according to the order to be fulfilled.

A huge variety is just not feasible within a single value chain. A value chain comes with a small number of stable partners. Business relationships just grow over time and some partners make it into preferred and established relationships. Instead, we now need to have access to a pool of thousands of partners that all are highly specialized experts in particular niches. This is because we need to select at any given time the then best fitting partner, even when this is going to be a partner for only a short period of time or for very low quantities of an asset. For the most part, supply chains of large businesses were not set up to deal with a world of thousands of partners, and they need to evolve into value networks (Fig. 3.3).

An ecosystem works differently. It requires partners to strictly focus on their unique expertise and allows them to become highly specialized. In an ecosystem, as purchaser, one depends on a huge pool of different offerings with finest differences mapped out. This allows purchasers to match the requirements of their customers

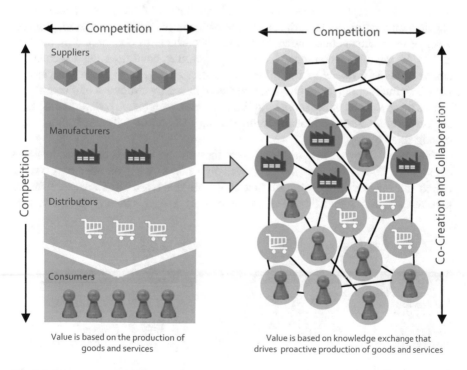

Value is based on the production of   Value is based on knowledge exchange that
goods and services   drives proactive production of goods and services

Fig. 3.3 Linear supply chains evolve into complex, dynamic and connected value networks that build on co-creation, collaboration and access to a pool of thousands of partners within an ecosystem (Source: own illustration based on [8]).

with products tailored to the orders. As a supplier, one depends on many different purchasers selecting the offered components or services in order to come up with high enough revenues. One just needs to be the best within one's narrow scope and then rely on others being the best in their niches. This is why an ecosystem as a whole is very agile in filling even the smallest niche and the smallest demand with dedicated offers. In an ecosystem, one does not depend on the small world of a particular value chain anymore. Instead, one reaches out to all the partners with opportunities to collaborate with all of them. With most of the partners, one collaborates only temporarily, but one always does so with enough partners to stay focused on one's particular expertise.

We call that way of splitting up work and distributing it across many different partners *separation of roles*. This is different to the way in which work is split up along a *product-line approach* [9] as *separation of roles* comes with a different granularity of management and with more responsibilities for a partner filling a role [10]. Of course, within our local scope, we can still run a product-line approach (see Chap. 1 for software product-line engineering).

Thus, the basic difference of a business ecosystem compared to a value-chain approach or compared to a product-line approach is its different way of managing the huge variety of partners (that all form their own organizational units) in order to better tackle a huge variety of tailored products. Instead of trying to manage all the required interactions and interfaces to cope with complexity, an ecosystem provides a set of mechanisms that allows thousands of partners to collaborate on demand and to do this under continually changing operating conditions. Management to ensure that the artifacts of the different partners fit together or that the partners smoothly interact in tightly coupled ways is replaced by structures and means that ensure those fit even under loose couplings between partners. A business ecosystem supports independent and locally managed work to end up with fits without extra management on top.

Figure 3.4 shows three different tiers that are typically found in a business ecosystem. The *uppermost tier* is driven by only a small number of partners, called the *ecosystem drivers*. At that level, all the basic structures and means that form sound foundations for the other tiers are provided. Tier 1 can also

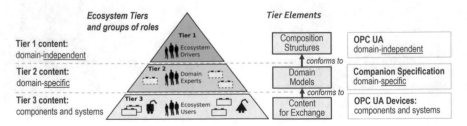

Fig. 3.4 Three tiers typical for any business ecosystem. Tier 3 is conforming to Tier 2 and Tier 2 is conforming to Tier 1. While there are only few ecosystem drivers forming Tier 1, there are already more domain experts at Tier 2 and finally all the ecosystem users are at Tier 3.

be called *foundations*. The *middle tier* is driven by domain experts that use the Tier 1 structures to come up with domain specifics. The middle tier is completely conforming to the uppermost tier. The separation between Tier 2 and Tier 1 is extremely important as Tier 2 allows to shrink the size of the domain under consideration such that agreements on structures become feasible. Tier 2 allows us to have many different domains in parallel and without conflicts in consistency, even coexisting and competing approaches for the same aspect. For example, the navigation domain and the vision domain might differ in the representation of coordinate systems. There might be different vocabularies for specifying trajectories all consistent in themselves. Of course, fragmentation without reason is not good, but Tier 2 allows for processes to resize scopes of domains and to sort out competing approaches where they are just different without any reason. As any Tier 2 domain is conforming to Tier 1, one can build links between different Tier 2 domains via Tier 1. Tier 2 is typically filled by domain-specific working groups. Tier 2 can also be called *definitions*. Finally, Tier 3 comprises all the concrete products and services that conform to at least one of the domains at Tier 2. As purchaser, one knows beforehand which parts can fit together. As supplier, one knows to what kind of structures one needs to adhere to so that the provided parts fit to others. Tier 3 can also be called *implementations*.

This is exactly the same structure that is also followed by *OPC UA* in the context of *industry 4.0* [11]. OPC UA provides the basic means for information models and accessing them (i.e. Tier 1). However, that still is not enough to make different industry 4.0 devices interoperable. Interoperable devices must use the same information models, that is, they must apply the generic OPC UA concepts in the same way. Thus, OPC UA introduces the so-called companion specifications (Tier 2) [12]. Finally, Tier 3 comprises all the conformant devices.

1.3   The Power of Composition

Complexity can be mastered by splitting up a complex problem into smaller ones, solve those and finally put these solutions together to form a solution for the original problem. Of course, that proved to be a reasonable approach in many engineering disciplines as we then can assign the work to different entities, let it be persons, teams, departments or even contracted other companies. This lets us do the work faster as we can get it done by a larger workforce. We can get it done concurrently and we can also identify whether there are already solutions for subproblems that we might want to reuse.

However, just splitting up the problem or splitting up the work is not enough. It needs to be done such that the subparts ultimately fit together and that they properly form the intended solution. Different approaches address exactly that challenge in significantly different ways [13].

A whole set of approaches is based on *integration*. The outcome of integration is an amalgamation of the previously separated subparts. It is quite difficult (and typically not foreseen) to again split apart the outcome afterward or even modify it. As everything is optimized for a set of preplanned configurations, any changes or variations interfere with the highly optimized setup that works like a clockwork. Basically, there is a huge management effort needed in keeping all the interactions between teams, and interfaces of parts and the work in general (i) up and running, (ii) in sync and (iii) orchestrated, so that at any point in time everything fits smoothly. Product-line approaches to face variations work well only within the same organization [9, 10] and as long as it is about foreseen variations of the same product for which the product line got already prepared. As soon as we need to cross organizational boundaries (i.e. when we are in need of managing resources not under our control, like other teams, other departments or external contractors), integration-based processes reach a level of complexity that cannot be managed anymore.

Another set of approaches is based on *composition*. It is the activity of selecting building blocks (modules, components, subsystems, etc.) *as they are* and putting them together. *Composability* is the ability to combine building blocks. *Compositionality* refers to the ability to understand a composite system by understanding its building blocks and how they are combined. *Modularity* is the degree to which a system's building blocks may be separated and recombined. Modularity comes with the challenge of designing building blocks with well-defined interfaces that can be used in a variety of contexts. Composition is one approach to achieve modularity. In contrast to an *integrated system*, a *composed system* can again be split apart into its building blocks. Thus, modifications and adaptations are possible requiring only *adequate* effort. *Adequate* means that effort and costs are in balance with the achieved result. That is exactly how we build complex Lego models out of given Lego bricks. Of course, as long as only regular bricks are used, many contours and shapes can only be approximated. Nevertheless, there is still room for most specialized bricks. These fit as long as they adhere to the standardized knobs. However, specialized bricks come with a price to pay in terms of extra costs, extra effort and less reuse in other contexts which does not pay off in case a solution based on standard bricks is adequate.

For example, *resource shares* and *reservation-based mechanisms* are composable. As long as there are resource shares available, these can be claimed without interfering with already assigned shares. However, *priorities* are not composable. Each new asset might require reassignments of priorities (see, e.g. priority assignment in rate monotonic scheduling).

To summarize, composition is a very powerful concept to build a huge range of diverse and modifiable systems in an economically feasible way, even down to lot-size one, as one does not start from scratch over and over again. Instead, one just reuses (compositions of) commodity building blocks as far as possible. This amortizes costs at a different level of granularity. It is not anymore by huge quantities of the same and difficult to modify finished product. It is at the level of huge quantities of commodity building blocks that become part of many different and diverse finished products.

1.4   Aiming for a Software Business Ecosystem for Robotics

Robotic systems need to become more flexible, more adaptable and still affordable even when it comes down to lot-size one for a particular robot. We select, combine, adjust and exploit those already available software engineering approaches that help us best in matching those needs. The approach is not limited to robotics.

A business ecosystem provides a lot of benefits in addressing the above challenges. In order to ramp up a business ecosystem for robotic software systems, one first needs to identify what structures need to be defined at which tier. One also has to answer the question of how those structures are presented to participants of an ecosystem. For them, it needs to be as simple as possible to adhere to the defined and agreed key structures and to apply those structures properly and in the intended way. This is of paramount importance in enforcing the principles of *separation of roles* and of *composition* as these form the pillars of a working and fruitful business ecosystem.

Figure 3.5 gives an overview of some core concepts. First of all, it shows different roles, such as a *component developer*, a *system builder* but also a *safety engineer*, a *domain expert*, a *system architect* and a *behavior developer*. For example, a component developer selects functional libraries from marketplaces of the ecosystem and uses them within software components. A system architect translates the requirements of a domain expert into a system design. A system builder selects those software components that fit best to a design and composes them to form the system. A behavior developer models action plots to achieve a certain task but in a way that is independent from a specific robotic system. Rather, a particular action plot needs to be executable with all robots that have the required capabilities. A safety engineer translates domain specific safety requirements into a set of constraints that are checked during the development as well as during the operation phase. There are many more roles, but it is important to understand that they all work concurrently and independently from each other and that there can be many partners for the same role at any time.

As shown in Fig. 3.5, the structures of all assets, whether these are functional libraries, software components or action plots, are represented via *blocks* with *ports*, where ports are linked by *connectors*. A block separates the internals from the externals. The only way to interact with a block is via its ports. Blocks can be nested as is the case, for example, of a software component that comprises functional libraries that are also blocks with ports. As soon as the experts of a particular domain agree on the kind of blocks, ports and connectors for the different assets of their domain (of course, structure *and* behavior), one can ensure that conforming assets finally fit together in the intended way.

Figure 3.5 also shows that each block comes with a *data sheet*. This proved to be an extremely successful mechanism to decouple suppliers and purchasers in a business ecosystem. A data sheet describes an outside view of an asset, including its foreseen variation points. A data sheet includes internals only as far as we need to know them for using the asset and for predicting its fit (behavior, structure) for our

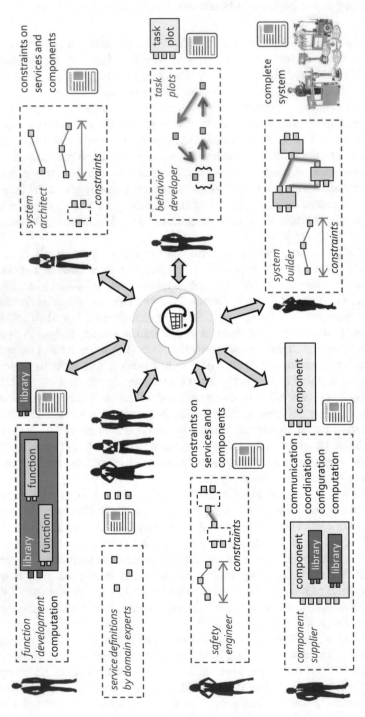

Fig. 3.5 Different roles in a robotics software ecosystem along with different kinds of assets exchanged via marketplaces. Each asset comes with a data sheet that represents the asset as a block with ports.

context. Data sheets are, by purpose, not rich enough for synthesizing the artifact. They are an *abstraction of the asset* and they are not in conflict with our needs in protecting our intellectual property. *Data sheets are models.* A data sheet is the minimum information we need to provide to users about our asset. The asset cannot be operated, and thus would be useless, if one does not know what to provide to its ports and how to set its variation points.

1.5 The Role of Model-Driven Software Engineering and of Data Sheets

It is important to understand what kind of roles we foresee for *models* and for *model-driven software engineering* in a business ecosystem for robotics [14]. An essential part of software engineering is managing interfaces between software assets. Despite the different levels of granularity of software artifacts, one ends up with different types of interfaces such as programming interfaces for libraries, service-oriented ports of software components and many others. These interfaces can all be represented by blocks, ports and connectors.

Models are considered as the most adequate way to explicate and link all the required and agreed structures in a consistent way. Models are not bound to a specific implementational technology. The represented insights are decoupled from the pace of new technologies. Models allow for early binding of semantics and for late binding of technology (in source-code-driven approaches, it often is exactly the other way around). Models form the baseline for role-specific model-driven tools that enforce the agreed structures and thus ensure that outcomes of independent work fit into the ecosystem.

Model-driven software engineering and associated *model-driven tools* are strong (1) in making structures with their semantics accessible and (2) in enforcing those structures with their semantics. In a robotics software business ecosystem, all those structures that implement the principles of *separation of roles* and of *composition* should be managed by a model-driven approach. Compliance to relevant structures and to their semantics is not anymore achieved by the developer's discipline. Instead, this compliance is ensured by the model-driven tools that generate (or just retrieve) the related source code for the asset's interfaces. Pregiven reference implementations for interfaces are a feasible way to provide their semantics (meaning and behavior of the structures). Model-driven tools relieve users from a huge cognitive load and from a huge responsibility while still giving all the freedom beyond the enforced structures.

We do not advocate fully replacing source code by models. Coding is to some extent modeling as well and models in the form of source code are even executable. Instead, we understand models as abstractions of artifacts, then called a *data sheet*. The more is expressed in the data sheet, the more details can be taken into account for prediction, analysis, etc. and the less there is a need to actually build all the variants in a costly way just to identify—by real-world testing—the wanted variant.

A model in the form of a data sheet does not cover everything to the finest detail, and it does not allow the synthesis of the represented artifact. Of course, the better a model can represent further properties of an artifact, the richer we can make the data sheet. Indeed, this can be a reasonable migration path to full-blown models and a full-blown model-driven approach. If it becomes both beneficial *and* feasible to avoid coding at all, then there might be a shift toward fully model-driven software engineering approaches. However, we might still end up with considering coding as the more efficient way of expressing some of the details. The reason is quite simple: software models are not just a level of indirection, they are an abstraction. A software model does not explicate all the complex details of its target platforms. Anyway, it is about a beneficial coexistence and about consistent links between different abstraction levels.

Figure 3.6 shows a fully model-driven approach where all the different models are composed to the final system (see ①). However, pure modeling without means to get models grounded in the real world is not a solution in robotics as robots finally need to act in the real world. Indeed, a model-driven approach in robotics makes sense only when there are no gaps between models and their grounding in real-world assets. After modeling is complete, one needs mechanisms to transform the models into an executable form. Up to now, such mechanisms exist only in a few domains. For example, *3D printing* is such a domain as a 3D printer is such a generator (see ②). As result, we get the real asset and a data sheet (see ③). The data sheet is an excerpt of the model used to generate the artifact and contains only information that the user needs to know for using the asset. Unfortunately, such a generator is still not there for complex software systems.

Figure 3.7 shows the approach we are favouring for a model-driven software approach in robotics. As a data sheet is an abstraction of a real asset, there is no data sheet without a related real asset. We can already compose a system at the model level by composing data sheets. This allows us to already check those system-level properties that are covered by the data sheet abstractions of the real assets (see also

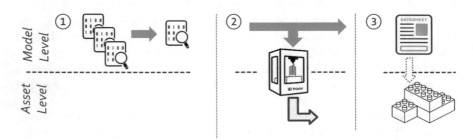

Fig. 3.6 A fully model-driven approach where we first model everything ① before we finally use a generator ② to get out of the model the real-world asset and an excerpt forming the data sheet ③.

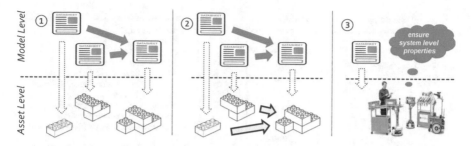

Fig. 3.7 Data sheets are *models* that are abstractions of real assets. A data sheet describes an outside view of an asset and it is grounded via the real asset.

Chap. 4 on testing, Chap. 7 on verifiable autonomy and in particular Chap. 8 on verification). For example, we can already perform various *what-if* analyses and we can trade off configurations of the variation points of the assets (see ①). This can all be done prior to buying any of the assets. Once we are fine with the predicted outcome, we can go for the according real assets (see ②). The transformation from the model level to the real-world level is by picking up the related assets and putting them together in the same way as we did at the data sheet level. This gives us a system with properties consistent to the predicted ones (see ③). Of course, all those properties of the system that are not covered by the data sheets can only be checked via the real assets.

This chapter is organized as follows. Section 2 describes the arrangement of software technologies such that a robotics software business ecosystem can get shaped. Section 3 goes in-depth into selected details of Tier 1 as this tier is decisive for the overall ecosystem approach. Section 4 gives selected insights into Tier 3 and into a world of model-driven tools from a user perspective as most ecosystem participants will operate at Tier 3. By that user-centric view, the role of Tier 2 and its links to Tier 3 and Tier 1 get clear as well. Section 5 relates the presented work to state of the art and concludes.

2 Structures for a Robotics Software Business Ecosystem

Separation of concerns is an important design principle to handle complex interdependent systems [15]. An example is the Internet protocol stack with its different well-defined layers. A good design defines *abstraction layers* where each layer can exist without the layers above it, but requires the layers below it to work. Otherwise, different layers just form levels of indirection. By definition, a layer cannot be defined on its own, since its semantics is the relationship between its artifacts and those of the neighbouring layers. The exercise to obtain well-defined relationships is a tough one. Of course, transitions between layers can be fluent and individual layers may also be split horizontally. There can also be implementations that just

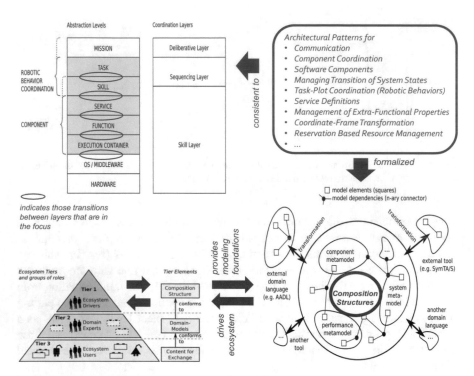

Fig. 3.8 Best practices and lessons learned are described in the form of architectural patterns. Experts translate architectural patterns into formal models using block-port-connector meta-models. Formalized and consistently arranged architectural patterns form the composition structures. Composition structures form the ecosystem foundations.

combine several adjacent layers into one, which means losing flexibility. However, it is not good practice at all to define relationships between non-adjacent layers.

Figure 3.8 shows widely agreed layers for robotic systems. The lowest layer is *hardware*, followed by the *operating system/middleware*. The *execution container* provides access to resources in a way independent of the implementational technology of the layer below. The *function* layer comprises the implementations of the algorithms. A *service* performs work based on contract and with its own resources and thereto interacts with other services. A *skill* arranges different services such that they become a particular robot capability. It translates between a generic name for a capability like *move* (from above) and configurations and parameterizations specific to the used services (downward). The *task* layer represents what steps and how these need to be executed in order to accomplish a particular job. Tasks are independent from a particular robot but can be executed with a particular robot only if that robot possesses matching skills. Finally, the *mission* layer represents the purpose of the robotic system.

The challenge is to define the relationships between the different layers. We need to find those patterns and structures that form the sweet spot between *freedom of choice* and *freedom from choice* [16, 17]. We need to support as much freedom as possible to not block progress and to address specific requirements in the most flexible way. On the other hand, we need to define at least those structures that ensure composability and separation of roles and that overcome fragmentation. The latter is about just different alternatives for the same concept but without any specific reason and without providing added value.

Obviously, it is not as simple as separating the overall robotic system along both the above abstraction layers and the different concerns such as *computation*, *communication*, *coordination* and *configuration*. This would end up in a granularity of assets that neither fits to a natural role in a robotics ecosystem nor forms marketable entities. In contrast, a software component is a good example for a reasonable building block as it is a marketable asset that fits to the well-understood role of a component developer. Of course, a software component comprises aspects of all the above concerns. Nevertheless, a well-defined software component model jeopardizes neither abstraction layers nor separation of concerns.

We introduce the notion of an *architectural pattern* to describe a particular recurring design [18]. An architectural pattern is a textual description comprising a specific design *context*, a recurring design *problem* and a well-proven *solution*. In our case, architectural patterns are driven by two fundamental objectives, that is, (1) to facilitate building systems by composition and (2) to support separation of roles. Architectural patterns allow us to compile knowledge and best practices from a community.

Via the description of the specific design context, it gets clear whether the pattern belongs to Tier 1 of an ecosystem, whether it is specific to a particular domain and thus belongs to Tier 2, or whether it is specific just to particular assets and thus belongs to Tier 3. Architectural patterns already allow in their textual form to identify competing approaches and overlap in scope. We can then initiate discussions, clarifications, classifications and decisions to ensure overall consistency across the different tiers. Afterward, all that can be formalized by using the fundamental modeling means as provided by Tier 1.

2.1 *(Meta-)Models and Tiers*

The concepts of *conform-to*, models, meta-models and tiers are key in understanding how an ecosystem gets organized. All the concepts at Tier 1 are prescriptive for Tier 1 and Tier 2. All the concepts at Tier 2 are conforming to the concepts at Tier 1 and they are prescriptive for Tier 2 and Tier 3. All the concepts at Tier 3 are prescriptive for Tier 3 and they are conforming to Tier 2.

At a particular tier, one can only operate in a way that is consistent with the scope of the already given prescriptive concepts. If there is already a prescriptive concept at a tier, one cannot go for an alternative one at a lower tier. The only option is to add at the tier comprising the prescriptive concept an alternative concept that is placed next to the already existing one. The alternative concept nevertheless still needs to be consistent with the concepts at this tier and with the prescriptive ones from the above tier. The more alternatives get introduced, the more we again end up with fragmentation that destroys composition. This is why it is so important to come up with very well thought-out concepts and place them at the proper tier.

For example, the key elements of the structure and the behavior of a software component are defined at Tier 1. This means that all the software components within that ecosystem adhere to that structure. The prescriptive parts of a software component are, for example, the basic states of its lifecycle automaton, the structure and behavior of its ports (such as a request-response or a publish-subscribe interaction), its behavior interface for its configuration and many more. The model of these key elements that constitute a well-formed software component is part of Tier 1. At Tier 2, a software component can still be tailored to a specific domain, for example, by the substates of the lifecycle automaton, the concrete data structures exchanged via the ports, etc. The prescriptive models of Tier 1 are used at Tier 2 to come up with a concrete set of domain-specific ports that are conforming in structure and behavior to the prescribed elements. The concrete set of domain-specific ports, which is defined at Tier 2, is again a set of models. This set of models is conforming to the related Tier 1 model and that Tier 1 model is a meta-model for it. Finally, a concrete software component is modelled at Tier 3, using the Tier 1 software component model as meta-model and the domain-specific refinements of Tier 2 as another consistent meta-model.

Tier 1 contains all structures that are relevant to shaping an ecosystem and that are independent from a specific domain. Nevertheless, these still provide all the hooks to get tailored at Tier 2 to domain-specific needs where necessary. For example, how to represent skills is defined at Tier 1, but the very concrete and domain-specific names of capabilities are defined at Tier 2. The structure and behavior of all the elements of a software component including the structure and behavior of its ports are fully defined at Tier 1, but the concrete set of instantiable domain-specific service ports are defined at Tier 2.

2.2 Tier 1: Foundations

This tier comprises all the sound foundations for expressing ecosystem structures. Tier 1 itself is organized into three different layers as shown in Fig. 3.9. The structures in Tier 1 are provided and driven by experts. The overall community compiles its body of knowledge via architectural patterns. Groups of experts moderate the consistent arrangement of the architectural patterns and then formalize them to become part of the bottom layer of Tier 1. As component developer, system builder, etc., one does not get directly in touch with those structures as one uses role-specific tools. The fully detailed description of the following concepts is available via the *RobMoSys Wiki* [19]. The Wiki is an effort of the robotics community that is driven and moderated by the *EU H2020 RobMoSys* project. It aims to explicate the body of knowledge of the robotics community for the purpose of ramping up a

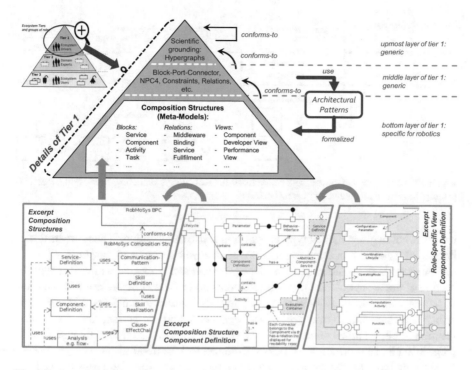

Fig. 3.9 A modeling view clusters related modeling concerns in one view. A view establishes the link between primitives in composition structures and ecosystem roles. Views enable roles to focus on their responsibility and expertise.

digital industrial platform for robotics. The *euRobotics AISBL* with its topic groups already adopted the role of *stewardship* for the robotics body of knowledge in order to ensure sustainability. Stewardship is about the responsibility to moderate, shepherd and safeguard the community's body of knowledge.

The uppermost layer provides a *hierarchical hypergraph model* as scientific grounding [20], [21, hypergraph-er]. A hierarchical (property) hypergraph [22, 23] is the modern, higher-order version of the *entity-relationship model*. In short, *hierarchical* means that every edge and every node can be a full graph in itself. In other words, any *relation* can be considered an *entity* in itself and can hence be used as an argument in another, higher-order relationship. *Hypergraph* means that every edge can join any number of nodes, that is, it is an n-ary *hyperedge*. *Property meta-data* means that every node and every edge in the graph has a property data structure attached to it (at least, a unique identifier and a meta-model identifier).

The middle layer contains the *entity-relationship model*, which is a specialization of the hypergraph model and that *conforms-to* a hypergraph. It comprises the concepts of an *entity*, a *relation* and a *property* and the basic set of standard relations *is-a*, *instance-of*, *conforms-to* and *constraints*. Their semantics is not specific to

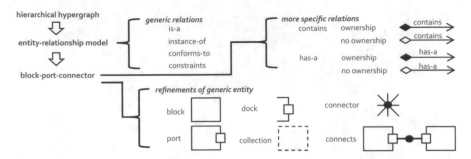

Fig. 3.10 Some selected entities and relations of the RobMoSys modeling foundations (see [21, principles:block-port-connector] for full details).

robotics, but is given by just typical common-sense descriptions. These are basically backed by generic modeling foundations such as *mereology* and *topology*.

This allows us to introduce the *block-port-connector* model, which is a specialization of the more abstract hypergraph model and of the entity-relationship model. It comes with two additional relations, namely, *contains* and *has-a*. The generic entity is refined into *block*, *port*, *dock*, *connector* and *collection*, and the generic relation is refined into *connects*. All these come with a description of their semantics (Fig. 3.10 and [21, principles:block-port-connector] for details).

We now have all the means to express structures and their relations by a *block-port-connector* model. Please note that while *blocks* and *ports* are semantically different, the structure of a port is represented as well by blocks. The kind of presentation is specific to a particular view, its purpose and its level of abstraction.

The bottom layer comprises the *composition structures*. The entry point for their detailed description is again the RobMoSys Wiki [21, composition-structures:start]. These are formed by the consistently arranged *architectural patterns* that are then formalized using the elements that are provided by the middle layer of this tier. This gives a set of meta-models [21, metamodels:start] that define the various elements and their relations. An excerpt of the composition structures is shown in the lower left-hand side of Fig. 3.9. The lower middle part of Fig. 3.9 shows a snippet of the meta-model of a *component* [21, metamodels:component] that comprises all the elements with all the internal and all the external ports that make the structure of a component and its links. In contrast thereto, the snippet on the lower right-hand side shows only those parts of the component meta-model that are relevant to the role of a *component developer*. A role-specific view is a *collection* out of those meta-model elements that are relevant for guiding in a consistent way the work of that role.

As we use blocks, ports and connectors to organize structure, *behavioral aspects* also get structured and separated along blocks, ports and connectors. The advantage now is that for describing the behavior of a block, it is sufficient to describe for all its ports (1) their behavior (just the behavior of the port as it is visible from outside the block) and (2) which input port influences which output port (connections inside a block and related transfer functions). This is fully in line with the way of thinking of the above-described data sheet approach. As long as not all relevant properties can be made available in a formalized way, and as long as one cannot deal yet with all the details due to overall complexity, one can already now offer a migration path toward this end by stepwise making data sheets richer and richer. A valuable approach that is already feasible now are compliance checks via running tests. Of course, there needs to be a mutual understanding and agreement about significance and coverage of the set of tests for properties of interest. In the same line are tests that run an artifact against a reference artifact to check whether there are deviations in properties defined as relevant. Another approach is to run artifacts in simulation to see whether the produced behavior is in line with the expectations (at least for the properties considered as relevant).

All the composition structures conform to the middle layer elements and, thus, either just use that semantics or they come with a consistent refinement that is further narrowing that semantics. Again, the semantics is specified either (1) by references to outside documentation, or (2) by referring to related reference implementations, or (3) by transformations into other representations that already come with a semantics. Of course, at least one mechanism, and sometimes even all three, is given for all the specific elements that are introduced in modeling the composition structures.

2.3   Model-Driven Tools to Access and Use Tier 1 Structures

The modeling mechanisms of Tier 1 are independent from any technology. Nevertheless, making them accessible and usable via model-driven tools has significant advantages in addressing what is called the semantic gap. Often, there are only textual descriptions of the semantics. Without model-driven tools, one depends on the users to properly interpret that semantics and to come up with a correct use in their particular context. There is a high risk of misinterpretations if different users do that individually. In a model-driven approach, highly specialized experts do this once and transform their reasoning and interpretation just into additional structural constraints. Pitfalls on the user side can be sorted out as some assets then fit only in the intended way or get directly linked to a trustworthy implementation providing the intended semantics without any chance to circumvent it. Structural constraints are easy to implement even with model-driven tools of limited expressiveness.

As illustrated in Fig. 3.11, the modeling mechanisms are represented in different ways in different model-driven tools. The modeling elements in the model-driven tools consistently refer back to the full set of Tier 1 structures by the unique identifiers that are used with all modeling elements. Unique identifiers are used, in particular, for elements of instances, elements of models and elements of meta-models.

Examples of specific technologies in model-driven tools are *Eclipse Ecore* and *UML profiles*. Although neither has enough expressiveness to directly represent the structures of Tier 1, one can still implement with them role-specific views. A role-

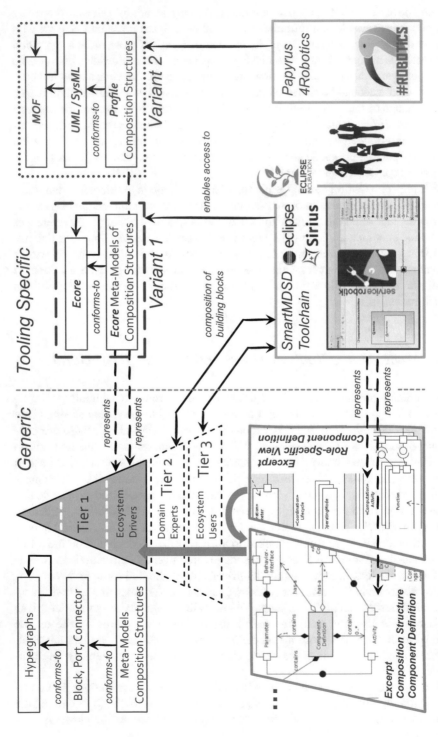

Fig. 3.11 The modeling mechanisms are independent from any technology. Tools make the structures accessible and usable in an easy and role-specific form.

specific view allows a particular person in its role to do its work within its context in a consistent way. Missing expressiveness can be circumvented much easier in the narrow scope of a particular view as measures can be specific. For example, an often-occurring challenge is to transform models about *what is not allowed* into proper mechanisms of model-driven tools. An approach often used for this is to structure the presentation of the view (guiding the user through particular steps) and to run checks for well-formedness before one can proceed with the next step.

2.4 Tier 2: Definitions

Tier 2 structures all the domains that are part of an ecosystem. A particular domain, such as vision, navigation, manipulation, etc., is shaped by the experts of that domain. Again, it is a community effort to structure a particular domain by creating domain-specific models. As a domain is limited in scope, it is much easier to come up with an agreement about, for example, particular vocabularies and data structures. At Tier 1, one would need to aim for an all-encompassing approach which several times already proved to be just infeasible (as evidenced by all the failed approaches to come up with a single ontology for a domain such as robotics). However, as all the domain-specific models in Tier 2 conform to Tier 1, these different models can be linked to each other in a consistent way.

By the way, that is exactly the role of *meta-models*. It is important to understand that the concept of a meta-model is a relative one. A meta-model is just a *model* as well. It becomes a meta-model only with respect to other models that conform to it.

As Tier 1 serves as meta-model for Tier 2, all models at Tier 2 conform to the models at Tier 1. Models at Tier 2 bind flexibility and narrow down semantics, but they are all consistent with the models in Tier 1. Of course, there can be different and coexisting domains, all with their own models. Although these are all consistent within their domain and although they all conform to Tier 1, concepts often cannot be easily linked one-to-one across domains. However, such links between domain-specific models are always available via the path through Tier 1.

Obviously, Tier 2 needs to be defined within the structures and semantics given by the bottom layer of Tier 1. In order to fulfill the relation *conform-to*, one can neither just pick out structures and ignore their semantics, nor can one just arbitrarily pick out elements from the top layer and the middle layer of Tier 1 as these layers do not form the interface to Tier 2. One cannot just go upward until it gets generic enough to then extend from there on. Of course, doing so would completely break the sound conceptual approach of layering abstraction levels and having them only interact with adjacent layers. It would break all the foundations that enable composition and that would again result in fragmentation.

If we cannot live with the framing of the adjacent superordinated layer, we cannot just put a different semantics next to it and ignore what already has been agreed. Instead, in case there are very good reasons why adjustments or extensions are needed at a particular layer, then we should just submit a convincing proposal that has a chance to get support from the community or expert group. As already mentioned before, that is why there is such a high responsibility in coming up with proper structures that match the sweet spot between freedom of choice and freedom from choice.

Very often, the above is misunderstood and it takes some effort explaining why it is not sufficient to just go for the structural approach of blocks, ports and connectors. Blocks, ports and connectors become of use only when one respects their concrete designs with semantics. By the way, that is exactly the challenge of applying *SysML*. It fits easily nearly everywhere in systems engineering as it basically just refers to blocks, ports and connectors. However, it gets useful only when there is another level of detailing to provide a consistent meaning and grounding for the kinds of blocks, ports and connectors in a particular domain.

2.5 Tier 3: Implementations

Tier 3 comprises all the content available in the ecosystem. In robotics, such content is libraries, software components, solution templates representing architectures as well as task plots that convey process knowledge independent of a particular robotic system, and many other artifacts. Tier 3 uses the domain-specific structures from Tier 2 in order to produce content to supply to the ecosystem and thereby using content from the ecosystem. For example, an expert in localization will adhere to the services defined for the navigation domain in building a software component for localization as it then fits to others conforming as well to the agreements for the navigation domain.

2.6 Coverage and Conformance

An ecosystem is viable and sustainable only when it finds the right balance between rigidity and plasticity. The further up a concept is in the hierarchy of the ecosystem structures, the more impact on the ecosystem comes with any modification of that concept and the better reasons have to be brought up for doing so. Thus, proposing an alternative to what is already defined (without having a real benefit, a real need or high enough added value) will hopefully not find enough supporters as this just fosters fragmentation. It is much simpler to make modifications at a lower tier of an ecosystem as the scope of the impact is much smaller. However, the lower a concept is in the hierarchy, the more constraints for new concepts come from the superordinated structures to which conformance is mandatory. In principle, there is always the option to put something next to what already exists (e.g. open up a new domain at Tier 2) as long as it is conforming to superordinated structures. This is an important means to take up new proposals and then perhaps let them even replace already existing structures.

Basically, the balance between rigidity and plasticity is dependent on the processes that manage the ecosystem. Of course, the three tiers already come with well-defined responsibilities (ecosystem drivers at Tier 1, domain experts at Tier 2 and all the ecosystem participants in Tier 3). This makes it easy to identify and then approach the according representatives whenever needed. However, that alone is not sufficient. In the ideal case, an ecosystem is managed in the form of a *meritocracy* for which the following key processes are decisive.

First, there need to be processes for the *vertical* interaction (i.e. between stakeholders at different tiers) and for the *horizontal* interaction (i.e. between stakeholders in different parts within the same tier). For example, there might be a concept that evolved within a particular domain and that might be of fundamental use for other domains as well. In such a case, one should discuss whether to move that concept up in the hierarchy. In case one detects coexisting approaches at lower tiers for the same concerns, one should stipulate a process to overcome this accidental fragmentation.

Second, there need to be processes to assign *coverage* and *conformance* labels to assets in the ecosystem. That enables us to explicate for any asset its link into the ecosystem. Assets include concepts, models, software components, tools and everything else of relevance in the ecosystem. This tremendously simplifies navigating through the ecosystem and selection of assets according to needs. Of course, there can be coexistence of assets that come with only partial conformance (relevant for uptake of new ideas in early stages or for establishing first links into neighbouring domains), but this always is explicated.

Basically, the outcome of processes to tackle aspects of coverage and of conformance are agreements (1) on content and format of data sheets for all kinds of assets in the ecosystem and (2) on procedures to state the properties in the data sheets. Of course, coverage of particular aspects with their particular degrees of conformance can be subsumed under dedicated labels. Some labels can even come with certification bodies.

3   Details of Selected Concepts at Tier 1

This section explains details of some selected concepts at Tier 1. These come with concrete structures and behaviors that are not specific to any domain in robotics. Exhaustive descriptions with lots of references to further documents are maintained via the *RobMoSys Wiki* [19].

A model at Tier 1 is called a *definition* meta-model when it defines the means for Tier 2 domain experts to model domain-specific aspects. It is called *realization* meta-model when it defines the means for Tier 3 ecosystem participants to model (and build) concrete assets. Otherwise, a model comes without any specific label. Of course, a model at Tier 1 directly serves as meta-model for Tier 3 in case there are no further refining models down to Tier 3.

3.1 The Software Component Model

A *software component* [24, 25] is one concrete example of a block with ports that needs to be specified (structures and behaviors) at the lowest level of Tier 1. A software component combines aspects of different concerns, namely, *computation*, *communication*, *coordination* and *configuration*. It comprises the abstraction levels ranging from the *execution container* to the *skill interface* but without diluting their layering and their interfaces.

Figure 3.12 shows the structure of a software component, that is, its elements and their relations. Of course, these elements all come with well-defined behavior, and the behavior of the software component model is defined by the behavior of these elements in the context of the interaction given by the explicated relations. Thus, it would be a complete misunderstanding if one conforms only to that structure but ignores the related specified behavior.

For example, the *Lifecycle* of a component is represented as a block with ports. It interacts with the *Activity* as an activity runs only in particular states. It interacts

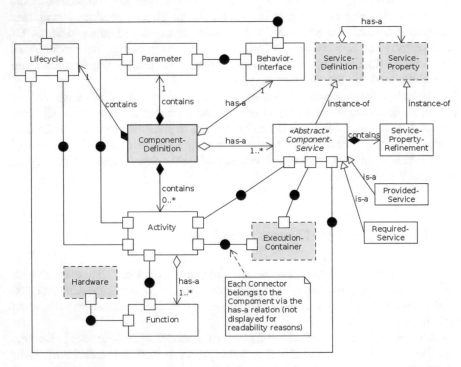

Fig. 3.12 The component definition meta-model shows the structure of a software component. A dashed gray box is a reference to another definition meta-model. The behavior of a block is defined by the behavior of its ports. A port is a block as well. A block refers to a behavior specification. A behavior specification can be in different forms (textual description, reference implementation, formalized model).

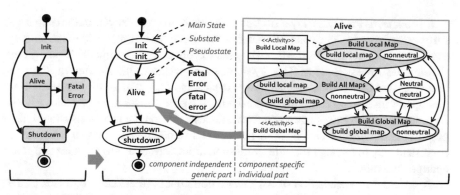

Fig. 3.13 The lifecycle of a software component with main states (visible outside), substates (inside coordination) and the pseudostate *Alive* (customizable).

with the *Services* of a component as their availability and their responses depend on the state of the component. It interacts with the *Behavior Interface* as this is the access from outside to the state of a component, its configuration and its monitoring (control flow interface).

The very concrete structure and behavior of the lifecycle of a software component are specified in detail in [26]. Figure 3.13 illustrates selected aspects. The lifecycle comes with the predefined main states *Init, Shutdown, Fatal Error* and *Alive*. At any point in time, a component can be in exactly one main state only. Main states come with substates. The main states *Init, Shutdown* and *Fatal Error* are created with exactly one substate each. As a convention, each name of a main state begins with a capital letter and each name of a substate begins with a small letter.

Activities inside a component refer to substates only and not to main states. An activity runs only when the required substates are active. From outside the component, only main states (and no substates) can be commanded. The reason for this approach is to decouple component internals from the outside view such that one can provide a standardized way of how to configure, control and monitor a component via its standard behavior interface. The approach becomes clear with the main state *Alive*, which is basically a pseudostate.

The main state *Alive* can be replaced by an individual state automaton that gets executed when the component is in the main state *Alive*. One can define an arbitrary set of substates for use by the activities within the component. The substates are clustered into user-defined main states. A substate can belong to any number of user main states. In this way, a user main state represents such a set of substates that can be consistently active at the same time and without conflicts. From the outside view, one needs to know neither the names of the individual substates nor the combinations of substates that can be consistently activated. One just needs to query the names of the user main states. One can set any of those user main states instead of the main state *Alive*, in any order.

Of course, as the lowest layer of Tier 1 already defines a lifecycle automaton for a software component, one cannot just define another one and put it next to the already existing one. Nevertheless, there can be ongoing discussions within the expert group shaping the lowest layer of Tier 1 whether one wants to have different kinds of lifecycles that then coexist (which means to explicitly agree to go for fragmentation, perhaps, as an interim solution). In particular, the pseudostate *Alive* provides a hook for other individual state automata. Nevertheless, alternative proposals should also properly deal with the challenge of providing a standardized behavior interface, a consistent way to manage activities and communication according to states and state changes and others.

A different representation of a software component (that fully conforms to the component definition meta-model) is shown in Fig. 3.14. It shows different role-specific views in a single figure. In the case of being a user of a software component, we see its outer view, which just comes as a block with ports as indicated by the solid surrounding line. These ports form the stable interface to other components and they are either a *provided service* or a *required service*. In the case of being a component developer, we see the stable interfaces of the inner part of the component

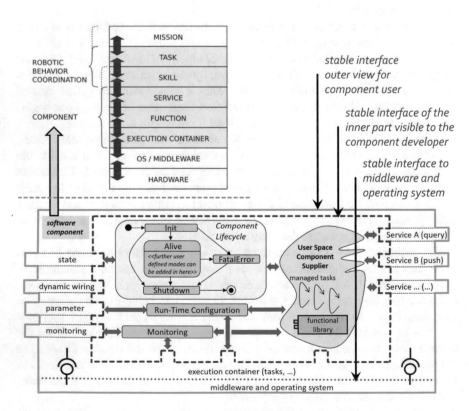

Fig. 3.14 Software component with communication (service-oriented ports), configuration (resources, parameters), coordination (modes, lifecycle) and computation (managed tasks).

(dashed line). Finally, the dotted line is the stable interface to the middleware and the operating system. This is an internal interface hidden from the user. It links the execution container to the middlewares and operating systems and ensures that the execution container is agnostic to middlewares and operating systems.

If we open the software component block, we see that its internals are again blocks with ports. In particular, a component service (which is a port of the software component block) itself becomes a block (the complete component service) with ports. Its first port forms the external dock of a component service (visible from outside the component as component service port—solid line). The second port forms the internal dock of a component service (visible inside the software component only and only by the component developer—dashed line). Other ports form the internal interface to the component lifecycle and to the execution container (that links to the OS/middleware).

As the basic structure and behavior of a software component is defined at Tier 1, these are prescribed for Tier 2 and for Tier 3. However, the software component model still provides enough freedom when it comes to filling in, for example, domain-specific resource management mechanisms inside the dashed block. Of course, when domain-specific refinements become of general use, they can be moved upward to Tier 1.

Basically, the model-driven tools are the means to come up with proper presentations of role-specific views. For example, the model-driven tools then ensure that we select ports of a component only from a palette that represents what has been agreed upon in this domain and that we cannot use others not specified for the selected domain. The guidance and consistency of what is allowed in which step by which role as it is given by the model-driven tools stems from the meta-models and the reference implementations linked to the elements.

3.2 Communication Patterns and Services

The link between the component internal ports (dashed interface in Fig. 3.14) and the component external ports (solid interface) is realized via *communication patterns*. Basically, they cover a request/response as well as a publish/subscribe interaction. Table 3.1 gives an overview of all the communication patterns used by a software component for communication and for coordination. The patterns for coordination are based on the patterns for communication. The meta-model for the small set of generic communication patterns is shown in Fig. 3.15.

A communication pattern always consists of two complementary parts. The outer docks of complementary communication patterns can be connected with each other in case the used communication objects are the same. Again, it is important to note that all patterns come with well-defined structures (i.e. the APIs for their docks visible inside a component and also the definition of what is a fit between external docks) and well-defined behavior (i.e. the fully specified behavior of their externally visible dock and of the API of their internally visible dock). That is given by both

Table 3.1 The *communication* patterns and the *configuration & coordination* patterns of the software component model (see [21, metamodels:commpattern])

Pattern name	Interaction	Description	Definition
Send	Client/server	One-way communication	[27, pp. 85–88]
Query	Client/server	Two-way request/response	[27, pp. 88–96]
Push	Pub/sub	1:n distribution	[27, pp. 96–99]
Event	Pub/sub	1:n asynchronous condition notification	[27, pp. 103–112]
Parameter	Master/slave	Runtime configuration of components	[17, 28]
State	Master/slave	Lifecycle management, modes	[26, 29]
Dynamic wiring	Master/slave	Runtime (re-)wiring of connections	[27, p. 112]
Monitoring	Master/slave	Runtime monitoring and introspection	[28, 30]

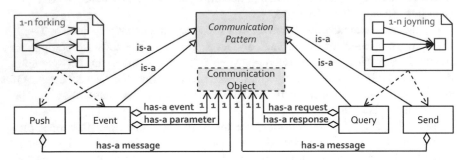

Fig. 3.15 The meta-model for the communication patterns (*Communication Pattern Meta Model*).

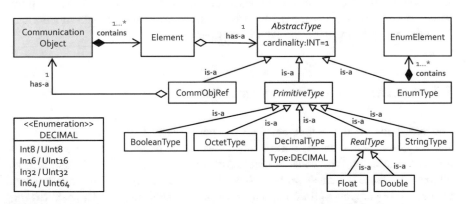

Fig. 3.16 The meta-model for a communication object (*Communication Object Meta Model*).

(1) descriptions in the form of documents [27] (that contain behavior specifications in the form of automata, sequencing diagrams, etc.) and in the form of API specs [31] and (2) reference implementations producing that very semantics (e.g. one such fully conformant implementation is given by SMARTSOFT/ACE [32]).

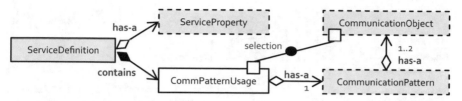

Fig. 3.17 The meta-model for defining services (*Service Definition Meta-Model*).

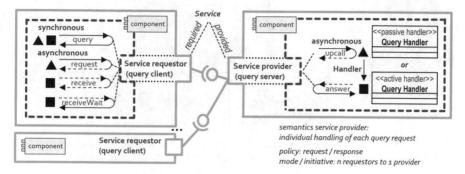

Fig. 3.18 A *service* based on the *query* communication pattern with the internal and external docks.

Figure 3.16 shows the meta-model for *communication objects*, and Fig. 3.17 shows the meta-model of a *service*. A *communication object* is a digital data representation that can be transferred *by-value* via communication patterns. A *service* is the combination of a particular communication pattern with selected communication objects. Thus, a service shows exactly the structure and behavior as defined by the used communication pattern, although it is customized via the used communication objects. A service is split up into a *provided service* and a *required service* according to the complementarity of the used communication pattern. Each port of a software component then is either such a provided service or a required one.

This is illustrated in a summary in Fig. 3.18. The shown *service* is based on the *query* communication pattern. The *provided* part of the service is based on the *query server* part of the query communication pattern. Accordingly, the *required* part of the service is based on the *query client* part of the query communication pattern. It shows also the interfaces of the internal docks that are visible to the component developer. The client-side access policy (synchronous, asynchronous) is completely decoupled from the server-side processing policy (asynchronous via handlers).

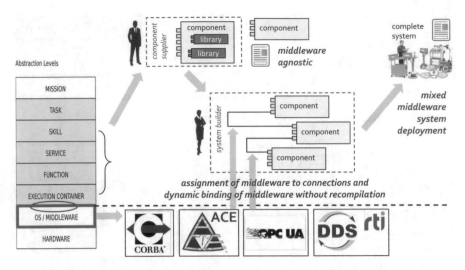

Fig. 3.19 The interface between the abstraction layer *execution container* and the abstraction layer *middleware/operating system* allows for middleware-agnostic software components.

3.3 Middleware-Agnostic Software Components

The structure and behavior of the communication patterns and of the communication objects, and thus of services as well, are completely middleware-agnostic. Their semantics is defined fully independently from any particular implementation technology. It is at the interface between the layer of the *execution container* and the layer of the *middleware/operating system* (Fig. 3.19) where the given and specified semantics of the communication patterns must be reproduced exactly as specified via the means of the selected underlying middleware technology. It is also at that interface where the elements of the communication object meta-model get mapped onto marshaling mechanisms of a middleware. The mapping between the execution container and a particular middleware/operating system is provided by experts on system-level software. Experts for model-driven tools make use of such mappings and reference to them from within their tools to generate artifacts for the according middleware/operating system.

By means of a common meta-model with a clearly specified semantics, one can replace a middleware by another one without impact on the overall behavior of the system. One can even mix different middlewares within a single system and one can decide on a middleware per connection. With dynamic linking, one can do this even at deployment time and without recompiling any of the components. A common meta-model also allows for the automatic transformation between different middleware-specific representations of the same communication object. Section 4.4 shows this via the *SmartMDSD* toolchain.

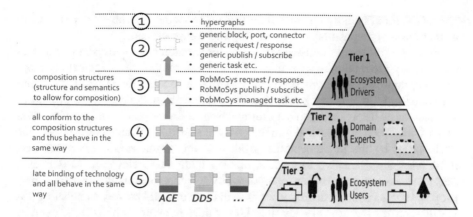

Fig. 3.20 Early binding of semantics and late binding of technology (not early binding of technology with lock-in into their semantics).

3.4 Early Binding of Semantics, Late Binding of Technology

Middleware-agnostic software components are a good example of the importance of the principle of *early binding of semantics but late binding of technology*. Only adhering to *structure* and ignoring the behavior destroys composition and thus results in fragmentation. The same happens when one just ignores the abstraction layers. Although that insight is a commonplace, there often is a tremendous lack in discipline when it comes to adhering to it.

Figure 3.20 illustrates this along the hierarchy of the tiers of an ecosystem. The composition structures (see ③, bottom level of Tier 1) are fully technology-agnostic (also agnostic to technologies of workbenches for model-driven tools). They ensure composition, composability and compositionality. These properties are available earliest at ③, which sets the frame for Tier 2 (see ④). Whatever domain-specific structures are introduced at ④, these are consistent with ③ and they do not destroy composition. The same holds true for all the assets (see ⑤) as these conform to ④. The behavior of a structure introduced at ③ frames the refinements at ④ and ⑤. Late binding of an implementational technology at ⑤ (such as a particular middleware) does not change the behavior given by ④ (and given by ③) as that binding is eligible only when it produces a behavior exactly in line with ④ (which is then also in line with ③). In this way, all the assets at ⑤ are not affected in their behavior when we move to another underlying implementational technology and there are nearly no migration costs.

In contrast to that is the approach of directly working with a particular implementational technology and let that subsequently drive structures and behaviors. Any change of the implementational technology would affect the behavior as well. This easily results in huge migration costs with any new technology and obviously is in conflict with the aim of composability. Therefore, it is also a misunderstanding that arbitrary solutions, which are available so far in source code only, can become part of the ecosystem by just modeling them. The models in Tier 1 are much more as they ensure composition, composability and compositionality.

Thus, the following often-seen proposal for extending the ecosystem structures does not help to attain the objectives of a business ecosystem. A typical but inappropriate approach is that an expert for a particular robotics domain or a particular software framework just picks out an arbitrary block from the ecosystem structures and ignores its behavior. A typical example is to select the generic *request/response* block from ② and to refine that into the *query* of a particular framework and provide its behavior by the implementation from that very framework. In consequence, it becomes an alternative to what is already there at ③. However, ③ defines a *query* that ensures composition, composability and compositionality. Either the new proposal comes with those properties as well (then it would be very astonishing if it is different from what is already there), or it does not come with those properties (then it should not be placed at ③ as then Tier 2 and Tier 3 cannot rely on composition anymore). It can also not be placed at ④ as this would not be consistent with ③. Just placing it at ④, skipping ③ and extending ② is also not an option as this again circumvents the composition structures and thereby jeopardizes the ecosystem foundations.

Of course, one can still propose to put another *query* at ③ that does not come with all the properties required for composition. However, such a query then does not help the ecosystem objectives. This is the same for all the other models such as the lifecycle of a software component, its behavior interface and many others. Again, model-driven tools can enforce compliance where necessary and it can circumvent relying on discipline only.

3.5   Horizontal Versus Vertical Composition

While *vertical composition* addresses the composition of parts located at different layers of abstraction, *horizontal composition* focuses on the composition of parts located at the same layer of abstraction. The concepts are illustrated in Fig. 3.21.

The difference becomes obvious in the context of control hierarchies. In a control hierarchy, resources assigned to a particular layer can be managed there. As long as one operates within the given boundaries (constraints on the assigned resources), there is no conflict with others. In particular, we can assign shares of our resources to entities that are under our control (vertical relationship). These then can again manage those resources within the given constraints and according to the given policy. It gets more complex in case we need access to resources outside our responsibility. In such a case, we need to come up with a contract that secures that horizontal relationship.

3.6   The Data Sheet

Figure 3.22 shows more details of a digital data sheet using an the example of a software component. The very same principles apply for all the assets in an ecosystem.

A digital data sheet comes in digital form and it contains different parts: at least a *technical part* that is generated from the technical models of the asset and a *descriptive part* that comprises manual annotations done by humans. The manual annotations in the descriptive part can range from free-form text fields to ontology-based labels. A data sheet is a model and thus its form, content and meaning are defined at Tier 1 (general parts) and Tier 2 (domain-specific parts).

The technical part is more like a technical product data sheet, operating manual, etc. It is technically binding and states guaranteed properties that can be claimed by customers. It supports aspects such as what-if analysis, prediction, composition, configuration and runtime adaptation. As can be seen in the lower right, the technical part may also contain an abstract representation of selected internals such as data flows and transfer functions needed for analysis, checking, predicting and configuration of system-level properties.

The descriptive part is more like a product description for presentation in a web store or in an advertisement. Its focus is on finding and preselecting assets from a marketplace. Thus, the descriptive part is more market-oriented, oriented to gain attention and awareness, and thus is more user-oriented. The statements in the product description are mapped to and grounded by the technical part of the data sheet.

The concept of a digital data sheet is fully consistent with the concept of an industry 4.0 asset administration shell (AAS). As shown in the upper right-hand side of Fig. 3.22, the digital data sheet can become just a submodel in the AAS and thus part of the AAS container.

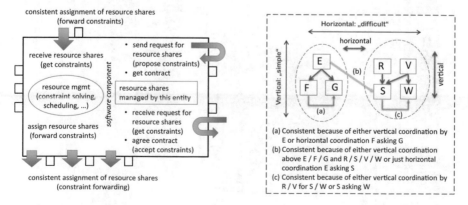

Fig. 3.21 Horizontal versus vertical composition: separation of control flow and data flow and responsibilities for resources.

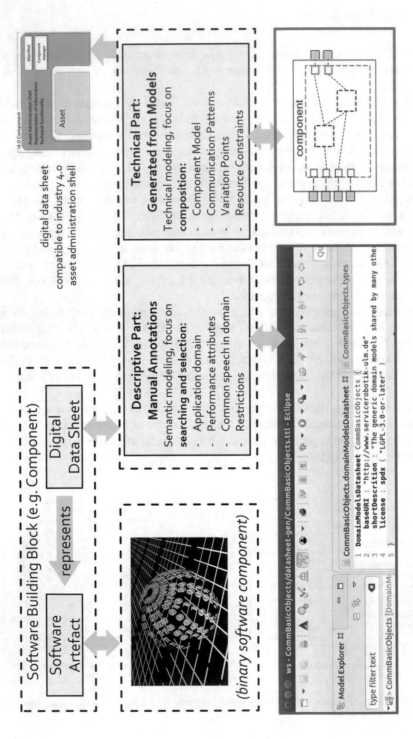

Fig. 3.22 A digital data sheet for the example of a software component.

3.7   *Dependency Graphs and Constraints on Services*

Dependency graphs can model system-level requirements that span across different components. Examples are properties along data flows, such as quality and ageing of data, but also consistency aspects, triggering along computational chains and arrival time analysis [33]. A dependency graph can also be used to express requirements on data privacy, for example, "there must not be any link between a service providing a raw camera image and a service connecting to the network outside the robot". A dependency graph can also express that, for example, the blur of a camera image depends on the speed of the robot. In order to keep the blur below a given threshold, one can derive the related maximum allowed velocity. A dependency graph can guide the system builder in selecting and configuring components according to the input from a system architect, a safety engineer and others. Thereto, the foreseen variation points, explicated in the data sheet, are exploited.

Dependency graphs can also be extracted from a virtual system composition (based on data sheets only; see Fig. 3.7, Sect. 1.5). Before building the real system, one can already check whether the dependency graphs in the virtual system comply with the constraints of the dependency graphs representing the requirements. Tools can be used to try different configurations and to check for matches. In that way, dependency graphs also help to implement traceability from requirements to fulfillment by configurations of variation points. Dependency graphs also serve at runtime as sanity checks (before finally implementing a runtime decision) and for monitoring integrity.

From a technical point of view, the meta-model of a dependency graph is again based on the entity-relation model provided in layer 2 of Tier 1, which refines into different variants at layer 3 of Tier 1. The dependency graph shown in Fig. 3.23 is a simple one that expresses dependencies between services (entity *service*, relation *uses*). It gets checked by a mapping between the dependency graph and the graph resulting from the data sheet composition.

Figure 3.24 shows two more examples of a dependency graph. The example on the left shows error propagation through a data flow across components. Transfer functions of a component specify what additional uncertainty comes on top. For example, the system builder can check whether the selected component is good enough to match the requirement at the end of the processing chain.

The dependency graph on the right is based on *name/value* pairs. The left component might be a service providing a camera image. That data is labelled as *critical* with respect to data privacy. The right component might provide a map based on the camera image (the transfer function for this port makes critical input to *uncritical* output) and the camera image enriched with pose information (critical input results in critical output at this port). Interval arithmetic and constraint solving are further mechanisms for dependency graphs. *Trigger chains* are another example that is illustrated in more detail in Sect. 4.6.

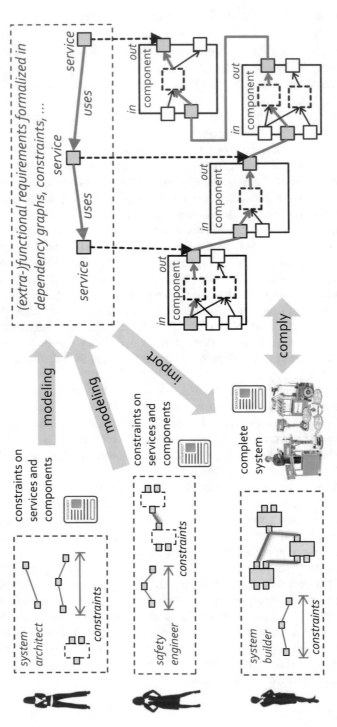

Fig. 3.23 Formalizing requirements by dependency graphs (e.g. safety engineer, system architect) and performing according configurations and checks (e.g. system builder).

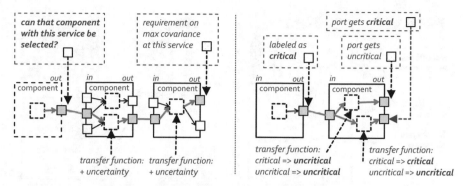

Fig. 3.24 Further examples of dependency graphs. Left: checking component selection (error propagation). Right: checking criticality of information (name/value pairs).

3.8   Tasks, Skills and the Behavior Interface

Robotics *behavior blocks* are separated into different levels of abstraction in order to allow for composition and for separation of roles. A *skill* provides access to functionalities realized by components. It is the bridge between generic descriptions of capabilities (independent of a particular implementation) and behavior interfaces (*configuration*—resources, parameters, wiring; *coordination*—modes, lifecycle, events) of a (set of) component(s) to achieve that capability (specific to the used components). A skill lifts the level of abstraction from a component-specific level to a generic level. Thus, different implementations for the same capabilities become replaceable as they are accessed in a uniform way.

Tasks describe via which steps (what, the ordering of steps) and in which manner (how, the kind of execution) to accomplish a particular job. This is done in an abstract manner independent from a particular robot as tasks refer to skills for their grounding (see also Chap. 12 on mission specification).

Tasks and skills are domain-specific as they refer to domain-specific vocabularies. Nonetheless, software components can be used in skill sets of different domains. Of course, one can also define a domain that holds generic tasks and skills that can then be used from within different domains.

We illustrate that for the domain of *intralogistics*. A *task* for *order fulfillment* arranges several other tasks such as *move to a shelf, locate and pick an item, follow the worker* and *deliver the item at a packing station*. Such tasks refer to skills for their execution, for example, to skills such as *navigate-to, recognize-object* and *grasp-object*. A particular task can be executed by a particular robot when the robot possesses matching skills for all the required skills. For example, a task might refine into a skill for grasping an object. From the execution context and its bindings, one knows that the object has a maximum weight of 1 kg. The robot can select any of its grasping skills that can handle at least 1 kg.

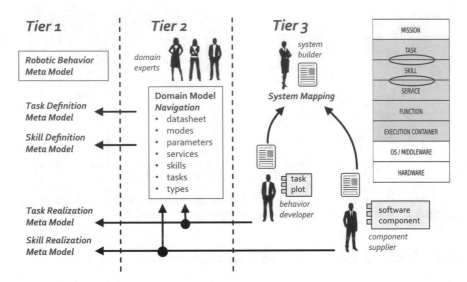

Fig. 3.25 The general context of meta-models for tasks and skills and their link to the behavior interface of software components.

Figure 3.25 illustrates the general context and selected roles, namely, *domain experts*, *component suppliers*, *system builders* and *behavior developers*. It also comprises different meta-models from Tier 1 and their use at Tier 2 and Tier 3. Skill definitions are being done by domain experts that belong to Tier 2 of an ecosystem. Those skill definitions are prescriptive for component developers that operate at Tier 3 of an ecosystem. A component developer adheres to the skill realization meta-model in making accessible the capabilities provided by its component (including references to other components needed for that capability) and thus does this in full conformance with the Tier 2 domain-specific specifications. Finally, the behavior developer is able to model task plots and ground them in domain-specific skill definitions. The skill definitions are the interface to software components providing the required functionalities and getting them executed. The components selected and composed by the system builder to form the robotic system all come with their individual skills. These skills get listed in the data sheet of the robotic system. Thus, the set of skills available on a concrete robotic system is part of and available via its accompanying data sheet.

Selected meta-models of Tier 1 for the behavior interface are shown in Fig. 3.26. The overall link between tasks, skills and services of a software component is organized by the *robotic behavior meta-model*. The *task definition meta-model* and the *skill definition meta-model* are the entry points to provide the means for the *modeling views* of the domain experts at Tier 2. They describe which elements form sound descriptions of a task and of a skill. They specify that input values and output values as well as results are described by *attribute-value* pairs. A result consists

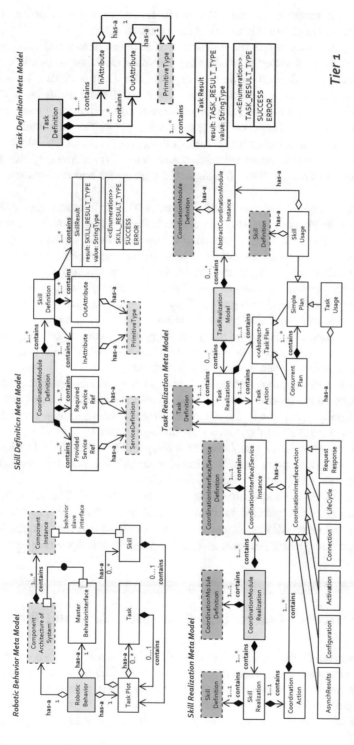

Fig. 3.26 Selected meta-models for tasks and skills and links to the elements of the software component meta-model.

of a list of string values, each labelled with either *success* or *error*. Task and skill definitions are grouped into sets to model their semantic cohesion.

Of course, the domain-specific definitions of attribute names and type assignments for input/output values as well as the domain-specific set of possible string values for results are defined by domain experts (Tier 2). It is at Tier 2 where all the tasks and skills that are possible within a particular domain are defined. That is done by modeling them in a way that is conforming to the definition meta-models for tasks and for skills. Based on the generic concepts at Tier 1, domain experts model for their domain the concrete forms of data sheets, tasks and skills. They model the concrete sets of communication objects and services, and they also define concretizations of configurations (configurable parameters and configurable main states).

The domain-specific task and skill definitions at Tier 2 ensure that different components with all their specifics present their capabilities in the same way. In consequence, skills belonging to the same domain represent the same capability when they use the same name and their input/output arguments as well as their result values follow the same domain-specific vocabulary and meaning.

The *task realization meta-model* and the *skill realization meta-model* put all those domain-specific models together. They form the foundation for presenting aspects of robot behavior in role-specific views at Tier 3. For example, software components comprise a *behavior interface*. Its generic structure and semantics is part of the *component definition meta-model* (Fig. 3.12, Sect. 3.1). It is used to configure control flows, even dynamically at runtime.

The *component supplier* needs to adhere to the domain-specific ports for software components (communication objects, communication patterns, services, parameterization, main states). The component supplier needs to model skills of its (set of) component(s) along the domain-specific agreements.

Task plots are modelled by the *behavior developer*. They refer to skill definitions, not to software components. The link between tasks/skills and skills/component instances on the robot is set up by the *system builder*.

More insights into role-specific views are given in Sect. 4.7. The differentiation between the *task/skill definition meta-models* and the *task/skill realization meta-models* is in the same vein as for communication objects and for services, where also the generic means are provided at Tier 1 and the domain-specific means are at Tier 2.

4   Links Between Composition Structures, Roles and Tools

This section gives further insights into some links between *definition meta-models*, *realization meta-models* and *roles* using views of them. This also gives insights into how the independent activities of different roles can result in consistent assets due to the overall guidance of the meta-models in the different tiers. Of course, the meta-models and role-specific views are independent from their implementation within a particular model-driven tool. However, examples are illustrated by screenshots from the open-source eclipse-based *SmartMDSD* toolchain [34], which implements the composition structures via *Ecore*. This is much more explanatory than going into details of their lengthy Ecore-mappings.

4.1 The Role of the Domain Expert (Tier 2)

Figure 3.27 show selected views for the Tier 2 domain experts. *Types* define communication objects, *services* define the combination of communication patterns and communication objects, *modes* define the agreed and configurable main states for software components, and *coordination services* define the coordination and configuration interface. Middleware-agnostic communication patterns require middleware-agnostic communication objects. The middleware-agnostic modeling of data structures is also used in a broader context and is then referred to as *digital data modeling*. In the context of communication objects, the digital data models are mapped in a semantics-preserving way onto the middleware marshaling mechanisms. *Skill definitions* define the task-level interface of skills and the *data sheet* refines the non-technical part of a data sheet (e.g. that license information is to be provided in that domain).

4.2 The View of the Component Developer (Tier 3)

Figure 3.28 shows a software component. The component developer selects from the domain-specific models at Tier 2 the ports (required and provided services) and the behavior interface (parameterization, configuration, main states of lifecycle, etc.) and models the skills of the component along a Tier 2 skill set. An example is shown in Fig. 3.29.

Skills always come with a component as a skill has a component-specific side (component behavior interface). A skill of a component can refer to skills of other components in order to allow for modeling skills that require a set of components. At the end, the set of models for a software component comprises the component model itself, the model for its parameters, its data sheet, its documentation and the skill realization.

4.3 The View of the Behavior Developer (Tier 3)

Figure 3.30 shows the *SmartTCL* version of the task realization model of a *transportation task*. *SmartTCL* is a task-coordination language for hierarchical task nets [35]. It comes with an interpreter, which at runtime dynamically expands tasks

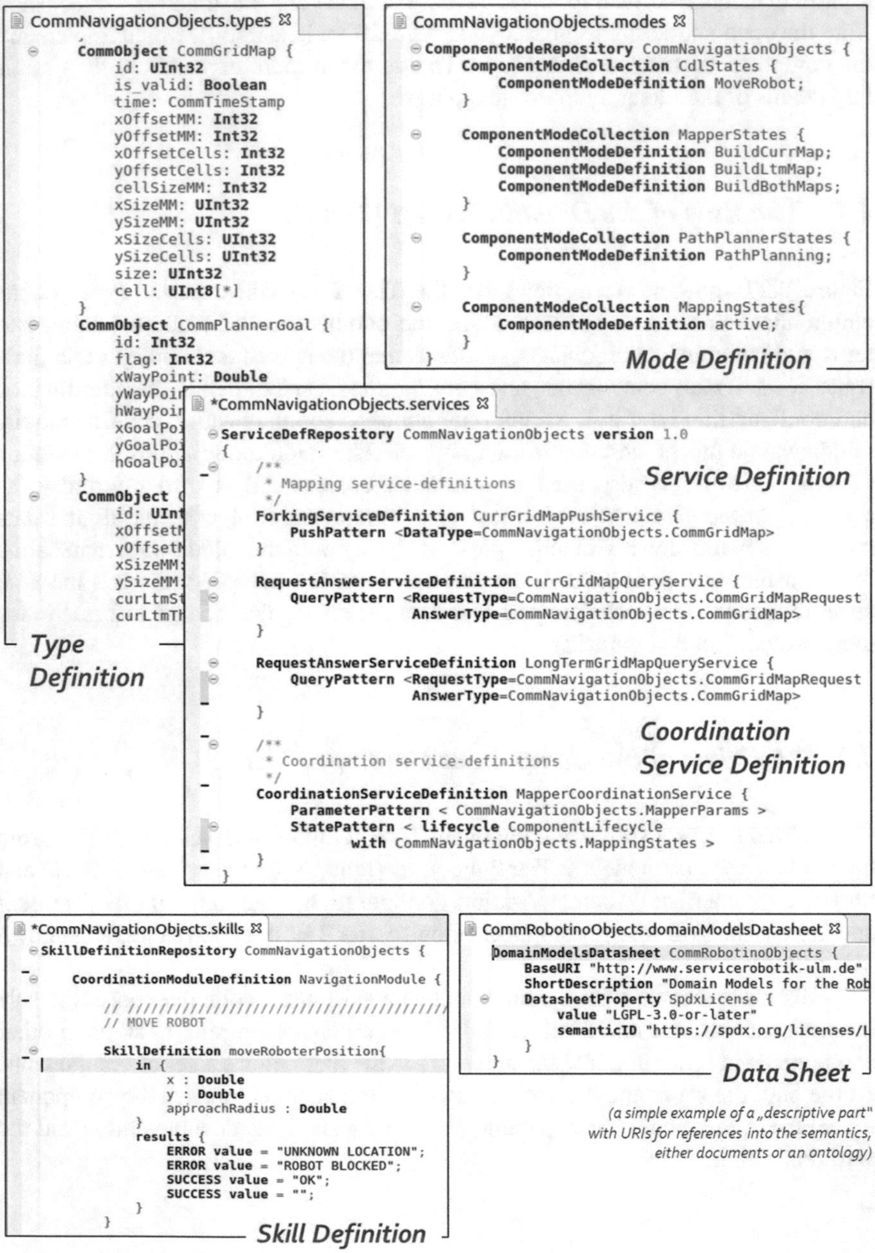

Fig. 3.27 Some views in the *SmartMDSD* toolchain for Tier 2 domain experts.

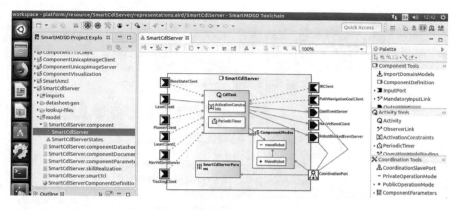

Fig. 3.28 The component developer view in the *SmartMDSD* toolchain (Tier 3).

```
 CommNavigationObjects.skills     CommNavigationObjects.services     SmartCdlServer.skillSmartTCL ⋈    CommNavigationObjects.modes

  SkillRealizationModel {
      CoordinationModuleRealization NavigationModule coordModuleDef CommNavigationObjects.NavigationModule
      uses {
          CommNavigationObjects.CdlCoordinationService instName cdl
          CommNavigationObjects.MapperCoordinationService instName mapper
          CommNavigationObjects.PlannerCoordinationService instName planner
      }(
          ;; realization of APPROACH LOCATION skill
          (define-skill-block (approachLocation ?locationid)
              (skillDefinition approachLocation)
              (precondition (equal '(region) (get-value (tcl-kb-query :kb-key '(is-a name) :kb-value '( (is-a location) (name
              (module "NavigationModule")
              (rules ( ruleUnknownLocation ruleLocationSuccess ruleRobotBlocked ))
              (plan ((moveRobotRegion ?locationid)))
              (abort-action (
                  (tcl-ci-state :server cdl :state Neutral)
                  (tcl-ci-state :server mapper :state Neutral)
                  (tcl-ci-state :server planner :state Neutral))))          Skill Realization
```

Fig. 3.29 Excerpt of the skill modeling for a software component (skill realization by the component developer in the *SmartMDSD* toolchain at Tier 3 using the SmartTCL representation).

into arrangements of other tasks/skills based on the current situation and context. Besides a hierarchical refinement (see the *tcl-push-plan* statement, in this example a refinement into skills), arrangements of tasks and skills can be *sequential, parallel, one-of* and others. The example nicely shows the binding of task variables via the knowledge base of the robot which, for example, contains the world model (linking names of rooms to regions in a particular map and a particular coordinate system) and also the data sheets of the used components. It is accessed as any other component via service ports (in our example, a service port for tell-and-ask expressions). The example also shows how a task model refers to skill realizations. Some more details on the relationship between design-time task and skill modeling and their runtime execution follow in Sect. 4.7.

```
▣ BehaviorNavigationScenario.taskSmartTCL ✕ │ BehaviorNavigationScenario.taskRealization │ CommNavigationObjects.tasks
  TaskRealizationModelTCL {
      AbstractCoordinationModuleInstance navigationInst coordModuleDef CommNavigationObjects.NavigationModule
      AbstractCoordinationModuleInstance kbModInst coordModuleDef CommBasicObjects.KBModule
      AbstractCoordinationModuleInstance localizationModInst coordModuleDef CommLocalizationObjects.LocalizationModule
      AbstractCoordinationModuleInstance base coordModuleDef CommNavigationObjects.MobileBaseModule
      AbstractCoordinationModuleInstance mpsInst coordModuleDef CommRobotinoObjects.MPSModule

      (define-task-block (transportationTask ?startStationID ?goalStationID)
          (taskDefinition CommNavigationObjects.transportationTask)

          (action ( (format t "============= >>> transportationTask from:~s to:~s ~%" ?startStationID ?goalStationID)
              (let*  ((start-station (tcl-kb-query :kb-key '(is-a id) :kb-value `((is-a station)(id ?startStationID))))
                      (start-station-approach-location (get-value start-station 'approach-location))
                      (goal-station (tcl-kb-query :kb-key '(is-a id) :kb-value `((is-a station)(id ?goalStationID))))
                      (goal-station-approach-location (get-value goal-station 'approach-location)))
                  (tcl-push-plan :plan `(
                      (localizationModInst.activateLocalization)
                      (navigationInst.approachLocation ,start-station-approach-location)
                      (mpsInst.mps_station_fetch_from ?startStationID)
                      (navigationInst.approachLocation ,goal-station-approach-location)
                      (mpsInst.mps_station_push_to ?goalStationID)
                      (localizationModInst.deactivateLocalization)))))))
                                                                                        Task Realization
  }
```

Fig. 3.30 Task model for a transportation task: realization by the behavior developer in the *SmartMDSD* toolchain at Tier 3 using *SmartTCL*.

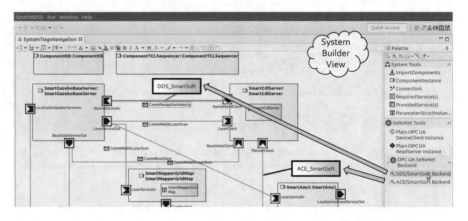

Fig. 3.31 Using different middlewares for different connections at the same time in a system.

4.4 Middleware-Agnostic Components and Mixed-Middleware Systems

The system builder view for selecting and composing different components is shown in Fig. 3.31. The system builder can select for each connection the middleware to be used. Different ports of a component can use different middlewares. Even different connections of the same port can use different middlewares. The components are not recompiled with a change of the underlying middleware as the links to the runtime executables of the middlewares are dynamically linked with the deployment process.

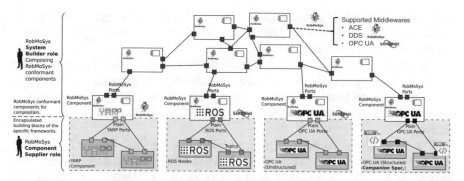

Fig. 3.32 Mixed-port-components serve as gateways to legacy systems and offer a migration path to the full-fledged composition structures.

4.5   The Mixed-Port Component as Migration Path

A smooth migration path from current systems to the full-fledged composition structures is very important to secure existing investments. Very often, we have to cope with a brownfield setting and thus need to be able to interact with our legacy parts. A mixed-port component shows two different sides of ports (Fig. 3.32). The first side consists of RobMoSys ports and the second side consists of ports of the legacy framework. A mixed-port component is like a gateway, which fully hides access to the legacy system. Thus, the resource management of the legacy system does not get in conflict with the RobMoSys composition structures (and their way of managing and configuring resource assignments). Relevant examples of mixed-port components are links to ROS systems and to industry 4.0 devices. The *SmartMDSD* toolchain comes with full support for ROS mixed-port components [36, start#lesson_6interfacing_with_ros_subsystems] and for OPC UA mixed-port components [36, start#lesson_5opc_unified_architecture_opc_ua].

It is important *not* to confuse mixed-port components with middleware-agnostic RobMoSys ports. For example, *OPC UA* can be used as middleware underneath RobMoSys ports. It is then fully hidden and just serves as a middleware to produce the semantics of the RobMoSys ports. In contrast, a mixed-port component for OPC UA offers at its second side native access to OPC UA devices (thereby, e.g. following OPC UA companion specifications).

4.6   Deployment-Time Configuration of Trigger Chains

In a robotics software business ecosystem, the development of a software component is first fully completed by a component developer before it is then offered *as-is* for use by others. In many approaches for robotics software components, the component developer ultimately decides how a particular task inside a component gets triggered (time-triggered such as *periodically* or event-triggered such as *by incoming data*) and then already invariably fixes that for the software component at the time of its implementation.

However, trigger chains (also called cause-effect chains) are a system-level property. They are a particular form of a *dependency graph*. Trigger chains occur in the context of data flows and typically map onto publish/subscribe communication. Trigger chains are part of the *information architecture* specified by the system architect.

The information architecture gets instantiated by a system builder by selecting software components that fit the information architecture and that match the dependency graphs. A software component will be extremely limited if it comes with prescribed trigger sources and thus only fits selectively. Thus, a software component must allow a system builder to configure the trigger sources for its activities. However, there should be no need at all to deliver the source code of a software component to the system builder and to have the software component recompiled. System builders neither want to go into internals of a software component nor are they expected to do so. Indeed, the data sheet of a software component shall contain trigger sources as variation points, which are then configurable without a need for recompilation.

For this use case, another abstraction layer is added inside a software component as refinement for an *activity*. This is just one example of how the internal structures of a software component can be refined or enriched with additional structures which form coexisting but consistent and dedicated offers for particular use cases. Further examples of such refining structures are scheduling mechanisms inside a software component or time-based approaches for functional composition.

The component developer now does not anymore write code that directly accesses the native API of ports providing incoming data. For example, the native API of a *push* client offers a *getUpdate* method (reading the latest available data without blocking) and a *getUpdateWait* method (blocking wait for the next update). Writing code using these access methods prescribes whether the access follows a register semantics (*getUpdate*) or a trigger semantics (*getUpdateWait*). Instead, the port is now accessed via a *get* method that forms a stable interface for programming independent from the subsequent trigger configuration. The *get* method is generated by the model-driven tools if the extension of trigger configurations is selected.

In a software component, the *activity* depends on a trigger. Thus, the additional abstraction layer is an extension of the *activity*. As shown in Fig. 3.33, the model of a software component comprises its ports and its activities, but also from which ports an activity reads its input and to which ports an activity forward its output. Model-driven tools generate from this model the component hull with its stable programming interfaces for use by the component developer.

An *activity* executes user code within its *on execute* method. The generated component hull includes all the mechanisms to configure an activity at any time to be either *without trigger* (calls the *on execute* method in a loop, user side self-trigger), *periodic* (calling the *on execute* method along the configured cycle rate and accessing all the ports with a *register* semantics) or *input port* (calling the *on*

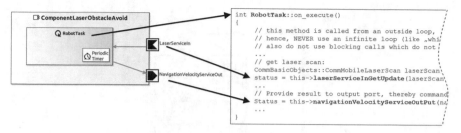

Fig. 3.33 The input port *LaserServiceIn* is of type *1-n forking* (see Sect. 3.2) for which the access method *laserServiceInGetUpdate* is generated.

execute method only when new data arrives on the port selected as trigger port). This configuration then also sets the according mapping of the generated *get* method to either the *getUpdate* or the *getUpdateWait* method. The trigger for a task can be specified at any point in time: (1) the component developer can specify a default value, (2) the system builder can set it according to system requirements, and (3) the task-coordination mechanism can set it even at runtime.

Figure 3.34 shows an application for tracing and checking the consistency of data flows, trigger chains, response times and more. For example, one can ask whether the response time for reacting to obstacles along the cycle A-B-C-A is fast enough, whether it is fine that *obstacle avoidance* gets triggered with each new laser scan, or whether the mapper with its configured update pace is good along the loop A-B-D-E-C-A.

4.7   Robotic Behavior Coordination: Skills, Tasks, World Model

Figure 3.35 shows the recurring principle of a block with ports and a data sheet in the context of runtime behavior coordination. In its simplest form, the complete robot is represented as a block that comes with a port. That port allows to call tasks or skills out of the data sheet of the robot and thereby have them executed. Examples for its use are user interfaces to command the robot but also fleet managers, or *MES* (manufacturing execution systems). Tasks and skills in that data sheet of the robot are a subset of those available on the robot as some are just internally used alternatives that are presented in a more generic way outside the robot. The skills of a robot come with the software components that the system builder selected and instantiated. The system builder also selects a task set that fits to the skills. Both the skills and the tasks end up on the robot so that they can be accessed by the *behavior executor*. This is a (set of) software component(s) providing the runtime execution mechanism for tasks and skills. At runtime, the *behavior executor* comprises at least the link from the task/skill models to the software components via their behavior interfaces (the *skill interface component*).

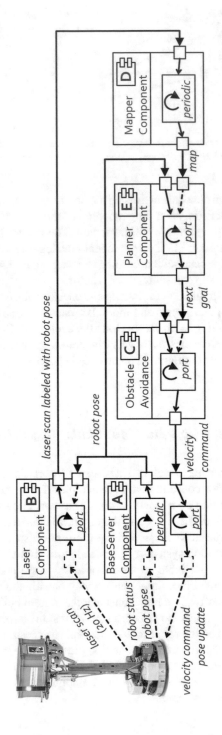

Fig. 3.34 A particular configuration of triggers for tasks. Inside a component, the input port selected as trigger for an activity with *port* as trigger is shown by a solid arrow.

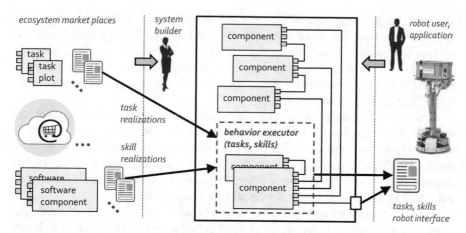

Fig. 3.35 Data sheets with tasks and skills: from the ecosystem marketplaces to their runtime execution by robots.

Push skill for execution / skill execution result message:

{ "msg-type" : "push-skill" , "id„ : <ID>, "skill" : { "name" : "<SKILLNAME>", "skillDefinitionFQN" : "<SKILL FQN>",
    "in-attribute" : { "<ATTRIBUTE>" : <VALUE> }, "out-attribute" : { "<ATTRIBUTE>" : <VALUE> }}}

{ "msg-type" : "skill-result" , "id" : <ID> , "result" : { "result" : "<SUCCESS|ERROR>", "result-value" : "<VALUE>" }}

Abort skill / abort current skill: **Optional information query:**

{ "msg-type" : "abort-skill", "id" : 2 }   { "msg-type" : "query" , "query" : { "type" : "<INFORMAION TO QUERY>" }}

{ "msg-type" : "abort-skill-result", "id" : 2, "result" : <"SUCCESS|ERROR>"}

{ "msg-type" : "abort-current-running-skill" }

{ "msg-type" : "abort-current-running-skill-result", "result" : "<SUCCESS|ERROR>"}

Fig. 3.36 The JSON format for commanding and querying a robot.

One implementation of the port for commanding tasks/skills to a robot uses a JSON representation (Fig. 3.36). The possible content and the parser in the *skill interface component* are directly derived from the data sheet of the robot.

A *behavior executor* can be as simple as allowing only one skill to be called at a time in order to avoid conflicts in concurrently executing skills. More advanced behavior coordination mechanisms are based on state automata or behavior trees. The tool *Groot* of the *RobMoSys ITP MOOD2Be* is one such example, which provides a dedicated graphical tool to support the behavior developer in arranging skills into behaviors. Skills can either be imported from the data sheet of a particular robotic system, or it can be the set of skills of a particular domain or even skills from different domains. The outcome is a *task* in the form of a *behavior tree*. That can be executed on any robotic system that comes with the skills referenced in the behavior tree. Typically, the *behavior-tree executor* is on-board the robot, becomes part of the *behavior executor* and uses its *skill interface component*. The robot becomes a block with a port where the port now forms a task-level interface.

Hierarchical task nets are another mechanism for behavior coordination. A powerful implementation is given by *SmartTCL* [35]. The behavior developer view within the *SmartMDSD* toolchain has already been introduced in Sect. 4.3. As shown in Fig. 3.37, hierarchical task nets allow for refinements at runtime taking into account the current situation and context. They can also deliberately include external solvers to decide between alternatives or to decide for a trade-off of different parameters. This is key when it comes to binding left-open variation points at runtime along given policies for executing tasks, even in open-ended environments, in an adequate and robust way.

Again, the robot comes with a port that allows to command tasks to the robot and the data sheet contains all the tasks that the robot accepts. All the tasks with their current execution status including their current level of refinement are present on the *agenda*. Tasks can also be generated from inside the robot. An example for this is a task that ensures that the robot drives to a charging station before it runs out of power. The agenda is processed by the so-called sequencer.

The core feature of the agenda-based mechanism (and of the hierarchical task nets) is the free-of-conflicts execution of concurrent activities. Hierarchical task nets go for resource reservations of their resources, and sub-tasks can only operate within the setting given by their superordinated task (horizontal/vertical interaction and composition of resource shares). For example, a task might reserve the gripper such that no other concurrent task can run concurrently that might release the carried object in the gripper. The self-model of the robot with its resources is part of the knowledge base. It gets initially filled from the data sheets of the components of the robot at the system building step.

5   State of the Art, State of the Practice and Conclusion

A comprehensive overview of the rich body of knowledge of different robotics software frameworks, of relevant software engineering approaches and of their applications in related domains (such as automotive, avionics, industry 4.0), of technologies (such as simulation in the loop, model-driven workbenches) and of advances in, e.g. formal methods (such as validation, verification, semantic modelling), is just impossible. Instead of just describing arbitrarily selected approaches, it is very revealing to think about how the relationship between robotics, programming, software engineering and the other just mentioned domains has changed over time.

There are all kinds of examples of *robot programming*. In earlier times, it was often related to bare-metal programming and could be done only by highly skilled programming experts. This approach neither scaled with respect to the complexity challenge nor allowed application domains to easily get access to and make use of robotics technology.

This was followed by a huge diversity of different robotics software frameworks, each favouring different needs and thus filling a particular niche [37, 38]. Indeed, these all can be best understood in their time since they have been heavily influenced by at least (1) the technology available and accessible at that time, (2) the kinds of robots and domains put into focus (mobile robots, manipulation, flying robots,

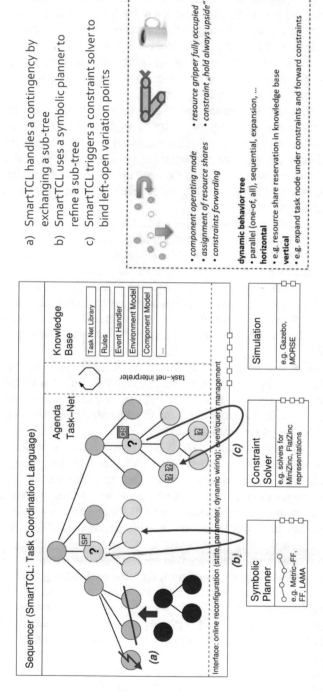

Fig. 3.37 The SmartTCL behavior execution, skills calling external experts and the interaction with the knowledge base.

etc.), (3) whether the focus has been more on functionalities (typically driven by academia) or on (application) scenarios (typically driven by industry and also by competitions like the RoboCup) and even by (4) the (non)availability and (non)affordability of standard robot platforms (mobile robot *Pioneer*, humanoid robot *NAO*, Willow Garage *PR2*, PAL Robotics *Tiago*, FESTO *Robotino*, the Universal Robots *UR-Arm* and many others), cheap sensors (laser rangers, RGB-D cameras, tablets for user interfaces), ubiquitous and always available Internet connections as well as enough and cheap computing power.

Robotics software has seen tremendous changes and shifts up to now. In the past, it has been left to the readers of a paper to try to implement a described algorithm on their own and it was nearly impossible to reproduce experiments. Nowadays, it is quite standard that papers are accompanied by at least prototypical and easy-to-access implementations. In particular, there are more and more graphical tools to support low-code programming (e.g. for parameterization of templates for particular tasks such as palletizing, etc.).

Latest developments in software frameworks for robotics now link cloud platforms and robotic systems. Examples are the *Microsoft Azure Cognitive Services* and the *Amazon Web Services RoboMaker*. Cloud services have the potential to boost object recognition and learning algorithms as well as inference algorithms and to add these to the variety of skills of robots. Cloud services have inherent advantages when it comes to cheap processing power, sharing of data sets for learning and even sharing of experiences between robots. This then allows a robot facing a new situation to take advantage from other robots that have dealt with that situation already.

In general, there is a long-standing fruitful relationship with mutual benefits between robotics and other related domains. For example, middleware systems are researched and pushed forward by a community in their own, and robotics makes use of their outcomes. On the other hand, the middleware community is stimulated by insights and demanding needs from robotics.

Nevertheless, all interdisciplinary fields, and thus robotics as well, come with a natural and deeply rooted challenge. We make this clear by exaggerated and catchy examples. Middleware experts provide generic middleware systems to try to cover as many application needs as possible. Some robotic experts consider those middleware systems to be of unnecessary complexity and with a too big footprint. In consequence, they implement message-based blackboard communication architectures on their own and thereby step-by-step discover all the challenges already well-known (and sometimes even solved) in the competent domain.

Software engineering experts come up with more and more generic offers for modeling aspects of software systems [39]. However, they either abstract away the steps to ground the models (as this is just domain-specific) or they do it in an exemplary manner only (e.g. by low-complexity examples with educational robots). Roboticists then neither see the benefits given their complex challenges nor do they see an easy-access path.

Another example relates to the question of how to generate trust into a robotics system. Roboticists are aware of formal methods for verification, validation, certification, runtime sanity checks, co-simulation (even at runtime by the robot before a decision is executed) and of many others [40]. However, it is about making reasonable matches for each technology: what kind of problem and what kind of complexity can it handle, what is the effort in using it, what is the coverage and reliability and is that within the needs and the economic constraints of the application domain. Unfortunately, there are again lots of mismatches. The reason is that at least advanced robotic systems have to face additional complexity due to open worlds.

Basically, after years of huge advancements within all the different silos relevant for robotics, it is now the time to again rethink matches between requirements and now available opportunities. Neither roboticists shall become experts for those silos, nor experts from those silos shall become experts for robotic systems. Rather, achieving mutual benefits requires (1) mutual respect for the body of knowledge of different silos and domains; (2) open and honest explications of the capabilities, but also the deficiencies of the state-of-the-art in a silo; and (3) neither blaming nor getting blamed for that state of the art, but rather seeking for possible hooks and interfaces between the silos and illustrating the benefits first along low-hanging use cases instead of starting with full-blown examples and missionary work first. Although that all should be quite standard, it always is a challenge because it is about establishing and providing opportunities for fruitful interactions between different silos for the mutual benefit of all.

We consider model-driven approaches, with their now-achieved maturity level and when applied as described in this chapter, to be a particular fit to serve as a moderator for this. As outlined, model-driven approaches are *not* foremost just modeling everything instead of coding. They are also *not* just about code generation. Instead, they are *the* means to provide consistent links between different domains and their assets and thus can establish links between so far isolated silos. They allow to stay with presentations specific to robotics but map those in a semantically correct way onto the huge variety of offered implementational technologies.

The broadest coverage of concepts, tools, implementations and applications of model-driven software engineering in robotics is represented by the EU H2020 project *RobMoSys* and can be found in the *RobMoSys Wiki* [19, 21]. In-depth presentations of core concepts are available via PhD theses (some still underway) [29, 41, 42]. *RobMoSys* started a movement toward model-driven software engineering approaches in robotics and is continuously updated with broadest community involvement based on discussions in the related forum [43]. It also contains lots of references beyond robotics.

The *euRobotics AISBL Topic Group on Software Engineering, System Integration, System Engineering* is another entry point for the community. It shapes the European road-mapping in software systems engineering for robotics. It will also take over stewardship for the body of knowledge managed and organized for the community by *RobMoSys*, heading for the sustainability of that starting point and the related movement and also ensuring its liveliness after the runtime of the project.

The community also gathers in the *Technical Committee on Software Engineering for Robotics and Automation (IEEE RAS TC-SOFT)*. Further currently active community services are the *MORSE* (Model-Driven Robot Software Engineering) workshops, the *SIMPAR* (Int. Conf. on Simulation, Modeling, and Programming for Autonomous Robots) conference, the *MODELS* (Int. Conf. on Model Driven Engineering Languages and Systems) conference but also *JOSER* (*Journal of Software Engineering for Robotics*). Further material is also available via the Dagstuhl Seminar 17071 *Computer-Assisted Engineering for Robotics and Autonomous Systems* [40]. Relevant software and modeling activities with impact on robotics are also driven by the *Reference Architecture Model Industrie 4.0 (RAMI)*, the *Asset Administration Shell* and related *OPC UA companion specifications*.

References

1. J. Stubbe, J. Mock, S. Wischmann, *The Acceptance of Service Robots: Tools and Strategies for the Successful Deployment in Companies*. Study commissioned by BMWi as part of the PAiCE Technology Programme (iit-Institut für Innovation und Technik in der VDI/VDE Innovation + Technik GmbH, Berlin, 2019)
2. D. Brugali (ed.) *Software Engineering for Experimental Robotics*. Springer Tracts in Advanced Robotics (Springer, Berlin, 2007). ISBN: 3540689494
3. M. Hägele, N. Blümlein, O. Kleine, *EFFIROB - Wirtschaftlichkeitsanalysen neuartiger Servicerobotik-Anwendungen und ihre Bedeutung für die Robotik-Entwicklung* (Eine Analyse der Fraunhofer-Institute IPA und ISI im Auftrag des BMBF, 2011)
4. SPARC - The Partnership for Robotics in Europe. Strategic Research Agenda (SRA) for Robotics in Europe 2014–2020. euRobotics aisbl (2013 & 2014)
5. D. Mourtzis, Challenges and future perspectives for the life cycle of manufacturing networks in the mass customisation era. Logist. Res. **9**, 2 (2016). https://doi.org/10.1007/s12159-015-0129-0
6. M. Teulieres, J. Tilley, L. Bolz, P.M. Ludwig-Dehm, S. Wägner, *Industrial Robotics – Insights into the Sector's Future Growth Dynamics* (McKinsey & Company, New York, 2019)
7. J.F. Moore, Predators and prey: a new ecology of competition. Harv. Bus. Rev. **71**(3), 75–83 (1993)
8. E. Kelly, *Business Ecosystems Come of Age*. Part of the Business Trends Series (Deloitte University Press, New York, 2015). DUP_1048-Business-ecosystems-come-of-age_MASTER_FINAL.pdf
9. S. Hallsteinsen, M. Hinchey, S. Park, K. Schmid, Dynamic software product lines. Computer **41**(4), 93–95 (2008)
10. J. Bosch, From software product lines to software ecosystems, in *Proceedings of the 13th Int. Software Product Line Conference* (2009), pp. 111–119. https://doi.org/10.1145/1753235.1753251
11. W. Mahnke, S.-H. Leitner, M. Damm, *OPC Unified Architecture* (Springer, New York, 2009). ISBN: 978-3-540-68898-3
12. OPC Foundation Companion Specifications. https://opcfoundation.org/about/opc-technologies/opc-ua/ua-companion-specifications/
13. J. Bosch, P. Bosch-Sijtsema, From integration to composition: on the impact of software product lines, global development and ecosystems. J. Syst. Softw. **83**(1), 67–76 (2010). ISSN:0164-1212. https://doi.org/10.1016/j.jss.2009.06.051

14. C. Schlegel, A. Lotz, M. Lutz, D. Stampfer, J.F. Inglés-Romero, C. Vicente-Chicote, Model-driven software systems engineering in robotics: covering the complete life-cycle of a robot. Inf. Technol. **57**(2), 85–98 (2015). De Gruyter, Oldenbourg
15. L. Andrade, J.L. Fiadeiro, J. Gouveia, G. Koutsoukos, Separating computation, coordination and configuration. J. Softw. Mainten. Evol. Res. Pract. **14**(5), 353–369 (2002)
16. E.A. Lee, S.A. Seshia, *Introduction to Embedded Systems: A Cyber-Physical Systems Approach*, 2nd edn. (MIT Press, Cambridge, 2017)
17. M. Lutz, D. Stampfer, A. Lotz, C. Schlegel, Service robot control architectures for flexible and robust real-world task execution: Best practices and patterns, in *Informatik 2014, Workshop Roboter-Kontrollarchitekturen*. LNI der GI (Springer, New York, 2014). ISBN:978-3-88579-626-8
18. F. Buschmann, R. Meunier, H. Rohnert, P. Sommerlad, M. Stal, *Pattern-Oriented Software Architecture, Volume 1, A System of Patterns* (Wiley Press, Hoboken, 1996). ISBN: 978-0-471-95869-7
19. RobMoSys Wiki. Cited 9. Aug 2020. https://robmosys.eu/wiki/
20. E. Scioni, N. Huebel, S. Blumenthal, A. Shakhimardanov, M. Klotzbücher, H. Garcia, H. Bruyninckx, Hierarchical hypergraphs for knowledge-centric robot systems: a composable structural meta-model and its domain specific language NPC4. JOSER - Spec. Iss. Domain-Spec. Lang. Mod. Robot. Syst. **7**(1), 55–74 (2016)
21. RobMoSys Wiki Modeling Section. Cited 9. Aug 2020. https://robmosys.eu/wiki/modeling:<title of subordinate document>
22. G. Engels, A. Schürr, Encapsulated hierarchical graphs, graph types, and meta types. Electron. Notes Theor. Comput. Sci. **2**, 101–109 (1995)
23. M. Levene, A. Poulovassilis, An object-oriented data model formalised through hypergraphs. Data Knowl. Eng. **6**, 205–224 (1991)
24. C.A. Szyperski, D. Gruntz, S. Murer, *Component Software - Beyond Object-Oriented Programming*. Addison-Wesley Component Software Series, 2nd edn. (Addison-Wesley, Boston, 2002)
25. I. Crnkovic, S. Sentilles, A. Vulgarakis, M.R.V. Chaudron, A classification framework for software component models. IEEE Trans. Softw. Eng. **37**(5), 593–615 (2011). https://doi.org/10.1109/TSE.2010.83
26. C. Schlegel, A. Lotz, A. Steck, SmartSoft - The State Management of a Component. Technical Report 2011/01. Hochschule Ulm, Germany (2011) ISSN:1868-3452. http://www.zafh-servicerobotik.de/dokumente/ZAFH-TR-01-2011-ISSN-1868-3452.pdf
27. C. Schlegel, Navigation and Execution for Mobile Robots in Dynamic Environments: An Integrated Approach. PhD thesis, Uni Ulm (2004)
28. D. Stampfer, A. Lotz, M. Lutz, C. Schlegel, The SmartMDSD toolchain: an integrated MDSD workflow and Integrated Development Environment (IDE) for Robotics Software. JOSER - Spec. Iss. Domain-Spec. Lang. Mod. Robot. Syst. **7**(1), 3–19 (2016)
29. M. Lutz, Model-Driven Behavior Development for Service Robotic Systems: Bridging the Gap between Software- and Behavior-Models (work in progress)
30. A. Lotz, A. Steck, C. Schlegel, Runtime monitoring of robotics software components: increasing robustness of service robotic systems, in *International Conference on Advanced Robotics (ICAR '11)*, Tallinn, Estonia (2011)
31. GitHub Repository with API specifications. Cited 9. Aug 2020. https://github.com/Servicerobotics-Ulm/SmartSoftComponentDeveloperAPIcpp
32. C. Schlegel, A. Lotz, ACE/SmartSoft - Technical Details and Internals. Technical Report 2010/01, Hochschule Ulm, Germany (2010). ISSN:1868-3452. http://www.zafh-servicerobotik.de/dokumente/ZAFH-TR-01-2010-ISSN-1868-3452.pdf
33. A. Lotz, A. Hamann, R. Lange, C. Heinzemann, J. Staschulat, V. Kesel, D. Stampfer, M. Lutz, C. Schlegel, Combining robotics component-based model-driven development with a model-based performance analysis, in *Proceedings of the IEEE International Conference on Simulation, Modeling, and Programming for Autonomous Robots (SIMPAR)* (2016), pp. 170–176

34. The Eclipse-based Open-Source SmartMDSD Toolchain. Cited 9. Aug 2020. https://wiki.servicerobotik-ulm.de/getting-started-guide. https://projects.eclipse.org/projects/modeling.smartmdsd

35. A. Steck, C. Schlegel, Managing execution variants in task coordination by exploiting design-time models at runtime, in *Proceedings of the IEEE/RSJ International Conference on Robotics and Intelligent Systems (IROS)*, San Francisco, USA, September (2011)

36. SmartMDSD Tutorials. Cited 9. Aug 2020 https://wiki.servicerobotik-ulm.de/tutorials:<title of subordinate document>

37. A. Elkady, T. Sobh, Robotics middleware: A comprehensive literature survey and attribute-based bibliography. J. Robot. (2012). Article ID 959013. https://doi.org/10.1155/2012/959013. https://core.ac.uk/download/pdf/52956509.pdf

38. A. Ramaswamy, B. Monsuez, A. Tapus, Model-driven software development approaches in robotics research, in *Proceedings of the 6th International Workshop on Modeling in Software Engineering (MiSE)*, June (2014), pp. 43–48. https://doi.org/10.1145/2593770.2593781

39. D. Akdur, V. Garousi, O. Demirors, A survey on modeling and model-driven engineering practices in the embedded software industry. J. Syst. Arch. (2018). https://doi.org/10.1016/j.sysarc.2018.09.007

40. E. Abraham, H. Kress-Gazit, L. Natale, A. Tacchella (organizers), Computer-Assisted Engineering for Robotics and Autonomous Systems. Dagstuhl-Seminar 17071, 12. - 17.02.2017. https://www.dagstuhl.de/17071

41. D. Stampfer, Contributions to System Composition using a System Design Process driven by Service Definitions for Service Robotics. PhD thesis, Technische Uni München (2018)

42. A. Lotz, Managing Non-Functional Communication Aspects in the Entire Life-Cycle of a Component-Based Robotic Software System. PhD thesis, Technische Uni München (2018)

43. The RobMoSys Discourse Forum. Cited 9. Aug 2020. https://discourse.robmosys.eu/

Chapter 4
Testing Industrial Robotic Systems: A New Battlefield!

Arnaud Gotlieb, Dusica Marijan, and Helge Spieker

Abstract Industrial robotics is a field that evolves very fast, with ever-growing needs in terms of safety, performance, robustness, and reliability. Nowadays, industrial robots are communicating cyber-physical systems that embed complex distributed multi-core software systems involving intelligent motion control, anti-collision, and advanced force or torque control. The increased complexity makes these robots more fragile and more error-prone than they were previously. Failures can originate from many sources, including system and software bugs, communication downtime, CPU overload, and robot wear and tear. Fortunately, advanced verification techniques such as constraint-based testing and validation intelligence are employed to cope with specification and development errors and ensure a better quality of delivered robots. In this chapter, we address the challenges of testing industrial serial robots and provide examples of artificial intelligence and constraint programming techniques being used to ease the automation of some parts of the robot testing processes. In particular, we present techniques for test generation, planning, and execution for industrial robots, as well as the deployment of this technology into the real-world continuous integration process of a large robot manufacturing company. The presented techniques are complementary to other strong formal verification techniques such as model checking and theorem proving, and only the combination of these techniques will lead to an industrial manufacturing world where robots are safer and more reliable.

1 Introduction

Robots are complex cyber-physical systems that are used nowadays in a range of safety-critical domains. In manufacturing, robots can perform tasks in full autonomy, for example, on-demand painting with colour and brush change [58], or

A. Gotlieb (✉) · D. Marijan · H. Spieker
Simula Research Laboratory, Lysaker, Norway
e-mail: arnaud@simula.no; dusica@simula.no; helge@simula.no

© Springer Nature Switzerland AG 2021
A. Cavalcanti et al. (eds.), *Software Engineering for Robotics*,
https://doi.org/10.1007/978-3-030-66494-7_4

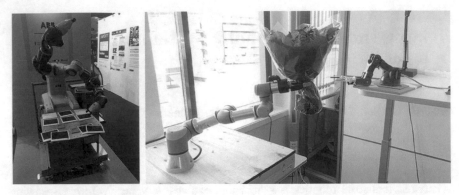

Fig. 4.1 Examples of collaborative industrial robots from ABB and Universal Robots (Picture credits: Arnaud Gotlieb, Mohit Kumar Ahuja).

they can collaborate with humans on the factory floor [57]. Robots are furthermore increasingly used in healthcare and medical operations [3, 36], education [42], and transportation [10]. Consequently, robots increasingly affect public safety [16], raising the interest of the community in their quality and safety assurance.

Due to the intrinsic complexity of robotic systems on the one hand [4] and the need of their error-proof interaction with the physical environment on the other [45], it is crucial to perform rigorous testing of these systems before their deployment into the field. This chapter focuses on industrial robots, which are robots used in factories to help improve worker productivity and safety by replacing or co-operating with humans. Examples of such robots are shown in Fig. 4.1. On the left, a collaborative dual-arm YuMi robot is shown where the robot automatically sorts objects. On the right, a UR3 robot is shown from our laboratory setup for testing of learning robots.

Industrial robotics is a field that advances quickly. Over the last few decades, industrial robots (IR) have gone from simple systems able to do easy pick-and-place tasks without any sensing capability to present-day automated and autonomous robots powered by artificial intelligence (AI), fully collaborative and able to dynamically adapt to their environment. While increasing manufacturing productivity by orders of magnitude, the underlying complex technology of modern IR introduces a vast set of challenges on IR robustness and safety. Comprehensive automated testing of IR is necessary both to ensure the early detection of potential robot faults before system deployment into operation and to keep maintenance costs low, once the system is in operation [5]. However, comprehensive testing of IR is challenging, owing to multiple reasons, discussed in more detail in the next section.

IR increasingly support high levels of human-robot interaction (HRI) [56] and high levels of robot autonomy, which in turn affects the nature of HRI [8]. Indeed, this context requires human imitation and demonstration capabilities [2, 84], rapid adaptation of the robots to changing environments, as well as safer robot behaviours [5]. Examples of industrial collaborative robots include single-arm robots that can work close to human workers (e.g. Universal Robot's UR3) and single-arm robots (e.g. CEA's SYBOT) or small-parts assembly robots (e.g. dual-arms ABB's YuMi) that cooperate with human co-workers.

To cooperate safely with humans, these robots have a special design where sensors and actuators are used to detect any human pressure or prevent dangerous robot actions. They are also equipped with perception systems that can detect obstacles and prevent collisions well in advance. However, the intelligent control systems that equip these robots have become more and more complex and they need to be frequently updated for maintenance. Hence, continuously controlling the safety and security of these systems is crucial [5, 61, 78].

Moreover, by using learning abilities, robots can optimize their trajectories to increase their speed and overall performance. Their actual behaviour is learnt instead of being carefully specified. Hence, controlling and testing the robot control systems has become very challenging as the precise expected behaviours of these collaborative robots are not known in advance.

IR can learn a range of skills, such as optimized trajectories, locomotion, grasping, obstacle detection and collision avoidance, or interactive abilities such as optimized object picking with human co-workers. Learning methods in IR use different theoretical frameworks of machine learning (ML), including imitation learning [1, 12, 52], where a robot-learner is trained to mimic human behaviours from demonstrations; reinforcement learning [44, 69], where a learner optimizes its actions by obtaining rewards or penalties from the environment; inductive programming [27], where a programming language allows learners to generalize rules from examples or instances; constraint acquisition [11], where constraint models are learnt by generalizing from given solutions and non-solutions; and deep learning (DL) [14], where multi-layers neural networks are trained over large corpus of data to efficiently recognize various type of signals. Learning skills can be given either through self-exploration of the space of trajectories or through the guidance of a human teacher.

Despite the considerable development of learning methods in the last two decades, testing IR faces specific challenges such as the limited availability of advanced robots to perform experiences, the unbearable cost of failures in robot deployment, the lack of systematic studies on how to evaluate robot learning, and the absence of appropriate test platforms. Furthermore, there is still no unifying learning framework for formalizing ML in robotics. This unifying framework could be helpful to understand how different learning methods can be combined to deal with difficult tasks.

This chapter discusses the prominent challenges of testing IR, including the robot autonomy and collaborative capabilities. In Sect. 2, we provide an overview of recent advances in this area, with a focus on test generation and validation, and test planning under resource constraints. These topics are further discussed in Sects. 3–5 with a more in-depth presentation of selected methods and techniques in these areas. The chapter closes with a summary of the current state of testing IR and open questions for future work in Sect. 6.

2   Testing Industrial Robots: Challenges and Recent Advances

Robots are getting smarter, in terms of reasoning, sensing, and adapting. They integrate a range of capabilities supported by AI and ML, to enable self-learning, advanced perception of the environment, synchronization with humans and other robots, and motion control, to name a few. These capabilities make IR able to perform complex tasks *autonomously*; however, robot autonomy induces issues regarding robot dependability [37] and trust [6]. Consequently, these advanced capabilities also lead to increased complexity of testing IR. As Guiochet et al. [37] argue, one of the core challenges for deploying robots on a large scale will be ensuring their dependability and safety.

IR are increasingly *collaborative*, intended to interact with humans in a shared workspace. In contrast to conventional IR, which can work autonomously but are caged or placed at a safe distance from a human, collaborative IR perform joint object manipulation, responding to the human-worker motion in real time. As such, these robots introduce the need to satisfy an increased level of safety requirements [45, 86], to ensure that human workers are not at risk when working collaboratively alongside robots.

A recent study [4] identified nine key challenges for automated testing of robots in practice. These challenges include: (1) unpredictable corner cases; (2) engineering complexity; (3) culture of testing; (4) coordination, collaboration, and documentation; (5) cost and resources; (6) environmental complexity; (7) lack of oracle; (8) software and hardware integration; and (9) distrust of simulation. The data for the study was collected from the interviews with 12 robotics practitioners from 11 robotics companies. While this study provides a comprehensive view of the practical and organizational challenges for testing robots, it does not consider several important technical challenges, which we discuss next.

Specifically, we discuss the challenges of automated testing of IR belonging to two broad categories: (1) test-input and test-output generation and (2) test planning under resource constraints.

Furthermore, we highlight the importance of continuous testing of IR, where robot code components are tested incrementally, as they are developed. Continuous testing for IR aims to make the process of developing IR faster and less expensive by detecting potential errors at early stages of development and thus avoiding expensive integration faults.

2.1   Test Generation

Test generation for IR is concerned with two basic tasks: test-input generation and test-output generation. Next, we discuss techniques for test-input and test-output generation.

2.1.1   Test-Input Generation

When testing IR, the goal is to specify realistic, diverse, and complete test inputs, because IR operate in complex, highly configurable, and nondeterministic environments. Conventional approaches to this end include combinatorial sampling [54], which can generate a test-input set that is complete for a certain n-wise coverage. N-wise coverage means that all combinations of values of n parameters are included. Another approach is model-based test input generation [5], where authors first generate abstract test sequences, which are concretized in a simulator execution.

The test-input space for typical IR is vast, as present-day robots integrate a range of sensors, dealing with a large amount of data. This consequently creates a significant challenge of selecting an adequate set of inputs. Answering the question "when to stop testing" leads to the definition of specific coverage criteria for the machine learning (ML) model of an IR, which can guide the selection of inputs able to reveal false classification or incorrect regression during testing. This problem has been encountered in the software testing domain, especially in the area of fuzzing [53]. Fuzzing is a testing technique where randomized inputs are provided to software under test with the goal of revealing vulnerabilities.

To address the challenge of test-input selection, several *test adequacy criteria* have been proposed. Source code coverage is one such metric, which makes it possible to decide when enough test cases have been collected and the test adequacy criterion is sufficiently fulfilled. Inspired by traditional code coverage metrics such as statement or decision coverage, neuron coverage [64] counts the number of activated neurons in the neural network when test inputs are submitted for classification. It is believed that a higher neuron activation coverage leads to a higher chance of detecting wrong classifications. Other studies have confirmed this initial result and also extended the proposed structural criteria by distinguishing neuron-level criteria from layer-level coverage criteria [50].

Instead of looking at the network structure, a different criterion called surprise adequacy has been introduced for measuring the diversity in training data [43]. The surprise of an input is defined as the difference between the input and the training dataset with respect to the behaviour of the ML model. It is then suggested that one should select inputs that are sufficiently surprising but still within the expected data distribution to compose an adequate test set. These criteria are specific to ML and can be used to guide the selection of test inputs, but additional techniques are needed to evaluate the correctness of results when test inputs are submitted to ML models.

2.1.2   Test-Output Generation

Expected test outputs are used in testing for checking the correctness of the actual outputs produced in a system execution. A mechanism for comparing the actual outputs against the expected outputs is known as a *test oracle*. The approaches to automated generation of test oracles can be broadly classified as *specified* and *derived* test oracles. The former exist if there is a specification (formal, semi-formal, or informal) of expected outputs. If not, test oracles are derived from various artefacts such as documentation, program executions, known properties of the system under test, or invariants inferred from system executions.

Both approaches face challenges when dealing with sophisticated systems such as IR. For IR integrating AI and ML, generating test oracles has become intrinsically complex. This is because ML introduces probabilistic reasoning that may give rise to nondeterministic system behaviour. This means, in simple terms, that a robot might produce different output given a certain input. Because IR learn to predict outputs from training data, the correctness of the output in testing IR cannot be easily determined. Therefore, it becomes crucial to *ensure the quality of training datasets*.

The quality of the training dataset largely affects the performance of the ML model (IR system). This is even more the case with deep learning (DL) models, which have become a popular technology used in IR [65]. While DL is advancing robot capabilities in general, there is a strong dependence of DL on the training data, which makes it vulnerable to adversarial attacks [31], where small modifications to input data can cause misclassifications. As Sunderhauf [80] pointed out, the application of DL in robotics introduces new challenges that are specific for robotic vision and distinct from those present in computer vision. For example, computer vision takes images and translates them into information, while robotic vision translates images into actions. Consequently, these new challenges need to be adequately addressed in testing.

The behaviour of DL models follows the examples given in the training data. When collecting a training dataset for a new application, the data must be diverse enough to cover all variances that are expected to be encountered by the deployed model, i.e. the data distribution for which and on which the model is trained.

There are multiple potential failures here. First, the collected training data might be insufficient to train a model that can generalize to the real-world data. In this case, the training data should be expanded and the test set needs to be enriched by examples that the model does not generalize to. This will make it possible to identify these issues before deployment of a new or improved model.

Second, the training data might lead to a capable model for the real-world data, but over time the data encountered by the deployed model gradually changes and no longer fits the initial training distribution. This reduces the model performance and increases false predictions. An approach to address this challenge is to monitor the actual data encountered by the deployed model and measure how it differs from the training data. This monitoring is then followed by frequent model fine-tuning or retraining with an adjusted dataset.

Other failures are caused by not carefully selecting the data and thereby introducing, for example, redundant data, adversarial data, or biases into the dataset. In case of biases, these can lead to a model that does not make fair predictions but is conditioned to repeat the biases encoded in the training data.

There are approaches inspired by *mutation testing* [41] proposed for evaluating the quality of test datasets for Deep Neural Network (DNN). DeepMutation [51] is one of the first works in this area. It specifies a set of mutation operators to inject faults into the training data. The idea is to, then, retrain ML models using the mutated training data, which will produce mutated models. In this way, faults are injected in the models, after which mutated models are tested using a test dataset. The quality of the test dataset is evaluated based on how many injected faults are detected. The drawback of DeepMutation is related to using basic mutation operators, which may seed faults that are not very representative of real faults [40]. Shen et al. [72] proposed MuNN as another approach for mutation testing of neural networks. In this work, it was shown that neural networks of different depth require different mutation operators. An extended discussion of mutation testing is given in Chap. 11, where its application towards the testing of robotic systems is demonstrated.

2.2 Test Planning Under Resource Constraints

When testing IR, it is common that the resources available for testing are constrained. The physical test agents, on which the test cases are executed, are limited since they require space, maintenance efforts, as well as initial acquisition costs. Therefore, it is often economically infeasible to remove this bottleneck in IR testing, especially when having to consider a variety of IRs with different feature sets. The availability of these test agents is further restricted in practice, albeit due to defects, maintenance windows, or usage from other projects. Furthermore, in some cases, test agents need exclusive access to shared additional resources, such as measurement devices, conveyor belts, or network equipment. Finally, the time windows for testing are limited as well and testing should often be completed at a fixed deadline or with minimal total execution time. All these constraints increase the complexity of the test organization, and test planning becomes necessary for efficient resource usage. While manual test planning is often the state of the practice, it is often also non-optimal and is less flexible as it has difficulty in adapting to varying scenarios where only subsets of test agents are available or the number of tests changes.

Following this description, test planning under resource constraints covers several smaller problems: deciding what to test, when to test, and where to test. Within the software testing community, these problems are referred to as test case prioritization, test case reduction (also referred to as minimization or selection), and test case scheduling.

This assumes that the execution environment is generic or decoupled from the software, and the test suite can be executed on any test environment. For testing robotic systems, it is further relevant to include the test environment into the test planning process as its resources need to be managed and considered, too. In 2012, Yoo and Harman published a major review on test selection and prioritization in software testing, and we refer the interested reader to this survey for an in-depth overview of the earlier literature [85], while we will discuss the specific challenges and developments in the industrial robotics context.

When creating a test plan, there are two distinct optimization goals. Either the test suite is fixed and the plan needs to minimize resource usage, e.g. total execution time, or the resource constraints are fixed and the test plan has to make the best use of these constraints, which can require to only select a subset of the test suite.

It is helpful in both cases to have a notion of relevance for each test case in the test suite to decide if it should be executed and how soon. The identification of this notion of relevance is called *test case prioritization* [25, 68], which has the goal to express an order of test case for high effectiveness, such that failing test cases are executed before passing test cases. The prioritization step can take a variety of information sources into account to prioritize a test case and establish a notion of how likely the test case might fail. These sources include source code coverage [25] and changes [30, 74] or historical test metadata [46, 78]. Shin et al. present a multi-objective test case prioritization approach for the acceptance testing of cyber-physical systems that includes both uncertainties in the test environment and potential hardware damages in the prioritization [75].

Once a priority has been assigned, a *test suite reduction* method selects the most relevant test cases for actual execution and thereby reduces the size of the test suite [34, 73, 83]. Reduction thereby performs the step to adjust the size of the initial test suite to the available, limited resources. For the case of sequential execution, the selection method can choose the most prioritized test cases that exhaust the available resources. Test suite reduction is also referred to as test suite minimization [39] as it minimizes the number of test cases from the full initial test suite to the test suite that is actually executed.

More complex approaches to test case reduction can consider additional constraints and requirements on the resulting test suite. For instance, they can demand certain coverage criteria to be fulfilled to cover the whole system even though the risk of failure in some subsystems has lower priority than in other subsystems. Additionally, it can be necessary to consider more resource constraints than just the available time, such as compatibility to available test agents or dependencies on external devices.

Finally, *test case scheduling* is the assignment of a test case to both a test agent and a time slot and is an application of constraint-based machine and project scheduling [7, 13, 38]. It differs from test case reduction, which selects a subset, by also making the additional assignment of time and execution location.

The scheduling model captures the constraints of the individual assignment between test cases and test agents as well as potential constraints on the execution orders of test cases. While test cases should generally be independent of each other, it can be necessary to group certain kinds of test cases to avoid costly setup times, or to avoid certain groupings of test cases because they rely on similar external resources. These constraints need to be developed together with domain experts, e.g. the quality engineers developing the tests, and need to be formalized and stored as metadata for the test cases.

3 Test Generation

This section presents two research leads that we pursue in the field of test generation for testing IR. The former subsection details the generation of test trajectories for IR and the latter presents our ongoing work to exploit metamorphic testing for testing robot's behaviours.

3.1 Stress Test-Trajectories Generation with Constraint Programming

Serial IR embed complex multi-core software systems with advanced motion control, collision avoidance, and intelligent torque control. As a result of the ever-increasing complexity of these systems and their interactions, IR have become more error-prone. Hence, generating tests that stress these robots to steer them to their limits is crucial to ensure quality and robustness. A goal in this area is to generate test trajectories of the end-effector of the robots that have the greatest potential to show deviations between the trajectory commanded to the robot and its actual trajectory. As an example, Fig. 4.2 illustrates such a deviation for a 3D trajectory specified to the robot. Generating test trajectories which exhibit such deviations is challenging as there is no model of deviation and the space of possible trajectories is unbounded.

To the best of our knowledge, there is no fully accurate analytic model that can solve the problem. The modelling of the robot's (inverse) kinematics involves

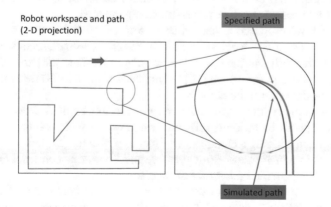

Fig. 4.2 2D projections of the robot specified test 3D trajectory (in orange) and its executed trajectory in simulation (in blue). Both are very close to each other (on the left), but, by zooming in the figure, one uncovers a deviation between the specified and simulated path (on the right). Those deviations reveal potential software or hardware failures which shall be discovered before deploying the robot in operational contexts.

solving a large number of multivariate polynomial equations. Typically, a 4-DoF[1] serial robot already requires to solve around 49 4th-degree polynomial equations with 49 variables [55]. That is why, even though careful design of the robot using simulators is possible, the automatic generation of stress trajectories for finding and testing deviations between specified and simulated paths is crucial, to ensure high quality of these industrial robots.

A possible approach for addressing this problem is to use *constraint programming*, which is a general-purpose framework [67] for solving combinatorial problems. Basically, these problems, called *constraint satisfaction problems (CSPs)* are defined by a set of variables V, a domain D, and a set of constraints C, which are relations over V. Each variable v_i takes its possible values in a finite (or continuous) domain D_i, and D is the Cartesian product of all D_i. The goal of solving CSPs is just to find an assignment of all the variables to a single value of their domain such that all the constraints in C are satisfied. Sometimes, a cost-objective function can be added, and when there are multiple solutions, the goal becomes to find one solution which optimizes that function. CP has been successfully used to solve many test generation problems [32], including applications in the IR domain [59–61].

The workspace of an IR is usually materialized by a subspace of a 3D space, a set of waypoints to be reached (in order to perform some dedicated tasks with the end-effector), and a set of obstacles. A valid trajectory in the workspace is a path that meets all or some of these waypoints at user-selected speeds, without hurting any of the obstacles. Generating trajectories to stress the robot involves finding valid trajectories that maximize the load of the robot, by changing directions and speed. The number of waypoints, obstacles, and the complexity of the load function to be maximized makes the problem hard to solve, as there are only a few valid trajectories in a huge search space of trajectories. In some cases, some waypoints cannot be reached. Hence, the problem is actually to find trajectories which hit some of the waypoints (not necessarily all) while maximizing the load.

In [21, 49], we proposed a method and a tool called Robtest to generate these trajectories by using CP. Robtest exploits a continuous-domain constraint solver called RealPaver [35] and the constraint language MiniZinc [63] with three backends solvers, namely, gecode [70], chuffed [20], and SICStus [17], for solving constraints over finite domains.

In CP, constraints with a non-fixed number of variables, called *global constraints*, are very successful in tackling difficult problems, as their solving is based on powerful domain-filtering techniques. Our model in Robtest exploits several global constraints such as INVERSE, SUBCIRCUIT, and TABLE to construct an effective approach, which generates stressful test trajectories.

Figure 4.3 shows the Robtest user interface where two cost-maximal trajectories have been automatically generated. Interestingly, these trajectories are automatically converted into scripts in a dedicated command language for the robot called Rapid. Robtest has been deployed and validated on real industrial robots and experimented

---

[1]Degree of freedom: typical industrial robots have 6-DoF.

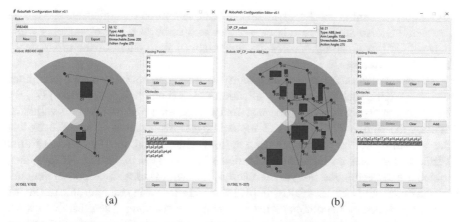

Fig. 4.3 Examples of Robtest interface, which shows the 2D projection of the robot workspace (in grey), waypoints and trajectories (in red), and obstacles (in blue). These optimal trajectories were generated by Robtest for 3D workspaces with two obstacles (**a**) and ten obstacles (**b**). Some waypoints are not reached but the generated trajectories are maximal w.r.t. the load function to optimize (Pictures by: Mathieu Collet).

with virtual workspaces containing more than 80 waypoints and 60 obstacles. For these workspaces, Robtest generated maximal trajectories in less than 5 min.[2] To the best of our knowledge, there is no other model able to deal with so many waypoints and obstacles.

The next subsection introduces the notion of metamorphic testing and how it can be used to also generate new test cases for IR.

3.2 Metamorphic Testing of Robots

Metamorphic testing is a testing paradigm, in which a source test case is transformed into a new follow-up test case for which the exact expected outcome is unknown, but a relation between the source and follow-up test case is available [19, 71]. By execution of the follow-up test case, it can be confirmed whether the system under test behaves according to the so-called metamorphic relation. If the relation is violated, a failure in the system has been identified. Metamorphic testing thereby addresses the oracle problem in software testing, where it is impossible or difficult to know the exact system output for a test case.

Examples of successful applications of metamorphic testing are search engines, where it is generally expected that additional keywords reduce the number of search results, or learning systems, where the exact reaction is unknown, but similar inputs

---

[2]The experimental benchmark and the CP model are publicly available at www.github.com/Makouno44/Robtest.

should reveal similar results, given an adequate definition of similarity. The oracle problem [4] also applies to robotic systems and some of their components when it is difficult to precisely specify the final state of the system or the effect every individual action has.

Recent works have applied metamorphic testing to robotics-related computer vision systems, such as image classification [23, 24] and object detection [77]. By using metamorphic relations, the test images are modified and can be validated against the output of the original input images. This makes it possible to test the variability of the computer vision system without additional labelling efforts. While it has received increasing attention in the context of testing ML systems, metamorphic testing has not yet been widely applied to industrial robots in general.

An initial study by Lindvall et al. [48] showed success in metamorphic testing of autonomous drones in combination with model-based testing. Their test framework generates test cases from a model based on the drone specifications, for example, a full flight scenario with obstacle avoidance. The applied metamorphic relations span five equivalences of the drone system: (1) The behaviour is expected to stay constant over multiple runs, days, and reboots. (2) The behaviour is rotation-invariant, i.e. rotating the world does not affect the outcome. (3) The behaviour is translation-invariant, i.e. against the movement to a different location in the environment. (4) The exact location of the obstacle does not affect the result. (5) The exact formation of the obstacles does not affect the result. The five equivalences can be combined and thereby create a wide variety of follow-up test cases from a single source test case, especially when further combined with model-based testing to even generate the source test cases.

To evaluate whether the behaviour matches the metamorphic relations, the authors outline different criteria, such as discretization of the sensor data to receive comparable values or the shape of the path the drone took, but the choice of this evaluation criterion is an open research question. Within their evaluation, they focus on the manual inspection of different flight paths for the generated scenarios, and the generated scenarios identified misaligned drone behaviour, leading to longer flight paths or irresponsible landing behaviour in some cases.

One potential outline for the application of metamorphic testing in robotics is related to the usage of domain randomization [81, 82] in the training of ML-based robotics. By introducing small randomizations in the training task, i.e. changing object colours, lightning conditions, or the physical model of the environment, the robot learns more robust behaviour. A similar approach can serve as a basis for metamorphic testing. The initial scenario is randomly modified in a similar way and expectations about the desired outcome are maintained. Maintaining the same expectations on the output, even though the test input has changed, is the identity function, i.e. the result stays the same independent of the modification. However, metamorphic relations make it possible to define more complex relations over the inputs and outputs of changed test cases and can formulate acceptable or mandatory changes in the outcome.

In conclusion, metamorphic testing is a flexible testing technique and has already shown to be applicable in some cases. We expect that the interest and attention on the combination of robotics and metamorphic testing will increase in future work.

4 Test Planning

This section discusses selected methods for test planning under resource constraints. In the first part (Sect. 4.1), we will present a technique for optimal test suite reduction, i.e. the selection of a smaller test suite that maintains a set of coverage properties. In the second part (Sect. 4.2), we then proceed to the scheduling and assignment of a test suite to a number of test agents.

4.1 Test Suite Reduction

Testing ABB's industrial robots is performed as part of a continuous integration (CI) process [60]. In CI, the goal is to build, test, and deploy software in frequent iterations and thus avoid software integration faults by detecting software errors at an early stage of development. Each development iteration, also called a CI cycle, deals with developing and testing only a part of the overall robot system source code.

Consequently, in each CI cycle, only relevant test cases are selected from a larger set of test cases that cover the overall functionality of the robot under test. Since the duration of a CI cycle is time-constrained, the relevant set of test cases needs to also be minimal, so that minimal time is needed for its execution. Selecting the smallest subset of test cases from a larger set, while preserving certain properties of the test suite (e.g. requirements coverage) is challenging, and some commonly proposed approaches address this problem only with approximated solutions [47, 66]. Finding such a smallest set of test cases (a.k.a optimal test suite) is NP-complete [22, 29], and the time required to find the optimal test set grows exponentially with the size of the problem, in the worst case.

In their previous work, the authors of this chapter have studied the problem of optimal test suite reduction using network maximum flows [33] and constraint programming [67]. As an example, we consider a simple test suite reduction problem consisting of a set of three test cases covering a set of seven test requirements, shown in Fig. 4.4. First, we encode the test suite reduction problem using a bipartite graph augmented by a source node S and a destination node D, shown in Fig. 4.4, with special capacity constraints. S and D are special vertices of a flow network, the source and the sink. Capacity is associated with each arc in a flow network, as the maximum limit of flow that the arc can receive. The bipartite graph is directed from a node S to a node D, with l/c values on each arc denoting flow values and capacities, respectively. Flow values of zero on all arcs mean that

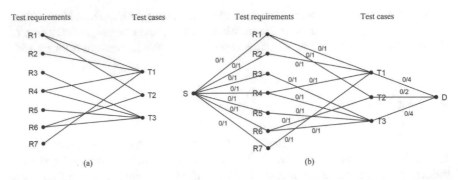

Fig. 4.4 Bipartite graph (**a**) and the corresponding flow network (**b**) for test suite reduction. The flow network consists of three test cases (T) covering seven test requirements (R). Flow values and capacities are denoted as l/c on each arc.

the flow is feasible but not maximum. To find maximum flows, the bipartite graph is first traversed using the Ford-Fulkerson method [28]. For the example test suite reduction problem of three test cases and seven test requirements, there are eight maximum flows, shown in Fig. 4.5. The maximum flows correspond to different solutions of the test suite reduction problem. However, not all maximum flows are optimal, as they do not reduce the original test suite. To find the optimal flows (there can be more than one), CP is used to search for the flow that maximized the number of zeros on arcs from nodes T to D nodes [33]. As the final solution of our example test suite reduction problem, there are two optimal flows found.

4.2 Test Scheduling and Distribution

In this section, we discuss two test scheduling and distribution problems. Both problems have in common that they require to assign test cases to test agents under consideration of additional constraints. The first problem are scheduling problems where test cases should be assigned to test agents while minimizing the total execution time of the final test plan. The second problem is to assign the test cases to the test agents such that in subsequent test cycles every test case is executed on a different test agent. Our discussion includes both a description of the problem and the presentation of solutions for these problems from the earlier work by the authors of this chapter.

4.2.1 Test Scheduling with Global Resources

We consider a scheduling problem where test cases should be assigned to test agents while minimizing the total execution time of the final test plan. This scheduling

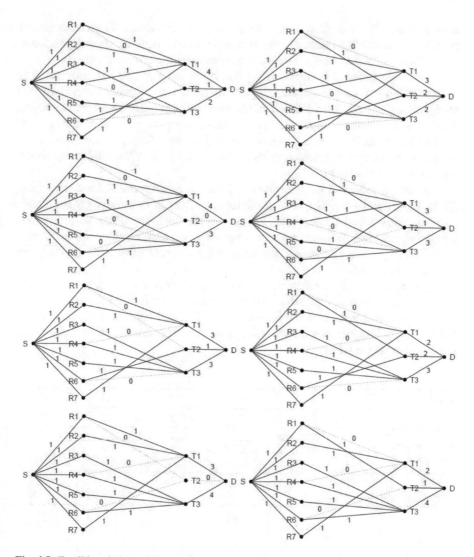

Fig. 4.5 Feasible solutions of the test suite reduction problem as maximum flows (from [33]).

problem is a variant of the generic machine scheduling problem [15] and covers a wide range of practical applications. Our discussion will be based on the work by Mossige et al. [62] and will contain the additional notion of exclusive access to shared global resources.

Some test cases may only be executable by a subset of the agents, e.g. a test agent might not provide a certain feature set, and some test cases can require exclusive access to a shared global resource. The final schedule will have to consider that only one test case at a time can make use of this global resource. It is furthermore necessary for the scheduling process to know the execution time of each test case. Because there can be some variation in the actual execution time, a practical approach is to maintain the execution time of earlier runs and provide an overestimation of the average execution time as a conservative estimate.

An example problem is shown in Fig. 4.6. A test suite of nine test cases has to be scheduled among three test agents (Fig. 4.6a). Most test cases are compatible with all agents, except test cases 6–9, which require some specific test agent functionality. Test cases 2–4 further require one specific global resource and test case 9 depends on the other global resource. The resulting optimal schedule (Fig. 4.6b) avoids the overlap of test cases 2–4, which leads to the critical path of the total test plan and the final test-execution time.

The model proposed by Mossige et al. [62] uses constraint-based scheduling [7] to model and solve the test scheduling problem in constraint programming. The cumulatives global constraint [9] supports the definition of the basic machine scheduling problem such that resource constraints on each machine, i.e. sequential execution of only a single task per time and non-preemptive execution without interruption, are followed.

By using an effective selection of global constraints, the CP solver can apply optimized filtering techniques that reduce the search space for the branch-and-bound tree search and thereby also the time until the first, and later the optimal, solution is found. Additionally, the search strategy for the tree search process affects the performance of the solver.

Test Number		1	2	3	4	5	6	7	8	9
Duration		2	3	2	4	2	1	2	3	5
Executable on Test Agent	1	X	X	X	X	X	X			X
	2	X	X	X	X	X		X		
	3	X	X	X	X	X			X	X
Use of Global Resource	1		X	X	X					
	2									X

(a)

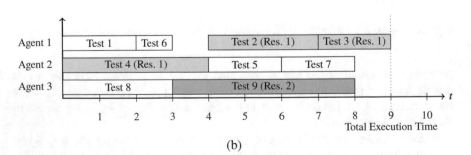

(b)

Fig. 4.6 Test scheduling with global resources: an example problem with nine test cases, three test agents, and two global resources (adapted from [62]). (**a**) Test suite overview. (**b**) Optimal schedule.

The search procedure attempts to assign values to variables such that all the constraints are satisfied and the cost function is minimized. Many different strategies can be used to explore the search space; however, it is known that the most effective approach is a search strategy depending on the characteristics of the scheduling problem to be solved [76].

In the case of test scheduling, the assignment is made by assigning first the earliest possible start time of the most demanding task, i.e., the task which requires the most global resources and take the longest time. This procedure is repeated for all test cases while trying to first distribute the test cases among the available machines. By initially preferring a different machine from the one that the previous test case was assigned to, we accept a compromise between the solving time of the schedule and the execution time of the schedule for the first found solution. Afterwards, branch-and-bound search minimizes the total execution time, but under resource constraints and with only short available time windows, it is often infeasible to solve every schedule to the global optimum rather than accepting a good enough, near-optimal solution.

The outlined approach is versatile and the general process persists for other variations of the scheduling. The constraint model and search strategy can be adapted to environment-specific conditions that require further constraints or requirements on the final schedule. However, the scheduling model is focused on a single test plan and cannot consider constraints and expectations that affect subsequent test schedules, e.g. from day to day. This issue will be discussed in the next section.

4.2.2   Test Scheduling with Rotational Diversity

Due to resource constraints in testing, trade-offs often have to be made between test coverage or completeness and resource usage. For example, each test is only executed once even though different test agents with different feature sets are available, and for a thorough and complete test result, it would be desirable to run each test on every compatible test agent.

One way to mitigate this trade-off is to follow the restrictions for a single test schedule, but address the distribution of test cases to different test agents over multiple test cycles, e.g. when running a form of continuous integration where the test process is repeated frequently. In every test cycle, the test schedule is automatically planned anew and the optimization function is adjusted so that it prefers the assignment of test cases to test agents that have not been recently made. This is especially relevant when the availability of test cases and test agents can change due to maintenance, new test cases, or restricted access to certain agents.

A solution to this problem of *rotational diversity* has been proposed in our earlier work [79]. As shown in Fig. 4.7, the main idea is to split the overall problem into two sub-problems. The inner problem is the actual test scheduling or test selection problem to be solved at every cycle to produce the test plan according to an optimization objective. The outer problem then adjusts the optimization weights of each test case in the schedule so that it incorporates information about the time

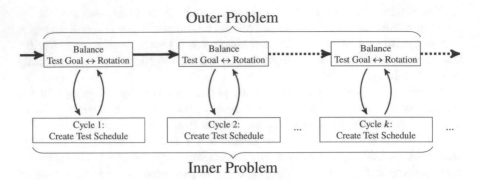

Fig. 4.7 Test scheduling with rotational diversity: the problem is divided into the actual scheduling problem per cycle and a balancing problem to steer the scheduling goal towards a trade-off between rotation and effectiveness (adapted from [79]).

since it has been assigned to each test agent, the so-called affinity. The higher the affinity, the better the solution if the assignment between these test cases and test agents is made.

The decomposed solution approach allows adopting the method for other problems or problems with a dedicated set of additional constraints, as long as the notion of an optimization objective and individual optimization weights per test case and test agent is given.

Several strategies for combining the original optimization weight with the affinity value have been proposed and compared [79]. Experimental results show that, in a scenario where a maximum time limit for testing should be exhausted, a product of both factors is a simple, but effective strategy. Such a strategy makes it possible to produce test plans that frequently rotate all test cases over their compatible test agents while only reducing the original optimization objective by on average less than 4%.

In this section, we discussed methods to distribute test cases onto physical test agents with the goal of making the best use of limited resources. The next section will look into the actual execution of these test cases on the physical hardware through the example of two case studies at our industrial partner ABB Robotics Norway.

5   Test Execution

In this section, we present an industrial test framework (Sect. 5.1) and a method for testing an integrated painting system (Sect. 5.2).

5.1   *Automated Regression Testing for Industrial Robots*

Test execution for IR is always challenging as it involves not only to perform an appropriate combination of testing in simulated environments and testing of real robots in safe environments but also to automate the testing process in continuous testing and delivery. Simulated environments for IR setup and testing include ABB's RobotStudio or Gazebo,[3] but the design and automation of input scenarios for these environments require skilled robotics engineers who are familiar with both the mechanics and electronic of the robots. Besides, these environments have usually not been created for repeated test execution, and they include neither test-scripting facilities nor test-results checking features.

Automating test execution for real robots is challenging as it involves robot motions (e.g. arm motions) that must be accurately synchronized with other subsystems, to perform specific tasks, such as painting or glueing. It is a usually labour-intensive step to prepare the robot environment for executing the tests, as strict safety regulations are enforced when humans are setting up the environment that involves moving machinery and sometimes dangerous paint fluids and gases [26]. Due to high cost, this type of test execution is usually performed during the final verification step and aims to detect subsystems synchronization problems or performance issues. Test-execution setups based on real robots have the advantage of providing very accurate results, but they require long labour-intensive preparation efforts.

Among the faults that are sought, software defects (as opposed to hardware defect) are usually prominent, and finding them is crucial before delivering the robots to their end-users. Fortunately, correcting software defects usually does not require long and costly procedures. Thus, there is a trend that consists in testing software systems independently from the physical robots, by using *testbenches*, such as the one shown in Fig. 4.8. These testbenches include layers of motherboards with actual subsystem CPUs, which enable extensive automated *regression tests execution*.

Regression tests are test scripts that are systematically executed for testing a new software release of an already deployed system. With these tests, if any of the test verdicts is failed, then the software release has detected a regression fault, meaning that some kind of added feature has broken previous correct behaviours of the system. The obvious advantage of using these testbenches for regression testing is that they test actual software components to be deployed on the robot (instead of simulated components), which allows test engineers to fully automate the test-execution process. These testing principles have been deployed at ABB Robotics with great success [60].

---

[3] http://gazebosim.org/.

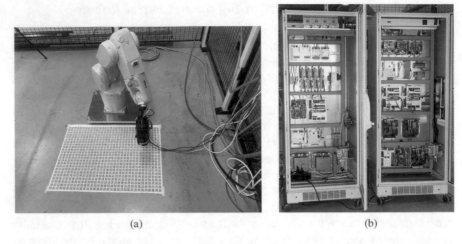

Fig. 4.8 Actual (**a**) robot testing versus (**b**) testbenches for testing IR software systems (Pictures by: Morten Mossige).

5.2   Generating and Executing Test Cases on an Integrated Painting System

ABB's Integrated Paint System (IPS) is a distributed control system that embeds various real-time embedded controllers for performing high-quality painting. These controllers are useful, for example, to control the air flow, the air pressure, the pump pressure in paint flow, or the high voltage for electrostatic charging used in painting car bodies. One of the main challenges encountered while testing IPS is related to the testing of timing characteristics of the distributed control system. Typically, the IPS is configured with embedded controllers that execute time synchronization protocols to keep their clocks synchronized, as shown in Fig. 4.9. As there are many possible configurations for the IPS, test configurations selection is required. Note that some configurations may involve more than 20 embedded controllers.

In [58, 59], we have introduced a constraint model that is based on input channels, which are responsible for controlling exactly one physical process, for example, air or paint, involved in generating a spray pattern, as shown in Fig. 4.10.

The IPS input is a sequence of paint-spray paths, that is, a sequence of timestamped events (B_i, t_i), denoting the i-th paint-spray path B_i and its application time t_i. The output of the model is given by variables that represent physical values for each channel j, $(P_{j,i}, t_{j,i})$. A detailed presentation of the constraint model and its constraint solving method can be found in [58, 60, 61].

Using this constraint model has been beneficial to generate test scenarios that can find subtle faults such as failing overlaps, burst scenarios, or invalid shutdowns. The model has been deployed at ABB Robotics for generating such test scenarios, while test execution is typically triggered by a build server on a testbench.

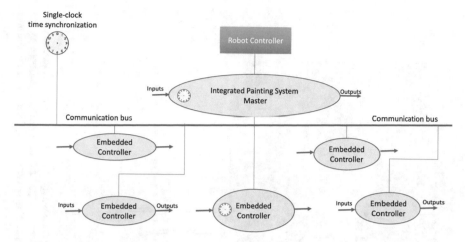

Fig. 4.9 ABB's Integrated Paint System as a distributed system (from [60]). Different subsystems are synchronized using a unique shared clock and communication bus. Implemented timing functions are used to trigger subsystems at the appropriate time-stamped events, but this needs to be heavily tested.

In brief, the IPS software system is built every night and all the embedded controllers are upgraded with the new build release. The IPS is then configured by using either customer configurations or on-purpose test configurations. A set of simple tests, so-called smoke tests, is then executed before solving the constraint model for new test-scenario generation. These tests are then executed by applying the generated time-stamped events sequence. As the constraint model also generates the expected outputs, the failed verdicts can be reported back to the test engineer. A detailed presentation of these steps is shown in Fig. 4.10.

The constraint model is solved with the finite-domain constraint solving library of SICStus Prolog, named `clpfd` [18]. The solver is called through a Python front-end layer that allows test engineers to easily integrate the test generator engine with existing build and test servers based on MS Team Foundation Server. A schematic overview of the architecture is shown in Fig. 4.11.

During the initial deployment of the constraint model, five critical bugs were discovered and immediately corrected, and several dozens of non-critical bugs were detected and progressively corrected. Today, the model is used on a daily basis and allows test engineers to continuously improve the quality of delivered IR.

6   Discussion and Conclusion

Continuous testing has emerged from the ever-growing complexity of industrial robots to become a crucial activity of the development process. An industrial robot is not only tested once before deployment, but its testing process involves frequently

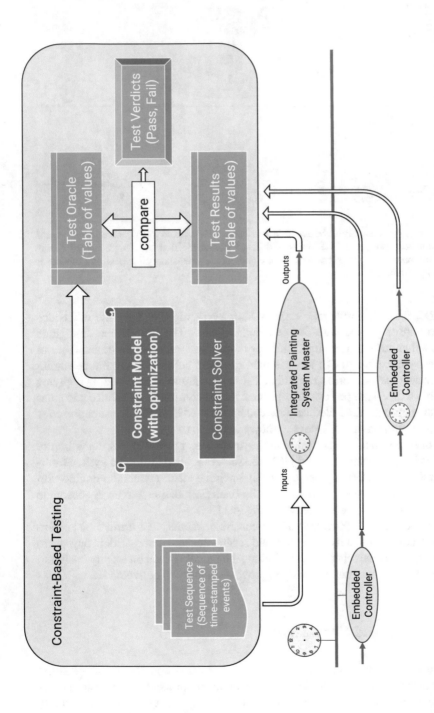

Fig. 4.10 A CP model for testing ABB's Integrated Paint System (from [60]). The *Constraint Model* generates inputs under the form of a sequence of time-stamped events (*Test Sequence*) and, based on its calculations, also produces expected outputs (*Test Oracle*). In parallel, the events are processed by the distributed system (*IPS Master*), which produces observed outputs (*Test Result*). By comparing expected outputs and observed outputs, failures are automatically detected and pointed out to the test engineer (*Compare Function*).

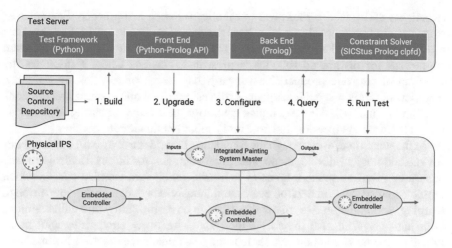

Fig. 4.11 Communication between the testbench and IPS (from [60]).

executing test suites to assess its robustness and performance and to quickly identify regression faults. One goal is therefore to identify or generate effective test cases that induce fault-revealing behaviour, and the other goal is to continuously improve the testing process for a better exploitation of available resources.

Throughout this chapter, we discussed techniques for these two aforementioned goals of testing industrial robots. We discussed approaches for the automated testing of industrial robots both in regard to the generation of diverse and complex test scenarios and to the test organization under resource constraints.

The view on testing robotic systems presented in this chapter is complementary to the traditional view of validation and verification using formal methods, as presented in Chaps. 7, 8, and 9. By designing a formal model of the system, model checking techniques can be used to formally prove or disprove some properties of the system under verification. Interestingly, such a formal model can also be used to generate test cases for finding faults in approaches such as simulation-based testing as explained in Chap. 5 or mutation-based test case generation as presented in Chap. 11.

In software and system testing, as discussed here, generated test cases are specifically defined scripts, often independent from an exact formal model, that are part of a continuous integration process, as described in Sect. 3.

Today, a solid methodological foundation exists for software and system testing and for how to design and implement tests for industrial robots, but the practical adoption of these techniques is not yet the state of the practice. Although the goals and challenges are acknowledged, both cultural and economic resistance against their adoption have been observed [4]. A direction for future work is therefore to maintain the connections and collaborations between the testing community, often based within the domain of software engineering, and the robotics community, especially with industrial partners, to present the benefits of automated testing techniques and establish them within robotics research and practice. Examples of existing initiatives towards the principled design and engineering of robotic systems are further discussed in the other chapters of this book. In Chap. 10 a software framework for distributed robotics is presented, whereas Chap. 1 discusses the adoption of software product lines towards the design of robotics software. An overview of domain-specific languages for designing robotic systems and missions is given in Chap. 12. All these topics have seen their origin or strong applications within software engineering and now find similar application in robotics.

At the same time, even though the foundation is solid, there are still opportunities for methodological developments and improvements. Additional dedicated focus should be given on modern types of industrial robots like learning robots, which employ trained ML models or even continuously adapt their behaviour through learning, and collaborative robots that work in close proximity and collaborations with humans. While ML-based components share many similarities with traditionally developed components, their testing has challenges that go beyond their basic functionality as measured through accuracy. The receptiveness for adversarial data, i.e. weaknesses due to manipulated inputs, needs to be considered as well as encountering situations outside the distribution of training data. Also traditional metrics for test completeness, such as code coverage or execution traces, are not as reliable anymore. Although there is work to adopt similar metrics for testing of ML systems, their meaning is not as intuitive or well-understood at this time and will need more consideration by the research community. Conclusively, the design of test cases and their efficient execution raises new challenges: How to construct efficient test cases that test robot components not only in isolation but also jointly? How to produce the necessary diversity in learning experiences to test the robot's learning capabilities and potential outcomes? How to model human interaction and human uncertainty?

In conclusion, the task of testing industrial robots includes open challenges for the future, even though several achievements have been observed in terms of test generation, test planning, and test execution.

Given the ongoing trend of flexible automation using lightweight, collaborative robots on the one hand and the continuing use of large-scale fixed industrial robots in automated assembly lines on the other hand, the variety of applications for testing increases. It will be the goal to transfer the existing body of knowledge and methods towards these open problems and to promote their adoption among practitioners.

Acknowledgements The authors would like to thank Morten Mossige, Mohit Kumar Ahuja, and Mathieu Collet for their tremendous contributions to the research works presented in this chapter. This work is supported by the Research Council of Norway under the T-Largo grant agreement no. 274786 (Testing of Learning Robots) and by the European Union under grant agreement no. 825619 (AI4EU).

References

1. P. Abbeel, *Apprenticeship Learning and Reinforcement Learning with Application to Robotic Control* (Stanford University, Stanford, 2008)
2. F. Abi-Farraj, T. Osa, N.P.J. Peters, G. Neumann, P.R. Giordano, A learning-based shared control architecture for interactive task execution, in *2017 IEEE International Conference on Robotics and Automation (ICRA)* (IEEE, Piscataway, 2017), pp. 329–335
3. L. Adhami, E. Coste-Maniere, Positioning tele-operated surgical robots for collision-free optimal operation, in *Proceedings 2002 IEEE International Conference on Robotics and Automation (Cat. No.02CH37292)*, vol. 3 (2002), pp. 2962–2967
4. A. Afzal, C. Le Goues, M. Hilton, C.S. Timperley, A study on challenges of testing robotic systems, in *Proceedings of the International Conference on Software Testing, Verification and Validation (ICST), ICST*, vol. 20 (2020)
5. D. Araiza-Illan, D. Western, A.G. Pipe, K. Eder, Systematic and realistic testing in simulation of control code for robots in collaborative human-robot interactions, in *Annual Conference Towards Autonomous Robotic Systems*, ed. by L. Alboul, D. Damian, J.M. Aitken (Springer International Publishing, Cham, 2016), pp. 20–32
6. A. Avizienis, J. Laprie, B. Randell, C. Landwehr, Basic concepts and taxonomy of dependable and secure computing. IEEE Trans. Dependable Secure Comput. **1**(1), 11–33 (2004)
7. P. Baptiste, C. Le Pape, W. Nuijten, *Constraint-Based Scheduling: Applying Constraint Programming to Scheduling Problems*, 1st edn. International Series in Operations Research & Management Science, vol. 39 (Springer US, New York, 2001)
8. J.M. Beer, A.D. Fisk, W.A. Rogers, Toward a framework for levels of robot autonomy in human-robot interaction. J. Hum. Robot. Interact. **3**(2), 74–99 (2014). https://doi.org/10.5898/JHRI.3.2.Beer
9. N. Beldiceanu, M. Carlsson, A New multi-resource cumulatives constraint with negative heights, in *Principles and Practice of Constraint Prog (CP'02)* (2002), pp. 63–79
10. M. Bernard, K. Kondak, I. Maza, A. Ollero, Autonomous transportation and deployment with aerial robots for search and rescue missions. J. Field Robot. **28**, 914–931 (2011)
11. C. Bessiere, F. Koriche, N. Lazaar, B. O'Sullivan, Constraint acquisition. Artif. Intell. **244**, 315–342 (2017)
12. A. Billard, S. Calinon, R. Dillmann, S. Schaal, *Robot Programming by Demonstration* (Springer, Berlin, 2008), pp. 1371–1394. https://doi.org/10.1007/978-3-540-30301-5_60
13. J. Blazewicz, J.K. Lenstra, A.R. Kan, Scheduling subject to resource constraints: classification and complexity. Discrete Appl. Math. **5**(1), 11–24 (1983)
14. K. Bousmalis, A. Irpan, P. Wohlhart, Y. Bai, M. Kelcey, M. Kalakrishnan, L. Downs, J. Ibarz, P. Pastor, K. Konolige, S. Levine, V. Vanhoucke, Using simulation and domain adaptation to improve efficiency of deep robotic grasping, in *2018 IEEE International Conference on Robotics and Automation (ICRA)* (IEEE, Brisbane, 2018), pp. 4243–4250. https://doi.org/10.1109/ICRA.2018.8460875
15. P. Brucker, S. Knust, *Complex Scheduling (GOR-Publications)* (Springer, New York, 2006)
16. B. Brumfield, Car assembly line robot kills worker in Germany (2015). http://www.cnn.com/2015/07/02/europe/germany-volkswagen-robot-kills-worker/
17. M. Carlsson, G. Ottosson, B. Carlson, An open-ended finite domain constraint solver, in *Proceedings of Programming Languages: Implementations, Logics, and Programs, 9th International Symposium, PLILP'97, Including a Special Trach on Declarative Programming Languages in Education, Southampton, 3–5 Sept 1997*, pp. 191–206
18. M. Carlsson, G. Ottosson, B. Carlson, An open-ended finite domain constraint solver, in *Proc. of the 9th Int. Symp. on Prog. Languages, Implementations, Logics, and Programs (PLILP '97)* (1997), pp. 191–206. https://doi.org/10.1007/BFb0033845
19. T.Y. Chen, F.C. Kuo, H. Liu, P.L. Poon, D. Towey, T.H. Tse, Z.Q. Zhou, Metamorphic testing: a review of challenges and opportunities. ACM Comput. Surv. **51**(1) (2018). https://doi.org/10.1145/3143561

20. G. Chu, P.J. Stuckey, A. Schutt, T. Ehlers, G. Gange, K. Francis, Chuffed, a lazy clause generation solver (2016)
21. M. Collet, A. Gotlieb, N. Lazaar, M. Carlsson, D. Marijan, M. Mossige, Robtest: a CP approach to generate maximal test trajectories for industrial robots, in *Proc. of the 26th Int. Conf. On Principles of Constraint Prog. (CP-20), Louvain-La-Neuve*. Lecture Notes in Computer Science, vol. 12333 (2020)
22. T. Cormen, C.E. Leiserson, R.L. Rivest, C. Stein, *Introduction to Algorithms*, 2 edn. (MIT Press, Cambridge, 2001)
23. J. Ding, X. Kang, X.H. Hu, Validating a deep learning framework by metamorphic testing, in *Proceedings - 2017 IEEE/ACM 2nd International Workshop on Metamorphic Testing, MET 2017* (2017), pp. 28–34. https://doi.org/10.1109/MET.2017.2
24. A. Dwarakanath, M. Ahuja, S. Sikand, R.M. Rao, R.P.J.C. Bose, N. Dubash, S. Podder, Identifying implementation bugs in machine learning based image classifiers using metamorphic testing, in *Proceedings of the 27th ACM SIGSOFT International Symposium on Software Testing and Analysis (ISSTA)* (2018), pp. 118–128. https://doi.org/10.1145/3213846.3213858
25. S. Elbaum, A. Malishevsky, G. Rothermel, Test case prioritization: a family of empirical studies. IEEE Trans. Softw. Eng. **28**(2), 159–182 (2002). https://doi.org/10.1109/32.988497
26. European Parliament and Council of the European Union, Directive 2006/42/EC on machinery (2006)
27. P. Flener, U. Schmid, *Inductive Programming* (Springer US, Boston, 2017), pp. 658–666. https://doi.org/10.1007/978-1-4899-7687-1_137
28. L. Ford, D. Fulkerson, *Flows in Networks* (Princeton University Press, Princeton, 1962)
29. M.R. Garey, D.S. Johnson, *Computers and Intractability; A Guide to the Theory of NP-Completeness* (W. H. Freeman, New York, 1990)
30. M. Gligoric, L. Eloussi, D. Marinov, Ekstazi: lightweight test selection, in *Proceedings of the 37th International Conference on Software Engineering*, vol. 2 (2015), pp. 713–716. https://doi.org/10.1109/ICSE.2015.230
31. I.J. Goodfellow, J. Shlens, C. Szegedy, Explaining and harnessing adversarial examples (2014). 1412.6572
32. A. Gotlieb, Constraint-based testing: an emerging trend in software testing, in *Advances in Computers*, vol. 99 (Elsevier, Amsterdam, 2015), pp. 67–101
33. A. Gotlieb, D. Marijan, FLOWER: optimal test suite reduction as a network maximum flow, in *Proc. of Int. Symp. on Soft. Testing and Analysis (ISSTA'14)* (2014), pp. 171–180
34. A. Gotlieb, M. Liaaen, P. Alexandre, Automated regression testing using constraint programming, in *Innovative Applications of Artificial Intelligence (IAAI)* (2016), pp. 4010–4015
35. L. Granvilliers, F. Benhamou, Algorithm 852: Realpaver: an interval solver using constraint satisfaction techniques. ACM Trans. Math. Softw. **32**(1), 138–156 (2006)
36. H. Gross, S. Mueller, C. Schroeter, M. Volkhardt, A. Scheidig, K. Debes, K. Richter, N. Doering, Robot companion for domestic health assistance: implementation, test and case study under everyday conditions in private apartments, in *2015 IEEE/RSJ International Conference on Intelligent Robots and Systems (IROS)* (2015), pp. 5992–5999
37. J. Guiochet, M. Machin, H. Waeselynck, Safety-critical advanced robots: a survey. Rob. Auton. Syst. **94**, 43–52 (2017). https://doi.org/10.1016/j.robot.2017.04.004
38. S. Hartmann, D. Briskorn, A survey of variants and extensions of the resource-constrained project scheduling problem. Eur. J. Oper. Res. **207**(1), 1–14 (2010). https://doi.org/10.1016/j.ejor.2009.11.005
39. R. Jabbarvand, A. Sadeghi, H. Bagheri, S. Malek, Energy-aware test-suite minimization for Android apps, in *Proceedings of the 25th International Symposium on Software Testing and Analysis - ISSTA 2016* (2016), pp. 425–436. https://doi.org/10.1145/2931037.2931067
40. G. Jahangirova, P. Tonella, An empirical evaluation of mutation operators for deep learning systems, in: *IEEE International Conference on Software Testing, Verification and Validation (ICST)* (IEEE, Piscataway, 2020)
41. Y. Jia, M. Harman, An analysis and survey of the development of mutation testing. IEEE Trans. Softw. Eng. **37**(5), 649–678 (2011). https://doi.org/10.1109/TSE.2010.62

42. A. Jones, G. Castellano, Adaptive robotic tutors that support self-regulated learning: a longer-term investigation with primary school children. Int. J. Soc. Robot. **10**, 357–370 (2018)
43. J. Kim, R. Feldt, S. Yoo, Guiding deep learning system testing using surprise adequacy, in *Proceedings of the 41st International Conference on Software Engineering, ICSE '19* (IEEE Press, Piscataway, 2019), pp. 1039–1049. https://doi.org/10.1109/ICSE.2019.00108
44. J. Kober, J.A. Bagnell, J. Peters, Reinforcement learning in robotics: a survey. Int. J. Robot. Res. **32**(11), 1238–1274 (2013)
45. P.A. Lasota, T. Fong, J.A. Shah, A survey of methods for safe human-robot interaction. Found. Trends Robot. **5**(3), 261–349 (2017). https://doi.org/10.1561/2300000052
46. C. Leong, A. Singh, M. Papadakis, Y.L. Traon, J. Micco, Assessing transition-based test selection algorithms at Google, in *Proceedings of the 41st International Conference on Software Engineering: Software Engineering in Practice, ICSE-SEIP '10* (IEEE Press, Piscataway, 2019), pp. 101–110. https://doi.org/10.1109/ICSE-SEIP.2019.00019
47. C. Lin, K. Tang, C. Chen, G.M. Kapfhammer, Reducing the cost of regression testing by identifying irreplaceable test cases, in *2012 Sixth International Conference on Genetic and Evolutionary Computing* (2012), pp. 257–260
48. M. Lindvall, A. Porter, G. Magnusson, C. Schulze, Metamorphic model-based testing of autonomous systems, in *Proceedings - 2017 IEEE/ACM 2nd International Workshop on Metamorphic Testing, MET 2017* (2017), pp. 35–41. https://doi.org/10.1109/MET.2017.6
49. M. Collet, A. Gotlieb, N. Lazaar, M. Mossige, Stress testing of single-arm robots through constraint-based generation of continuous trajectories, in *IEEE Int. Conf. On Artificial Intelligence Testing (AITest'19), San Francisco, CA* (2019)
50. L. Ma, Y. Liu, J. Zhao, Y. Wang, F. Juefei-Xu, F. Zhang, J. Sun, M. Xue, B. Li, C. Chen et al., Deepgauge: multi-granularity testing criteria for deep learning systems, in *Proceedings of the 33rd ACM/IEEE International Conference on Automated Software Engineering - ASE 2018* (2018) https://doi.org/10.1145/3238147.3238202
51. L. Ma, F. Zhang, J. Sun, M. Xue, B. Li, F. Juefei-Xu, C. Xie, L. Li, Y. Liu, J. Zhao, Y. Wang, DeepMutation: mutation testing of deep learning systems, in *2018 IEEE 29th International Symposium on Software Reliability Engineering (ISSRE)* (IEEE, Piscataway, 2018), pp. 100–111. https://doi.org/10.1109/ISSRE.2018.00021
52. M. Malekzadeh, J. Queißer, J.J. Steil, Imitation learning for a continuum trunk robot, in *Proceedings of the 25th European Symposium on Artificial Neural Networks, Computational Intelligence and Machine Learning, ESANN 2017* (2017)
53. V.J.M. Manès, H. Han, C. Han, S.K. Cha, M. Egele, E.J. Schwartz, M. Woo, The art, science, and engineering of fuzzing: a survey. arXiv: Cryptography and Security (2018)
54. D. Marijan, A. Gotlieb, S. Sen, A. Hervieu, Practical pairwise testing for software product lines, in *Proceedings of the 17th International Software Product Line Conference, SPLC '13* (Association for Computing Machinery, New York, 2013), pp. 227–235. https://doi.org/10.1145/2491627.2491646
55. J.P. Merlet, Interval analysis and reliability in robotics. Int. J. Reliab. Saf. **3**(1/2/3), 104–130 (2009)
56. A. Mohammed, T. Viola, C.S. Tucker, J. Duarte, Towards co-robot navigation in manufacturing environments through machine learning of human movement patterns. Chall. Technol. Innov. **189**(194), 189–194 (2017)
57. M. Morioka, S. Sakakibara, A new cell production assembly system with human–robot cooperation. CIRP Ann. **59**(1), 9–12 (2010). https://doi.org/10.1016/j.cirp.2010.03.044
58. M. Mossige, A. Gotlieb, H. Meling, Testing Robotized paint system using constraint programming: an industrial case study, in *IFIP International Conference on Testing Software and Systems*, ed. by M.G. Merayo, E.M. de Oca. Lecture Notes in Computer Science (Springer, Berlin, 2014), pp. 145–160. https://doi.org/10.1007/978-3-662-44857-1_10
59. M. Mossige, A. Gotlieb, H. Meling, Using CP in automatic test generation for ABB robotics' paint control system, in *Principles and Practice of Constraint Programming*. Lecture Notes in Computer Science, vol. 8656 (Springer International Publishing, Cham, 2014), pp. 25–41. https://doi.org/10.1007/978-3-319-10428-7_6

60. M. Mossige, A. Gotlieb, H. Meling, Testing robot controllers using constraint programming and continuous integration. Inf. Softw. Technol. **57**, 169–185 (2015). https://doi.org/10.1016/j.infsof.2014.09.009

61. M. Mossige, A. Gotlieb, H. Meling, Generating tests for robotized painting using constraint programming, in *Proceedings of the Twenty-Fifth International Joint Conference on Artificial Intelligence, IJCAI 2016, 9–15 July 2016*, ed. by S. Kambhampati (IJCAI/AAAI Press, New York, 2016), pp. 4200–4204

62. M. Mossige, A. Gotlieb, H. Spieker, H. Meling, M. Carlsson, Time-aware test case execution scheduling for cyber-physical systems, in *Proceedings of the 23rd International Conference on Principles and Practice of Constraint Programming*. Lecture Notes in Computer Science, vol. 10416 (2017), pp. 387–404. https://doi.org/10.1007/978-3-319-66158-2_25

63. N. Nethercote, P.J. Stuckey, R. Becket, S. Brand, G.J. Duck, G. Tack, MiniZinc: towards a standard CP modelling language, in *Proceedings of Principles and Practice of Constraint Programming - CP 2007, 13th International Conference, CP 2007, Providence, RI*, 23–27 Sept 2007, pp. 529–543. https://doi.org/10.1007/978-3-540-74970-7_38

64. K. Pei, Y. Cao, J. Yang, S. Jana, Deepxplore, in *Proceedings of the 26th Symposium on Operating Systems Principles* (2017). https://doi.org/10.1145/3132747.3132785

65. H.A. Pierson, M.S. Gashler, Deep learning in robotics: a review of recent research. Adv. Robot. **31**(16), 821–835 (2017). https://doi.org/10.1080/01691864.2017.1365009

66. X. Qu, M.B. Cohen, G. Rothermel, Configuration-aware regression testing: an empirical study of sampling and prioritization, in *Proceedings of the 2008 International Symposium on Software Testing and Analysis, ISSTA '08* (Association for Computing Machinery, New York, 2008), pp. 75–86. https://doi.org/10.1145/1390630.1390641

67. F. Rossi, P.V. Beek, T. Walsh, *Handbook of Constraint Programming (Foundations of Artificial Intelligence)* (Elsevier Science, New York, 2006)

68. G. Rothermel, R.H. Untch, C. Chu, M.J. Harrold, Prioritizing test cases for regression testing. IEEE Trans. Softw. Eng. **27**(10), 929–948 (2001). https://doi.org/10.1145/347324.348910

69. S. Schaal, J. Peters, J. Nakanishi, A. Ijspeert, Learning movement primitives, in *Robotics Research. The Eleventh International Symposium*, ed. by P. Dario, R. Chatila (Springer, Berlin, 2005), pp. 561–572

70. C. Schulte, G. Tack, M.Z. Lagerkvist, Modeling and programming with Gecode (2018)

71. S. Segura, G. Fraser, A.B. Sanchez, A. Ruiz-Cortes, A survey on metamorphic testing. IEEE Trans. Softw. Eng. **42**(9), 805–824 (2016). https://doi.org/10.1109/TSE.2016.2532875

72. W. Shen, J. Wan, Z. Chen, Munn: mutation analysis of neural networks, in *2018 IEEE International Conference on Software Quality, Reliability and Security Companion (QRS-C)* (2018), pp. 108–115

73. A. Shi, A. Gyori, S. Mahmood, P. Zhao, D. Marinov, Evaluating test-suite reduction in real software evolution, in *Proceedings of the 27th ACM SIGSOFT International Symposium on Software Testing and Analysis - ISSTA 2018* (2018), pp. 84–94. https://doi.org/10.1145/3213846.3213875

74. A. Shi, P. Zhao, D. Marinov, Understanding and improving regression test selection in continuous integration, in *International Symposium on Software Reliability Engineering* (2019), pp. 228–238

75. S.Y. Shin, S. Nejati, M. Sabetzadeh, L.C. Briand, F. Zimmer, Test case prioritization for acceptance testing of cyber physical systems: a multi-objective search-based approach, in *Proceedings of the 27th ACM SIGSOFT International Symposium on Software Testing and Analysis - ISSTA 2018* (2018), pp. 49–60. https://doi.org/10.1145/3213846.3213852

76. H. Simonis, B. O'Sullivan, Search strategies for rectangle packing, in *Principles and Practice of Constraint Programming*. Lecture Notes in Computer Science, vol. 5202 (Springer, Berlin, 2008), pp. 52–66

77. H. Spieker, A. Gotlieb, Adaptive metamorphic testing with contextual bandits. J. Syst. Softw. **165**, 110574 (2020). https://doi.org/10.1016/j.jss.2020.110574

78. H. Spieker, A. Gotlieb, D. Marijan, M. Mossige, Reinforcement learning for automatic test case prioritization and selection in continuous integration, in *Proceedings of 26th ACM SIGSOFT International Symposium on Software Testing and Analysis (ISSTA 2017)* (2017), pp. 12–22. https://doi.org/10.1145/3092703.3092709
79. H. Spieker, A. Gotlieb, M. Mossige, Rotational diversity in multi-cycle assignment problems, in *Thirty-Third AAAI Conference on Artificial Intelligence* (2019), pp. 7724–7731. https://doi.org/10.1609/aaai.v33i01.33017724
80. N. Sunderhauf, O. Brock, W. Scheirer, R. Hadsell, D. Fox, J. Leitner, B. Upcroft, P. Abbeel, W. Burgard, M. Milford, P. Corke, The limits and potentials of deep learning for robotics. Int. J. Robot. Res. **37**(4–5), 405–420 (2018). https://doi.org/10.1177/0278364918770733
81. J. Tobin, L. Biewald, R. Duan, M. Andrychowicz, A. Handa, V. Kumar, B. McGrew, A. Ray, J. Schneider, P. Welinder, W. Zaremba, P. Abbeel, Domain randomization and generative models for robotic grasping, in *2018 IEEE/RSJ International Conference on Intelligent Robots and Systems (IROS)* (2018), pp. 3482–3489. https://doi.org/10.1109/IROS.2018.8593933
82. J. Tremblay, A. Prakash, D. Acuna, M. Brophy, V. Jampani, C. Anil, T. To, E. Cameracci, S. Boochoon, S. Birchfield, Training deep networks with synthetic data: bridging the reality gap by domain randomization, in *Proceedings of the IEEE Conference on Computer Vision and Pattern Recognition Workshops* (2018), pp. 969–977
83. R. Wang, Y. Lu, B. Qu, Empirical study of the effects of different profiles on regression test case reduction. IET Softw. **9**(2), 29–38 (2015). https://doi.org/10.1049/iet-sen.2014.0008
84. H. Yin, P. Alves-Oliveira, F.S. Melo, A. Billard, A. Paiva, Synthesizing robotic handwriting motion by learning from human demonstrations, in *Proceedings of the 25th International Joint Conference on Artificial Intelligence, CONF* (2016)
85. S. Yoo, M. Harman, Regression testing minimization, selection and prioritization: a survey. Softw. Test. Verif. Reliab. **22**(2), 67–120 (2012). https://doi.org/10.1002/stvr.430
86. A. Zacharaki, I. Kostavelis, A. Gasteratos, I. Dokas, Safety bounds in human robot interaction: a survey. Saf. Sci. **127**, 104667 (2020). https://doi.org/10.1016/j.ssci.2020.104667

Chapter 5
Gaining Confidence in the Trustworthiness of Robotic and Autonomous Systems

Kerstin Eder

Abstract Because no single technique is adequate to cover a whole system in practice, a variety of complementary techniques will be needed to gain confidence in the trustworthiness of robotic and autonomous systems. In this chapter, we argue that demonstrable trustworthiness can be achieved by design, through transparency and by rigorous verification and validation. We then concentrate on ensuring correctness during system development, with specific focus on how to achieve implementations that are free from runtime errors and how design flaws can be detected prior to implementation. Our discussion is illustrated on two examples, robot navigation code and Simulink controllers. In both cases, the power of fully automatic theorem proving combined with test-based techniques can deliver strong confidence in system correctness. Attention is then turned to simulation-based verification, focusing in particular on the test generation challenge. We motivate the benefits of introducing agency into the test environment and show how exploiting multi-agent systems for test generation can lead to robust tests that achieve a high level of coverage as part of a coverage-driven verification methodology. Finally, some challenges for robotic and autonomous system verification are discussed, including specification, (more) automation, discipline, innovation and creativity in combining techniques and exploiting AI for verification.

1 Introduction

Ultimately, we would like to create *flawless* systems, but this is not enough. Systems need to be built in such a way that this flawlessness can be demonstrated. In the absence of this ideal, however, systems should at least be trustworthy. Trust is a broad concept that varies over time as a function of context and many other factors [32]. Everyone knows how hard it can be to trust, whether this is trust in

K. Eder (✉)
Trustworthy Systems Laboratory, Department of Computer Science, University of Bristol, Bristol, UK
e-mail: kerstin.eder@bristol.ac.uk

© Springer Nature Switzerland AG 2021
A. Cavalcanti et al. (eds.), *Software Engineering for Robotics*,
https://doi.org/10.1007/978-3-030-66494-7_5

people or technology, how easily trust can be lost and how hard it is to regain trust. Acknowledging the subjective nature of trust, the focus in this chapter will instead be on demonstrable trustworthiness [18].

This chapter concentrates on design and verification techniques that support engineers in establishing the evidence required to justifiably claim that the system they are building can be trusted. Here, verification is understood to mean *the process used to gain confidence in the correctness of a system with respect to its specification* [8]. This interpretation is commonly used in the hardware design verification community, i.e. by the verification engineers responsible for signing off ever more complex semiconductor designs for manufacture.

It is noteworthy that the above definition of verification does not stipulate any particular technique that should be used in order to gain such confidence. Indeed, there is no single technique that is adequate to cover a whole system in practice. For this reason the use of a variety of complementary verification techniques is typically required to gain confidence in the correctness of complex real-world systems. This has long been widely accepted in the hardware design verification community and is also true for the domain of robotic and autonomous systems (RAS). However, some members of the software community tend to use verification synonymous to formal verification, despite the fact that established standards like ISO/IEC 29119-1 clearly show (see Annex A of ISO/IEC 29119-1 in Fig. 5.1) that both formal methods and testing are at the same level in a classification hierarchy, and next to verification and validation (V&V) analysis, all three being activities that contribute to V&V.

The chapter is structured as follows. The next section gives a general overview of techniques that can be used to gain confidence in the trustworthiness of a system. We will briefly consider design techniques and the principle of transparency, illustrating each with an example. Our focus will then shift to V&V techniques and how a variety of complementary techniques can be used in combination. Section 3 is focused on correctness from specification to implementation, promoting the use of formal methods. It introduces reconvergence models, a conceptual representation that shows how the verification process relates to the system development process.We then concentrate on the implementation level, specifically on eliminating runtime errors in robot navigation code. By programming in SPARK, a language used mainly in the safety-critical systems domain, we show how to gain the benefits of fully automatic formal proof while achieving performance comparable to that of conventional C or C++ implementations. This demonstrates that verifiability can be promoted to a first-class software development objective for RAS and that the choice of programming language is instrumental in this process. Next, we study the design level, with the aim of detecting design flaws in Simulink controller designs prior to implementation. We show how high-level control system requirements can be decomposed, captured as assertions and verified using simulation combined with automated formal proof in order to gain confidence in the system design meeting its specification. Section 4 shifts our attention to simulation-based verification, presenting the architecture of a contemporary testbench together with coverage-driven verification on the example of a human-robot interaction scenario. To address the test generation challenge, we exploit multi-agent systems for test generation and

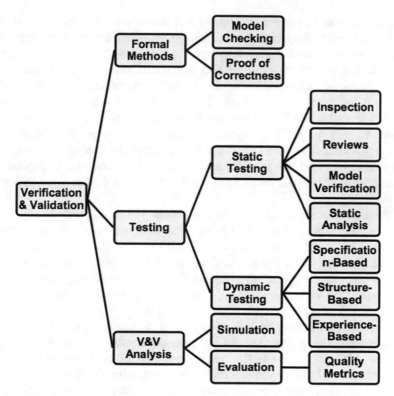

Fig. 5.1 Annex A of ISO/IEC/IEEE 29119-1:2013. Adapted and reprinted with permission from IEEE. Copyright IEEE 2013. All rights reserved.

show how this leads to robust tests that achieve a high level of coverage as part of a coverage-driven verification methodology. The chapter closes with an outlook on challenges for robotic and autonomous system verification, including specification, (more) automation, discipline, innovation and creativity in combining techniques and exploiting AI for verification in Sect. 6.

2 How to Gain Confidence in the Trustworthiness of a System

Confidence in a system's trustworthiness can be gained in many different ways; these include by design, through transparency and through V&V. With respect to design, systems that are simple are often significantly easier to understand and verify than complex systems, an ancient principle termed *Occam's razor*. Likewise, systems that are designed for verification are significantly easier to verify than systems for which verification has been left to after the design stage.

An example of this is the well-established *design-for-test* principle, a trade-off that has been made for decades in semiconductor manufacture. Designing for test means observing several design constraints and dedicating part of the silicon area to an infrastructure that facilitates testing after manufacture. This infrastructure, the *scan chain*, allows automated testing of whether the manufacturing process has been successful, i.e. whether all the basic electronic elements and connections work as expected, irrespective of the functionality of the silicon chip. The scan chain is a trade-off, or, from a different point of view, an investment, that makes testing the manufactured chips tractable. It saves a significant amount of time, allows test automation to be used and provides confidence in the manufacturing process. Any remaining issues can thus be attributed to the pre-silicon design stage, rather than manufacturing.

With respect to transparency, systems that allow us an insight into how they make decisions, why they are acting in a certain way or how they use resources become understandable, predictable, verifiable and, on that basis, to some degree worthy of our trust. An example of this is recent research into enabling energy transparency, a new concept that requires information on energy usage of programs to be made available, ideally without running the programs and across the various layers in the system stack from machine code to source code [19].

To achieve this, the energy consumed by a given hardware platform is measured for various workloads and captured in an energy consumption model. Figure 5.2 shows two heat maps that visualize the power dissipation over time for two machine instructions, a multiplication and a logic AND, processing 8-bit data. The data can be used to construct energy consumption models that can be used by static analysis tools to predict the energy consumed by a program at compile time [22, 23, 33].

Alternatively, the models can be used to predict the energy consumption of a program based on its execution trace [22]. Although not static, this method allows software developers to gain an insight into their program's energy consumption without having to explicitly instrument and measure the energy consumed by the program during execution. Providing resource consumption transparency to end users, developers and the tool chain allows them to analyze and understand how resources are used during a computation. This understanding can foster confidence and trust in the system's behavior.

Finally, V&V—rigorous proof complemented by simulation-based testing using intelligent test generation methods as well as experimental validation of selected scenarios in dedicated test environments—can provide convincing evidence of a system's trustworthiness. In fact, it is important to realize that by combining these techniques, the shortfalls of one technique, when applied in isolation, can often be compensated for by other techniques [40].

For example, while formal methods like model checking [14] are exhaustive in their exploration of a system's state space, they suffer from scalability issues. Abstraction is used to overcome this limitation. The models for formal verification are typically representations of the system's behavior at a very high abstraction level, where many implementation details cannot be represented. However, high-

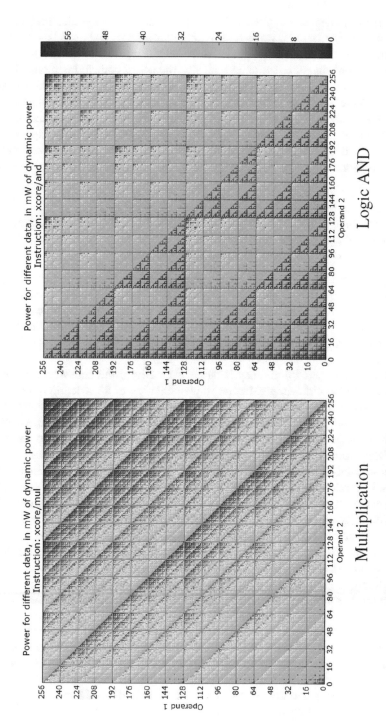

Fig. 5.2 Every calculation performed by a processor consumes energy by dissipating power over time. These 2D heat maps visualize power dissipation when a calculation is performed on two integers, each ranging from 0 to 255, with multiplication shown on the left and the logic AND operation on the right.

level formal properties that capture safety requirements and essential functional requirements can often still be verified using these abstract formal models.

In contrast, the models used for simulation-based testing are often much closer to those in the system's target environment. Ideally, the actual code to be executed on RAS can be tested in simulation, interacting with a simulated physical environment based on realistic stimulus. However, simulation-based verification cannot be performed exhaustively for real-world systems. Instead, the state space can only be sampled and the system's behavior checked against a set of formally expressed properties, termed assertions, to determine whether or not requirements are met. When used in isolation, each approach has clear shortfalls giving rise to questions such as:

- How does the formal model relate to the actual code and RAS?
- When all formal properties have been proven using a high-level model, what is known about the properties of the actual code?
- Which parts of RAS have been exercised during simulation?

To answer these questions and to bridge the gap, the different abstraction levels and verification techniques can be linked in different ways. One approach is to refine the high-level formal properties to lower-level assertion monitors that operate during simulation. This seeks to confirm that properties established on the formal models also hold in simulation. If they are found not to hold, then any discrepancies need to be investigated.

For instance, the actual code may behave differently to the behavior captured in the formal model, indicating either an implementation or a modeling fault. Or, a condition assumed to hold for a formal proof is violated during simulation, indicating a potential misunderstanding of the operational design domain, which can prompt re-evaluation of the formalization and the constraints used to ensure validity of the simulation stimulus.

Another technique that links formal models to simulation-based verification techniques is model-based testing [39]. Formal models of the system and its environment are systematically traversed, e.g. using search-based techniques guided by heuristics or formal methods such as model checking, in order to generate traces that serve as abstract templates for stimulus generation. Full coverage of the formal model, measured, for example, in the number of paths traversed, states visited or transitions explored, can be used as a measure of extensiveness in this context. The set of test templates is then refined and parameters are instantiated, potentially by pseudo-randomly sampling from a given range, to obtain concrete test instances at the abstraction level of the simulation.

By exploiting the formal model for test generation, the functional behavior of the model is being reflected in the test stimulus and thereby transferred into the simulation. Again, any discrepancies in behavior require investigation and potentially corrections of the formal model, the simulation environment, the code under test or the test generation, especially the step that refines the high-level test template to concrete stimulus.

This concludes our general overview of techniques to gain confidence in the trustworthiness of a system. In the next section our focus is on correctness from specification to implementation, with special emphasis on how formal methods can be integrated into the system development process.

3   Ensuring Correctness from Specification to Implementation

Conventional system design starts from user requirements and high-level specifications. These are then refined in a series of transformations into lower-level designs that can be analyzed and optimized before being implemented in full detail in software, hardware or both. The need for verification arises from the lack of confidence in the transformative processes that are used at each refinement step from specification down to implementation.

Reconvergence models [8] illustrate the role of verification in confirming that the result of such a transformation preserves the intent. A transformation such as coding may start from a specification and produce source code. Verification is used to establish that the source code meets the specification. A wide variety of transformations are used in practice: some are manual, such as coding or modeling; others are automated by tools, such as compilation, where executable code is produced from high-level language source code, or high-level synthesis, where implementable hardware is automatically generated from high-level language descriptions of algorithms. Unless it can be shown that these tools work correctly, a process referred to as tool qualification, it is generally advisable to check that the output of the transformation is preserving the intent of the input, irrespective of whether the transformation is manual or automatic.

Figure 5.3 shows a simple two-stage reconvergence model, comprising two transformations, the first from user requirements to design and the second from design to implementation. Identification of the start and end points of each transformation, i.e. its input and output, is critically important for verification, which is the process used to gain confidence in the correctness of the output of the transformation with respect to the input. The transformation can of course also be transitive, motivating verification of whether the implementation satisfies the specification.

In the following two sections the discussion will first focus on the implementation level, investigating

What can be done to ensure implementations are free from runtime errors?

and second, on the design level, asking

What can be done to identify design flaws before coding?

In answering these questions, we seek to promote verifiability to a first-class system design objective. In addition, we show how to combine test-based with

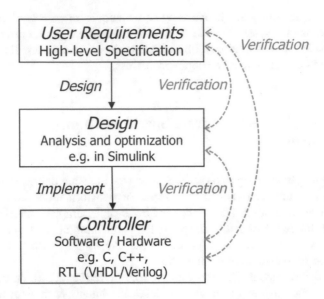

Fig. 5.3 Reconvergence model illustrating the stepwise transformative process from user requirements via a Simulink design to an implementation on the example of a controller. The start and end points of each transformation are critically important for verification.

formal techniques using examples from two domains, robot navigation code and Simulink controller designs.

3.1 What Can Be Done at the Code Level?

Mobile robotics relies heavily on navigation algorithms. While much research has been invested into proving robotic navigation algorithms correct, little has been done to demonstrate that the implementations are free from runtime errors. It is easy to find open-source robot navigation code online, yet convincing correctness arguments or ready-to-use test environments are not typically made available alongside the implementations, many of which are written in C or C++, languages with features that are considered unsafe. Verification is often not the top priority for robotic software developers. The situation is worsened by the perceived high cost of manual verification and the lack of tool support.

How much the choice of programming language limits the verifiability of robotic code, and whether the limitations can be overcome by a different choice, has been studied in [38]. The research was focused on eliminating runtime errors in robot navigation code, including access to arrays and vectors beyond their bounds; runtime exceptions; arithmetic errors including division by zero, overflows and underflows for integers and floating point numbers; errors resulting from calling mathematical functions with arguments outside their domains; null pointer dereferencing; dynamic memory management issues; blocking inter-thread communication; accessing uninitialized variables; and operations that are not specified as part of the programming language definition. Considering the vast space that testing would need to cover, the confidence that can be obtained from the use of formal methods in this context is stronger than that obtainable only from testing. Thus, the focus in [38] was on the feasibility of formal proof complemented by testing and the availability of tools to automate this process.

The SPARK language was selected in [38] for its comprehensive verification support, combining formal proof with testing. SPARK [6] is an imperative programming language with strong typing. It offers a verifiable subset of the Ada language and is designed for high-integrity, safety-critical application domains. In addition, the SPARK language specification and associated tools are freely available for the development of free and academic software.

Three popular open-source robot navigation implementations were selected and encoded in SPARK; these were implementations of the VFH+ (Vector Field Histogram) [10], ND (Nearness Diagram) navigation [34] and SND (Smooth Nearness Diagram) navigation [17] algorithms. Rather than using more recent code, the research intentionally focused on finding potential weaknesses in code that has been available for a number of years and could therefore be considered stable. An initial attempt to verify these implementations directly using state-of-the-art tools for C/C++ was abandoned due to insufficient tool support for the formal proofs and the substantial effort that would have been required to annotate and transform the source code for it to become suitable for automated verification.

Instead, the workflow included manually rewriting the original C/C++ source code of the respective navigation algorithm into functionally equivalent SPARK code. As part of this process, explicit annotations were introduced to declare pre- and post-conditions of subprograms and functions; these capture what constitutes valid input and the conditions the output should satisfy, respectively. In addition, loop invariants and variants were added to the SPARK code. Loop invariants express properties that remain stable for each iteration of the loop, i.e. they hold before and immediately after each loop iteration. In contrast, loop variants are expressions that can be shown to monotonically decrease in value at each loop iteration, which permits reasoning about the termination of a loop. Encoding in SPARK also required the definition of numeric subtypes to capture valid data ranges, formal data containers as well as contracts to enable formal verification to be performed for each subprogram independently of other subprograms.

Thus, coding in the SPARK subset of Ada naturally enabled verifiability of the code and promoted verification to a primary software development objective. The resulting source code was on average 30% longer than the original encoding of the algorithms in C/C++, with C/C++ accounting for 2.7 kSLOC (1000 Source Lines of Code) and SPARK for 3.5 kSLOC in total. This increase in the number of source code lines can be attributed to Ada being more verbose than C/C++, but also accounts for verifiability as the extra annotations enable verification.

After translation of the source code to SPARK, the workflow moved to veri-fication, focusing first on runtime verification. This involved compiling the Ada code with runtime checks enabled. Validation of the executable was performed using the navigation scenarios available from the simulation environment of the original code, allowing a comparison between the SPARK and the original C/C++ implementations both in terms of functionality and performance. The runtime checks injected by the Ada compiler exposed several runtime errors, such as calling a mathematical function with arguments outside the function's domain. It was also discovered that for VFH+, an array was accessed out of bounds at algorithm initialization. While this does not necessarily lead to a runtime error, it is a vulnerability that could be exploited to trigger undesirable behavior as part of a potential cyberattack.

The second stage of verification, formal proof, is supported by the static analysis tool GNATprove [26]. After statically checking that the Ada source code satisfies the SPARK language restrictions, verification conditions are created in WhyML to catch potential runtime errors, exceptions or assertion violations. WhyML is the specification and programming language of the deductive program verification platform Why3 [21]. The proof obligations arising from the verification conditions can then be discharged with the help of a variety of automatic provers or, if necessary, interactive proof assistants available via the Why3 platform. The formal proof establishes, whether or not a given verification condition is satisfied for all possible states of the program variables. In practice, due to the high level of expertise required to guide interactive proof assistants, fully automatic proofs are a much better fit to the skills of most software developers.

Regarding the three SPARK versions of the navigation algorithms, well over 90% of the proof obligations could be proved automatically, with several hundreds discharged within 6–48 min by employing the SMT solvers Alt-Ergo [15] and Z3 [16], both available with Why3. The remaining few proof obligations are either straightforward to examine, and can be discharged manually, or require more sophisticated interactive proof techniques such as inductive proof, which would benefit from automatic discovery of type invariants. Beyond runtime safety, for two algorithms, VFH+ and SND, termination could be confirmed by demonstrating that the loop variants hold for all the while loops in the code.

Translating to SPARK has clearly increased the verifiability of the code, but at what cost? Our performance assessment compared the average runtimes of the C/C++ code with the SPARK code with and without runtime checks enabled. In the best case, when runtime checks are disabled, the SPARK code runs as fast as the original code (ND algorithm), with the worst case incurring a performance penalty of 2.7 times (SND algorithm). Enabling runtime checks comes with a performance drop of between 3.8 and 4.6 times. It would be for the software developers and roboticists to decide whether this is a price worth paying.

In conclusion, the example of re-coding three robot navigation algorithm imple-mentations in SPARK has shown that, by coding for verification, runtime errors can be identified and eliminated almost fully automatically. The runtime checks injected automatically by the Ada compiler attract a performance penalty when left in the production code. This may be a cost worth paying or even a regulatory assurance requirement where robots operate in safety-critical application areas. However, without these runtime checks, performance comparable to that of the original buggy implementations can be achieved. Thus, the choice of programming language can significantly increase the verifiability of source code as well as enable access to powerful verification techniques and tools, ultimately leading to provable runtime safety.

3.2   What Can Be Done at the Design Level?

When a controller is not operating as expected, this could be due to an implementation fault, manifesting itself as a runtime error as discussed in the previous section, or perhaps due to a design flaw. The ability to distinguish a design flaw from an implementation fault is important in practice. For control software, a patch can possibly be provided to quickly fix a fault in the implementation. Correcting a design flaw, however, is likely to require a far more costly re-design and, if the application is safety critical, potentially even a complete product recall.

In the case of control systems, designs are often written in Simulink before they are implemented in a programming language or in hardware. The design process starts with the identification of high-level requirements on the behavior of the control system. Typically, one of these requirements is stability of the controller. In practice, control engineers often translate the requirements directly into Simulink designs for the target application in order to perform numerical simulation, without capturing the control laws used and assumptions made in the process.

It can be particularly challenging to verify control systems because many of the signals and parameters are from the domain of real numbers, giving rise to undecidability, thus rendering verification techniques such as model checking [14] unusable in this context. On the other hand, and as already discussed in the previous section, theorem proving is a deductive verification method that allows symbolic reasoning over entire domains and state spaces. How can this powerful technique be made more accessible to control engineers?

In [2], an approach is presented to verify the stability of Simulink controller designs using assertion checks and theorem proving in combination. This approach was inspired by assertion-based verification (ABV) methods that have successfully been used in hardware design verification since the early 2000s [7].

ABV was first enabled by a library, the Open Verification Library (OVL), which has been an Accellera standard[1] since 2007. When OVL was introduced, it offered hardware designers and verification engineers access to more than 40 assertion monitors, expressed in widely used Hardware Description Languages (HDL). The fact that engineers did not need to learn a new language to specify formal properties

---

[1] https://www.accellera.org/activities/working-groups/ovl.

made OVL easy to use and very popular. This significantly contributed to the uptake of ABV. The current version of OVL also supports dedicated property specification languages such as SystemVerilog Assertions (SVA) [11] and the Property Specification Language (PSL) [20], paving the way to the wide use of model checking in hardware design verification as SVA and PSL properties can be directly read into HDL model checking tools. Thus, once formalized, verification can be performed using simulation-based as well as formal methods. Can a similar approach be developed for the design of control systems?

The properties to be verified first need to be identified and formalized. How this can be done has been shown in [2] on the example of the discrete, linear, time-invariant system given by the state-space model in Eq. 5.1, where the system parameters $\mathbf{A}$ and $\mathbf{B}$ are a constant matrix and a constant vector, respectively, $\mathbf{x}$ is the system state, $\mathbf{u}$ are the inputs and k a discrete time variable.

$$\mathbf{x}(k+1) = \mathbf{A}\mathbf{x}(k) + \mathbf{B}\mathbf{u}(k) \tag{5.1}$$

The control requirement is stability. To meet this requirement, it is proposed to replace $\mathbf{u}(k)$ in Eq. 5.1 with the feedback controller $-\mathbf{K}\mathbf{x}(k)$, where $\mathbf{K}$ is a constant vector. This substitution yields the following results:

$$\mathbf{x}(k+1) = \mathbf{A}\mathbf{x}(k) + \mathbf{B}(-\mathbf{K}\mathbf{x}(k))$$
$$= (\mathbf{A} - \mathbf{B}\mathbf{K})\mathbf{x}(k) \tag{5.2}$$

The high-level stability requirement can be broken down into a set of sub-requirements by using Lyapunov's second method. Accordingly, a Lyapunov function, $V(\mathbf{x}(k))$, $V : \mathbb{R}^n \to \mathbb{R}$, needs to be found such that the sub-requirements L1, L2 and L3 listed below are satisfied.

(L1) $V(\mathbf{x}(k)) = 0$ if and only if $\mathbf{x}(k) = 0$.
(L2) $V(\mathbf{x}(k)) > 0$ for all $\mathbf{x}(k) \neq 0$.
(L3) $V(\mathbf{x}(k))$ must be strictly decreasing for all $\mathbf{x}(k) \neq 0$.

In general, the existence of a Lyapunov function implies stability. A candidate Lyapunov function as shown in Eq. 5.3, where the parameter $\mathbf{P}$ is a constant matrix,

$$V(\mathbf{x}(k)) = \mathbf{x}(k)^{\mathrm{T}}\mathbf{P}\mathbf{x}(k) \tag{5.3}$$

can be obtained by solving the discrete Lyapunov equation given in Eq. 5.4.

$$(\mathbf{A} - \mathbf{B}\mathbf{K})^{\mathrm{T}}\mathbf{P}(\mathbf{A} - \mathbf{B}\mathbf{K}) - \mathbf{P} = -\mathbf{I} \tag{5.4}$$

It now needs to be shown that the candidate Lyapunov function in Eq. 5.3 satisfies the sub-requirements L1, L2 and L3.

- L1 is a straightforward numeric check.

- L2 can be further refined to the requirement that the constant matrix $\mathbf{P}$ in Eq. 5.3 is positive definite. This gives rise to requirement R1.1.1 in [2].
- L3 can be formalized as $V(\mathbf{x}(k)) < V(\mathbf{x}(k-1))$ for all $\mathbf{x}(k) \neq 0$, which can be decomposed again into two sub-requirements:

  - the requirement that specifies the decrease of $V(\mathbf{x}(k))$ to be

  $$V(\mathbf{x}(k+1)) - V(\mathbf{x}(k)) = \mathbf{x}(k)^{\mathrm{T}}[(\mathbf{A} - \mathbf{BK})^{\mathrm{T}}\mathbf{P}(\mathbf{A} - \mathbf{BK}) - \mathbf{P}]\mathbf{x}(k) \qquad (5.5)$$

  for all $\mathbf{x}(k) \neq 0$, derived from Eqs. 5.2 and 5.3, and giving rise to requirement R1.2.1 in [2], and,
  - to ensure this decrease is strict, $V(\mathbf{x}(k)) - V(\mathbf{x}(k-1)) < 0$ for all $\mathbf{x}(k) \neq 0$, the requirement that the matrix shown in Eq. 5.6 is positive definite,

  $$\mathbf{P} - (\mathbf{A} - \mathbf{BK})^{\mathrm{T}}\mathbf{P}(\mathbf{A} - \mathbf{BK}) \qquad (5.6)$$

  giving rise to requirement R1.2.2 in [2].

The final sub-requirements can then expressed as assertions in Simulink, using `require` blocks following the technique introduced in [1]. Each `require` block contains the built-in `assert` block and specifies pre- and post-conditions as part of the Simulink model. Formalizing such assertions in the same language as the controller design makes them an inherent part of the design and a first-class objective of the design process. Thus, there is no need for control engineers to learn a new language in order to capture the fundamental control laws and assumptions their Simulink designs are expected to satisfy.

Figure 5.4 shows the original control system (top left) together with the added assertions in the dashed box. A sizeable part of the Simulink design is now occupied by assertions. However, it is beneficial that the design intent and the relevant control laws have been formally captured and that the resulting assertions are now part of the design. Thus, in the first instance, the new assertions can be checked via numerical simulation in Simulink; this fits well into the traditional Simulink design flow.

Linking back to ABV in hardware design, this stage is very similar to how OVL assertion monitors were first used. Historically, the next step toward gaining higher confidence in the hardware design community has been to formally prove assertions.

While confidence gained from numerical simulation may be sufficient for some areas, safety-critical applications, for example, require stronger verification results. We observe that the assertions derived from the high-level stability requirement can be divided into two groups. First, there are those that specify properties such as the positive definiteness of matrices: R1.1.1 and R1.2.2. Both matrices are ground, i.e. they do not contain any variables. These assertions can therefore be checked using numerical simulation.

Requirement R1.2.1, however, specifies that $V(\mathbf{x}(k))$ must decrease according to Eq. 5.5 for all $\mathbf{x}(k) \neq 0$. Satisfaction of such a requirement necessitates demonstrating that it holds for all values $\mathbf{x}(k) \neq 0$. Numerical simulation, by its very nature, can only establish this for a small subset of values and may be subject

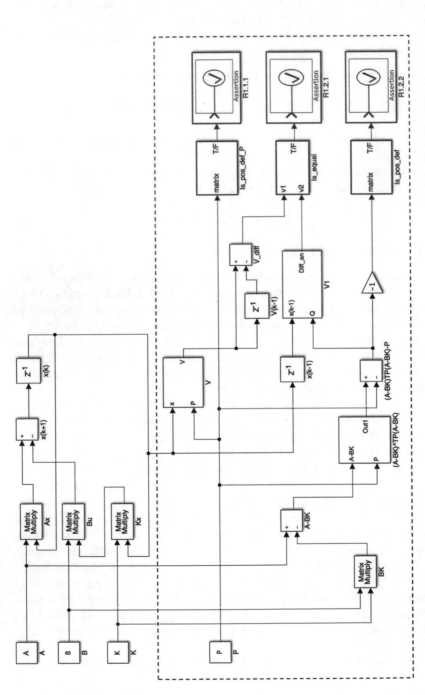

Fig. 5.4 Simulink design with assertions to capture the fundamental control laws that the system design is expected to satisfy. The dashed box surrounds the part of the Simulink design that is occupied by the assertions. The remaining part on the top left is the original Simulink controller design.

to floating point errors. Formal proof tools, on the other hand, perform a symbolic analysis. Thus, the proof, if established, holds for all values.

Thus, to gain the benefit of formal proof, the signals and blocks in the Simulink design and also the assertions were translated into higher-order logic predicates, targeting the input logic language of the deductive program verification platform Why3 [21]. The details of how this translation has been automated using MATLAB are described in [1]. The translation relies on the provision of a library of theories in Why3 to specify axioms that formalize the input-output behavior of the Simulink blocks included in the design. Establishing such theories has already been challenging for the simple controller verified in [2]. To verify more complex Simulink designs using this approach, the development of suitable theories would first be required.

Once formalized, theorem proving can be performed in Why3, which provides interfaces to the input languages of a variety of SMT solvers. Using the assertions as proof goals for formal verification, the SMT solvers Z3 and CVC4 automatically confirmed the validity of requirement R1.2.1 for a correct version of the controller, but failed to find proofs for some other versions into which errors had been injected that invalidate this requirement. Full automation of these proofs is a necessity for widespread use of this method in practice.

In conclusion, to find design flaws early, the fundamental control laws and assumptions made during the design of a controller can be systematically refined into a set of sub-requirements that specify the design intent. These sub-requirements can then be formalized and encoded in the form of assertions that become an inherent part of the Simulink design. This allows verification either by traditional numerical simulation or via fully automatic theorem proving using the power of SMT solvers, provided that suitable translation tools and theories are available. Code generation should then only be performed once sufficient confidence has been gained that the Simulink design satisfies its high-level requirements. As with the runtime checks that were automatically injected by the Ada compiler into the SPARK code in the previous section, control engineers can decide to remove the Simulink assertions before code generation to obtain code performance that is not affected by runtime assertion checks.

4   What Can Be Done to Increase the Productivity of Simulation-Based Testing?

Although the previous two sections have strongly promoted formal methods, once systems reach a certain level of complexity, rigorous proof is typically not tractable. In hardware design verification, for example, automatic formal apps are now routinely used, e.g. to verify design properties at the unit level, to check connectivity between sub-systems and at the chip level, to determine whether coverage is reachable, etc. However, simulation-based testing is still considered the major workhorse in this domain.

It has been recognized that simulation will also be a major tool for V&V of RAS [13], where the complexity arising from interacting hardware and software, high degrees of concurrency and communication within the system and with its environment, including with people, is posing particularly difficult challenges. In addition, experimental testing is typically expensive and unsafe. It also does not scale well—recent research has estimated that 275 million miles, equivalent to 12.5 years of driving assuming a fleet of 100 vehicles, would be required to achieve an acceptable level of confidence in the safety of an autonomous vehicle [29].

When faced with challenges of this nature, simulation-based testing is often the only viable approach. However, far more automation is needed to ensure that simulation is efficient and effective. Coverage-driven verification is a methodology that achieves a very high level of automation; it is introduced in the next section together with the architecture of testbenches that support a coverage-driven verification methodology. Section 4.2 then proposes the use of agency to address the test generation challenge for simulation-based verification of RAS.

4.1 Increasing the Degree of Automation in Simulation-Based Testing of RAS

Sophisticated simulation environments, termed testbenches, can be built to gain insight into the behavior of very complex systems at different levels of abstraction. A testbench is a completely closed system that models "the universe" from the perspective of the system under test (SUT). Our research has shown how coverage-driven verification (CDV), a simulation-based testing technique that is well established in hardware design verification, can be used to verify code that controls robots in application areas where robots directly interact with humans [3].

The components of a CDV testbench support a fully automatic workflow, including test generation, driving stimulus into the SUT, checking and coverage collection. Figure 5.5 introduces the architecture of a CDV testbench for code that controls a robotic assistant working with a human co-worker as part of a human-robot collaborative manufacturing task as shown in Fig. 5.6. Further details on the components of a CDV testbench will be provided later on in this section.

Our research concentrated on the most critical part in this human-robot interaction, the handover. The robot is tasked to hand work pieces to a human co-worker, needing to decide when it is safe to let go based on the human's gaze, the pressure sensed on the robot's hand and the location of the human's hand [24]. The success of each handover interaction critically depends on the robot making the right decision in a timely manner.

The test generator in a CDV testbench produces tests, i.e. test input data and potentially also execution conditions, that, ideally, are effective at reaching verification targets. These tests are then driven into the SUT by a dedicated driver component in the testbench. The driver provides test inputs to the SUT, observing

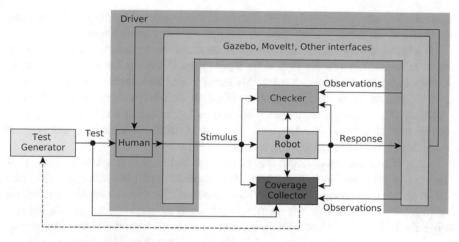

Fig. 5.5 The architecture of a testbench that enables a coverage-driven verification methodology, first introduced in [3] for simulation-based testing of code that controls a robotic assistant as part of a human-robot collaborative interaction scenario.

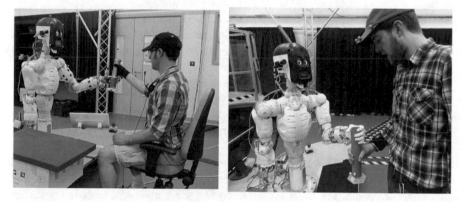

Fig. 5.6 Scenes from a human-robot collaborative manufacturing task where the robot hands over work pieces to a human co-worker.

any existing protocols on interfaces. It may issue test stimuli in interaction with the SUT, controlling and monitoring the execution of the test and reacting to the SUT's response as part of a test sequence. Drivers also bridge the abstraction gap between high-level test input and the low-level physical stimulus the robot senses as well as the changes observable as a consequence of the robot's actions in response to the test stimulus. In Fig. 5.5 the interaction between a simulated human co-worker and the robotic code is strictly through a physical layer enabled by the physics engines in Gazebo[2] and MoveIt![3] motion planning as well as inverse kinematics.

For simulations to be effective, manual test creation or pseudo-random sampling from what is in practice a very large input space is clearly not sufficient. Yet, in mainstream industrial development, this seems to still be common practice. Instead, more guidance is needed to reach verification objectives faster and with more certainty. Note that, to ensure repeatability, pseudo-random sampling should be used in this context instead of random sampling, so that simulations can be repeated when a test is generated using the same random seed(s).

Constrained pseudo-random test generation and search-based techniques [25] have been proposed to achieve this. Furthermore, model-based testing [39] can be employed to target specific scenarios, especially when complex interaction sequences between the environment and the robot are required to achieve the test objectives. A directed test usually comprises a set of predefined test inputs that have been designed to exercise specific functionality in the design together with the expected design behavior specifically for this test. The automation of test generation, however, makes it necessary to introduce automatic checkers that are independent of the test inputs and to collect coverage during simulation.

Checkers determine whether a test passed or failed. The expected behavior is formalized and expressed as assertions that continuously monitor the SUT [3]. Checking can either be performed during simulation or after, based on data collected throughout a simulation run. Checkers are always active, irrespective of the test, which may have been generated using some form of random sampling.

To understand the quality and diversity of the tests generated and how testing is progressing over time, it is important to collect and analyze coverage of different types. It is generally considered good practice to use different coverage models in combination [36].

The most common coverage models are code coverage, for example, statement or branch coverage; structural coverage, such as finite state machine coverage; and functional coverage. The latter includes requirements coverage, assertion coverage and cross-product functional coverage [30]. Assertion coverage measures how often assertions have been activated during simulation. Cross-product functional coverage is focused on specific signals identified as important in a verification plan. Combinations of these signals are systematically enumerated based on the Cartesian product of their value domains. Each tuple in this Cartesian product is termed a coverage task. Cross-product functional coverage measures how often each of these coverage tasks occurs during simulation. More specifically, a cross-product functional coverage model is composed of four parts [30]:

- A semantic description of the model, often referred to as the *story*, to capture the coverage model's intent as given in the verification plan

---

[2]http://gazebosim.org/.

[3]https://moveit.ros.org/.

- A list of attributes mentioned in the story—these typically map to signals in the SUT
- The set of possible values for each attribute, i.e. the attribute value domains—these typically correspond to the types of the selected signals in the SUT
- A list of restrictions on the legal combinations in the Cartesian product of the attribute value domains

The detailed coverage data and analysis presented in [5] illustrate clearly how different test generation techniques complement each other; full coverage was achieved most efficiently when constrained pseudo-random and model-based techniques were used in combination.

CDV is a systematic, goal-directed, simulation-based testing methodology that uses coverage strategically to direct the verification process to bias test generation toward coverage closure, e.g. by tightening constraints or using model-based techniques. CDV offers a high level of automation and is capable of exploring systems of realistic detail under a broad range of environmental conditions.

Once the verification environment has been set up, engineers can focus on test generation and coverage. Constraining test generation toward achieving coverage targets requires significant engineering skill and SUT knowledge. While model-based test generation allows targeting outstanding coverage more effectively than constrained pseudo-random test generation, a suitable model needs to be available. Unfortunately, this is unlikely unless such models are an inherent part of the system design process, as is the case for model-based system engineering [27]. Thus, test generation remains a critical and time-consuming part of simulation-based testing.

The high degree of automation enabled by CDV does not only increase verification productivity. Large amounts of data are created in the process, mostly *labelled* data such as coverage data over time as well as data collected by assertion monitors. This provides a basis for exploiting artificial intelligence (AI), especially machine learning (ML), to further automate test generation toward coverage-directed test generation (CDTG).

Automating the feedback loop from coverage to test generation, indicated by a dashed line in Fig. 5.5, by embedding ML into the CDTG process, has been the focus of research for more than a decade [28]. Researchers have explored the benefits of a broad variety of ML techniques, including evolutionary algorithms, Bayesian networks and Markov models as well as data mining and inductive logic programming. Yet, none of these approaches have to date matured for them to be found in professional test generation tools. Hence, the test generation challenge remains.

4.2 Introducing Agency into Test Environments for Simulation-Based Verification of RAS

Testing autonomous systems faces an additional challenge that goes well beyond the conventional test generation challenge. Autonomous systems are designed to adapt and change over time; their functionality evolves during operation. They are expected to interact intelligently with open environments, to be responsive, proactive, cooperative and, potentially, even social. In short, they exhibit *agency*. This agency, the freedom these systems have to react differently to what may appear externally to be the same situation, means that the same stimulus may trigger different system behavior over time. Thus, testing can no longer rely on a predefined response in an interaction sequence with the SUT, as would be the case when two interacting components use a standardized communication protocol.

Moreover, enumerating all possible response options is not practical. Thus, the execution of tests that rely on a specific response from the SUT, e.g. as part of an interaction sequence, may get stuck waiting for this response while the SUT has already reacted in a different, perfectly legitimate way. In verification terminology, the SUT can be considered a *responder*, i.e. a component that *responds* to the stimulus it senses from its environment, i.e. from the *agents* it interacts with, whether these are humans; other systems, including robots; or simply traffic lights and signs.

Thus, agency in the test environment is required so that a test can adapt its reaction to the observed behavior of the SUT in an interaction sequence while steering the interaction toward reaching verification objectives. How can agency be introduced into a test environment?

In [4] a first study investigated the use of BDI agent models [9] for test generation, whether ML can automate the agent-based test generation process and how BDI agent models compare to conventional automata-based techniques for model-based test generation. The behavior of BDI agents is driven by their beliefs, i.e. what the agent knows about the world; desires, i.e. the goals the agent aims to achieve; and intentions, i.e. the desires the agent has committed to perform and, where applicable, the means to achieve them, e.g. via specific plans or actions depending on the situation. The execution of an action can change the agent's beliefs and enable further actions toward eventually achieving the agent's overall goals.

Research in [4] focused on the human-robot collaborative manufacturing scenario introduced in the previous section and depicted in Fig. 5.6. For agent-based test generation, a multi-agent system was set up containing BDI agents for the human co-worker; the sensors involved, including gaze; pressure and location; and the robot. Interactions between agents are controlled by beliefs: 38 beliefs in this use case, giving a total of 2^{38} belief subsets. The identification of belief subsets that enable the generation of a good variety of meaningful interaction sequences is challenging due to the vast number of options.

To address this challenge, a verification agent is introduced in the first instance. The verification agent is a meta-agent with full visibility of the state of the multi-agent system. This agent can influence the beliefs and thereby the actions of individual agents, thus biasing or directing the interactions between agents toward verification targets. Options to select belief subsets for test generation include manual selection, pseudo-random selection and learning an optimal belief set using ML, e.g. reinforcement learning.

For evaluation, belief subsets were obtained using each of these techniques. The belief subsets were subsequently deployed for agency-directed test generation, where the actions of the verification environment, i.e. the test stimuli, were directly derived from the interactions of the multi-agent system. The results, measured in terms of coverage of the code that controls the robot, show that manual selection is most effective, leading to a fast and steep increase in coverage with a relatively low number of tests. Pseudo-random selection is also effective, but does not achieve the cumulative coverage achieved by manual selection. The tests generated from learning reach the highest coverage of around 90%, using the largest number of tests to achieve this.

As a baseline, conventional pseudo-random test generation never reached more than 66% coverage, while all agent-based testing achieved coverage over 85%. When evaluating performance, it is important to take into account the cost of belief-set selection. While manual selection requires significant domain knowledge and engineering effort, reinforcement learning achieved convergence within 300 iterations, taking less than 3 h. Seeing that the learning-based method reached the highest coverage fully automatically, it clearly outperforms the other techniques in terms of efficiency and effectiveness for the given example.

In summary, generating tests from multi-agent systems is both efficient and effective. Furthermore, ML can automate the selection of belief sets so that agent-based testing can be biased toward maximizing coverage. Compared to conventional model-based test generation, multi-agent systems are more intuitive to set up as the agency in the test environment can naturally be expressed in this paradigm, smaller in size as well as more predictable to explore, and equal if not better in terms of achieving verification goals.

More recently, agency-directed test generation was applied to create tests that activate assertions in simulation-based verification of autonomous vehicles [12]. In this context, the agents represent pedestrians crossing a road in a way that allowed the autonomous vehicle sufficient reaction time to avoid a collision. Thus, an agent's perceptions of the changes within the test environment, i.e. the autonomous vehicle approaching, enabled the agent to act in a natural manner and at the right time for the assertion to be activated. Testing assessed whether the autonomous vehicle indeed managed to avoid a collision, e.g. by braking or manoeuvring. Agency-directed test generation was found to be efficient, robust and twice as effective in achieving the verification goal than pseudo-randomly generated tests.

Using BDI agents for test generation constitutes a new model-based testing technique, where the model is given by the multi-agent system and agency is being used for model traversal [39]. Comparing the agency-directed test generation to conventional model-based test generation using model checking timed automata, considering coverage, the vast majority of agency-directed tests achieve higher coverage, with maximum coverage often achieved by agent-based tests. Furthermore, the agent-based approach has several advantages, including significantly smaller models when measured by source lines of code, and a stable model exploration time, while model checking times were found to differ by up to three orders of magnitude.

5 Conclusions

On the example of robot navigation code and control system designs, we have seen how verification techniques can be combined to gain stronger confidence in the correctness of the implementations and the designs of RAS, respectively. In both cases, automatic theorem proving was enabled by the latest advances in SMT solver technology, making deductive formal verification usable by software developers without specific training in formal verification. The key insight from both these examples is that verifiability can be achieved by considering verification early, i.e. as an inherent part of the system design and implementation process. The work presented in this chapter directly complements Chaps. 8 and 11 in this book.

It is perhaps remarkable to observe that AI and ML are now commonly used to make RAS smarter in various different ways, yet testing these smart systems is still mainly a manual process in industry. This calls for change!

A test environment needs to be at least as smart as, ideally a lot smarter than, the SUT. No matter how smart the SUT, the testbench needs to reflect the agency and intelligence the SUT can reasonably be expected to meet in its target environment, which in many cases will include people. Agent-based testing exploits the agency that is a natural part of multi-agent systems to generate tests that interact with and react to the observable behavior of the SUT while targeting verification goals as part of a coverage-driven verification methodology. Agency-directed test generation is a first step toward the development of novel AI-based test automation techniques for simulation-based verification of RAS.

6 Challenges for the Verification and Validation of RAS

Many challenges remain; the most urgent is the specification challenge. Traditionally, system development and verification rely on a specification that fully defines the functional and increasingly also the non-functional behavior of the system to be designed and verified. However, RAS are developed so that they can adapt to the circumstances they encounter in their environment. They are expected to sense, understand, decide how to respond and act within their environment, without this being prescribed in full detail.

It is by intent that the RAS specification and design process does not explicitly enlist a specific response for every possible scenario RAS may face; there usually are far too many possible scenarios for this to be practical. Instead, we need flexible specifications [35] expressed in terms of acceptable and required behavior with associated precise limits for critical properties complemented by vague and imprecise indications of desired actions permitting flexibility when assessing the degree of requirement satisfaction.

For ML-based systems, whether end-to-end learning or ML-based components that are part of a conventional system, more innovative solutions need to be found. The development of such systems departs from the traditional refinement and composition-based system development methods. Instead, the development process can be described as data-driven software synthesis. An explicit specification in the traditional sense is not part of this process, as clearly identified in [37]. Instead, the specification is implicit in the data set, and more research is required to understand what properties these data sets must have as well as how to verify them.

Automation is a challenge that is currently being addressed. Significant advances have been made for more than a decade in SMT solver technology, offering the potential to revolutionize problem solving and mathematical reasoning in complex domains through an unprecedented degree of automation. Now is the time to exploit these innovations to further automate those parts of the verification process that require specialist skills, excessive resources or high engineering effort.

A big obstacle to achieving more automation is potentially the huge variety of different development approaches, modeling paradigms, programming languages and environments available to the RAS community. In contrast, the choice is far more limited in hardware design, where the entire design process is constrained in order to ensure that the designed system can be manufactured with confidence in its functional correctness.

It is this *freedom from choice* [31], the discipline exercised by engineers to meet the set constraints, that has made it possible to automate much of the hardware design process. The Electronic Design Automation industry, a $6.3 billion sector, provides sophisticated tools and methodologies that support both design and verification. As a result, hardware designers and verification engineers have tools and techniques to manage the complexity inherent in billion transistor designs.

For the RAS community to enjoy the same benefits, more design discipline will be required. The bottom-up design and verification approach described in Chap. 8 and the RoboStar technology introduced in Chap. 9 offer promising approaches in this direction. Likewise, Chap. 3 promotes the benefits of more discipline in Software Engineering for Robotics.

Finally, more innovation and creativity are needed. Novel methodologies that boost V&V may employ innovative combinations of existing techniques or inject paradigms from other fields that can be harnessed for verification. For instance, how the power of AI can be utilized in V&V is an area that is now attracting more attention of the research community. Chapter 4 covers very recent work on using AI and constraint programming to automate the testing process.

Acknowledgments While conducting this research, the author was in part supported by the following funded research projects: EPSRC RIVERAS "Robust Integrated Verification of Autonomous Systems" (EP/J01205X/1), EPSRC ROBOSAFE "Trustworthy Autonomous

Assistants" (EP/K006320/1), ROBOPILOT and CAPRI, both part-funded by the Centre for Connected and Autonomous Vehicles (CCAV), delivered in partnership with Innovate UK under grant numbers 103703 (CAPRI) and 103288 (ROBOPILOT), respectively. Finally, the author would like to thank Greg Chance, Abanoub Ghobrial, Kevin McAreavey, Mohamed Salem and Anas Shrinah as well as the anonymous reviewers for suggestions, comments and corrections of earlier versions of this chapter.

References

1. D. Araiza-Illan, K. Eder, A. Richards, Formal verification of control systems' properties with theorem proving, in *2014 UKACC International Conference on Control (CONTROL)* (2014), pp. 244–249. https://doi.org/10.1109/CONTROL.2014.6915147
2. D. Araiza-Illan, K. Eder, A. Richards, Verification of control systems implemented in Simulink with assertion checks and theorem proving: a case study, in *2015 European Control Conference (ECC)*, pp. 2670–2675 (2015). https://doi.org/10.1109/ECC.2015.7330941
3. D. Araiza-Illan, D. Western, A. Pipe, K. Eder, Coverage-driven verification—an approach to verify code for robots that directly interact with humans, in *Hardware and Software: Verification and Testing*, ed. by N. Piterman (Springer International Publishing, Cham, 2015), pp. 69–84. https://doi.org/10.1007/978-3-319-26287-1_5
4. D. Araiza-Illan, A.G. Pipe, K. Eder, Intelligent agent-based stimulation for testing robotic software in human-robot interactions, in *Proceedings of the 3rd Workshop on Model-Driven Robot Software Engineering, MORSE'16* (Association for Computing Machinery, New York, 2016), pp. 9–16. https://doi.org/10.1145/3022099.3022101
5. D. Araiza-Illan, D. Western, A.G. Pipe, K. Eder, Systematic and realistic testing in simulation of control code for robots in collaborative human-robot interactions, in *Towards Autonomous Robotic Systems*, ed. by L. Alboul, D. Damian, J.M. Aitken (Springer International Publishing, Cham, 2016), pp. 20–32. https://doi.org/10.1007/978-3-319-40379-3_3
6. J. Barnes, SPARK: The Proven Approach to High Integrity Software. Altran Praxis (2012)
7. L. Bening, H. Foster, *Principles of Verifiable RTL Design*, 2 edn. (Springer, Berlin, 2001)
8. J. Bergeron, *Writing Testbenches: Functional Verification of HDL Models*, 2 edn. (Springer, Berlin, 2003)
9. R. Bordini, J. Hubner, M. Wooldridge, *Programming Multi-Agent Systems in AgentSpeak using Jason* (Wiley, Chichester, 2007)
10. J. Borenstein, Y. Koren, The vector field histogram-fast obstacle avoidance for mobile robots. IEEE Trans. Robot. Autom. **7**(3), 278–288 (1991)
11. E. Cerny, S. Dudani, J. Havlicek, D. Korchemny, *SVA: The Power of Assertions in SystemVerilog*, 2 edn. (Springer, Berlin, 2015)
12. G. Chance, A. Ghobrial, S. Lemaignan, T. Pipe, K. Eder, An agency-directed approach to test generation for simulation-based autonomous vehicle verification, in *IEEE International Conference On Artificial Intelligence Testing (AITest)* (IEEE Computer Society, Washington, 2020), pp. 31–38. https://doi.org/10.1109/AITEST49225.2020.00012. Preprint: https://arxiv.org/abs/1912.05434
13. J. Clark, J. McDermid, Software Systems Engineering Initiative (SSEI)—Predictable Complex Systems Via Integration. Tech. Rep. SSEI-TR-000020, The University of York, 2011 (unclassified)
14. E.M. Clarke, O. Grumberg, Peled, D.A.: *Model Checking* (MIT Press, Cambridge, 2000)
15. S. Conchon, M. Iguernelala, A. Mebsout, A collaborative framework for non-linear integer arithmetic reasoning in Alt-Ergo, in *2013 15th International Symposium on Symbolic and Numeric Algorithms for Scientific Computing* (2013), pp. 161–168
16. L. De Moura, N. Bjørner, Z3: an efficient SMT solver, in *Proceedings of the Theory and Practice of Software, 14th International Conference on Tools and Algorithms for the*

Construction and Analysis of Systems, TACAS'08/ETAPS'08 (Springer, Berlin, 2008), pp. 337–340

17. J.W. Durham, F. Bullo, Smooth nearness-diagram navigation, in *IEEE/RSJ International Conference on Intelligent Robots and Systems* (2008), pp. 690–695

18. K. Eder, C. Harper, U. Leonards, Towards the safety of human-in-the-loop robotics: challenges and opportunities for safety assurance of robotic co-workers', in *The 23rd IEEE International Symposium on Robot and Human Interactive Communication (ROMAN)* (2014), pp. 660–665. https://doi.org/10.1109/ROMAN.2014.6926328

19. K. Eder, J.P. Gallagher, P. López-García, H. Muller, Z. Banković, K. Georgiou, R. Haemmerlé, M.V. Hermenegildo, B. Kafle, S. Kerrison, M. Kirkeby, M. Klemen, X. Li, U. Liqat, J. Morse, M. Rhiger, M. Rosendahl, Entra: whole-systems energy transparency. Microprocess. Microsyst. **47**, 278–286 (2016). https://doi.org/10.1016/j.micpro.2016.07.003. http://www.sciencedirect.com/science/article/pii/S0141933116300862

20. C. Eisner, D. Fisman, *A Practical Introduction to PSL* (Springer, Berlin, 2006)

21. J.C. Filliâtre, A. Paskevich, Why3—Where programs meet provers, in *Programming Languages and Systems (ESOP)*, ed. by M. Felleisen, P. Gardner, no. 7792 in Lecture Notes in Computer Science (Springer, Berlin, 2013), pp. 125–128. https://doi.org/10.1007/978-3-642-37036-6_8

22. K. Georgiou, S. Kerrison, Z. Chamski, K. Eder, Energy transparency for deeply embedded programs. ACM Trans. Archit. Code Optim. **14**(1) (2017). https://doi.org/10.1145/3046679

23. N. Grech, K. Georgiou, J. Pallister, S. Kerrison, J. Morse, K. Eder, Static analysis of energy consumption for LLVM IR programs, in *Proceedings of the 18th International Workshop on Software and Compilers for Embedded Systems, SCOPES'15* (Association for Computing Machinery, New York, 2015), pp. 12–21. https://doi.org/10.1145/2764967.2764974

24. E.C. Grigore, K. Eder, A.G. Pipe, C. Melhuish, U. Leonards, Joint action understanding improves robot-to-human object handover, in *2013 IEEE/RSJ International Conference on Intelligent Robots and Systems* (2013), pp. 4622–4629. https://doi.org/10.1109/IROS.2013.6697021

25. M. Harman, S.A. Mansouri, Y. Zhang, Search-based software engineering: trends, techniques and applications. ACM Comput. Surv. **45**(1) (2012). https://doi.org/10.1145/2379776.2379787

26. D. Hoang, Y. Moy, A. Wallenburg, R. Chapman, SPARK 2014 and GNATprove. Int. J. Softw. Tools Technol. Transfer **17**, 695–707 (2014). https://doi.org/10.1007/s10009-014-0322-5

27. T. Huldt, I. Stenius, State-of-practice survey of model-based systems engineering. Syst. Eng. **22** (2018). https://doi.org/10.1002/sys.21466

28. C. Ioannides, K. Eder, Coverage-directed test generation automated by machine learning – a review. ACM Trans. Des. Autom. Electron. Syst. **17**(1) (2012). https://doi.org/10.1145/2071356.2071363

29. N. Kalra, S.M. Paddock, Driving to safety: How many miles of driving would it take to demonstrate autonomous vehicle reliability? Transp. Res. A Policy Pract. **94**, 182–193 (2016)

30. O. Lachish, E. Marcus, S. Ur, A. Ziv, Hole analysis for functional coverage data, in *Proceedings of the 39th Annual Design Automation Conference, DAC'02* (Association for Computing Machinery, New York, 2002), pp. 807–812. https://doi.org/10.1145/513918.514119

31. E.A. Lee, Freedom from choice and the power of models: in honor of Alberto Sangiovanni-Vincentelli, in *Proceedings of the 2019 International Symposium on Physical Design, ISPD'19* (Association for Computing Machinery, New York, 2019), p. 126. https://doi.org/10.1145/3299902.3320432

32. J.D. Lee, See, K.A.: Trust in automation: designing for appropriate reliance. Hum. Factors **46**(1), 50–80 (2004). https://doi.org/10.1518/hfes.46.1.50_30392. PMID: 15151155

33. U. Liqat, K. Georgiou, S. Kerrison, P. Lopez-Garcia, J.P. Gallagher, M.V. Hermenegildo, K. Eder, Inferring parametric energy consumption functions at different software levels: ISA vs. LLVM IR, in *Foundational and Practical Aspects of Resource Analysis*, ed. by M. van Eekelen, U. Dal Lago (Springer International Publishing, Cham, 2016), pp. 81–100. https://doi.org/10.1007/978-3-319-46559-3_5

34. J. Minguez, L. Montano, Nearness diagram (ND) navigation: collision avoidance in trouble-some scenarios. IEEE Trans. Robot. Autom. **20**(1), 45–59 (2004). https://doi.org/10.1109/TRA.2003.820849
35. J. Morse, D. Araiza-Illan, K. Eder, J. Lawry, A. Richards, A fuzzy approach to qualification in design exploration for autonomous robots and systems, in *IEEE International Conference on Fuzzy Systems (FUZZ-IEEE)* (2017), pp. 1–6. https://doi.org/10.1109/FUZZ-IEEE.2017.8015456
36. A. Piziali, *Functional Verification Coverage Measurement and Analysis* (Springer, Berlin, 2008). https://doi.org/10.1007/b117979
37. R. Salay, K. Czarnecki, Using machine learning safely in automotive software: an assessment and adaption of software process requirements in ISO 26262 (2018). https://arxiv.org/abs/1808.01614
38. P. Trojanek, K. Eder, Verification and testing of mobile robot navigation algorithms: a case study in SPARK, in *2014 IEEE/RSJ International Conference on Intelligent Robots and Systems* (2014), pp. 1489–1494. https://doi.org/10.1109/IROS.2014.6942753
39. M. Utting, A. Pretschner, B. Legeard, A taxonomy of model-based testing approaches. Softw. Test. Verif. Reliab. **22**(5), 297–312 (2012). https://doi.org/10.1002/stvr.456
40. M. Webster, D. Western, D. Araiza-Illan, C. Dixon, K. Eder, M. Fisher, A.G. Pipe, A corroborative approach to verification and validation of human-robot teams. Int. J. Robot. Res. **39**(1), 73–99 (2020). https://doi.org/10.1177/0278364919883338

Chapter 6
Robot Accident Investigation: A Case Study in Responsible Robotics

Alan F. T. Winfield, Katie Winkle, Helena Webb, Ulrik Lyngs, Marina Jirotka, and Carl Macrae

Abstract Robot accidents are inevitable. Although rare, they have been happening since assembly line robots were first introduced in the 1960s. But a new generation of social robots is now becoming commonplace. Equipped with sophisticated embedded artificial intelligence (AI), social robots might be deployed as care robots to assist elderly or disabled people to live independently. Smart robot toys offer a compelling interactive play experience for children, and increasingly capable autonomous vehicles (AVs) offer the promise of hands-free personal transport and fully autonomous taxis. Unlike industrial robots, which are deployed in safety cages, social robots are designed to operate in human environments and interact closely with humans; the likelihood of robot accidents is therefore much greater for social robots than industrial robots. This chapter sets out a draft framework for social robot accident investigation, a framework that proposes both the technology and processes that would allow social robot accidents to be investigated with no less rigour than we expect of air or rail accident investigations. The chapter also places accident investigation within the practice of responsible robotics and makes the case that social robotics without accident investigation would be no less irresponsible than aviation without air accident investigation.

A. F. T. Winfield (✉)
Bristol Robotics Lab, UWE Bristol, Bristol, UK
e-mail: alan.winfield@brl.ac.uk

K. Winkle
School of Electrical Engineering and Computer Science, KTH Royal Institute of Technology, Stockholm, Sweden
e-mail: winkle@kth.se

H. Webb · U. Lyngs · M. Jirotka
Department of Computer Science, University of Oxford, Oxford, UK
e-mail: helena.webb@cs.ox.ac.uk; ulrik.lyngs@cs.ox.ac.uk; marina.jirotka@cs.ox.ac.uk

C. Macrae
Nottingham University Business School, University of Nottingham, Nottingham, UK
e-mail: Carl.Macrae@nottingham.ac.uk

© Springer Nature Switzerland AG 2021
A. Cavalcanti et al. (eds.), *Software Engineering for Robotics*,
https://doi.org/10.1007/978-3-030-66494-7_6

1   Introduction

> *What could possibly go wrong?*
> Imagine that your elderly mother, or grandmother, has an assisted living
> robot to help her live independently at home. The robot is capable of
> fetching her drinks, reminding her to take her medicine and keeping in
> touch with family. Then one afternoon you get a call from a neighbour
> who has called round and sees your grandmother collapsed on the floor.
> When the paramedics arrive, they find the robot wandering around
> apparently aimlessly. One of its functions is to call for help if your
> grandmother falls or stops moving, but it seems that the robot failed
> to do this.
>
> Fortunately, your grandmother makes a full recovery. Not surpris-
> ingly you want to know what happened: did the robot cause the
> accident? Or maybe it didn't but made matters worse, and why did it
> fail to raise the alarm?

Although this is a fictional scenario, it could happen today. If it did we would be
reliant on the goodwill of the robot manufacturer to discover what went wrong.
Even then we might not get the answers we seek; it is entirely possible that neither
the robot nor the company that made it is equipped with the tools and processes to
undertake an investigation.

Robot accidents are inevitable. Although rare, they have been happening since
assembly line robots were first introduced in the 1960s. First wave robots include
factory robots (multi-axis manipulators), autonomous guided vehicles (AGVs)
deployed in warehouses, remotely operated vehicles (ROVs) for undersea exploration and maintenance, teleoperated robots for bomb disposal and (perhaps surprisingly) Mars rovers for planetary exploration. A defining characteristic of first
wave robots is that they are designed for jobs that are dull, dirty or dangerous; these
robots are typically either dangerous to be around (and therefore enclosed in safety
cages on assembly lines) or deployed in dangerous or inaccessible environments.

In contrast, second wave robots are designed to operate in human environments
and interact directly with people. Those human environments include homes,
offices, hospitals, shopping malls and city or urban streets, and—unlike first wave
robots—many are designed to be used by untrained, naive or vulnerable users,
including children and the elderly. These are robots in society, and hence social
robots.[1] Equipped with sophisticated embedded artificial intelligence (AI), social
robots might be deployed as care robots to assist elderly or disabled people to live

---

[1] Noting that we take a broader view of social robotics than usual.

independently. Smart robot toys offer a compelling interactive play experience for children, and increasingly capable autonomous vehicles (AVs) offer the promise of hands-free personal transport and fully autonomous taxis.

Social robots by definition work with and alongside people in human environments; thus, the likelihood and scope of robot accidents are much greater than with industrial robots. This is not just because of the proximity of social robots and their users (and perhaps also bystanders), it is also because of the kinds of roles such robots are designed to fulfil, and further exacerbated by the unpredictability of people and the unstructured, dynamic nature of human environments.

Given the inevitability of social robot accidents, it is perhaps surprising that no frameworks or processes of social robot accident investigation exist. This chapter addresses that deficit by setting out a draft framework for social robot accident investigation: a framework which proposes both the technology and processes that would allow social robot accidents to be investigated with no less rigour than we expect of air or rail accident investigations.

This chapter proceeds as follows. We first survey the current practices and frameworks for accident investigation, including in transport (air, rail and road) and healthcare, in Sect. 2. In Sect. 3 we survey robot accidents, including both industrial and social robot accidents, then analyse the scope for social robot accidents to understand why social robot accidents are more likely (per robot deployed) than industrial robot accidents. Section 4 then places accident investigation within the practice of responsible robotics, by defining responsible robotics within the broader practice of responsible innovation; the section also briefly surveys the emerging practices of values-driven design and ethical standards in robotics. Section 5 sets out both the technology and processes of our draft framework for social robot accident investigation, then illustrates the application of this framework by setting out how an investigation of our fictional accident might play out. Finally, in Sect. 6, we conclude by outlining work currently underway within project RoboTIPS to develop and validate our framework with real-robot mock accident scenarios, before considering some key questions about who would investigate real-world social robot accidents.

Since the embedded AI is a critical software component of any intelligent robot, it follows that the specification, design and engineering of robot software should take account of the needs of robot accident investigation, hence the relevance of this chapter to the present volume.

2 The Practice of Accident Investigation

Investigating accidents and incidents is a routine and widespread activity in safety-critical industries such as aviation, the railways and healthcare. In healthcare, for instance, over two million safety incidents are reported across the English National Health Service each year, with around 50,000 of those incidents causing moderate to severe harm or death [29]—such as medication errors or wrong-site surgery. Accident investigation also has a long history. In aviation, the world's first national air accident investigation agency was established over a century ago in 1915 [1], and the international conventions governing air accident investigation were agreed in the middle of the last century [15]. The practices and methods of accident investigation are therefore well-established in many sectors and have been developed and refined over many years. As such, when considering a framework for the investigation of social robot accidents, it is instructive to examine how accident investigation is conducted in other safety-critical domains.

First, it is important to emphasise that the core principle and fundamental purpose of accident investigation is learning. Whilst investigations primarily focus on examining events that have occurred in the past, the core purpose of an accident investigation is to improve safety in the future. Accident investigations are therefore organised around three key questions [24]. The first is factual: what happened? The second is explanatory: why did those things happen and what does that reveal about weaknesses in the design and operation of a particular system? The third is practical: how can systems be improved to prevent similar events happening in future? The ultimate objective for any accident investigation is to develop practical, achievable and targeted recommendations for making safety improvements.

Conducting an accident investigation involves collecting and analysing a range of evidence and data from a variety of sources to understand what happened and why. This can include quantitative data, such as information from 'black box' Flight Data Recorders on aircraft that automatically collect data on thousands of flight parameters. It can also include qualitative data in the form of organisational documentation and policies, and in-depth interviews with witnesses or those directly or indirectly involved in an accident such as a pilot or a maintenance engineer. Accident investigators therefore need to be experts in the methods and processes of accident investigation and also need to have a deep and broad understanding of the industry and the technologies involved. However, accident investigations are also typically collaborative processes that are conducted by a diverse team of investigators who, in turn, draw on specific sources of expertise where required [21].

A variety of methods have been developed to assist in the collection and analysis of safety data, from cognitive interviewing techniques [11] to detailed human factors methods [37] and organisational and sociotechnical systems analysis techniques [38]. Importantly, to understand why an accident has occurred and to determine how safety can be improved in future, accident investigations typically apply a systemic model of safety that helps identify, organise and explain the factors involved and how they are interconnected.

One of the most widely applied and practical accident models is the 'organisational accident' framework—commonly referred to as the 'Swiss cheese' model [34]. This has been adapted and applied in various ways [3], but at its core, this provides a simple framework that conceptualises system safety as dependent on multiple layers of risk controls and safety defences that span from the front-line to organisational and regulatory levels—such as redundant systems, emergency alarms, operator training, management decisions or regulatory requirements. Each of these safety defences will inevitably be partial or weak in some way, and accidents occur when the 'holes' in these defences all line up in some unexpected way—thus the eponymous image of multiple slices of 'Swiss cheese', each slice full of holes. The focus of accident investigations is to examine the entire sociotechnical system to understand the safety defences, the weaknesses in those defences, why those weaknesses arose and how they can be addressed. Similar premises underpin a variety of different accident investigation models, methods and tools.

Safety-critical industries have developed sophisticated systems to support these activities of investigation and learning at all levels. These range from lengthy investigations of major accidents that are coordinated by national investigative bodies to relatively rapid local-level investigations of more minor incidents or near-miss events that are conducted by investigators within individual organisations [22].

A good deal of media attention understandably focuses on the investigations into high-profile accidents that are conducted by national investigation bodies, such as the US National Transportation Safety Board's investigations of the various accidents involving semi-automated Tesla vehicles [27] and Uber's autonomous test vehicle [28]. However, much of the more routine work of investigation occurs within individual organisations, like airlines and hospitals, which regularly conduct hundreds or thousands of investigations each year. These local-level investigations examine more minor safety incidents as well as near-miss events—where there was no adverse outcome but some sort of safety-relevant failure was detected, such as a poorly specified maintenance procedure in an airline leading to a technical failure that causes a rejected take-off [22]. Local-level investigations employ similar methods and approaches to those conducted at a national level but are often much more rapid, lasting days or weeks rather than months and years. They are also typically limited to examining sources of risk within one single organisation, unlike national-level investigations which can consider regulatory factors and interactions between different organisations across an industry.

At all these levels of investigative activity, accident and incident investigation is always focused on learning. One of the main implications of this focus on learning is that accident investigation activities are typically entirely separated from other investigative activities that seek to attribute blame or determine liability. In aviation, for example, the information collected during major accident investigations may only be used for safety investigation and improvement—and may not be used, for instance, to punish individuals or pursue legal claims against organisations [10]. This is also the case within individual organisations in safety-critical industries, where it is common to have a safety department or team that is entirely separated from operational processes or production pressures and whose sole focus is monitoring safety, investigating incidents and supporting improvement [23]. The information that is collected by these safety teams is usually only used for safety improvement purposes and purposefully not used for line management or disciplinary processes. This ensures that staff feel they can openly and honestly provide safety-relevant information to support ongoing efforts to investigate and understand failures and continuously improve safety.

3   Robot Accidents

Robert Williams is believed to be the first person killed by a robot in an industrial accident, in January 1979, at a Ford Motor Company casting plant.[2] Given that there are over 2.4M industrial robots in use worldwide [16], it is surprisingly difficult to find statistics on robot accidents. The US Department of Labor's Occupational Safety and Health Administration (OSHA) maintains a web page listing industrial accidents since 1984, and a search of this listing with the keyword 'robot' returns records of 43 accidents;[3] all resulted in injuries of which 29 were fatal. The US National Institute for Occupational Safety and Health (NIOSH) found 61 robot-related deaths between 1992 and 2015, noting that 'These fatalities will likely increase over time because of the increasing number of conventional industrial robots being used by companies in the United States, and from the introduction of collaborative and co-existing robots, powered exoskeletons, and autonomous vehicles into the work environment'.[4]

Finding data on accidents involving robots that interact with humans (HRI) is also difficult. One study on the safety of interactive industrial robots in Finland [25] notes that 'International accident-report data related to robots is scarce'. The study reports that the majority of the 6000 robots in Finland are 'used in small robot work cells rather than large automation lines' and that 'a natural feature of the production is that it requires substantial (mainly simple) HRI inside the robot working envelope'. The study analyses a total of 25 severe robot or manipulator-related accidents between 1989 and 2006, of which 3 were fatal, noting also that most of the accidents occurred towards the end of this period. Key characteristics of these accidents were:

- The cause of an accident is usually a sequence of events, which have been difficult to foresee.
- Operator inattentiveness and forgetfulness may lead them to misjudge a situation and even a normal robot movement may surprise them.
- Most of the accidents involving robots occurred when (the) robot moved unexpectedly (from worker's point of view) against the worker within the robot working area.
- Inadequate safeguarding featured significantly as a cause of accidents.
- Many accidents are associated with troubleshooting disturbances.
- Only about 20% of accidents occurred during undisturbed automated runs [25].

Although now somewhat dated, Chapter 4 'Robot Accidents' of [8] sets out a comprehensive analysis of industrial robot accidents, including data from accidents in Japan, Western Europe and the United States. Noting that 'human behavior plays

---

[2]https://en.wikipedia.org/wiki/Robert_Williams_(robot_fatality).

[3]https://www.osha.gov/pls/imis/AccidentSearch.search?acc_keyword=%22Robot %22&keyword_list=on.

[4]https://www.cdc.gov/niosh/topics/robotics/aboutthecenter.html.

an important role in certain robot accidents', the paper outlines a set of typical human behaviours that might result in injury. A number of these are especially instructive to the present study:

- Humans often incorrectly read, or fail to see, instructions and labels on various products.
- Many people carry out most assigned tasks while thinking about other things.
- In most cases humans use their hands to test or examine.
- Many humans are unable to estimate speed, distance, or clearances very well. In fact, humans underestimate large distances and overestimate short distances.
- Many humans fail to take the time necessary to observe safety precautions.
- A sizeable portion of humans become complacent after a long successful exposure to dangerous items.
- There is only a small percentage of humans which recognise the fact that they cannot see well enough, either due to poor illumination or poor eyesight [8].

Finding data on non-industrial robot accidents is even more difficult, but one study reports on adverse events in robotics surgery in the United States [2]; the paper surveys 10,624 reports collected by the Food and Drug Administration between 2000 and 2013, showing 8061 device malfunctions, 1391 patient injuries and 144 deaths, during a total of more than 1.75M procedures. Some robot accidents make headline news, for example, car accidents in which semi-automated driver-assist functions are implicated; a total of six 'self-driving car fatalities' have been reported since January 2016.[5] We only know of one accident in which a child's leg was bruised by a shopping mall security robot because it was reported in the national press.[6]

A recent study in human-robot interaction examines several serious accidents, in aviation, the nuclear industry and autonomous vehicles, to understand 'the potential mismatches that can occur between control and responsibility in automated systems' and argues that these mismatches 'create moral crumple zones, in which human operators take on the blame for errors or accidents not entirely in their control' [9].

3.1 The Scope for Social Robot Accidents

As we have outlined above, industrial and surgical robot accidents are—thankfully—rare events. It is most unlikely that social robot accidents will be so rare. Several factors lead us to make this forecast:

1. Social robots, as broadly defined in this chapter, are designed to be integrated into society at large: in urban and city environments, schools, hospitals, shopping

---

[5]https://en.wikipedia.org/wiki/List_of_self-driving_car_fatalities.

[6]https://www.wsj.com/articles/security-robot-suspended-after-colliding-with-a-toddler-1468446311.

malls and in both care and private homes. Unlike industrial robots operating behind safety cages, social robots are designed to be among us, up close and personal. It is this proximity that undoubtedly presents the greatest risk.

2. Operators of industrial robots are required to undertake training courses, both on how to operate their robots and on the robot's safety features, and are expected to follow strict safety protocols. In contrast, social robots are designed to benefit naive users—including children, the elderly and vulnerable or disabled people—with little more preparation than might be expected to operate a vacuum cleaner or dishwasher.

3. Industrial robots typically operate in an environment optimised for them and not humans. Social robots have no such advantage: humans (especially young or vulnerable humans) are unpredictable and human environments (from a robot's perspective) are messy, unstructured and constantly changing. Designing robots capable of operating safely in human environments remains a serious challenge.

4. The range of serious harms that can arise from social robots is much greater than those associated with industrial robots. Social robots can, like their industrial counterparts, cause physical injury, but the responsible roboticist should be equally concerned about the potential for psychological harms, such as deception (i.e. a user coming to believe that a robot cares for them), over-trust or over-reliance on the robot or violations of privacy or dignity. Several *ethical hazards* are outlined in Sect. 4.[7]

5. The scope of social robot accidents is thus enlarged. The nature of the roles social robots play in our lives—supporting elderly people to live independently or helping the development of children with autism, for instance [5]—opens up a range of ethical risks and vulnerabilities that have hitherto not been a concern of either robot designers or accident investigators. This factor also increases the likelihood of social robot accidents.

If we are right and the near future brings an increasing number of social robot accidents, these accidents will need to be investigated to address the three key questions of accident investigation outlined in Sect. 2. What happened? Why did it happen? And what must we change to ensure it doesn't happen again? [24].

4 Responsible Robotics

In essence, responsible robotics is the application of responsible innovation (RI) to the field of robotics, so we first need to define RI. A straightforward definition of RI is a set of good practices for ensuring that research and innovation benefit society and the environment. There are several frameworks for RI. One is the EPSRC AREA

---

[7] An ethical hazard is a possible source of ethical harm.

framework,[8] built on the four pillars of Anticipation (of potential impacts and risks), Reflection (on the purposes of, motivations for and potential implications of the research), Engagement (to open up a dialogue with stakeholders beyond the narrow research community) and Action (to use the three processes of anticipation, reflection and engagement, to influence the direction and trajectory of the research) [31]. The more broadly framed Rome Declaration on Responsible Research and Innovation[9] is built around the six pillars of open access, governance, ethics, science communication, public engagement and gender equality.[10]

We define responsible robotics as follows:

> Responsible robotics is the application of responsible innovation in the design, manufacture, operation, repair and end-of-life recycling of robots that seeks the most benefit to society and the least harm to the environment.

Robot ethics—which is concerned with the ethical impact of robots, on individuals, society and the environment, and how any negative impacts can be mitigated—and ethical governance both have an important role in the practice of responsible robotics. Responsible robotics must also be built upon a foundation of best practice in the engineering of robots and their subsystems, including software components (such as the robot's embedded AI), which must be engineered to very high standards of reliability, robustness and transparency.

In recent years, many sets of ethical principles have been proposed in robotics and AI—for a comprehensive survey, see [18]—but one of the earliest is the EPSRC Principles of Robotics, first published online in 2011[11] [4]. In [42] we set out a framework for ethical governance that links ethical principles to standards and regulation and argue that when such processes are robust and transparent, trust in robotics (or, to be precise, the individuals, companies and institutions designing, operating and regulating robots) will grow.

Responsible social robotics [39] is the practice of responsible robots in society with a particular ambition of creating robots that bring real benefits in both quality of life and safeguarding to the most vulnerable, within a framework of values-based design based upon a deep consideration of the ethical risks of social robots.

There are several approaches and methods available to the responsible social roboticist, including emerging new ethical standards and an approach called ethically aligned design, which we will now review in Sects. 4.1 and 4.2.

---

[8]https://epsrc.ukri.org/research/framework/area/.

[9]https://ec.europa.eu/research/swafs/pdf/rome_declaration_RRI_final_21_November.pdf.

[10]http://ec.europa.eu/research/science-society/document_library/pdf_06/responsible-research-and-innovation-leaflet_en.pdf.

[11]https://epsrc.ukri.org/research/ourportfolio/themes/engineering/activities/principlesofrobotics/.

4.1   Ethically Aligned Design

In April 2016, the IEEE Standards Association launched a Global Initiative on the Ethics of Autonomous and Intelligent Systems.[12]   The initiative positions human well-being as its central tenet. The initiative's mission is 'to ensure every stakeholder involved in the design and development of autonomous and intelligent systems (AIS) is educated, trained, and empowered to prioritize ethical considerations so that these technologies are advanced for the benefit of humanity'.

The first major output from the IEEE global ethics initiative is a document called Ethically Aligned Design (EAD) [14]. Developed through an iterative process over 3 years, EAD is built on the three pillars of Universal Human Values, Political Self-Determination and Data Agency and Dependability, alongside eight general (ethical) principles covering Human Rights, Well-being, Data Agency, Effectiveness, Transparency, Accountability, Awareness of Misuse and Competence. EAD sets out more than 150 issues and recommendations across its 10 chapters. In essence, it is both a manifesto and roadmap for a values-driven approach to the design of autonomous and intelligent systems. Spiekermann and Winkler [36] detail the process of ethically aligned design within a broader methodological framework they call value-based engineering for ethics by design. Responsible social robotics must be values-based.

4.2   Standards in Social Robotics

A foundational standard in social robotics is ISO 13482:2014 *Safety requirements for personal care robots*. ISO 13482 covers mobile servant robots, physical assistant robots and person carrier robots, but not robot toys or medical robots [17]. The standard sets safety (but not ethical) hazards, including hazards relating to battery charging, robot motion, incorrect autonomous decisions and actions and lack of awareness of robots by humans.

A new generation of explicitly ethical standards is now emerging [30, 40]. Standards are simply 'consensus-based agreed-upon ways of doing things' [6]. Although all standards embody a principle or value, explicitly ethical standards address clearly articulated ethical concerns and—through their application—seek to remove, reduce or highlight the potential for unethical impacts or their consequences [40].

The IEEE ethics initiative outlined above has initiated a total of 13 working groups to date, each tasked with drafting a new standard within the 7000 series of so-called human standards. The first of these to reach publication is IEEE 7010-2020

---

[12]https://standards.ieee.org/content/dam/ieee-standards/standards/web/documents/other/ec_about_us.pdf.

Recommended Practice for Assessing the Impact of Autonomous and Intelligent Systems on Human Well-Being.[13]

Almost certainly the world's first explicitly ethical standard in robotics is BS8611-2016 *Guide to the ethical design and application of robots and robotic systems* [7]. 'BS8611 is not a code of practice, but instead guidance on how designers can undertake an ethical risk[14] assessment of their robot or system, and mitigate any ethical risks so identified. At its heart is a set of 20 distinct ethical hazards and risks, grouped under four categories: societal, application, commercial and financial, and environmental. Advice on measures to mitigate the impact of each risk is given, along with suggestions on how such measures might be verified or validated' [40]. Societal hazards include anthropomorphisation, loss of trust, deception, infringements of privacy and confidentiality, addiction and loss of employment, to which we should add the uncanny valley[26], weak security, lack of transparency (for instance, the lack of data logs needed to investigate accidents), unrepairability and unrecyclability. Ethical risk assessment is a powerful and essential addition to the responsible roboticist's toolkit, as it can be thought of as the opposite face of accident investigation, seeking—at design time—to prevent risks becoming accidents.

5 A Draft Framework for Social Robot Accident Investigation

We now set out a framework for social robot accident investigation outlining first the technology in Sect. 5.1, then the process in Sect. 5.2, followed by an illustration of how the framework could be applied in practice, in Sect. 5.3.

5.1 Technology

We have previously proposed that all robots should be equipped with the equivalent of an aircraft Flight Data Recorder (FDR) to continuously record sensor inputs, actuator outputs and relevant internal status data [41]. We call this an ethical black box[15] (EBB) and argue that the EBB will play an important role in the process of discovering how and why a robot caused an accident.

All robots collect sense data, then—based on that sense data and some internal decision-making process (the embedded artificial intelligence)—send commands to actuators. This is of course a simplification of what in practice will be a complex set

---

[13] https://standards.ieee.org/standard/7010-2020.html.

[14] An ethical risk is possible ethical harm arising from exposure to a hazard.

[15] Because it would be unethical not to have one.

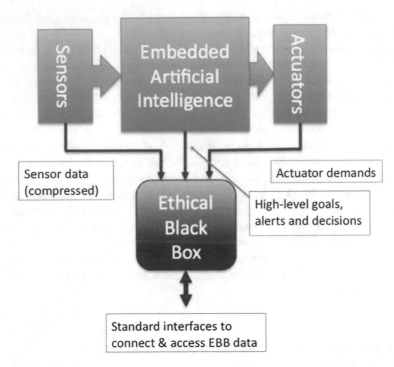

Fig. 6.1 Robot subsystems with an ethical black box and key data flows (amended from Winfield and Jirotka [41]).

of connected systems and processes, but, at an abstract level, all intelligent robots will have the three major subsystems shown in blue, in Fig. 6.1. It is important to note that the robot's embedded AI is a software component, the correct functioning of which is critical to the safety and reliability of the robot.

Our EBB will have much in common with its aviation and automotive counterparts, the flight data recorder [13] and event data recorder (EDR) [12], in particular: data is stored securely in a rugged unit designed to survive accidents; stored records are time and date stamped; and storage capacity is limited and organised such that only the most recent data are saved—overwriting older records (an FDR typically records 17–25 h of data, an automobile EDR as little as 30 s).

The technical specification for an EBB for a social robot is beyond the scope of this chapter. It is, however, important to outline here the kinds of data we should expect to be recorded in the EBB. Consider the Pepper robot, as an exemplar of an advanced social robot [32]. The Pepper is equipped with 25 sensors, including 4 microphones, 2 colour cameras, 2 accelerometers and various proximal and distal sensors. It has 20 motors, a chest-mounted touch display pad, loudspeakers for speech output and Wi-Fi connectivity. We would therefore expect an EBB designed for the Pepper robot to be able to record:

1. Input-sense data, including sampled (and compressed) camera images, audio sampled by the microphones, accelerometers and touch screen touches.
2. Actuator demands generated by the robot's control system along with sampled position of joints, from which we can deduce the robot's pose.
3. Synthesised speech and touch screen display outputs.
4. Periodically sampled battery levels.
5. Periodically sampled status of the robot's Wi-Fi and Internet connectivity.
6. The robot's position (i.e. x, y coordinates) within its working environment (noting that this data might be available from a tracking system in a 'smart' environment, or if not, deduced via odometry from the robot's known start position at, say, its re-charging station and subsequent movements).

The EBB should also record high-level decisions generated by the robot's AI (see the data flow in Fig. 6.1); thus, it is important that the AI software is designed both to provide such outputs to the EBB and to be transparent and traceable in order to support post-accident analysis. Given also that social robots are likely to accept speech commands, we would, ideally, be able to keep both the raw audio recorded by the microphones and the text sequence produced by the robot's speech recogniser.

5.2 Process

Conventionally, accident investigation is performed in four stages: (1) information gathering, (2) fact analysis, (3) conclusion drawing and—if necessary—(4) implementation of remedies. Stage (2) might typically take the form of causal analysis, and we propose to adopt here the method of why-because analysis developed by Ladkin et al. [19, 20, 35].

Why-because analysis (WBA) is a method of causal analysis of complex sociotechnical systems and hence well suited to social robot accidents. WBA lends itself to a simple quality check: whether one event is a causal factor in another can be determined by applying a counter-factual test. The primary output of WBA is a why-because graph (WBG) and a further advantage of WBA is that—if necessary—the correctness of the WBG can be formally proven. Figure 6.2 shows, in flow chart form, the process of why-because analysis.

We next elaborate briefly on some of the steps in Fig. 6.2. 'Gather information' requires collecting all available information on the accident. This will include witness testimony, records from the EBB, any forensic evidence and contextual information on the setting of the accident. The next step 'determine facts' requires the investigation team to sift the information and glean the essential facts. Noting that witness testimony can be unreliable, any causal inferences from that testimony should ideally be corroborated with forensic evidence, so that—even if not certain—the team can have high confidence in those inferences.

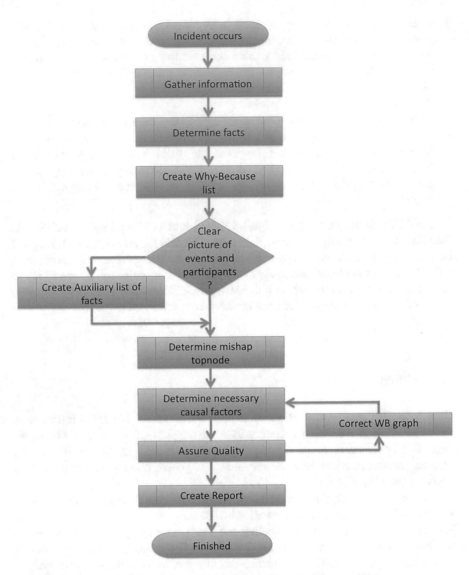

Fig. 6.2 An overview of why-because analysis (adapted from Ladkin et al. [20]).

The third step: 'create why-because list' links the facts of events—including both things that happened and things that should have happened but did not (unevents)—to outcomes. If they give the team a clear picture of the sequence of events and participants in the accident, then the team agree on the 'mishap topnode(s)' of the why-because graph, i.e. the accident—or perhaps multiple accidents. Then the why-because graph is created, top-down. This is likely also to require several iterations and—if necessary—quality checking using counter-factual tests or formal proof of completeness. For a complete explanation of the method, refer to the introduction to WBA in [35].

5.3   The Application of the Framework

To understand how this framework would operate in practice, we return to the fictional scenario outlined at the start of the chapter. As described in the scenario, an elderly lady 'Rose' has a fall in her home. She is found, still on the floor, sometime later by her neighbour, Barbara, who calls for medical help as well as alerting Rose's family. Whilst Rose receives hospital treatment, an investigation team is formed, who begin to collect evidence for the investigation. Photos of Rose's flat are taken and information about her home setup is collected; for instance, Rose lives in a smart home with various sensors to detect her movements and communicate with the robot as necessary. Preliminary observation of the robot also reveals details about its design and use in the home. The robot can fetch drinks, provide reminders (such as for Rose to take medication) and act as an interface so that Rose can control her television and other devices through it with voice commands. The robot will also respond to commands sent via a smartphone or tablet application.

Barbara, the paramedics and Rose herself are interviewed to provide witness statements. These statements combine with the initial observations to provide important early findings that shape the ongoing investigation. Of key importance are the following: (1) Rose did not put on her fall alert bracelet on the day of the accident, and (2) from her condition (as observed by the paramedics), it seems that after her fall Rose was too weak to call out to the robot if it was more than two metres away from her.

In her witness testimony, Rose states that she had climbed on a chair and was reaching for something in an upper cupboard in her kitchen, but then slipped and fell on the floor. She has no memory of what happened after that and does not recall where the robot was when she fell. Barbara states that she does not recall seeing the robot when she entered the flat, and feels that the shock of finding Rose on the floor meant she did not notice much else around her. The paramedic states that he noticed the robot moving about around the flat—between the living area and the kitchen. It did not seem to be moving for the accomplishment of any particular action so he described the robot as acting 'aimlessly'.

The investigation team gather further information. They interview the manager of the retirement complex that Rose lives in; she provides details of the organisational structure of the complex including the infrastructure that enables residents to set up their homes as smart homes and have assistance robots. They also talk to others who saw Rose on the day of her accident.

The last person to see Rose before her fall was Michelle, a cleaner who works for the retirement complex. Michelle saw Rose whilst she was in Rose's flat for an hour to do her regular weekly clean. Michelle reported that Rose seemed very cheerful and chatty and did not complain of feeling ill or mention any worries or concerns. Michelle said that she did her usual clean in its usual order as requested by the retirement complex: collect up rubbish to take outside, wipe bathroom surfaces and floor, wipe kitchen work surfaces and clean floor, polish wooden surfaces, hoover carpeted areas and use disinfectant wipes on door handles and all over the robot for infection control. When asked by the investigation team, she said she thought the robot was in the living room for most of the time she was there but could not be sure. She didn't notice anything unusual about what the robot was doing.

The team also gets a report from Rose's general practitioner about her overall health status before the accident. This states that Rose had some limited mobility and needed to take daily medication to help her arthritis. She had also recently complained of forgetting things more and more easily. However, she was generally healthy for her age and had no acute concerns.

Finally, the team extracts data from the ethical black box. These are in CSV format and contain timestamped information regarding (1) the location and status of the robot and Rose and others within the apartment, (2) actions undertaken by the robot and (3) sampled records of all other robot inputs and outputs over the previous 24 h period. It enables the team to conclude that the robot lost connection to the central 'home connection hub' intermittently over the past 24 h, coinciding with Rose's fall. In addition, processing of the camera feed and other sensors used for navigation appear to be producing erroneous data. The records showed no log of Rose's fall but did log that the robot made several 'requests for help'— by speaking out loud—regarding its inability to connect to the home connection hub.

Having collected and analysed all of the material, the investigation team identify key relevant factors. At the individual level, certain actions by Rose—forgetting to put on her accident bracelet and reaching to a high cupboard—certainly increased the safety risk. Aspects of the local environment are likely to have also contributed to this risk and influenced the technical problems that occurred—for instance, the repeated disinfecting of the robot, as required by the retirement complex, has almost certainly impaired its sensors. The robot's response to losing connection to the home hub, i.e. 'asking for help' was not effective in getting the problem addressed, most likely because Rose did not understand the problem.

Concerning the robot's standard functionalities, it failed to detect Rose's fall and therefore raise an alert following the fall. The robot's fall detection system relies, in part, on data collected by distributed sensors placed around the smart home. This data is delivered to the robot via the home connection hub, so the intermittent connectivity issues prevented the robot's fall detection functionality from operating as intended. The team make use of these key facts to construct the why-because graph shown in Fig. 6.3.

The first thing to note in Fig. 6.3 is that there are two mishap topnodes and hence two sub-graphs. On the left is the sub-graph showing the direct causal chain that resulted in Rose's fall (Accident 1), following her unwise decision to try and reach a high cupboard. The sub-graph on the right shows the chain of factors—processes,

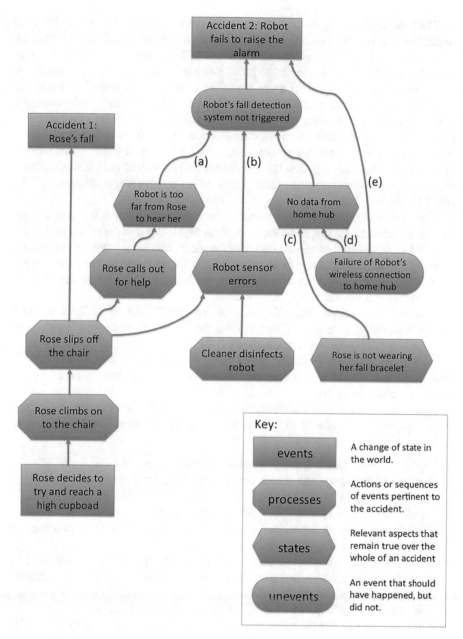

Fig. 6.3 Why-because graph for Rose's accident.

states and unevents (things that should have happened but did not)—that together led to Accident 2: the robot failing to raise the alarm.

The sub-graph leading to Accident 2 shows that the four ways in which the robot might have raised the alarm all failed, for different reasons. The first (a) is that the robot was too far away from Rose to hear her calls for help (most likely because the failure of its connection with the home hub means that the robot did not know where she was). The second (b) is that the robot's sensors that should have been able to detect her fall (together with data from the smart environment) were damaged, almost certainly by cleaning with disinfectant, and the third (d) was the failure of the wireless communication between the robot and the home hub, which meant there was no data from the home's smart sensors. A fourth reason (c) is due to two factors: (1) Rose had forgotten to put on her fall alarm bracelet, but (2) even if she had been wearing it, the bracelet would have been ineffective as it too communicates with the robot via the home hub.

The failure of communication between the robot and home hub is particularly serious because, as the graph shows, even if the first two pathways (a) and (b) to the robot's fall detection system had operated correctly, the robot would still not have been able to raise the alarm, indicated by path (e). To use the Swiss cheese metaphor from Sect. 2, over-reliance on communication with the home hub represents a set of holes in the layers of safety which all line up.

The key conclusions from this analysis are that (1) the robot did **not** cause Rose's accident, (2) the robot failed to raise the alarm following Rose's fall—one of its primary safeguarding functions and (3) failures and shortcomings of the smart home's infrastructure contributed significantly to the robot's failure. The robot's failure might have had very severe consequences had a neighbour not called upon Rose and raised the alarm.

As a consequence of their investigation, the team make a set of recommendations to prevent similar accidents happening in future. These recommendations are, in order of priority:

1. Equip the robot with a backup communications system, in case the Wi-Fi fails. A recommended approach might, for instance, be to integrate a module allowing the robot to send text or data messages via the public cellular phone network.
2. Equally important is that if the robot detects a Wi-Fi connectivity failure, it should not be alerting its user (Rose) but instead sending an alert to a maintenance engineer via its backup communication system.
3. Equip the home hub with the ability to send an emergency call directly—via a landline, for instance—when the fall bracelet is triggered, so that this particular alarm is not routed via the robot.
4. Improve the sensitivity of the robot's microphones to increase their range.
5. Add a new function to the robot so that it reminds Rose to put on her fall bracelet every day.
6. Advise the cleaner not to use disinfectants on the robot.

6 Concluding Discussion

We conclude this chapter with a brief introduction to the project within which the work of this chapter falls, followed by a discussion of both the composition of robot accident investigation teams and the important question of which agency should be responsible for investigating social robot accidents.

6.1 RoboTIPS

The work of this chapter is part of the 5-year programme RoboTIPS: Developing Responsible Robots for the Digital Economy.[16] RoboTIPS has several themes, two of which are of relevance to this chapter. The first is the technical specification and prototyping of the EBB, including the development of model EBBs, which will be used as the basis of implementations and trials by project industrial partners.

There will be two model EBBs, one a hardware implementation and the other a software module, and their specifications, code and designs will be published as open-source to support and encourage others to build EBBs into their robots. Ultimately we would like to see industry standards emerge for EBBs, noting that we would need separate specifications for different domains: one EBB standard for AVs, another for healthcare robots, a third for robot toys/educational robots and so on.

Second, we are designing and running three staged (mock) accidents, and we anticipate one in the application domain of assisted living robots, one in educational (toy) robots and another for autonomous vehicles. We believe this to be the world's first study of accident investigation in the field of social robots. For each of these staged scenarios, we will be using real robots and will invite human volunteers to act in three roles, as

1. subject(s) of the accident,
2. witnesses to the accident, or
3. members of the accident investigation team.

One key aim of these staged accidents is to trial, develop and refine the framework for social robot accident investigation outlined in this chapter.

Thus, we aim to develop and demonstrate both technologies and processes (and ultimately policy recommendations) for effective social robot accident investigation. And as the whole project is conducted within the framework of Responsible Research and Innovation, it is a case study in responsible robotics.

---

[16]https://www.robotips.co.uk/.

6.2 The Bigger Picture

There are two important details that we omitted from the accident scenario outlined in Sect. 1, then developed in Sect. 5.3. The first is who needs to be on a robot accident investigation team. And the second—and perhaps more fundamental question—which agency should investigate social robot accidents?

Concerning the makeup of a social-robot accident-investigation team, if we follow best practice, it would be a multi-disciplinary team. One report, for instance, described multi-disciplinary teams formed to investigate sleep-related fatal vehicle accidents as 'consisting of a police officer, a road engineer, a traffic engineer, a physician, and in certain cases a psychologist' [33]. Such teams did not require the involvement of vehicle manufacturers, but more recent fatal accidents involving AVs have needed to engage the manufacturer, to provide both expertise on the AV's autopilot technology and access to essential data from the vehicle's proprietary data logger [27].

Robot accident investigations will similarly need to call upon the assistance of robot manufacturers, both to provide data logs and advice on the robot's operation. We would therefore expect social robot accident investigation teams to consist of (1) an independent lead investigator with experience of accident investigation, (2) an independent expert on human-robot interaction, (3) an independent expert on robot hardware and software, (4) a senior manager from the environment in which the accident took place and (5) one of the robot manufacturer's senior engineers. Depending on the context of the accident, the team might additionally need, for instance, a (child-)psychologist or senior healthcare specialist.

Consider now the question: who does one call when there has been a robot accident?[17] At present, there is no social robot equivalent of the UK Air Accidents Investigation Branch[18] or Healthcare Safety Investigation Branch (HSIB).[19] A serious AV accident would of course be attended by a police road traffic accident unit, although they would almost certainly encounter difficulties getting to the bottom of failures of the vehicle's autopilot AI. The US National Transport Safety Board[20] (NTSB) is the only investigation branch known to have experience of AV accidents, having investigated five to date (notably the NTSB is the same agency responsible for air accident investigation in the United States and thus able to bring that considerable experience to bear on AV accidents).

For assisted living robots deployed in care-home settings, such as in our example scenario, accidents could be reported to both the Care Quality Commission[21] (CQC)—the regulator of health and social care in England—and the Health and

---

[17]After the paramedics, that is.

[18]https://www.gov.uk/government/organisations/air-accidents-investigation-branch.

[19]https://www.hsib.org.uk/.

[20]https://www.ntsb.gov/Pages/default.aspx.

[21]https://www.cqc.org.uk/.

Safety Executive[22] (HSE), since care homes are also workplaces. Again it is doubtful that either the CQC or HSE would have the expertise needed to investigate accidents involving robots. Accidents involving workplace assistant robots, or robots used in schools—including near misses—would certainly need to be reported to the HSE. It is clear that as the use of social robots in society increases, regulators such as the CQC and HSE will need to create the capacity for robot accident investigation branches, as would the HSIB for surgical or healthcare robots. Even more urgent is the need to record all such accidents—again including near misses— so that we have, at the least, statistics on the number and type of such accidents.

Until such mechanisms exist, or for robot accidents in settings that fall outside those outlined here, the only recourse we have is to contact the robot's manufacturer, thus underlining the importance of clear labelling of the robot's make and model alongside contact details for the manufacturer of the robot itself.[23] Even if the robot and its manufacturer do not yet have data logging technologies (such as the EBB) or processes for accident investigation in place, we would hope that they would take accidents seriously. A responsible manufacturer would both investigate the accident—drawing in outside expertise where needed—and effect remedies to correct faults. Ideally, social robot manufacturers would adopt the data-sharing philosophy that has proven so effective in aviation safety, summed up by the motto 'anybody's accident is everybody's accident'.

Acknowledgements The work of this chapter has been conducted within EPSRC project RoboTIPS, grant reference EP/S005099/1 RoboTIPS: Developing Responsible Robots for the Digital Economy. CM's contribution to this work was supported by the Wellcome Trust [213632/Z/18/Z].

References

1. Air Accident Investigation Branch, *AAIB Centenary Conference* (2015)
2. H. Alemzadeh, J. Raman, N. Leveson, Z. Kalbarczyk, R.K. Iyer, Adverse events in robotic surgery: a retrospective study of 14 years of FDA data. PloS One **11**, 4 (2016)
3. ATSB, Analysis, causality and proof in safety investigations. Technical Report, Canberra: Australian Transport Safety Bureau (2007)
4. M. Boden, J. Bryson, D. Caldwell, K. Dautenhahn, L. Edwards, S. Kember, P. Newman, V. Parry, G. Pegman, T. Rodden, T. Sorrell, M. Wallis, B. Whitby, A. Winfield, Principles of robotics: regulating robots in the real world. Connect. Sci. **29**, 124–129 (2017)
5. E. Broadbent, Interactions with robots: the truths we reveal about ourselves. Ann. Rev. Psychol. **2017**(68-1), 627–652 (2017)
6. J. Bryson, A. Winfield, Standardizing ethical design for artificial intelligence and autonomous systems. Computer **50**(5), 116–119 (2017)
7. BSI, *BS8611:2016 Robots and Robotic Devices, Guide to the Ethical Design and Application of Robots and Robotic Systems*. British Standards Institute (2016)

---

[22]https://www.hse.gov.uk/.

[23]See EPSRC Principle of Robotics #5 in [4].

8. B.S. Dhillon, Robot accidents, in *Robot Reliability and Safety* (Springer, New York, 1991)
9. M.C. Elish, Moral crumple zones: cautionary tales in human-robot interaction. Engag. Sci. Technol. Soc. **5**, 40–60 (2019)
10. EU, Regulation No 996/2010 of the European Parliament and of the Council of 20 October 2010 on the investigation and prevention of accidents and incidents in civil aviation and repealing Directive 94/56/EC. *Official Journal of the European Union*, 12.11.2010 (2010)
11. R.P. Fisher, R.E. Geiselman, *Memory Enhancing Techniques for Investigative Interviewing: The Cognitive Interview* (Charles C Thomas Publishe, Springfield, 1992)
12. H. Gabler, C. Hampton, J. Hinch, Crash severity: A comparison of event data recorder measurements with accident reconstruction estimates. *SAE Technical Paper 2004-01-1194* (2004)
13. D.R. Grossi, Aviation recorder overview, national transportation safety board [NTSB]. J. Accid. Investig. **2**(1), 31–42 (2006)
14. IEEE, The IEEE global initiative on ethics of autonomous and intelligent systems. ethically aligned design: A vision for prioritizing human well-being with autonomous and intelligent systems, first edition. Technical Report, IEEE (2019)
15. International Civil Aviation Authority, *Annex 13 to the Convention on International Civil Aviation, Aircraft Accident and Incident Investigation* (ICAO, Montreal, 2007)
16. International Federation of Robotics (IFR), *Executive Summary World Robotics 2019 Industrial Robots* (2019)
17. International Standards Organisation, *ISO 13482:2015: Robots and robotic devices - Safety requirements for Personal Care Robots* (ISO, Geneva, 2014)
18. A. Jobin, M. Ienca, E. Vayena, The global landscape of AI ethics guidelines. Nat. Mach. Intell. **1**, 389–399 (2019)
19. P.B. Ladkin, *Causal System Analysis* (Springer, Heidelberg, 2001)
20. P.B. Ladkin, J. Sanders, T. Paul-Stueve, *The WBA Workbook* (Causalis in der IIT GmbH, Bielefeld, 2005). https://rvs-bi.de/research/WBA/TheWBACaseBook.pdf
21. C. Macrae, Making risks visible: Identifying and interpreting threats to airline flight safety. J. Occup. Organ. Psychol. **82**(2), 273–293 (2010)
22. C. Macrae, *Close Calls: Managing Risk and Resilience in Airline Flight Safety* (Palgrave, London, 2014)
23. C. Macrae, The problem with incident reporting. BMJ Qual. Saf. **25**(2), 71–75 (2016)
24. C. Macrae, C. Vincent, Investigating for improvement. building a national safety investigator for healthcare. clinical human factors group thought paper. Technical Report (2017)
25. T. Malm, J. Viitaniemi, J. Latokartano, et al., Safety of interactive robotics - learning from accidents. Int. J. Soc. Rob. **2**, 221–227 (2010)
26. R. Moore, A bayesian explanation of the uncanny valley - effect and related psychological phenomena. Sci. Rep. **2**, 864 (2012)
27. National Transportation Safety Board, *Collision Between a Car Operating With Automated Vehicle Control Systems and a Tractor-Semitrailer Truck Near Williston, Florida.* Washington (2016)
28. National Transportation Safety Board, *Preliminary Report for Crash Involving Pedestrian.* Washington (2018)
29. NHS, NaPSIR quarterly data summary April-June 2019. Technical Report, NHS (2019)
30. C. O'Donovan, Explicitly ethical standards for robotics. Technical Report, Working paper for the international symposium: Post-automation, democratic alternatives to Industry 4.0 SPRU - Science Policy Research Unit, University of Sussex, 11–13 September, 2019 (2020)
31. R. Owen, The UK engineering and physical sciences research council's commitment to a framework for responsible innovation. J. Res. Innov. **1**(1), 113–117 (2014)
32. A.K. Pandey, R. Gelin, A mass-produced sociable humanoid robot: pepper: the first machine of its kind. IEEE Rob. Autom. Mag. **25**(3), 40–48 (2018)
33. I. Radun, H. Summala, Sleep-related fatal vehicle accidents: characteristics of decisions made by multidisciplinary investigation teams. Sleep **27**(2), 224–227 (2004)
34. J.T. Reason, *Managing the Risks of Organisational Accidents* (Ashgate, Aldershot, 1997)

35. J. Sanders, *Introduction to Why Because Analysis* (Bielefeld University, 2012). http://rvs.uni-bielefeld.de/research/WBA/WBA_Introduction.pdf
36. S. Spiekermann, T. Winkler, Value-based Engineering for Ethics by Design (2020). https://arxiv.org/abs/2004.13676
37. N.A. Stanton, P.M. Salmon, L.A. Rafferty, G.H. Walker, C. Baber, D.P. Jenkins, *Human Factors Methods: A Practical Guide for Engineering and Design* (Routledge, London, 2013)
38. P. Underwood, P. Waterson, Systems thinking, the Swiss cheese model and accident analysis: a comparative systemic analysis of the Grayrigg train derailment using the ATSB, AcciMap and STAMP models. Accid. Anal. Prev. **68**, 75–94 (2014)
39. H. Webb, M. Jirotka, A.F. Winfield, K. Winkle, Human-robot relationships and the development of responsible social robots, in *Proceeding of the Halfway to the Future Symposium 2019 (HTTF 2019)*, NY, 2019. Association for Computing Machinery Article 12, pp. 1–7
40. A. Winfield, Ethical standards in robotics and AI. Nat. Electron. **2**(2), 46–48 (2019)
41. A.F. Winfield, M. Jirotka, The case for an ethical black box, in *Towards Autonomous Robotic Systems (TAROS 2017) Lecture Notes in Computer Science Vol. 10454*, ed. by Y. Gao, S. Fallah, Y. Jin, C. Lekakou (Springer, Cham, 2017), pp. 262–273
42. A.F. Winfield, M. Jirotka, Ethical governance is essential to building trust in robotics and artificial intelligence systems, phil. Trans. R. Soc. A **376** (2018). https://doi.org/10.1098/rsta.2018.0085

Chapter 7
Verifiable Autonomy and Responsible Robotics

Louise Dennis and Michael Fisher

Abstract The move towards greater autonomy presents challenges for software engineering. As we may be delegating greater responsibility to software systems and as these autonomous systems can make their own decisions and take their own actions, a step change in the way the systems are developed and verified is needed. This step involves moving from just considering what the system does, but also why it chooses to do it (since decision-making may be delegated). In this chapter, we provide an overview of our programme of work in this area: utilising hybrid agent architectures, exposing and verifying the reasons for decisions, and applying this to assessing a range of properties of autonomous systems.

1 Introduction

Autonomy is surely one of the key technological themes for the twenty-first century. We expect to soon see fully autonomous vehicles, robots, and software, all of which will need to be able to make their own decisions and take their actions, without direct human intervention. Depending on our levels of trust in such technologies, this potential future may be exciting or frightening, and possibly both! Exciting in that autonomous systems can allow us to devise support and tackle tasks well beyond current possibilities; frightening in that the direct control of these *autonomous* systems is now taken away from us.

How do we know that they will work? How do we know that they are safe? And how can we trust them? All of these are currently impossible to answer since we cannot be sure that such systems will never deliberately cause harm. Yet, without such guarantees, increased autonomy could be unlikely to be accepted by engineers, allowed by regulators, or trusted by the public.

L. Dennis · M. Fisher (✉)
Department of Computer Science, University of Manchester, Manchester, UK
e-mail: louise.dennis@manchester.ac.uk; michael.fisher@manchester.ac.uk
http://www.cs.man.ac.uk/~dennisl; https://web.cs.manchester.ac.uk/~michael

© The Author(s) 2021
A. Cavalcanti et al. (eds.), *Software Engineering for Robotics*,
https://doi.org/10.1007/978-3-030-66494-7_7

As we can never precisely describe, or model, the real world, we are never able to say that a system will be absolutely safe or will definitely achieve something; instead, we can only say that it *tries* to be safe and tries to carry out its tasks to the best of its ability. This distinction is crucial: we can only prove that an autonomous system never decides to do the wrong thing and always aims to act safely, we cannot guarantee that accidents will never happen.

However this step, in itself, is important—if an accident occurs, did the system intentionally cause the accident, or did it just make a mistake? Examining the decision-making process, and particularly the *reasons* for such decisions, becomes crucial.

Traditionally, practical autonomous systems comprise complex and intricate controllers, often involving open and closed-loop control systems of many varieties (parametric, neural, fuzzy, etc.). However, since the reasons for any choice of action are often opaque, such systems are hard to develop, control, and analyse.

We utilise a *hybrid agent architecture*, wherein we can separate the discontinuous changes (i.e. 'deliberate' choices) into a distinct 'agent', leaving the low-level control aspects to traditional methods. These low-level elements range across adaptive control, image classification, sensor fusion techniques, manipulation, etc. A further aspect is that this agent must have, and be able to explain, explicit reasons for making the choices it does [21, 31].

Once we have such a hybrid agent architecture, the reasons for particular choices are exposed and transparent. This is vital to regulators who need to be able to interrogate the exact motivations for decisions, for engineers who need clear and precise structure in systems decisions, and the public who may be convinced of a robot's trustworthiness if they can access the internal reasoning processes. Using this approach we can also directly tackle complex questions of safety, ethics, legality, etc. [10, 54]. Without this fundamental step, no amount of testing or simulation is likely to satisfy system producers, the public, or government regulators. This will involve formally verifying the agent's (discrete, high-level, finite) decision-making and that decisions are taken for the right reasons.

In this chapter, we provide an overview of our programme of work in this area: utilising hybrid agent architectures, exposing and verifying the reasons for decisions, and applying this to assessing a range of properties of autonomous systems. Our long-term aim is to ensure that practical and useful (effective) autonomous systems can be constructed that can truthfully (verifiably) explain their intentions. Since truly autonomous robots will have at least some, and possibly quite significant, responsibilities, this will be crucial. This chapter has close links with, and relevance to, Winfield's work (Chap. 6). Indeed, we have collaborated closely in developing formal verification for his ethical governors; see Sect. 8.2. Furthermore, the ability of our systems to explain, reliably and before acting, what they will do and why they make these choices solves one of the core issues in explainability, as highlighted by Winfield and Jirotka's work on an *ethical black box* [56]; see Sect. 7.1.

In what follows, we will begin by exposing the key issues with autonomous systems. We follow this with a description of our approach and then explanations of how this solves many of the issues highlighted earlier.

2 What Is the Problem?

2.1 Concerns About Autonomy

Once the decision-making process is taken away from humans, even partially, can we be sure what autonomous systems will do? Will they be safe? Can we ever trust them? What if they fail? This is especially important once robotic devices, autonomous vehicles, etc., increasingly begin to be deployed in safety-critical situations.

Example from Popular Culture We have seen in science fiction films such as *Terminator* and *I, Robot* typical examples of robots with sinister intent. In such films, there is an assumption that, just as with humans, we cannot accurately expose the intentions that these robots have. However, if we construct our robot's software carefully, then we certainly can assess the *reasons* for making certain choices.

2.2 No Psychiatrists for Robots?

As we move towards increased autonomy, we need to assess not just *what* the robot will do, but *why* it chooses to do it. As above, the common view is that we cannot reason about why the robot makes the decisions it does, and specifically cannot capture its *intent*. However, with an *autonomous system* that we have built, we should, if we are careful with its construction, be able to examine its internal programming and find out exactly

1. *what* it is 'thinking',
2. what *choices* it has, and
3. why it *decides* to take particular options.

Consequently, we explicitly program autonomous systems, such as robots, using an intentional approach and ensure that these are both transparent and exposed. Returning to the *Terminator* example above, think how different (and perhaps uninteresting) such films would have been if robots had to truthfully explain their intentions. If we could guarantee, through proof for example, that decisions were always made for the right reasons and that these reasons were truthfully displayed, then think how quickly you would become confident, and possibly more trusting, of such robots.

3 Background

Although robots have been in use for many years, these have been typically part of automated systems (such as production lines) that have very fixed tasks to undertake. Furthermore, these robotic systems are regularly monitored by human operators. But, this situation will change. Increasingly, there is a strong need for robotic systems that can not only continue working for long periods of time but can do so without close human supervision. It is only when these systems become autonomous that we will see significant change and efficiency improvements.

However, current robotic systems, such as driver less cars, unmanned air vehicles, domestic robots, manufacturing robots, etc., are far from being autonomous. They remain closely controlled by human operators, drivers, or pilots in most crucial situations.

Why is this? Essentially, there is a step change that is needed to move from analysis of standard cyber-physical systems, typically controlled by humans, to truly autonomous systems that make their decisions and have their own responsibilities. This often presents a barrier to the development of technology.

We would like to have autonomous home-care robots that can significantly improve social care for the elderly, driver less cars that fundamentally improve accident rates on our roads, nuclear robots that can reliably decommission a complex power-station without danger to humans, etc. All of these types of systems can strongly impact, and enhance, both societies and economies.

The main barriers to the development of such systems are not technological, but ethical, legal, and social. Again understanding why decisions are made in autonomous systems is central.

3.1 Autonomy

We begin with a straightforward definition of *autonomy*:

- the ability of a system to make its own decisions and to act on its own, and to do both without direct human intervention.

With this definition, we are particularly focussing on systems where *software* has a key role in decision-making.

Who Makes the Decisions? Even within the above definition of 'autonomy', there are important variations concerning how this autonomy (or autonomous decision-making) is achieved.

- **Automatic:** involves several fixed, and prescribed, activities; there may be options, but these are generally fixed in advance and rely on fairly minimal input from the environment—for example, to check the positioning of a target is within acceptable bounds.

- **Adaptive:** improves its performance/activity based on feedback from its environment—typically developed using tight continuous control and optimisation (e.g. a feedback control system) and generally modelled using differential equations.
- **Autonomous:** decisions made based on the system's (belief about its) current situation at the time of the decision—the environment may still be taken into account, but the motivations and beliefs within the system are important.

Especially when we come to (formal) verification, distinguishing *between* these variations is often crucial. If we think about who or what makes the decisions, then we have very different forms of, and targets for, verification:

- in *automatic systems*, the decisions are encoded by the system developer and so the exact options can be verified before deployment;
- in *adaptive systems*, the decisions are essentially made by the environment with tight feedback control driving the system through environmental interaction, to which the system merely reacts, and so verification here requires a model (always imprecise) of the environment; and
- in *autonomous systems*, decisions are taken by the software based on internal goals and motivations while taking into account beliefs about the environment, and so here verification should be applied particularly to the core decision-making software (agent).

Our system architectures will combine these elements and so analysis will require verification of all these different forms. So, we next revisit a range of suitable verification techniques that we might use for the above components.

3.2 Verification

The aim of verification is to ensure that a system matches its requirements; formal verification takes this further, not only having precise formal requirements but carrying out a comprehensive mathematical analysis of the system to 'prove' whether it corresponds to these formal requirements.

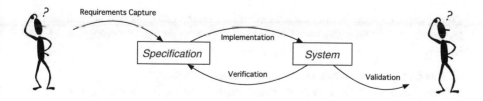

Essentially, verification is the process of assessing, sometimes formally, that we have built the system satisfying its requirements/specification.

As we will see later, we capture the system's core autonomous behaviour as a *rational agent*. Once we do this, formal verification techniques for autonomous decision-making become possible. Any verification technique here needs to take into account not only what the agent does but why it chooses to do it [31]. In this case, a logical proof about the agent's choices is required. Note that, in legacy cases, where a distinguished decision-making agent is *not* a core part of the architecture, we might be able to formally verify any governor component usually used to regulate the safety/ethics of decisions within a robotic system [5, 58].

Varieties of Formal Verification Verifying that an agent, A, matches a (formal) requirement, R, for that agent can be carried out in many ways.

- *Proof:* the behaviour of the agent is itself formally described by the logical formula, φ_A, and verification involves *proving* that this behaviour, φ_A, always implies that the requirement, R. (Essentially, $\vdash \varphi_A \Rightarrow R$.)
- *Traditional Model-Checking:* where R is checked against a representation of all possible execution paths of the agent, A.
- *Synthesis:* from R generate an automaton implementing A that guarantees to only produce runs satisfying R.
- *Dynamic Fault Monitoring (aka Runtime Verification):* where executions *actually* generated by the agent, A, as it is running are checked against R.
- *Program Model-Checking:* all execution paths the program implementing the agent, A, can take are checked against R.

We also note that *Testing*, whereby R is checked on a subset of the possible executions of the agent A, is also a popular, though not exhaustive, route. As we will see later, the above varieties of verification will be used for the different components within the system.

Formal verification, through model-checking, of agent-based autonomous systems is still at a relatively early stage [7, 8, 30, 46] though the techniques and tools reported in this chapter are at the forefront of this work, specifically the MCAPL framework and AJPF program model-checker [20, 21, 28].

3.3   Trustworthiness

The concept of trust in autonomous systems is quite complex and subjective [33]. However, the *trustworthiness* of an autonomous system usually comprises at least two aspects:

1. *Reliability*—does system work reliably and predictably?
2. *Beneficiality*—is the system working for our benefit, for example, is it trying to help us, not harm us?

Even for (non-autonomous) cyber-physical systems, providing evidence of reliability is problematic. This often involves mathematical models of the system and its environment, both of which will necessarily be wrong, in the sense that they do not capture the physical world. In the case of autonomous systems, we are particularly unclear about exactly 'how wrong' these models are, especially once we deploy them in unknown and dynamic environments. To assess the reliability of a system, we will also need to be clear about exactly what we expect it to do (and so what its requirements are). While this captures the trustworthiness of a standard cyber-physical system, once we move to autonomous systems, then the concept of *beneficiality* becomes important.

Since we are delegating activity and responsibility to an autonomous system, and especially where this might involve some critical activities, we must be sure that the system will behave as we would want even when not directly under our control. It is here that 'why' the system will choose to make certain decisions becomes crucial. In complex environments we cannot predict all the decisions that must be made (and so cannot pre-code all the 'correct' decisions), but we can ensure that, in making its decisions, an autonomous system will carry them out 'in the right way'.

Example A domestic robot helper might have a standard mode that prohibits contact with humans. Thus this robot in our home keeps out of our way. In the (uncommon) case where the single home occupant is unconscious and a fire has started in the house, then the robot should carry out reasoning (and assess the risks) hopefully deciding to (gently) carry the home occupant out of the house.

Clearly, for this example we could hard-wire a policy but there will always be scenarios that we have not considered (and so set rules for). In making its decisions, the robot's primary goal is to keep the occupant safe.

Providing evidence for the beneficial activities of the system is a challenge, particularly for systems built from opaque components, such as deep-learning modules. However, as we will see later in this chapter, if we build the system appropriately, exposing the reasons for decisions, and then apply formal verification techniques, we can have much greater confidence in the system, especially its beneficiality.

4 Autonomous Robotics

Robots are already here: machines that clean our houses are well established [49] (see also Fig. 7.1), and factory automation has been available for a while with robots making our cars, our clothes, even our food [35]! But now a new generation of robots is being designed that have the potential to affect safety. Unlike factory robots, they have no cages around them, and unlike robot floor cleaners, they are multi-functional with a range of sophisticated capabilities.

These robots (e.g. Care-O-bot, Fig. 7.2) are increasingly *autonomous*—they are designed to make their own decisions, and even act, *without* human intervention. These decisions may be essential in critical situations: maybe the robot is assisting a person at home who has an accident, or is assisting a person at work who does

Fig. 7.1 Roomba vacuum cleaner. Source: iRobot: https://www.irobot.com/for-the-home/vacuuming/roomba.

Fig. 7.2 Care-O-bot 4 robotic home assistant.
Source: Fraunhofer IPA: http://www.care-o-bot.de/en/.

not see imminent danger. These are not an isolated class, since unmanned vehicles (see Figs. 7.3 and 7.4) and broad AI decision systems might all be required to take critical decisions.

Fig. 7.3 Parrot Bebop 2 drone. Source: http://www.parrot.com.

Fig. 7.4 Waymo self-driving car. Source: Waymo: http://waymo.com.

As such systems become increasingly autonomous, so the need to capture the core autonomous decision-making (where possible) becomes crucial. This enhances explainability, but, if we can then apply formal verification techniques, then we can impact upon trustworthiness of these new classes of robotic system.

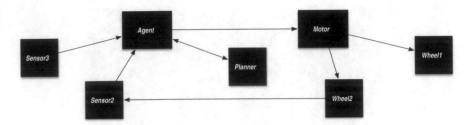

Fig. 7.5 Modular architecture. Copyright Dennis and Fisher, printed with permission in this chapter.

4.1 Architectures

In constructing autonomous systems such as autonomous robots, software engineering is a key aspect since the 'autonomous behaviour' is primarily encapsulated within software. There is a range of (software) architectural styles used for these systems, but with several underlying themes as described below.

4.1.1 Robot Architectures: Modularity

First, *modularity* has now been widely adopted, both because of developing international standards (ISO/DIS 22166)[1] and through the widespread use of de facto standards such as the *Robot Operating System* (ROS) [45]. Such modularity separates key architectural elements and permits improved design, analysis, and maintenance. See Fig. 7.5 for a representation of a typical modular architecture.

In addition to permitting and supporting *inter-operability* amongst sub-components, such modularity provides two specific benefits for us. First, it provides a natural basis for compositional verification and so improves viability and efficiency of, for example, formal verification in that we need only verify smaller components. A secondary benefit is that modular architectures allow heterogeneous components to be employed. These may range across agents incorporating symbolic reasoning, deep learning-based object recognition, adaptive control for grasping, etc. All these components will have very different verification techniques applied to them, so the modularity of the architecture also supports the heterogeneous nature of verification across complex autonomous robotics [27].

4.1.2 Robot Architectures: Transparency

Second, there is a strong move towards *transparency* to ensure that designers, users, and assessors have a clear idea of what is in each component. This is

---

[1] https://committee.iso.org/home/tc299.

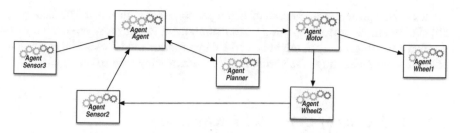

Fig. 7.6 Transparent modular architecture. Copyright Dennis and Fisher, printed with permission in this chapter.

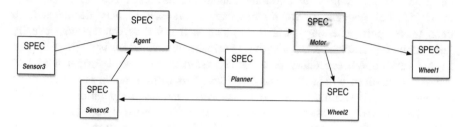

Fig. 7.7 Architecture with module specifications. Copyright Dennis and Fisher, printed with permission in this chapter.

particularly captured by standards such as the IEEE's forthcoming P7001 standard on *Transparency of Autonomous Systems* [37] and through research on *Explainable AI* (see also Sect. 7.1). So, rather than just imposing modularity, we do not want many, or even any, of our modules to be *black boxes*. We require modular components to be *transparent*, i.e. we must be able to inspect and assess the internal behaviour of the module; see Fig. 7.6.

4.1.3 Robot Architectures: Verifiability

Being able to see each component's code still might not help us—understanding adaptive behaviour just from a component's implementation can be hard/impossible. For example, seeing the coding of a complex deep-learning network will likely not aid us in understanding exactly what it is doing. Instead, we usually require a concise, and ideally *formal*, description of the anticipated behaviour of each component.

In Fig. 7.7, each component's specification describes what it is expected to do. Once we have this, we can select the most appropriate verification technique to assess how well it does, indeed, achieve its specification.

Software Engineering for Robotic systems is a complex task, and so modularity is often enhanced with descriptions (even metadata) of the intended behaviour, for example, Genom [34] (see Chap. 8). The formalisation of these descriptions paves the way for compositional reasoning, for example, through assume-guarantee techniques [39]. Furthermore, our use of heterogeneous techniques is useful here [27, 55]. For example, a deep-learning vision component might most productively be verified by testing, an agent might be verified by program model-checking, a motion planning component might be verified by theorem proving, and so on.

4.2   Verification of Cyber-Physical Systems

Before turning to the formal verification of the discrete 'agents' of the type we described above, we provide a comment about the verification of more traditional cyber-physical systems. If we do not care about exactly why decisions have been made, only assessing what the system will do, then our systems usually fall within the automatic or adaptive categories described earlier. To carry out formal verification, especially of adaptive systems, we need a formal model of the environment in which the system resides. This formal model is usually quite complex with differential equations mixed with discrete computations (and will, of course, be wrong). This explains why the majority of formal verification in this area utilises hybrid system verification [3, 11, 43, 44] which in turn often relies on approximation [1, 2, 40].

5   Verifiable Autonomy

In this section, we will describe both our architectural basis and the verification techniques, in order to provide stronger verification of autonomous systems, especially their decision-making.

5.1   Robot Architectures: Responsibility

Following on from the discussion of software architectures in the last section, there is one further architectural aspect that will be crucial to us: the identification of key decision-making responsibilities, usually into identifiable and verifiable components.

As our systems become increasingly autonomous, it is important to be clear *where* key decisions are made and who (whether human or software) is *responsible* for these decisions. Furthermore, how is responsibility distributed, shared, and reinforced? This leads us to further architectural requirements wherein areas of responsibility, and mechanisms for changing them, are explicitly highlighted in decision-making components (agents), as in Fig. 7.8.

Autonomous systems must make their own decisions about what to do, and so such decisions are made in the software that controls the system. In our approach,

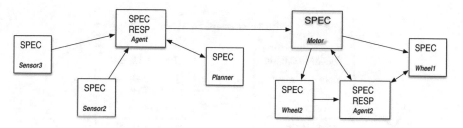

Fig. 7.8 Architecture with agent responsibilities. Copyright Dennis and Fisher, printed with permission in this chapter.

the high-level decision-making is captured as a *rational agent* [31, 60]. These software agents reason in a simplified, but structured, manner.

They have a set of 'goals' they wish to achieve, a set (often quite large) of 'beliefs' about the state of the world, and a set of 'plans' describing their known routes to achieving such goals. On top of these, agents typically have deliberation mechanisms that help decide which goals can be tackled, how to tackle them, and which are the most important to tackle at present [59]. All these aspects are then modified by the agent's evolving context and its dynamic beliefs about the world. Such agents are the core of our approach to responsible autonomous systems.

5.2 Hybrid Agent Architectures

The requirements for *reasoned* decisions and explanations, as well as self-awareness (see later), have led on to our development of *hybrid agent architectures* [42] combining:

1. one or more *rational agents* for *high-level* autonomous decisions, and
2. a range of traditional *feedback control systems* for *low-level* activities,

These can be more *understandable*, easier to *program* and *maintain*, and, often, more *flexible* so providing benefits over monolithic control architectures [19]. An overview is provided in Fig. 7.9, and in subsequent sections we refine this basic concept to provide transparency and verifiability, self-awareness, fault tolerance and reconfigurability, and generality.

Our approach is that

- we should be certain what the autonomous system **intends** to do and how it **chooses** to go about this

and so the core *rational agent*

- must have explicit **reasons** for making the choices it does, and should expose/-explain these when needed.

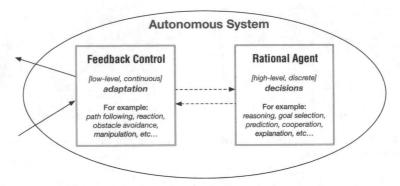

Fig. 7.9 Overview of hybrid agent architecture. Copyright Dennis and Fisher, printed with permission in this chapter.

Example: From Pilot to Rational Agent Within a modern aircraft, the *Autopilot* software can essentially fly the aircraft 'on its own'

- keeping on a particular path,
- keeping flight level and steady under environmental conditions,
- planning routes around obstacles, etc.

In many cases the *human* pilot will then be concerned with high-level decisions, such as

- where to go to,
- when to change route,
- what to do in an emergency, etc.

Once we think of constructing a fully autonomous version of this, the *rational agent* now makes the decisions the pilot used to make.

5.3 Verifying (Rational) Agents

As we saw in Sect. 3.2, we have many options of how to carry out formal verification, but for rational agents, we commonly choose to use *program model-checking* [52] whereby a logical specification is checked against the *actual* agent code that is used in the robot/system. We have developed a toolkit for such verification [16], based on the Java Pathfinder program model-checker [38, 51].

In Java Pathfinder, Java code is symbolically executed (through a modified Java virtual machine), and, as all the possible executions are explored, each is checked against the agent requirements. In Fig. 7.10, we can see (on the left) a backtracking interpreter exploring all possible branches in a program's execution while (on the right) an automaton assesses the required properties on each execution path

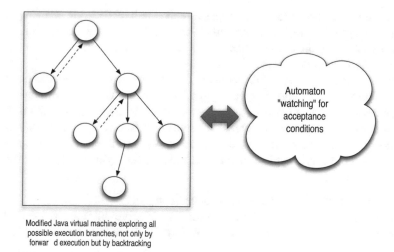

Modified Java virtual machine exploring all
possible execution branches, not only by
forwar d execution but by backtracking

Fig. 7.10 Representation of Java Pathfinder exploration, from [29]. Copyright Dennis and Fisher, printed with permission in this chapter.

explored. This combination of (backtracking) symbolic execution and a monitoring automaton in parallel is termed *'on the fly' model-checking* [36].

5.4 Agent Java Pathfinder (AJPF)

There are many different agent programming languages and agent platforms based, at least in part, on the rational agent view above. These languages are essentially rule-based, goal-reduction languages, with the extra aspect that deliberation, the ability to change between goals and change between plan selection strategies at any time, is a core component. Consequently, they allow the representation and programming of high-level reasoning mechanisms that have many similarities with basic practical reasoning in humans [9]. The basic agent programming language we use is called Gwendolen [15], a language particularly built on the concepts of beliefs, goals (alternatively, *desires*), and intentions [48].

Gwendolen is built on Java and so Java Pathfinder (JPF) can be enhanced to verify Gwendolen programs rather than just Java programs. This requires the extension of the program model-checking approach with the additional agent aspects of Gwendolen (beliefs, plans, goals, etc., as above).

In addition to exposing the structure of the system, we need formal properties to specify. While standard formal specifications often utilise algebraic or temporal specifications, we now need to extend these with agent aspects. To describe these extensions, we will begin with standard temporal logics [29]. While classical propositional logic is typically used to capture *static* and unchanging situations, we turn to propositional temporal logics to describe how the state changes as time progresses. As an example, the formula '$this \land \bigcirc that \land \Diamond the\_other$' means that *this* is true now (in the current state), *that* is true in the next state, and *the_other* is true at *some* state in the 'future'.

Clearly, for describing requirements of our rational agent, we should choose an appropriate way to extend temporal logic. However, there are very many options here, such as: *logics of knowledge, real-time temporal logics, probabilistic logics, cooperation or coalition logics*, etc. No simple logic exactly captures the behaviours of rational agents, and so, in realistic scenarios, we need to *combine* several logics. In our case (and this is typical of many rational agent logics [47]), we, at least in principle, combine:

- a temporal logic, to represent basic dynamic activity;
- a modal logic of *belief* (of the KD45 variety), to represent information [26]; and
- a modal logic of *intention* (of the KD variety), to represent motivation [47].

Combining logics can become very complex, very quickly, so, in practice, we actually use a restricted combination of simple linear temporal logic combined with (non-nested) representations of belief and intention, both referring to goals and plans that the agent has.

Example Specification A typical statement we might have concerning human-robot interaction:

- If a robot 'believes' that a human is directly in front of it, and it has a goal to get to a room beyond the human, then it should *never* choose a plan that brings it into proximity with the human, unless there is no alternative, in which case the robot may even decide to wait for the person to move or instead may decide to drop or revise its goal.

This can be formalised in the logic combing time, intention, and belief (as well as some formalisation of plans and actions).

Formal requirements can then be formally verified of our rational agent, allowing us to analyse the core decision-making in our autonomous system.

5.5   *Heterogeneous Verification*

Recall that our hybrid agent-based architecture is as shown in Fig. 7.11. Now, we can employ different verification techniques to different components within our architecture.

- We can *formally verify* the agent's decision-making and so can be certain about the way decisions are made.
- We can *simulate and test* or alternatively *monitor* the feedback control components. This either provides us with a probabilistic estimate (testing) or informs us when 'bad' behaviour occurs (monitoring).

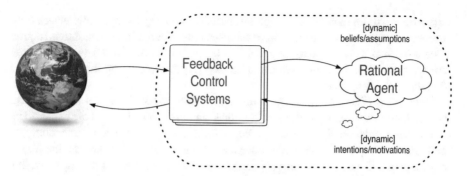

Fig. 7.11 Hybrid agent architecture interacting with environment. Copyright Dennis and Fisher, printed with permission in this chapter.

- We can *practically test* the whole system. Although this gives relatively weak verification evidence, it provides important confidence to the developers, regulators, or users that the system *can* work in realistic scenarios.

So, when we verify autonomous robotic systems constructed using our hybrid agent architecture, we can say *for certain* what it will decide to do once presented with a particular scenario, but cannot guarantee that such scenarios will be recognised accurately. Similarly, once the agent decides what to do, we cannot guarantee that this will be carried out effectively.

Our agent essentially replaces the high-level, human decision-making so formal verification here has an important role in certification.

5.6 Problems: Requirements

As described earlier, if we are to carry out verification, and particularly *formal* verification, of a component, we need to be clear what specification we are verifying against. This leads us to the ongoing problem of *where* these requirements come from. There are two variations on this. In one version the requirements we are given of a system or component may be *extremely* vague and under-specified:

- Please verify the system is safe!
- Please verify the system is secure!
- Please verify the system is usable!
- Please verify the system is nice!

Not only are many of these aspects subjective ('nice', 'usable', etc.) and so more relevant to (user) validation techniques, but no further detail of terms such as 'safe' or 'secure' is given. There might be a link to a relevant standard (where this exists), but such standards are also vague and ambiguous.

A second version of the problem is where we just have *far too many* requirements. There may be precise *safety* requirements and precise *security* requirements.

Furthermore, there may be local regulations to incorporate or, especially where fully autonomous systems are used, *ethical* requirements. Then, if an autonomous robotic system (such as a domestic robot or a driver less car) is owned by a person, then they will likely require their preferences to be incorporated and so on through a variety of additional constraints.

Here the problem is that the number and size of requirements become vast. Especially when we carry out formal verification, this can be beyond the capabilities of contemporary verification tools. Furthermore, we might need quite complex mathematical apparatus even to represent all of these (originally, natural language) requirements. If we are carrying out formal verification on logical descriptions, this often means combining a wide range of logics, and this can easily push verification systems beyond their limits in terms of computational complexity of the combined logics considered or practical complexity of the size of the targets.

6 Applications: 'Act as Humans Should'

A typical sequence of moving from automation to autonomy is to:

1. begin with an automated system (often with fixed behaviour);
2. move to human (and, probably, remote) control of the system;
3. and then gradually replace human decisions with autonomous ones.

It is step (3) where we particularly need verification of the autonomous behaviours. Specifically, we need to ensure that the decisions the autonomous system makes correspond to what the human operators would have done in step (2).

Once we can expose the high-level system decisions, we can match these against a range of 'expected' behaviours. In particular, we can match against specific legal requirements we might have. This is not always possible, especially if we have no prescribed behaviours for human operators, drivers, or pilots, but in some cases can be useful. Below we outline obvious example areas from the realm of autonomous vehicles.

6.1 Towards UAS Certification

This work, from [12, 53] and particularly [54], summarised below shows how we might formally verify that an agent controlling an unmanned air system makes the same (high-level) decisions that a human pilot would (or at least *should*). The 'Rules of the Air' [13] describe what a human pilot should do when in control of an air vehicle. If we replace human control by a rational agent, then clearly the agent should follow the rules prescribed for a human pilot; see Fig. 7.12. Note that the aircraft's *autopilot* will take care of low-level flight skills, with the agent taking care of the higher-level decisions, e.g.

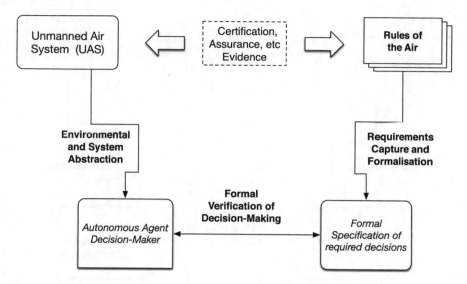

Fig. 7.12 Providing certification evidence for UAS. Copyright Dennis and Fisher, printed with permission in this chapter.

- when two aircraft are approaching head-on ... and there is danger of a collision, each shall alter its course to the right [13].

This type of requirement is then translated to a formal specification. For example, in our temporal logic of belief and intention (see Sect. 5.4, where '□' means 'it is always the case', ◇ means 'at some future moment', and **B** means 'the agent believes'):

$$\Box(\mathbf{B}\ \text{detected\_aircraft} \Rightarrow \Diamond \mathbf{B}\ \text{engage}(\text{emergency\_avoid}))$$

ensuring that emergency_avoid will be engaged. This can then be verified of the agent. Note that the question of how effective emergency_avoid is in ensuring the aircraft turns to the right is assessed differently (through testing) as is the reliability of aircraft detection. Note also that the '**B**' modality is necessarily weak; we cannot be certain about any aspect of the environment so belief is the best we can hope for.

There are many 'Rules of the Air', with many being ambiguous or imprecise (after all they are intended for human pilots) meaning that formalisation can be quite difficult. However, we would expect a trained pilot to adhere to these, and so if we have an autonomous system, it is our rational agent that should be responsible for following these rules.

In [54] we have developed a Gwendolen agent that can control, at a high level, a (simulated) autonomous air vehicle and have formalised a selection of the 'Rules of the Air'. We used AJPF to verify that the agent's decisions match the formal requirements. Note that, while verifying the above rule is relatively simple,

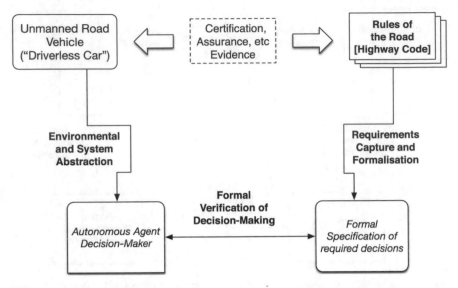

Fig. 7.13 Assessing 'Rules of the Road' in vehicles. Copyright Dennis and Fisher, printed with permission in this chapter.

increasingly complex rules together with a more sophisticated agent will lead to complex verification, requiring us to improve the verification efficiency [57].

6.2 'Driver less' Car Analysis?

In road vehicles, technology is moving beyond adaptive cruise control, lane-keeping, and automated parking, towards more autonomous capabilities. Once we introduce software to control the vehicle, for example, when a driver is taking a rest, we will need much more sophisticated software analysis. Just as a rational agent controlling an autonomous air vehicle might be verified with respect to the 'Rules of the Air' (see the previous section), the rational agent controlling an autonomous road vehicle can be verified against the 'Rules of the Road' [4]. This follows a similar approach to the UAS work, as shown in Fig. 7.13.

In fully autonomous vehicles, the reliability of the agent will be crucial. This reliability is under-specified at present, but must eventually be able to cope with unexpected events, failures, aggressive road users, etc. Thus, the verification of suitable properties concerning the decision-making will be vital. (As ever, capturing these requirements, beyond prescribed 'Rules of the Road', will be difficult.)

7 Applications: Self-Awareness

7.1 Explainability

In some systems, especially those with opaque components, explainability can be difficult. Hence the popularity of the field of Explainable AI [25]. However, if we have constructed our autonomous system using a hybrid agent architecture, then the rational agent

1. has symbolic representations of its motivations (goals, intentions) and beliefs
2. reasons about these to decide what to do, and
3. records all the other options explored (and reasons for not taking them).

And so, it essentially has a trace of reasoned decisions. Then with a little work [41], we can provide explainability and transparency in decision-making. This allows us to provide

- recording facilities—such as Winfield/Jirotka's *ethical black box* [56] (see also Chap. 6);
- interactive capabilities—such as a 'why did you do that?' button [41]; and
- advanced capabilities—such as a 'what will you do next, and why?' button.

In our approach, we do not use any *learning* components for high-level decision-making. Although it may be possible to incorporate them, work remains to be done on explaining new, learned behaviours.

7.2 Self-Awareness

Just as in psychology, where *self-awareness* involves internal models, we can enhance and implement introspection and self-awareness in autonomous systems by providing a range of internal models; see Fig. 7.14. The elements here comprise the following models:

- An *Interaction Model* describing the form of, and requirements for, interaction with operators, humans, inspectors, etc.
- A *Safety Model* capturing the assumptions and guarantees extracted from the system certification process, involving how techniques such as safety cases and fault identification identify safety aspects.
- A *Self Model* describing the components, architecture, and effectiveness of the system itself.
- A *Task Model* describing the goals, tasks, and schedules for the robotic system's activities.

All of these are available to the agent and all are useful for different aspects.

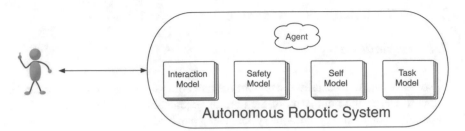

Fig. 7.14 Internal models accessible from the agent; from [32]. Copyright Dennis and Fisher, printed with permission in this chapter.

Example Robots that must continue working for long periods will have failures in some of their subsystems. These failures could range from minor to severe, and, in the former case, we require the system to continue operating, though perhaps at reduced efficiency. In order to recognise such failures and, potentially, provide replacement behaviours, the agent must

- be aware of system components and their expected behaviours,
- be able to monitor its own performance, and
- be able to autonomously *reconfigure* its software.

All of these elements depend on an up-to-date and accurate *Self Model*. Each model could be a separate architectural component, or multiple models of these forms might be directly accessible by the rational agent. Furthermore, this form of internal structure provides possibilities for carrying out formal verification of activities related to self-awareness [18].

7.3 Awareness of Acceptable Boundaries

If we have a clear view of the envelope of behaviour for which our systems approved, then this can be captured within the *Safety Model*. Monitoring system behaviour can highlight when the system steps beyond (or is close to stepping beyond) legal limits.

A more general version of this involves capturing the behaviour boundaries and assessing when the agent moves outside these boundaries. When we carry out formal verification of the agent, there will be some assumptions about the environment in which the agent is embedded. In [23, 28] we showed the use of *Runtime Verification* to monitor whether our system is operating within the bounds where it had been verified. Specifically, when the original formal verification is carried out, a runtime monitor is also generated that can be used by the agent to allow it to check if it is reaching (or has passed) its verified envelope.

8 Applications: Beyond the Predictable

While conforming to legal requirements may be sufficient for many autonomous systems, a further question, particularly for systems deployed in domestic settings, is what happens in unpredictable situations or in situations where some violation is unavoidable.

Recall the example from Sect. 3.3, where a robot should rescue a person from a burning house. Although there may be a very strict rule about human contact, the robot helper decides to 'break' this and save the human. We are here beginning to veer into ethical, rather than straight-forwardly legal, issues.

Cointe et al. [14] integrate rational agents and ethical reasoning into a comprehensive framework in which agent reasoning determines sets of *desirable*, *feasible*, and *moral* actions and then uses context-sensitive ethical principles to select one action from these sets. We are not so comprehensive but, in the sections below, highlight how some ethical decision-making can be implemented and verified within our approach.

8.1 Simple Ethical Ordering

In [22] we considered a situation, where ethical reasoning is only invoked when none of the system's existing plans apply, or a plan is being applied but is not achieving the robot's goal—this follows from the agent having some self-awareness of the effectiveness of its actions and the options it has available. In this situation, we considered an architecture where a route planning system is invoked to produce a wider range of options and they are annotated with the ethical consequences of selecting that option.

We considered examples from the domain of unmanned air systems and an ethical theory based on prima facie duties in which the system has a preference order over its ethical duties (e.g. its duty to minimise casualties takes precedence over its duty to obey the Rules of the Air). In this system, we were able to verify that the aircraft, if forced into an emergency landing, would always land in a field rather than on a road.

This approach used straight-forward ethical priority ordering such as

$$save\ life \gg save\ animals \gg save\ property$$

Then the agent simply assesses the options and takes the most *ethical* one with respect to that ordering. Once the agent decisions take these *ethical concerns* into account, then we can extend formal verification to also assess these. For example, we can formally verify that

if a chosen course of action violates some substantive ethical concern, *A*

then   the other available choices all violated some concern that was equal to, or
more severe than, A.

Though simple, this shows, at least for some simple ethical views, that the
combination of self-awareness ('what decisions are made, and why') and formal
verification ('are all decisions made in the right way') gives us a mechanism for
exploring verifiable robot/machine ethics.

8.2 Ethical Governors (See also Chap. 6)

In [10] we explored the ethical issues further. We used Asimov's Laws of Robotics
as a simple (and well known) example of an ethical theory that could be used to
decide courses of action in an agent.

In experiments, a robot had a goal to move to a particular location, but through
monitoring of its environment, it became aware that a 'human' (also represented by
a robot) was moving towards a dangerous area. The robot could continue moving to
its desired location (as ordered) or choose to intercept the human (and potentially in
some situations could do both). Where the goal-based reasoning did not produce
an ethically acceptable outcome (i.e. where harm befell the human), the moral
decision-making could override the default choices and would select the option for
intercepting the human.

This approach was based on the use of an ethical governor utilising an internal
simulation-based model [50]. As the ethical governor is essentially a rational agent,
we verify this agent against ethical requirements and properties. In Fig. 7.15, we see
how the ethical governor intercepts the actions the robotic system is deciding upon
and either prunes or orders this set of options before returning them to the robot
to choose. In this way, the ethical governor can remove any particularly unwanted
options. It is important, however, for the ethical governor to have mechanisms for
analysing the outcomes of various options and it is here that Winfield et al. utilise
internal simulations [50].

Fundamental to this work was both the self-awareness involved in monitoring the
robot's environment and predicting the outcomes of its actions (through the internal
simulation), and the explicit internal representation of Asimov's laws that allowed
it to pick the most ethically acceptable option.

8.3 Multiple Ethical Theories

Finally, we just mention our work in [6, 24]. Here, we allow and verify the
modification of ethical theories and priorities as our autonomous systems change
contexts. We provide hybrid agent architecture capable of ethical reasoning that
allows agents to change their reasoning as contexts (both physical and social)
change.

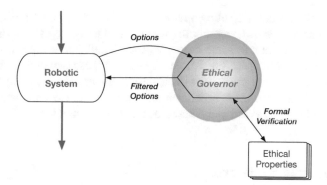

Fig. 7.15 Ethical governor and its verification. Copyright Dennis and Fisher, printed with permission in this chapter.

9 Responsibility

As we have discussed, an *autonomous* system must be able to make its own decisions, and even act, without direct human control. At this point, it becomes important to be sure what the system will choose to do, as it may make this choice without human input. We must not only be able to assess the autonomous decision-making but, in doing so, should be able to convince ourselves that the system will always endeavour to act responsibly and ethically. We term this *responsible autonomy*.

Irresponsible Autonomy Although we might aim for trustworthy and responsible systems, in many cases we cannot ensure these. Either:

1. we know for certain what the system is aiming to do and this will be unacceptable in some form (typically, unethical or irresponsible); or
2. we do not know what the system is exactly doing, what it is aiming to do, or why it makes the choices it does.

Both are bad! In the first case, we are clear about what the system is doing but we have chosen to program it to act badly. In the second case, we do not even know for certain what the system is doing. To distinguish these, we might term the first variety as *unethical* and the second as *irresponsible*.

A related aspect concerns how much we can, or should, rely on these components. Especially with autonomous systems in safety-critical scenarios,

- it is vital that we use strong verification techniques for any component we rely on.

We can be sure what it is trying to do (as above). As a corollary,

- we should never fully **rely** on any component that has not been formally verified.

It is vital that we design and build autonomous systems in ways that are more amenable to strong verification and provide transparency about intention.

Example [17] A robot is holding a hammer, but then drops it on a person's head. Either this was a *deliberate* act and the robot aimed to injure the person, or it was an accident and was due to some failure in, for example, its gripper control. From the viewpoint of public trust, the difference between these two options is *huge*. So, if we can formally verify the robot's decision-making processes to show that it will *never* decide to take an action it believes will cause harm, then this is a significant step on the path to increasing public confidence in this technology.

10 Concluding Remarks

The key, new aspect in *autonomous systems* is that the system is able to *decide for itself* about the best course of action to take. This is a significant change in the roles of engineered systems and requires very different approaches to those traditionally used for (human-controlled) cyber-physical systems. Our approach to this problem is to advocate systems architectures in which the core, high-level decision-making capabilities can be captured as a *rational agent*. Then, by *formally verifying* the rational agent, we verify not what the system does, but what it *tries* to do and why it decided to try! This issue of 'why' decisions are made is crucial. It becomes essential to be able to identify exactly *why* the system took this choice as this may be important in verification or certification; society may be more forgiving of entities that at least *try* to do the right thing.

Once we have delegated decision-making to the agent, then being able to ensure that it will make its future decisions for the right reasons is vital. Our work described here represents initial steps towards providing stronger guarantees of autonomous systems behaviour and so helping trustworthiness and responsibility.

Acknowledgements The work described in this chapter has been supported by EPSRC, over several projects on reconfigurable autonomy (EP/J011770), verifiable autonomy (EP/L024845), sensor systems (EP/N007565), and robotics for hazardous environments, specifically nuclear (EP/R026084), offshore (EP/R026173), and space (EP/R026092). This work was also supported by a Royal Academy of Engineering *Chair in Emerging Technology*. Finally, we would like to thank the organisers of, and participants in, the RoboSoft meeting at the Royal Academy of Engineering in London in November 2019 for their encouragement and feedback.

Apart from Figures 7.1, 7.2, 7.3, and 7.4, all images are copyright Dennis and Fisher, reproduced with permission in this chapter.

References

1. A. Abate, M. Prandini, J. Lygeros, S. Sastry, Probabilistic reachability and safety for controlled discrete time stochastic hybrid systems. Automatica **44**(11), 2724–2734 (2008)
2. A. Abate, J.-P. Katoen, A. Mereacre, Quantitative automata model checking of autonomous stochastic hybrid systems, in *Proceedings of the 14th ACM International Conference on Hybrid Systems: Computation and Control (HSCC)* (ACM, New York, 2011), pp. 83–92

3. R. Alur, C. Courcoubetis, N. Halbwachs, T.A. Henzinger, P.-H. Ho, X. Nicollin, A. Olivero, J. Sifakis, S. Yovine, The algorithmic analysis of hybrid systems. Theor. Comput. Sci. **138**(1), 3–34 (1995)

4. G.V. Alves, L. Dennis, L. Fernandes, M. Fisher, Reliable decision-making in autonomous vehicles, in *Validation and Verification of Automated Systems: Results of the ENABLE-S3 Project*, ed. by A. Leitner, D. Watzenig, J. Ibanez-Guzman (Springer International Publishing, Cham, 2020), pp. 105–117

5. R.C. Arkin, Governing lethal behavior: embedding ethics in a hybrid deliberative/reactive robot architecture, in *Proc. 3rd ACM/IEEE International Conference on Human Robot Interaction (HRI)* (ACM, New York, 2008)

6. M. Bentzen, F. Lindner, L. Dennis, M. Fisher, Moral permissibility of actions in smart home systems, in *Proceedings of the FLoC 2018 Workshop on Robots, Morality, and Trust through the Verification Lens* (2018)

7. R.H. Bordini, M. Fisher, W. Visser, M. Wooldridge, Model checking rational agents. IEEE Intell. Syst. **19**(5), 46–52 (2004)

8. R.H. Bordini, M. Fisher, W. Visser, M. Wooldridge, Verifying multi-agent programs by model checking. J. Auton. Agent. Multi-Agent Syst. **12**(2), 239–256 (2006)

9. M.E. Bratman, *Intentions, Plans, and Practical Reason* (Harvard University Press, Cambridge, 1987)

10. P. Bremner, L.A. Dennis, M. Fisher, A.F.T. Winfield, On proactive, transparent, and verifiable ethical reasoning for robots. Proc. IEEE **107**(3), 541–561 (2019)

11. M.L. Bujorianu, J. Lygeros, Toward a general theory of stochastic hybrid systems, in *Stochastic Hybrid Systems: Theory and Safety Critical Applications*. Lectures Notes in Control and Information Sciences (LNCIS), vol. 337 (Springer, Berlin, 2006), pp. 3–30

12. N. Cameron, M. Webster, M. Jump, M. Fisher, Certification of a civil UAS: a virtual engineering approach, in *AIAA Modeling and Simulation Technologies Conference, AIAA-2011-6664, Portland*, 2011

13. Civil Aviation Authority, CAP 393 air navigation: the order and the regulations (2010). http://www.caa.co.uk/docs/33/CAP393.pdf

14. N. Cointe, G. Bonnet, O. Boissier, Ethical judgment of agents' behaviors in multi-agent systems, in *Proceedings of the 2016 International Conference on Autonomous Agents & Multiagent Systems (AAMAS)* (ACM, New York, 2016), pp. 1106–1114

15. L.A. Dennis, Gwendolen semantics: 2017. Technical Report ULCS-17-001, University of Liverpool, Department of Computer Science, 2017

16. L.A. Dennis, The MCAPL framework including the agent infrastructure layer and agent java pathfinder. J. Open Source Softw. **3**(24), 617 (2018)

17. L.A. Dennis, M. Fisher, *Verifiable Autonomous Systems—Using Rational Agents to Provide Assurance about Decisions made by Machines* (2021, in preparation)

18. L.A. Dennis, M. Fisher, Verifiable self-aware agent-based autonomous systems. Proc. IEEE **108**(7), 1011–1026 (2020)

19. L.A. Dennis, M. Fisher, N. Lincoln, A. Lisitsa, S.M. Veres, Reducing code complexity in hybrid control systems, in *Proc. 10th International Symposium on Artificial Intelligence, Robotics and Automation in Space (i-Sairas)* (2010)

20. L.A. Dennis, M. Fisher, M. Webster, R.H. Bordini, Model checking agent programming languages. Autom. Softw. Eng. **19**(1), 5–63 (2012)

21. L.A. Dennis, M. Fisher, N.K. Lincoln, A. Lisitsa, S.M. Veres, Practical verification of decision-making in agent-based autonomous systems. Autom. Softw. Eng. **23**(3), 305–359 (2016)

22. L.A. Dennis, M. Fisher, M. Slavkovik, M.P. Webster, Formal verification of ethical choices in autonomous systems. Rob. Auton. Syst. **77**, 1–14 (2016)

23. L.A. Dennis, A. Ferrando, D. Ancona, M. Fisher, V. Mascardi, Recognising assumption violations in autonomous systems verification, in *Proc. International Conference on Autonomous Agents and Multiagent Systems (AAMAS)* (2018)

24. L.A. Dennis, M. Bentzen, F. Lindner, M. Fisher, Verifiable machine ethics in changing contexts, in *Proc. 35th AAAI Conference on Artificial Intelligence (AAAI)* (2021)

25. Explainable AI, https://www.darpa.mil/program/explainable-artificial-intelligence
26. R. Fagin, J. Halpern, Y. Moses, M. Vardi, *Reasoning About Knowledge* (MIT Press, Cambridge, 1996)
27. M. Farrell, M. Luckcuck, M. Fisher, Robotics and integrated formal methods: necessity meets opportunity, in *Proc. 14th International Conference on Integrated Formal Methods (iFM)*, ed. by C.A. Furia, K. Winter. Lecture Notes in Computer Science, vol. 11023 (Springer, Berlin, 2018), pp. 161–171
28. A. Ferrando, L.A. Dennis, D. Ancona, M. Fisher, V. Mascardi, Verifying and validating autonomous systems: towards an integrated approach, in *Runtime Verification*, ed. by C. Colombo, M. Leucker (Springer, Berlin, 2018), pp. 263–281
29. M. Fisher, *An Introduction to Practical Formal Methods Using Temporal Logic* (Wiley, Chichester, 2011)
30. M. Fisher, M. Singh, D. Spears, M. Wooldridge, Guest editorial: logic-based agent verification. J. Appl. Log. **5**(2), 193–195 (2007)
31. M. Fisher, L.A. Dennis, M. Webster, Verifying autonomous systems. ACM Commun. **56**(9), 84–93 (2013)
32. M. Fisher, E. Collins, L.A. Dennis, M. Luckcuck, M. Webster, M. Jump, V. Page, C. Patchett, F. Dinmohammadi, D. Flynn, V. Robu, X. Zhao, Verifiable self-certifying autonomous systems, in *Proc. 8th IEEE International Workshop on Software Certification (WoSoCer), Memphis* (2018)
33. M. Fisher, C. List, M. Slavkovik, A. Weiss, Ethics and trust: principles, verification and validation (Dagstuhl Seminar 19171). Dagstuhl Rep. **9**(4), 59–86 (2019)
34. S. Fleury, M. Herrb, R. Chatila, GenoM: a tool for the specification and the implementation of operating modules in a distributed robot architecture, in *International Conference on Intelligent Robots and Systems* (IEEE, Piscataway, 1997), pp. 842–849
35. Food Machinery Company — Automated Ready Meal Production line. www.foodmc.co.uk
36. R. Gerth, D. Peled, M.Y. Vardi, P. Wolper, Simple on-the-fly automatic verification of linear temporal logic, in *Proc. 15th IFIP WG6.1 International Symposium on Protocol Specification, Testing and Verification XV, London* (Chapman & Hall, London, 1996), pp. 3–18
37. Institute of Electrical and Electronics Engineers. P7001—transparency of autonomous systems, 2019
38. Java PathFinder, http://javapathfinder.sourceforge.net
39. C.B. Jones, Tentative steps toward a development method for interfering programs. ACM Trans. Program. Lang. Syst. **5**(4), 596–619 (1983)
40. A.A. Julius, G. Pappas, Approximations of stochastic hybrid systems. IEEE Trans. Autom. Control **54**(6), 1193–1203 (2009)
41. V. Koeman, L.A. Dennis, M. Webster, M. Fisher, K. Hindriks, The 'Why did you do that?' Button: answering why-questions for end users of robotic systems, in *Proc. of the 7th International Workshop on Engineering Multi-Agent Systems (EMAS)* (2019)
42. N. Lincoln, S.M. Veres, L.A. Dennis, M. Fisher, A. Lisitsa, An agent based framework for adaptive control and decision making of autonomous vehicles, in *Proc. IFAC Workshop on Adaptation and Learning in Control and Signal Processing (ALCOSP)* (2010)
43. C. Livadas, N.A. Lynch, Formal verification of safety-critical hybrid systems, in *Hybrid Systems: Computation and Control*. Lecture Notes in Computer Science, vol. 1386 (Springer, Berlin, 1998), pp. 253–272
44. A. Platzer, J.-D. Quesel, KeyMaera: a hybrid theorem prover for hybrid systems, in *Proceedings of 4th International Joint Conference on Automated Reasoning (IJCAR)*, ed. by A. Armando, P. Baumgartner, G. Dowek. Lecture Notes in Computer Science, vol. 5195 (Springer, Berlin, 2008), pp. 171–178
45. M. Quigley, K. Conley, B.P. Gerkey, J. Faust, T. Foote, J. Leibs, R. Wheeler, A.Y. Ng, ROS: an open-source robot operating system, in *Proc. ICRA Workshop on Open Source Software* (2009)
46. F. Raimondi, A. Lomuscio, Automatic verification of multi-agent systems by model checking via ordered binary decision diagrams. J. Appl. Log. **5**(2), 235–251 (2007)

47. A.S. Rao, Decision procedures for propositional linear-time belief-desire-intention logics. J. Log. Comput. **8**(3), 293–342 (1998)
48. A.S. Rao, M.P. Georgeff, An abstract architecture for rational agents, in *Proc. 3rd International Conference on Principles of Knowledge Representation and Reasoning (KR&R)* (Morgan Kaufmann, San Francisco, 1992), pp. 439–449
49. Vacuum Cleaning Robot, http://www.irobot.co.uk/home-robots/vacuum-cleaning
50. D. Vanderelst, A.F.T. Winfield, An architecture for ethical robots inspired by the simulation theory of cognition. Cogn. Syst. Res. **48**, 56–66 (2018)
51. W. Visser, P.C. Mehlitz, Model checking programs with Java PathFinder, in *Proc. 12th International SPIN Workshop*. Lecture Notes in Computer Science, vol. 3639 (Springer, Berlin, 2005), p. 27
52. W. Visser, K. Havelund, G.P. Brat, S. Park, F. Lerda, Model checking programs. Autom. Softw. Eng. **10**(2), 203–232 (2003)
53. M. Webster, M. Fisher, N. Cameron, M. Jump, Formal methods and the certification of autonomous unmanned aircraft systems, in *Proc. 30th Int. Conf. Computer Safety, Reliability and Security (SAFECOMP)*. Lecture Notes in Computer Science, vol. 6894 (Springer, Berlin, 2011), pp. 228–242
54. M. Webster, N. Cameron, M. Fisher, M. Jump, Generating certification evidence for autonomous unmanned aircraft using model checking and simulation. J. Aerosp. Inf. Syst. **11**(5), 258–279 (2014)
55. M. Webster, D. Western, D. Araiza-Illan, C. Dixon, K. Eder, M. Fisher, A.G. Pipe, A corroborative approach to verification and validation of human-robot teams. Int. J. Rob. Res. **39**(1), (2020). https://doi.org/10.1177/0278364919883338
56. A.F.T. Winfield, M. Jirotka, The case for an ethical black box, in *Proc. 18th Annual Conference Towards Autonomous Robotic Systems*. Lecture Notes in Computer Science, vol. 10454 (Springer, Berlin, 2017), pp. 262–273
57. M. Winikoff, L.A. Dennis, M. Fisher, Slicing agent programs for more efficient verification, in *Proc. 6th International Workshop on Engineering Multi-Agent Systems (EMAS)*. Lecture Notes in Computer Science, vol. 11375 (Springer, Berlin, 2019), pp. 139–157
58. R. Woodman, A.F.T. Winfield, C.J. Harper, M. Fraser, Building safer robots: safety driven control. Int. J. Rob. Res. **31**(13), 1603–1626 (2012)
59. M. Wooldridge, *An Introduction to MultiAgent Systems* (Wiley, New York, 2002)
60. M. Wooldridge, A. Rao (eds.), *Foundations of Rational Agency*. Applied Logic Series (Kluwer Academic, Dordrecht, 1999)

Chapter 8
Verification of Autonomous Robots: A Roboticist's Bottom-Up Approach

Félix Ingrand

Abstract Autonomous robots may one day be allowed to fly or to drive around in large numbers, but this will require their makers and programmers to show that the most critical parts of their software are robust and reliable. Moreover, autonomous robots embed onboard deliberation functions. This is what makes them autonomous but open for new challenges. There are many approaches to consider for the V&V of AR software, e.g. write high-level specifications and derive them through correct implementation, deploy and develop new or modified V&V formalisms to program robotics components, etc. One should note that learned models aside, most models used in deliberation functions are already amenable to formal V&V. Thus, we rather focus on functional-level components or modules and propose an approach that relies on an existing robotics specification and implementation framework (G^{en}M), in which we harness existing well-known formal V&V frameworks (UPPAAL, BIP, FIACRE-TINA). G^{en}M was originally developed by roboticists and software engineers who wanted to clearly and precisely specify how a reusable, portable, middleware-independent, functional component should be specified and implemented. As a result, G^{en}M has a rigorous specification and a clear semantics of the implementation, and it provides a template mechanism to synthesize code that opens the door to automatic formal-model synthesis and formal V&V (offline and online). This bottom-up approach, which starts from components implementation, is more modest than top-down ones which aim at a larger and more global view of the problem. Yet, it gives encouraging results on real implementations on which one can build more complex high-level properties to be then verified and validated offline but also with online monitors.

Félix Ingrand (✉)
LAAS/CNRS, University of Toulouse, Toulouse, France
e-mail: felix@laas.fr
http://homepages.laas.fr/felix

1   Introduction

Validation and Verification (V&V) of autonomous systems (AS)[1] software is not a "new problem". More than 20 years ago, a seminal work [31] started a study on how to guarantee robustness, safety and overall dependability of software. Yet, for several reasons, the robotic and AI communities have mostly been focussed on other problems with respect to safe dependable autonomous robots. Meanwhile, there are other fields where bugs and errors can lead to catastrophic events (e.g. aeronautic, nuclear industry, rail transportation) where there is already a large corpus of research, and also successfully deployed tools and frameworks [78], whose goal is to improve the trust we can put in the software controlling these complex, although not autonomous, systems.

The fast and recent developments of autonomously driving cars have put the spotlight on the dramatic consequences of unverified software. Unfortunately, there is no doubt that autonomous vehicles will cause deadly accidents. They will only become "acceptable" if carmakers deploy all reasonably applicable and available techniques to ensure trust and robustness, and if as a result of these techniques, they outperform a regular human driver by one or two orders of magnitude. The decision to deploy these techniques may come from the carmakers themselves, as a commercial argument to safety or as an incentive from car insurances, or they may come from government certification agencies (see Chap. 14) representing the general public's concerns about safety (as this is already the case for aeronautic, railway, etc.). In any case, despite the somewhat human-biased argument that we are "all" good drivers, autonomous cars will probably prevail if 5, 10 or 20 years from now, statistics show that they are indeed safer than human-driven cars.

One original aspect to consider with respect to AS is that, unlike most critical systems in other domains, they exhibit and use deliberative functions (e.g. planning, acting, monitoring, etc.) [44]. If one considers the models used by these deliberative functions, some are explicitly written by humans, while others are learned [4, 46]. Similarly for functional components, some also use learned models. We will see that if the explicit models are amenable to formal verification, the learned ones pose a new challenge to the V&V community.

For now, most of the trust we put in the AS Software (ASS) is acquired through test [47, 71]. Moreover, there are a number of "good" practices, architectures design, software development methodologies, model-based techniques [17, 58] and specification tools that all contribute to establish this trust. Still, formal V&V, when applicable, has the potential to bring a level of confidence unreachable by other practices.

We also need to focus on approaches addressing realistic applications with real implementations and experiments. We already have reached the point where AS with tens of sensors and effectors, executing millions of lines of code, running tens of programs on multiple CPUs are being deployed. The time when one could

---

[1]We consider here autonomous systems at large, i.e. including autonomous robots, autonomous vehicles, drones, cyber physical systems, etc.

illustrate an approach with an example, oversimplifying reality, adding limited value to the field has passed. Proving that a Lego Mindstorm will not crash into a wall, thanks, e.g., to its LTL-generated controller, is not quite the same problem than showing that an autonomous car is not going to do the same.

Moreover, the dependability of the software should be considered as a whole, looking at the entire system. When it comes to V&V, unless you take adapted protective containment measures between components, the overall system will be, at best, as strong as your weakest link or component. The way the components are organized and communicate and share resources in the architecture is as critical as the components themselves.

There are some recent valuable surveys available, each with its own reading grid and focus. Their coverage of the decisional components may be limited [54], or they are limited to the decisional components [66], or they present a larger safety picture [37] (and Chap. 7), but no formal V&V. Still, they are a good source of information in this fast-growing field.

Last, our perspective is definitely from a roboticist's point of view. First, we want to rely, as much as possible, on automatic synthesis of models and code. Second, we are aiming at proving properties which are "useful" for the ASS programmer. Can we guarantee that the plan produced by the planner is safe and that it will be properly executed by the acting system? Can the CPU resources available on the robots guarantee that all components will run fast enough? Can we guarantee that the robot will stop in time when an obstacle has been properly detected, that the initialization sequence will not deadlock, etc.? Overall, starting from some real ASS implementations, what is the current status with respect to V&V and how can we improve it?

This chapter is organized as follows. Section 2 presents the V&V models and techniques that we think are relevant to ASS, while Sect. 3 reviews the various situations with respect to availability of formal models in an AS. We then present in Sect. 3.6 some of the robotic software specification frameworks which could be transformed toward a formal model. Section 4 introduces G$^{en}_o$M, a tool to specify and deploy software functional component, and its template mechanism. A complex robotics example using G$^{en}_o$M is introduced in Sect. 5, followed by Sect. 6 which presents the four formal frameworks for which G$^{en}_o$M can synthesize a model. Section 7 illustrates some of the V&V results we obtain both offline and online. Section 8 concludes the chapter and presents some possible future topics of research.

2 Formal Models and V&V

Our goal here is not to survey such a large field of research. We point the reader to [14, 30, 78] for overviews of formal methods in software development.[2] Formal methods use mathematical or logical models to analyze and verify programs. The

---

[2]Note that none of these three recent surveys mention robots or AS.

key point here is that these models can then be used rigorously to prove properties of the modeled program. Of course, formal methods can cover various parts of the program life cycle, from the specifications down to the code. Here, we mostly focus on approaches that are close to the deployed and running code.

2.1 Models and Methods

There are many formal models available, and none of them cover all the needs. Some models are grounded in simple yet powerful primitives. Automata and state machines [15, 52, 74] are often put forward as they easily capture the various states of the subsystems and their transitions. Petri nets [24, 50] are also often used, as they easily model coordination, together with their time extension, e.g. time Petri nets [12]. Time is also at the core of Timed Automata widely used in UPPAAL, Kronos and more recently in the latest BIP version. Other models are provided as languages defined at a higher level of abstraction, such as the synchronous system family, but can be translated to mathematical or logical representation. Such a category also includes temporal logic [49], situation calculus [22, 51] and interval temporal logic deployed on robots in various components. There are also methods geared toward hybrid systems [73].

From a roboticist's point of view, we should consider the models and methods that seem the most appropriate to represent the type of behavior we have to model and to prove the type of properties we want to check.

2.2 V&V Approaches

Among the various techniques available to the V&V community, we focused on four different families.

State Exploration and Model Checking These approaches, given an initial state and a transition function, explore offline the reachable state of the system. The search space can sometimes be studied without being completely built (e.g. by finding a finite set of equivalence classes), but most often, these approaches suffer from state explosion and face scalability issues.

Statistical Model Checking (SMC) These approaches, instead of exploring the complete state space, sample it using a probability model of the transition function and thus evaluate the properties to be verified with a resulting probability. Indeed, there are many properties which are desirable, but not required 100% of the time. For example, if your drone flight controller, running at 1 kHz, loses 1 cycle every 100 cycles, the consequences are probably not as dramatic as if were to lose one cycle every other cycle. So SMC approaches allow one to "explore" a states space whose size blows the regular state exploration techniques.

Logical Inference These approaches work offline by building an overapproximation of a set of reachable states as a logical statement (e.g. obtained by combining invariants of components). So the set is explored by checking logical properties, not its individual element. If these approaches can potentially address the state explosion, which breaks model checking, they face another problem as the logical invariants can be too general and too loosely fit the real reachable states set.

Last, *runtime verification* is more an online approach where the state transition model is given to an engine that monitors and checks properties and consistency of the model on the fly. This approach requires specification of what needs to be done when a property is violated but allows verification of models that would not scale to or fit the three previous approaches.

We shall now consider the current availability of these models and V&V approach in ASS.

3 Autonomous System Software and Formal Models

With respect to the availability of formal models within ASS, here we examine the overall situation (i.e. the architecture) but also the various types of software components and how they rely (or not) on formal models: programmed with formal models (knowingly or not), learned models, no model and some models.

3.1 Software Architecture

ASS needs to be organized along a particular chosen architecture framework. See [48] for a survey of the different architectures available for AS. Some architectures help and ease the V&V of the overall system, by precisely defining how the layers are organized and how the components interact. To keep things simple with respect to architecture, we consider two levels: the Decisional Level and the Functional Level (see Fig. 8.1). As a result, we often find two types of software components:

- Those performing one of the deliberative functions, which often rely on models (e.g. planning actions, fault detection models, acting skills) that are then used to explore the problem space and find a solution (e.g. a planner combines action models to find a plan, a fault detection component monitors the state of the system to find discrepancies with its model, etc.)
- Those performing some data and information processing to solve a problem or provide a service through a regular algorithm implementing the function (image processing, motion planning, etc.)

In the first category, algorithms involve heuristic searches in a space that cannot be reasonably explicitly computed, while the second involves searching for the

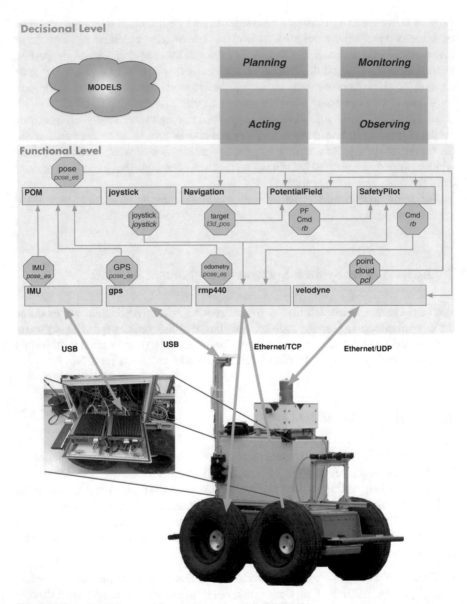

Fig. 8.1 Overall architecture of an autonomous robot (RMP440).

solution explicitly through deterministic algorithms. The algorithm in the first category may take an a priori unbounded amount of time, while the ones in the second usually have a predictable global computation time. The software in the first category often rely on high-level abstraction models, while the ones in the second are often programmed in classical programming languages.

As a result of this dichotomy, the V&V of these different components may rely on different approaches. We now examine the different situations with respect to the availability of models suitable for V&V.

3.2 Directly Programming with Formal Models

We consider here systems or software components that are programmed "directly" using some well-established formal frameworks, which may be used in other domains.

For example, the "synchronous approach" [10, 11] has been instantiated in a number of languages (e.g. Lustre, Esterel, Signal, etc.) and commercial frameworks. It has been deployed in critical domains such as, but not limited to, aeronautics (Scade, Lustre and Esterel) and electronics (Signal). Esterel [16] is used in [68], which presents a framework (Orccad) to deploy robotics systems programmed in the Maestro language (then translated in Esterel) to specify *Robot Tasks* and *Robot Procedures*. This was used to successfully program a complete AUV and prove some logical and temporal properties on the system. Yet, the Maestro language was probably too abstract for robotic system programmers, who did not adopt it.

Similarly, there are decisional components that are programmed using situation calculus. Such a formalism has been used in Golog (for planning) [22, 51] and Golex [38] (for acting) within a museum guide robot. However, the language is quite cumbersome and has hardly been used for other deployments.

There are numerous other formalisms available in the V&V community (e.g. BIP, LTL, UPPAAL, etc.) which could have been used to directly program some robot components, but apart from toy examples, or localized and limited functionalities, this was not attempted. Similarly, there are formal tools to help the specification and the analysis of ASS (e.g. ALTARICA [19]), but they are not easily linked to the real code running on the robot.

RoboChart [20, 57, 61] (and Chap. 9) proposes an interesting approach where the programmer explicitly models the robotics application in a formal framework based on timed communicating sequential process. So the model is provided by the programmer who must have some familiarity with formal models and languages. In Chap. 11, they extend this framework to automatically generate test cases, while Chap. 13 proposes an extension to model uncertainty using probability.

Overall, in this category we find some interesting attempts to bridge the gap between ASS and V&V, but clearly, the languages proposed and the required knowledge and skill to make good use of them are a challenge robotic programmers must overcome.

3.3 "Hidden" Formal Models

In this category we consider the software components that are programmed using some formalisms on their own, but that have not been explicitly used with V&V methods.

For example, PDDL models widely used in automated *planning* have all the right features to be considered formal models. Even if PDDL expressiveness can be somewhat limited when it comes to real robotics application, the semantic is clear and unequivocal, and most heuristic searches used to find plans are correct (if not complete, nor optimal). Similarly, there are other planning formalisms (ANML, HTN, TLPlan, TalPLannner, situation calculus, temporal interval logic, etc.) [44] that all fall in the same category. They were all designed for planning but they have all the right features to be used for V&V. Some works, e.g. [1, 9], explicitly study the link between V&V and planning and scheduling.

Similarly, there are many studies on the *acting* component, which is more concerned with the operational model (skill) of how to execute an action (as opposed to planning, which is more concerned with a model of how to use it). ASPiC [50] is an acting system based on the composition of *skill Petri nets*. [67] show that TDL-based acting components can be verified using NuSMV. RMPL [76] (which relies on an Esterel-like language) is also used for acting and monitoring. RAFCON also provides mechanisms to model and execute plans produced by high-level planners. Similarly, in Chap. 12 the authors propose to program the acting part of the robotics system with a DSL (domain-specific language), relying on LTL, which expresses various control structures and event-handling templates.

So there are many components (mostly deliberative within the decisional layer) that rely on some models that can be linked or transformed to some of the formal models used for V&V. Such a transformation is seldom used, but is an option that could be easily activated if needed.

3.4 Learned Models

Models acquired through learning are of a different kind. They are popular because they successfully tackle problems that resisted analytical modeling and solving. In an AS, one can learn a skill for *acting* (e.g. using reinforcement learning [46] or DBN [43]), an action model for *planning* (e.g. using MDP) or a perception classifier (e.g. using deep learning and convolutional neural network), but from a V&V point of view, these learned models are mostly black boxes.[3]

Still there are some attempts to improve the dependability of learned models. In [3], the authors identify five design pitfalls that can lead to "negative side effects",

---

[3]Note that if learning itself is considered a deliberative function, the learned models can be used within different components, functional or decisional.

and they propose some guidelines to prevent them. Unfortunately, none of them rely on formal methods, and if we can expect better models following them, there is no guarantee that false-positives will not slip in.

Cicala et al. [21] present three areas of ASS where they propose an automatic approach to do V&V. One of the areas is Safe Reinforcement Learning for which they propose to deploy probabilistic model checking on discrete-time model Markov chains. Similarly [66] identify five challenges, to formally verify systems that use AI and Machine Learning: Environment Modeling; Formal Specification; Modeling Systems That Learn; Computational Engines for Training, Testing; and Verification and Correct-by-Construction Intelligent Systems. But overall, researchers are just starting to look at these issues, and the inclusion, or not, of learned models in ASS will depend on their success.

For now, we think that these models must be confined to non-critical components, and if not, their results should be merged, combined and checked for consistency with other results before being considered as input in a critical decisional process.

An interesting work is proposed by Feth et al. [32] where they learn a model to help identify situations which require a higher level of awareness of the system.

3.5 No Model

In many situations, the code and the programs are written following some hopefully good programming practices, but overall, there is no model of what it does. The code is the model. Still, there are a number of tools that make a thorough checking of the code with static analysis and even some invariants extraction [30]. Furthermore, if formal V&V is really required, one can deploy an approach such as the one presented in [72], which requires one to annotate all the functions in the program with logical preconditions, assertions and effects, which will then be checked and inferred by the formal tool (Isabelle). This is a rather tedious process and can only be done by people that are both familiar with the formal frameworks used *and* understand the algorithms being implemented. Nevertheless, the results are very encouraging.

3.6 Some "Specification" Models

In this category, we consider all the components that are somehow specified with some languages, or model-driven frameworks that are not formal. For example, most robotic domain-specific languages (DSLs) are seldom formal. One often uses the term "semi-formal" in the sense that they have a clear syntax, but their semantics is often ambiguous, which prevents them from being directly fed to some V&V engine.

There exist numerous frameworks to develop and deploy robotic software [17, 59]. Some offer specification languages or rely on well-known specification framework (e.g. UML, AADL, etc.). Some just focus on providing tools and API libraries to ease integration of different components (Chap. 1). Orocos [18] focuses on real-time control of robots. [28, 79] present RobotML, a robotic domain-specific language (Papyrus) and toolchain (based on Eclipse Modeling Project) that facilitates the development of robotics applications. Smartsoft [65] (Chap. 3) provides a framework to also specify the complete architecture of a robotics system, while [53] provide a meta-model to separate user programmer concerns and system integrators issues. But if these tools greatly ease the overall architecture and analysis of the system, they remain short of connecting to a formal model.

Despite its success, ROS provides little support when it comes to ease V&V of the software with formal models. There are some efforts to model its communication layer [39] or to verify some simple properties [23, 56, 77], but overall, the lack of structure required to write ROS nodes makes it rather difficult to extract anything worth verifying. [5] propose to model the robot software in AADL and then synthesize code in ROS. There are also systems that build some "runtime" verification of properties on the top of ROS [42, 70], but they hardly rely on ROS itself. [45] propose model transformation using MontiArcAutomaton from a high-level specification down to ROS code. Last we should point out that the SMACH [15] component can be used to control ROS nodes with state machine models.

MAUVE [29, 35] is an interesting framework that allows to extract temporal information from runs and verify them with an LTL checker, but can also perform some runtime verification of temporal properties. Similarly, [27] propose Drona, to perform model checking and runtime verification to check and enforce properties on a model using a Signal Temporal Logic.

3.7 Discussion

Overall, the situation is not completely hopeless. Many components (in particular the deliberative ones) use formal models (hidden but present), which can be used for V&V. Providing the algorithm they use are sound, and that the models are valid w.r.t. the specifications, the results will be consistent w.r.t. the models. On the other end of the spectrum, we have systems using code without any model at all. From our point of view, such practice should not be allowed in critical systems. All the running code should be developed with some level of specification and structure to enable some verification. As for learned models, we need to consider a change of paradigm, and not so much prove that a model is validated and verified, but that it will be used and deployed in such a way that the trust we put in the system is not jeopardized.

Last, one can also expect encouraging results for components developed using a robotics framework, for which a DSL can be derived toward a formal language. In the following sections, we shall examine in detail a particular instance in this category.

We have seen that most robotic tools and frameworks do not provide any formal models per se, and if they do, it is usually up to the programmers to write both the program to run on the robot and the formal model.

Bjørner and Havelund [14] write:

> We will argue that we are moving towards a point of singularity, where specification and programming will be done within the same language and verification tooling framework. This will help break down the barrier for programmers to write specifications.

Following this advice, we advocate that the best way to introduce formal V&V in robotic components is to rely on existing robotic specification languages and frameworks and to offer some automatic translation to formal models. For this, we need to ensure that the semantics of the specification is correct and is properly modeled in the targeted formalism.

We shall now present an existing robotic specification language and its versatile template mechanism.

4 The G^{en}₀M Tool

G^{en}₀M (GENerator Of Modules) [55] is a tool to specify and implement robotic functional components: modules (see the nine module "boxes" of the functional level in Fig. 8.4). These modules provide services in charge of functionalities that may range from simple low-level driver control (e.g. VELODYNE or IMU modules to control Velodyne HLV32 or XSens IMU, respectively) to more integrated computations (e.g. POM for localization with an Unscented Kalman Filter or POTENTIALFIELD for navigation). G^{en}₀M proposes a language to completely specify the functional component down to (but not including) the C or C++ functions (codels) that implement the different stages and steps of the implemented services. This language fully specifies the shared ports (the green octagons in Fig. 8.4) between components (in and out), as well as the shared variables in a component, and the periodic tasks (i.e. threads) in which the services run. For each service, one defines the arguments (in and out) and the automata specifying the steps to follow to execute the codels, as well as their arguments. From a specification point of view, there is a clear semantics of what should be done and how it should be properly implemented.

We now briefly present G^{en}₀M specification and its template mechanism.

4.1 G^{en}ₒM *Specification*

Figure 8.2 presents a generic G^{en}ₒM component, composed of:

Control Task: A component always has an implicit cyclic *control task* that
  manages the control flow by processing *requests* and sending *reports* (from and
  to external clients); it also runs control (attribute and function) *services* and
  activates or interrupts activity *services* in *Execution Tasks.*
Execution Task(s):   Aside from the *control task,* whose reactivity must remain
  high, one may need one or more cyclic *execution tasks,* aperiodic or periodic,
  in charge of longer computations needed by activity *services.*
Services: The core algorithms needed by the component are encapsulated within
  services. Services are associated with *requests* (with the same name). The
  algorithm executed by these services may require a *short* computation time
  or a *long* one. *Short* services are known as *control services* and are directly
  executed by the control task. Control services are in charge of quick computations
  and may be *attributes* (setters or getters) or *functions. Longer* services are
  known as *activities* and they are executed by *execution tasks.* Activities ensure
  longer computations and are modeled with an automaton that breaks down the
  computation into different *states* (see an example in the lower right part of
  Fig. 8.2). Each state is associated with a *codel,* which specifies a C or C++
  function (top right part of Fig. 8.2). The execution of that codel leads to the next
  state in the automaton, to execute immediately, or in the next period if this next
  state name is prefixed with pause.

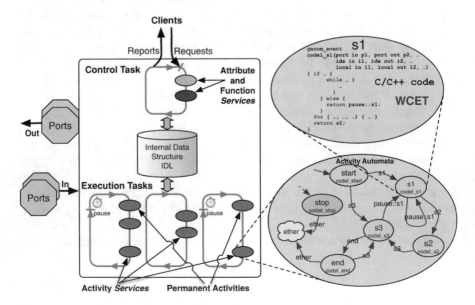

Fig. 8.2 A G^{en}ₒM generic functional component (module).

IDS: A local *internal data structure* is provided for all the *services* to share parameters, computed values or state variables of the component. A codel which needs to access (in or out) fields from the IDS specifies them in its argument list, and G^{en}ₒM will ensure proper locking of the IDS fields during computation.

Ports: They specify the shared data, in and out, the component needs and produces from and for other components. Access to ports is mutually exclusive and is also specified in the *codels* arguments list.

This generic component is then instantiated for each specific component specification with a template mechanism.

4.2   G^{en}ₒM *Templates*

```
void
genom_<"$comp">_activity_report(
  struct genom_component_data *self,
  struct genom_activity *a)
{
  switch(a->sid) {
    case -1: return; /* permanent activity reports nothing */
<'foreach s [$component services] {'>
    case <"$comp">_<"[$s name]">_RqstId:
      genom_<"$comp">_<"[$s name]">_activity_report(
        self,
        (struct genom_<"$comp">_<"[$s name]">_activity *)a);
      return;
<'}'>
  }
```

Listing 1 A simple template code snippet.

A G^{en}ₒM template is a set of text files that include Tcl code, whose evaluation in the context of a G^{en}ₒM call on a specification file will produce the target of this particular template. The target can be as simple as one file with the list of the name of the services specified in the module (in which case the template file will just include a loop over all services and print their name), or it can be the C code which controls the execution of an activity automaton, or it can be all the codes which implement the module itself in ROS. Templates are the building blocks of any output of G^{en}ₒM.

The template mechanism was initially introduced to deal with the *middleware independency* problem [55]. Indeed, the specifications presented above do not subsume any specific middleware. In short, the components are specified in a generic way using G^{en}ₒM, and different templates are then used to automatically synthesize the components for different middleware, which are then linked to the codels library for the considered module.

A template, when called by G^{en}ₒM on a given module specification, has access to all the information contained in the specification file such as services names and types, ports and IDS fields needed by each codel, execution tasks periods, etc. Through the template interpreter (using Tcl syntax), one specifies what they need the template to synthesize. For instance, Listing 1 shows an excerpt of a template code and Listing 2 the C code it produces when called together with the NAVIGATION component specification file. The interpreter evaluates anything enclosed in *markers* <' '> without output, while on the code between <" ">, variables and commands substitution is performed and the result is output in the destination file, together with the text outside of the markers. For example, <'foreach s [$component services] {'> ... <"[$s name]"> ... <'}'> iterates over the list of services of the component contained in the $component variable, while <"[$s name]"> is replaced by the name of the service contained in the $s variable bound by the foreach statement.

```
void
genom_Navigation_activity_report(
  struct genom_component_data *self,
  struct genom_activity *a)
{
  switch(a->sid) {
    case -1: return; /* permanent activity reports nothing */
    case Navigation_connect_port_RqstId:
      genom_Navigation_connect_port_activity_report(
        self,
        (struct genom_Navigation_connect_port_activity *)a);
      return;
    ...
    case Navigation_GotoPosition_RqstId:
      genom_Navigation_GotoPosition_activity_report(
        self,
        (struct genom_Navigation_GotoPosition_activity *)a);
      return;
    case Navigation_GotoNode_RqstId:
      genom_Navigation_GotoNode_activity_report(
        self,
        (struct genom_Navigation_GotoNode_activity *)a);
      return;
  }
  ...
```

Listing 2 Excerpt of the synthesized C code for the PocoLibs NAVIGATION component corresponding to the template in Listing 1 (note how the C code is synthesized for all the services of the component).

There are already templates to synthesize: the component implementation for various middleware (e.g. PocoLibs,[4] ROS-Com [60], Orocos [18]), client libraries to control the component (e.g. JSON, C, OpenPRS), etc. Among the available middleware, we rather focus on PocoLibs as it is the most suitable for real-time

---

[4]https://git.openrobots.org/projects/pocolibs.

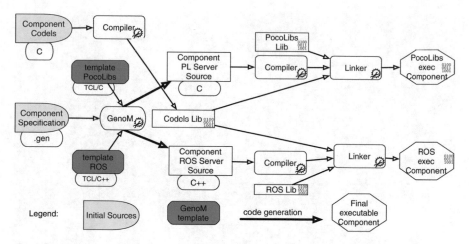

Fig. 8.3 Workflow to synthesize regular PocoLibs or ROS G^{en}₀M components.

applications (notably UAVs). Yet, its implementation, as efficient as it can be, cannot guarantee crucial properties such as schedulability of periodic tasks, for which we need formal V&V.

Figure 8.3 shows the workflow to synthesize regular PocoLibs or ROS G^{en}₀M components.

We now present a non-trivial real-world experiment implemented using G^{en}₀M. As we shall see, it is not an autonomous car, but it shares a lot of common sensors and effectors with one.

5   Not A Toy Example

To illustrate our approach, we present a complete navigation experiment for an RMP440 LE robot (called Minnie). Minnie has the following sensors: an XSens MTi IMU, a KVH DSP-5000 fiber optic gyro and a Novatel GPS, all connected through serial or USB lines, and an HDL-32E Velodyne lidar (on an Ethernet UDP interface). The RMP440 platform comes with a low-level controller (accessed through the Ethernet interface), which allows controlling the robot with a speed (x-linear and z-angular) command, and returns the platform wheel odometry. There is also a wireless Sony PS2 joystick connected to a USB port. Finally, the platform includes a Nuvis 5306RT i7-6700 CPU with 16 Gb RAM and a 256 Go SSD drive, running Ubuntu 16.04.

These hardware components are controlled through their respective G^{en}₀M modules (see Fig. 8.4).[5] They produce their respective ports (e.g. pose estimation,

---

[5]The gyro does not appear as a separate module, because it is managed inside the RMP440 module.

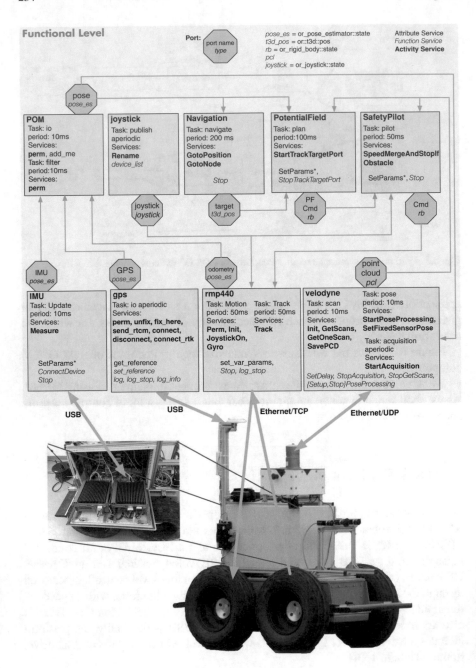

Fig. 8.4 Functional level of the Minnie RMP440 experiment.

point cloud, buttons and axes pushed, etc.). On top of these, POM uses an UKF to merge pose estimations from GPS, IMU and RMP440 (gyro and odometry) and to provide the position of the platform in the **Pose** port. NAVIGATION offers services to navigate in a graph of positions in a topological map of the environment and produces a port with the next **Target** to reach. This **Target** is used by POTENTIALFIELD as the current goal to reach, while avoiding obstacles in the **point cloud** using a Potential Field method inspired by [36] (the points in the cloud are collected in an occupancy grid which is then used to provide obstacles' positions in the local map). The speed reference **PF Cmd** is then read by SAFETYPILOT which, as a last resort, checks that no obstacles are too close to the robot, and stops the robot if needed. It also considers the **joystick** port and uses it as a speed command producer if the proper joystick buttons are pushed (which is a way to gain back control of the robot platform in case something goes astray). The final speed **Cmd** produced is then read by RMP440, which pushes it to the low-level controller of the robot. The goal of this chapter is not to discuss the overall localization and navigation implemented on Minnie, but to give a reasonable idea of the overall complexity entailed by a non-trivial robotic experiment.[6]

Figure 8.4 indicates how many execution tasks each module has, the associated activity services (in bold), the function services (in italic) and the attribute services. Ports, represented by octagons, have a name and the data type they hold. For example, VELODYNE has three execution tasks: scan and pose running at 100 Hz and acquisition aperiodic. scan has four services, *Init*, *GetScans*, *GetOneScans* and *SavePCD*. To further illustrate the G^{en}₃M specification, Listing 3 presents the *GetScans* activity service of the VELODYNE. Note the automata specification, which is also presented in Fig. 8.5.

Overall, the Minnie experiment includes 9 modules, 9 ports, (13 + 9) tasks, 38 activity services (with automata), 41 function services, 43 attribute services, 170 codels over 14k loc (lines of codes) and their respective WCET. The synthesized G^{en}₃M modules amount to 200k loc for all to which one must add external libraries (middleware, PCL, Euler, etc.).

We now briefly present the formal frameworks for which we have written G^{en}₃M templates.

6   Synthesized BIP, FIACRE, UPPAAL Formal Models

The template mechanism used to synthesize the G^{en}₃M modules from their specifications and codels can also synthesize models for three formal frameworks. These templates also use temporal and statistical information obtained by running regular modules with proper probes. In particular, for all formal models, we include the

---

[6]The complete code of the Minnie experiment is available at: https://redmine.laas.fr/projects/minnie.

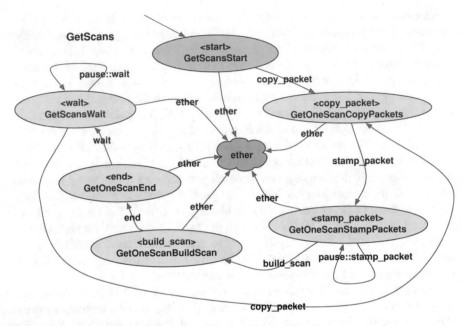

Fig. 8.5 Finite-state machine of the activity GetScans (Listing 3).

extracted Worst-Case Execution Time of each codel, as well as the distribution of state transitions in the service automata for the UPPAAL-SMC model. We now briefly present the three formal frameworks.

6.1 RT-BIP

RT-BIP is a framework[7] that allows modeling of embedded real-time systems, within components, including automata with guards (logical and temporal) and ports for synchronization (rendezvous and broadcast) with other components. RT-BIP is not to be confused with the original BIP template used in [8], which did not include time information. RT-BIP can be used offline with RTD-Finder [7] to prove some properties. For this, it automatically extracts an invariant, analyzing the automata, the component interactions and the clock's histories. This invariant is then used as an overapproximation of the reachable states of the system. It then tries to prove with a SAT solver that this invariant entails a desired property. More interestingly, it can also be used online, and then use the RT-BIP Engine [69] to run the model itself, linked to the codels, and enforce the properties at runtime. This runtime verification can also be augmented with more complex properties the roboticist may want to enforce.

---

[7]http://www-verimag.imag.fr/RSD-Tools.html.

```
 1    activity GetScans(
 2        in double firstAngle = :"First angle of the scan (in degrees)",
 3        in double lastAngle = :"Last angle of the scan (in degrees)",
 4        in double period = :"Time in between two scans",
 5        in double timeout = :"Timeout used when stamping packets")
 6    {
 7      doc "Acquire full scans from the velodyne sensor periodically";
 8      task scan;
 9
10      validate GetScansValidate(in firstAngle, in lastAngle, in period);
11
12      codel<start> GetScansStart(in acquisition_params)
13                          yield copy_packets;
14      codel<copy_packets> GetOneScanCopyPackets(in acquisition_params,
15                               out mutex_buffer)
16                          yield stamp_packets;
17      codel<stamp_packets> GetOneScanStampPackets(in acquisition_params,
18                               out mutex_pose_data, in timeout)
19                          yield pause::stamp_packets, build_scan;
20      codel<build_scan> GetOneScanBuildScan(in acquisition_params,
21                               in firstAngle, in lastAngle)
22                          yield end;
23      codel<end> GetOneScanEnd(in acquisition_params,
24                               port out point_cloud, inout usec_delay)
25                          yield wait;
26      codel<wait> GetScansWait(in period)
27                          yield pause::wait, copy_packets;
28
29      interrupts GetOneScan, SavePCD, GetScans;
30    };
```

Listing 3 The G^{en}oM specification of the *GetScans* activity (VELODYNE).

6.2 UPPAAL and UPPAAL-SMC

UPPAAL is an integrated tool environment[8] for modeling, validation and verification of real-time systems modeled as networks of timed automata, extended
with data types. Unlike BIP, it uses model checking to verify simplified TCTL
(timed computation tree logic) properties in the modeled systems. The latest
UPPAAL version (UPPAAL-SMC[9]) addresses the state explosion limit by offering
a Statistical Model Checking extension. The timed automata can then be enriched
with transition frequencies to perform sampling of the reachable states consistent
with the transition frequencies. The properties are then proven with a confidence,
which is the ratio of state in which they are true over all the sampled states.

---

[8]http://www.uppaal.org/.

[9]http://people.cs.aau.dk/~adavid/smc/.

6.3 FIACRE

FIACRE[10] is a formal language[11] for specifying concurrent and real-time systems also based on automata (behavior), ports and transitions, which can be guarded and sensitized over a time interval (similar to time Petri nets). The semantics is different from the timed automata used in BIP and UPPAAL. FIACRE provides a rich model to represent behavior and timing aspects of concurrent systems, using complex types, functions and externals. The produced model can then be analyzed with a model checking tool (see the TINA (TIme petri Nets Analyzer) toolbox[12]) but can also be deployed and executed using the Hippo model execution engine.

Table 8.1 sums up the various formal frameworks for which $G^{en}_b M$ templates are available, and the corresponding tools used. The resulting formal models are automatically synthesized for the experiment, such as the one presented in Fig. 8.4. The offline versions abstract codels with their WCET and are enriched with a client model (specific or generic) that specifies the sequence of requests to execute. The resulting model is then fed to the respective V&V tools (see some results in [33]). The online versions (RT-BIP Engine and FIACRE-Hippo) provide an API to allow for asynchronous external events handling to receive requests from the real clients, codels execution in separate threads and reporting results to the clients. So the online versions run the real components (in place, e.g., of the PocoLibs implementation of the component); we will show that we can add some formal properties in the model and thus perform runtime verification.

Writing these templates is tedious. It requires a very good knowledge of the $G^{en}_b M$ specification and implementation and of course a good knowledge of the formal frameworks used. But an interesting side effect is that writing the formal version of a synthesized implementation (e.g. the PocoLibs implementation of the module) requires to also clarify when the specification and the implementation are subject to ambiguities.

We shall now examine how the synthesized formal models for the Minnie experiment can be used to provide the Minnie $G^{en}_b M$ module programmers with more confidence and proof of what the robot is doing.

Table 8.1 Existing formal framework templates for $G^{en}_b M$

Formal frameworks	Offline	Online PocoLibs [40]	Online ROS-Comm [60]
RT-BIP [69]	RTD-Finder [7]	RT-BIP Engine [2]	RT-BIP Engine
FIACRE [13]	Tina [25]	Hippo[41]	Hippo
UPPAAL [6]	OK	NA	NA
UPPAAL-SMC [26]	OK	NA	NA

---

[10]"Format Intermédiaire pour les Architectures de Composants Répartis Embarqués", French for "Intermediate Format for Embedded Distributed Component Architectures".

[11]http://projects.laas.fr/fiacre/.

[12]http://projects.laas.fr/tina/.

7   Putting These Formal Models to Use

The three frameworks presented in the previous section have been deployed and tested online and offline with the Minnie experiment.

7.1   Online Runtime Verification with BIP

The BIP model of the Minnie experiment (Sect. 5) is 27k lines long and is linked to all the codels of the experiment. It handles requests, reports and ports exactly like the regular G^{en}₀M module. Thus, the BIP Engine can run the whole experiment, in place of the regular G^{en}₀M module implementation, but also check and enforce properties. Thus, it will check that the tasks' specified periods are respected, but it can also check more complex properties. For example, we may want to stop the robot when the **point cloud** port has not been properly published. For such temporal properties, we can add a monitor (Listing 4) such that:

- The **scan()** BIP port will be connected to the GetOneScanEnd codel (Listing 3, line 23) termination (which corresponds to the writing of the **point cloud** G^{en}₀M port).
- The **report()** BIP port will be connected to the BIP model of the *Track* service of the RMP440 modules, and force a transition to the *stop* state of this automaton which makes an emergency stop of the robot (i.e. bringing both linear and angular speed to 0).

To test this scenario, we introduced a *SetDelay* service in VELODYNE that artificially delays the execution of the GetOneScanEnd codel, and indeed, after 2 seconds, one can see the robot make an emergency stop.[13]

Running the BIP Engine, in the Minnie experiment, incurs a 15% increase in the CPU load, which remains acceptable considering the added safety. The offline version of this model has also been tested and run with RTD-Finder, but did not bring any noticeable results.

7.2   Offline Verification with UPPAAL

The synthesized UPPAAL model for the whole Minnie experiment is 9k lines. This model can be used for offline verification, but we need to add a "client" model, which represents the overall procedure, including the initialization sequence, of request to start and then run the experiment. Indeed, in robotics applications, the initialization sequence is rather critical: many subsystems need to start at once, and

---

[13]BIP Engine traces and videos are available at this url https://redmine.laas.fr/projects/minnie/gollum/index which demonstrate the expected behavior.

```
atom type monitor_timeout()

        clock c unit millisecond
        export port Port scan()
        export port Port report()

        place idle, busy // Automata state

        initial to idle //Initial state

        on scan
        from idle to busy
        do {c = 0;} // reset clock

        on scan // scan interact, we stay in
        from busy to busy // busy state
        provided (c <= 2000) // providing it took less than 2 sec
        do {printf("monitor_timeout <= 2000.\n"); c = 0;} // reset clock

        on report // report interact
        from busy to idle // when in busy (transit to idle)
        provided (c > 2000) // if it took more than 2 sec
        do { printf("monitor_timeout > 2000.\n");}

    end
```

Listing 4 A BIP monitor atom type.

race conditions and deadlock are to be avoided. Such a "client" model is also written in UPPAAL, translated from the regular TCL script used to run the experiment. We can then check offline that no **port** is ever read before it has been written at least once. The UPPAAL property for the **Pose** port is:

A☐ (**not** port_reading[Pose] **or** port_inited[Pose])

Then it is just a matter of making a conjunction for all the ports. If the UPPAAL model checker finds this formula to be false, it will report the sequence which led to this state.

In fact, this very property was initially false because of a race condition in the initialization sequence of the robot where the NAVIGATION module may have tried to navigate before the first **pose** position of the robot had been produced by the POM module.

7.3 Offline and Online Verification with FIACRE

The FIACRE model of the Minnie experiment is around 34k lines and includes 9 modules, 13 execution tasks and 38 activity services calling 170 codels. It is the most detailed formal model we have and provides an execution fully equivalent to the regular G^{en}oM version.[14] The biggest advantage of this model is that it is exactly the same when used offline with the TINA tools and online with the Hippo runtime execution engine. The differences between the two methods are minimal:

- The codel-returned values, which define the next state in the activity automata, in the model executed online are replaced with nondeterministic "select" choices in the model used offline.
- Real code execution time in Hippo (dispatched in a separate thread) is handled with transitions over the time interval of [0, *wcet*] in the offline version.
- We use external ports in the online version to receive requests from the client, while the offline version deploys a generic or specific FIACRE process to produce them and get reports.

In fact, the G^{en}oM template[15] that produces the formal model is the same, there is just a flag to pass at synthesis time to select which version to produce (online or offline).

As a result, we are able to show properties similar to the ones we have shown in UPPAAL, even if the two approaches use two different time presentations (timed automata and global clock versus local clocked guard and sensibilization). Listing 5 shows the code to check for Uninitialized Port Read in FIACRE with Hippo; at runtime an error can be reported when the UPR state is reached. A similar version (without `start/sync`) can be used offline searching if there is a path leading to UPR.

Similarly, we can compute the maximum time it takes between a *stop* request sent to NAVIGATION and the writing of the zero speed on the robot HW controller by RMP440 from the **Cmd** port of SAFETYPILOT. Given a proper model of the scheduler, which can also be written in FIACRE, we can check whether the number of cores on the CPU is sufficient or not [34]. The model includes an `overshoot` state for each periodic task, and if this state is reached, it means that the task has taken more time than its specified period. The number of cores available in an experiment can be considered in the model, and one can adjust the value or the period or the activity services automata, or the codel to change the WCET, to ensure the schedulability of all the tasks. Note that this property is checked offline (considering WCET and the number of cores) but also at runtime during the real execution.

So not only are these properties checked offline, but as a result, they are also true of the same model which executes them in a monitor online with the Hippo engine. Like BIP, we can write monitors in FIACRE that enforces more complex properties at runtime.

---

[14]Without going into the details, the BIP and UPPAAL models are too abstract compared to the FIACRE one; they oversimplify the control task part of requests execution, which handles interruption and precedence execution, and do not properly model the interleaving of activities within one task (activities run completely to ether or pause before the next activity gets called).

[15]https://redmine.laas.fr/projects/genom3-fiacre-template.

```
from Navigation_start
    on (navigate_turn = my_index);
    if (navigate_activities[my_index].state = Navigation_stop) then
        to Navigation_stop
    end;
    on ((mutex_ports[Pose_port] = no_codel) and /* guard to check both */
        (mutex_ports[Target_port] = no_codel)); /* ports are available */
    mutex_ports[Pose_port] := SetTargetToPoseStart_port_codel; /* we lock */
    mutex_ports[Target_port] := SetTargetToPoseStart_port_codel; /* them */
    navigate_running_codel := SetTargetToPoseStart;
    if (not write_ports[Pose_port]) then /* Pose has not been written yet */
        to UPR /* the UPR state will report the error, */
    end; /* and take actions */
    /* The codel is called (start) in its own thread */
    start Navigation_SetTargetToPose_start(navigate_activities[my_index]);
    to Navigation_start_sync

from Navigation_start_sync
    /* waiting (sync) for the codel to return */
    sync Navigation_SetTargetToPose_start state;
    mutex_ports[Pose_port] := no_codel; /* ports are released */
    mutex_ports[Target_port] := no_codel;
    write_ports[Target_port] := true; /* Target port is marked as written */
    navigate_running_codel := 0; /* navigate task has no running codel now
        */
    to Navigation_start_dispatch
```

Listing 5 A FIACRE code snippet handling Uninitialized Port Read of the **Pose** port by the *SetTargetToPose* service from the NAVIGATION module.

Overall, even if model checking techniques suffer from state-space explosion, the results obtained here on fairly complex robotic experiments are still encouraging.

All these results are obtained with models automatically synthesized and with automatic verification tools. One still needs to understand how to express properties in the corresponding query language (e.g. LTL and patterns for FIACRE, TCTL for UPPAAL) and how to interpret the results. Still, it is a big step forward in providing V&V tools to roboticists. To formally validate the obtained model, the semantics of G^{en}₀M has been first specified in Timed Transition Systems and then transformed into Timed Automata with Urgency and Data [34]. Still this is subject to the correctness of the semantics of G^{en}₀M in TTS. If there is a flaw in this semantics, the flaw will also end up in the verified model. From our roboticist's point of view, having exactly the same model (produced from the same template file) for execution and offline verification is a much stronger argument for validation of the model, as it exhibits a behavior at runtime perfectly similar to the regular PocoLibs implementation. Finally, one should note that all implemented modules in G^{en}₀M get all these equivalent formal models for free, and their programmer can run the various V&V tools associated with them.

8 Conclusion and Future Work

In the proposed architecture, we distinguish between functional and decisional components. We have seen that some of these components already provide formal models (Sect. 3.3); therefore, for these components, future research may focus on verifying the correctness of the search algorithms deployed. Some rely on DSL and specific frameworks (Sect. 3.6) which could be extended to automatically provide formal models on which one can perform V&V, as we have shown in Sect. 4. The G^{en}M formal frameworks templates should not be seen as the only possible path to infusing V&V in robotics. It is an example of such a possible path, and there are numerous robotics frameworks that could do a similar automatic transformation of their specification toward formal models. We invite other robotic framework programmers to reach out and look at possible formal frameworks they can connect to. Learned models (Sect. 3.4) have already been identified as outliers, so they probably need a "special" architectural setup for now and until we are satisfied with the confidence we can put in them. As for components with no model at all (Sect. 3.5), there are tedious solutions; if one cannot deploy them, one should at least consider reorganizing the code in such a way that it can rely on existing DSL or robotics frameworks.

On a different topic, the same way AI and robotics have to take into account human presence and interaction, V&V must also integrate it in the process. So we need to consider models of human behavior to introduce them in the V&V process. This can be models of users of the AS itself (e.g. passenger of an autonomous car) but also models of people around the AS (e.g. pedestrian or drivers of a regular car)[75]. Of course this adds another layer of variability and expands again the size of the models to explore, but we should also consider this as an opportunity to "close" the model and keep the reachable states at a reasonable size.

Another topic which needs to be considered is how these different models coexist and complement each other when it comes to proving a property of all the ASS. For example, the link between the different layers and components must also be verified. If one can rely on "compatible" formalisms between these layers, the better, as it will ease checking properties which are defined over more than one layer. The communication and middleware should also be properly modeled to be part of the V&V process. Shared memory is rather straightforward to model with locks, but publish and subscribe over XML-RPC (e.g. in ROS) is a completely different story. One needs to take into account network latencies, size of the queue, etc. We also need to guarantee consistency over the various models deployed.

Last, we should keep in mind that specifications, even if they produce formally equivalent models that can be fed to V&V tools, need to correctly capture the intent of the system designer. As pointed out by [62], "there is no escaping the 'garbage in, garbage out' reality". For this problem, we think that for now, we should rely on good old testing (see in this book Chap. 4 and 5) of the system to check that specifications are correct and synthesize the proper formal model.

Acknowledgments This work has been supported by the European Union's Horizon 2020 research and innovation programme under grant agreement No. 825619 (AI4EU) and the Artificial and Natural Intelligence Toulouse Institute – Institut 3iA (ANITI) under grant agreement No: ANR-19-PI3A-0004.

References

1. Y. Abdeddaim, E. Asarin, M. Gallien, F. Ingrand, C. Lesire, M. Sighireanu, Planning robust temporal plans: A comparison between CBTP and TGA approaches, in *Proceedings of the International Conference on Automated Planning and Scheduling* (2007). https://hal.archives-ouvertes.fr/hal-00157935
2. T. Abdellatif, J. Combaz, J. Sifakis, Model-based implementation of real-time applications, in *International Conference on Embedded Software* (2010). http://dl.acm.org/citation.cfm?id=1879052
3. D. Amodei, C. Olah, J. Steinhardt, P. Christiano, J. Schulman, D. Mané, Concrete Problems in AI Safety (2016). http://arxiv.org/abs/1606.06565v2
4. B.D. Argall, S. Chernova, M.M. Veloso, B. Browning, A survey of robot learning from demonstration. Rob. Auton. Syst. **57**(5), 469–483 (2009)
5. G. Bardaro, A. Semprebon, M. Matteucci, A use case in model-based robot development using AADL and ROS, in *ACM/IEEE Workshop on Robotics Software Engineering* (ACM Press, New York, 2018), pp. 9–16. https://doi.org/10.1007/978-3-319-10783-7_13. http://dl.acm.org/citation.cfm?doid=3196558.3196560
6. G. Behrmann, A. David, K.G. Larsen, A Tutorial on Uppaal 4.0. Technical Report, Department of Computer Science, Aalborg University, Denmark (2006). https://www.uppaal.com/uppaal-tutorial.pdf
7. S. Ben Rayana, M. Bozga, S. Bensalem, J. Combaz, RTD-finder - A tool for compositional verification of real-time component-based systems, in *International Conference on Tools and Algorithms for the Construction and Analysis of Systems* (2016). http://link.springer.com/chapter/10.1007/978-3-662-49674-9_23
8. S. Bensalem, L. de Silva, F. Ingrand, R. Yan, A verifiable and correct-by-construction controller for robot functional levels. J. Softw. Eng. Rob. **1**(2), 1–19 (2011). http://arxiv.org/abs/0908.0221v1
9. S. Bensalem, K. Havelund, A. Orlandini, Verification and validation meet planning and scheduling. Int. J. Softw. Tools Technol. Trans. **16**(1), 1–12 (2014). https://doi.org/10.1007/s10009-013-0294-x. http://link.springer.com/10.1007/s10009-013-0294-x
10. A. Benveniste, G. Berry, The synchronous approach to reactive and real-time systems. Proc. IEEE **79**(9), 1270–1282 (1991)
11. A. Benveniste, P. Caspi, S. Edwards, N. Halbwachs, P. Le Guernic, R. de Simone, The synchronous languages 12 years later. Proc. IEEE **91**, 64–83 (2003). https://dblp.org/rec/journals/pieee/BenvenisteCEHGS03
12. B. Berthomieu, M. Diaz, Modeling and verification of time-dependent systems using time petri nets. IEEE Trans. Softw. Eng. **17**(3), 259–273 (1991). http://gateway.webofknowledge.com/gateway/Gateway.cgi?GWVersion=2&SrcAuth=mekentosj&SrcApp=Papers&DestLinkType=FullRecord&DestApp=WOS&KeyUT=A1991FE66100005
13. B. Berthomieu, J.P. Bodeveix, P. Farail, M. Filali, H. Garavel, P. Gaufillet, F. Lang, F. Vernadat, Fiacre: An intermediate language for model verification in the topcased environment, in *Embedded Real-Time Software and Systems, HAL - CCSD, Toulouse* (2008). http://hal.inria.fr/docs/00/26/24/42/PDF/Berthomieu-Bodeveix-Farail-et-al-08.pdf
14. D. Bjørner, K. Havelund, 40 Years of Formal Methods - Some Obstacles and Some Possibilities? FM (2014). https://dblp.org/rec/conf/fm/BjornerH14

15. J. Bohren, S. Cousins, The SMACH high-level executive. IEEE Rob. Autom. Mag. **17**(4), 18–20 (2010). https://doi.org/10.1109/MRA.2010.938836. http://ieeexplore.ieee.org/lpdocs/epic03/wrapper.htm?arnumber=5663871

16. F. Boussinot, R. de Simone, The ESTEREL language. Proc. IEEE **79** , 1293–1304 (1991)

17. D. Brugali, Model-Driven Software Engineering in Robotics. IEEE Rob. Autom. Mag. **22**(3), 155–166 (2015). https://doi.org/10.1109/MRA.2015.2452201. http://ieeexplore.ieee.org/lpdocs/epic03/wrapper.htm?arnumber=7254324

18. H. Bruyninckx, Open robot control software: The OROCOS project, in *IEEE International Conference on Robotics and Automation* (2001)

19. F. Cassez, C. Pagetti, O.H. Roux, A timed extension for ALTARICA. Fundam. Inform. **62**, 291–332 (2004). https://dblp.org/rec/journals/fuin/CassezPR04

20. A. Cavalcanti, Formal methods for robotics: RoboChart, RoboSim, and more, in *Formal Methods: Foundations and Applications* (Springer International Publishing, Cham, 2017), pp. 3–6. https://doi.org/10.1145/1592434.1592436. http://link.springer.com/10.1007/978-3-319-70848-5_1

21. G. Cicala, A. Khalili, G. Metta, L. Natale, S. Pathak, L. Pulina, A. Tacchella, Engineering approaches and methods to verify software in autonomous systems, in *International Conference on Intelligent Autonomous Systems* (2016).http://link.springer.com/chapter/10.1007/978-3-319-08338-4_121

22. J. Claßen, G. Röger, G. Lakemeyer, B. Nebel, Platas—integrating planning and the action language golog. KI-Künstliche Intell. **26**(1), 61–67 (2012). http://link.springer.com/article/10.1007/s13218-011-0155-2

23. D. Come, J. Brunel, D. Doose, Improving code quality in ROS packages using a temporal extension of first-order logic, in *IEEE International Conference on Robotic Computing* (IEEE, Piscataway, 2018), pp. 1–8. https://doi.org/10.1109/IRC.2018.00010. http://ieeexplore.ieee.org/document/8329874/

24. H. Costelha, P.U. Lima, Robot task plan representation by Petri Nets: modelling, identification, analysis and execution. Auton. Rob. **33**(4), 337–360 (2012). https://doi.org/10.1142/3376. http://link.springer.com/10.1007/s10514-012-9288-x

25. S. Dal Zilio, B. Berthomieu, D. Le Botlan, Latency analysis of an aerial video tracking system using fiacre and tina, in *FMTV Verification Challenge of WATERS 2015, LAAS-VERTICS* (2015). http://arxiv.org/abs/1509.06506v1

26. A. David, K.G. Larsen, A. Legay, M. Mikučionis, D.B. Poulsen, UPPAAL SMC tutorial. Int. J. Softw. Tools Technol. Trans. **17**, 1–19 (2015). https://doi.org/10.1007/s10009-014-0361-y. http://dx.doi.org/10.1007/s10009-014-0361-y

27. A. Desai, T. Dreossi, S.A. Seshia, Combining model checking and runtime verification for safe robotics, in *International Conference on Runtime Verification RV* (2017). https://dblp.org/rec/conf/rv/DesaiDS17

28. S. Dhouib, S. Kchir, S. Stinckwich,T. Ziadi, M. Ziane, RobotML, a domain-specific language to design, simulate and deploy robotic applications, in *IEEE International Conference on Simulation, Modeling, and Programming for Autonomous Robots* (2012). http://link.springer.com/chapter/10.1007/978-3-642-34327-8_16

29. D. Doose, C. Grand, C. Lesire, MAUVE runtime: A component-based middleware to reconfigure software architectures in real time, in *IEEE International Conference on Robotic Computing* (IEEE, Piscataway, 2017), pp. 208–211. https://doi.org/10.1109/IRC.2017.47. http://ieeexplore.ieee.org/document/7926540/

30. V. D'Silva, D. Kroening, G. Weissenbacher, A survey of automated techniques for formal software verification. IEEE Trans. Comput. Aided Design Integr. Circuits Syst. **27**(7), 1165–1178 (2008). https://doi.org/10.1109/TCAD.2008.923410. http://ieeexplore.ieee.org/lpdocs/epic03/wrapper.htm?arnumber=4544862

31. B. Espiau, K. Kapellos, M. Jourdan, Formal verification in robotics: Why and how?, in *International Symposium on Robotics Research* (1996). http://citeseerx.ist.psu.edu/viewdoc/download?doi=10.1.1.54.3091&rep=rep1&type=pdf

32. P. Feth, M.N. Akram, R. Schuster, O. Wasenmüller, Dynamic Risk Assessment for Vehicles of Higher Automation Levels by Deep Learning (2018). http://arxiv.org/abs/1806.07635v1
33. M. Foughali, Formal Verification of the Functional Layer of Robotic and Autonomous Systems. PhD Thesis, LAAS/CNRS, 2018
34. M. Foughali, B. Berthomieu, S. Dal Zilio, P.E. Hladik, F. Ingrand, A. Mallet, Formal verification of complex robotic systems on resource-constrained platforms, in *FormaliSE @ The International Conference on Software Engineering ICSE* (ACM Press, New York, 2018), pp. 2–9. https://doi.org/10.1016/S1571-0661(05)80435-9. https://hal.laas.fr/hal-01778960
35. N. Gobillot, F. Guet, D. Doose, C. Grand, C. Lesire, L. Santinelli, Measurement-based real-time analysis of robotic software architectures, in *IEEE/RSJ International Conference on Intelligent Robots and Systems* (IEEE, Piscataway, 2016), pp. 3306–3311. https://doi.org/10.1109/IROS.2016.7759509. https://ieeexplore.ieee.org/xpl/articleDetails.jsp?tp=&arnumber=7759509&contentType=Conference+Publications
36. M. Guerra, D. Efimov, G. Zheng, W. Perruquetti, Avoiding local minima in the potential field method using input-to-state stability. Control Eng. Pract. **55**(C), 174–184 (2016). https://doi.org/10.1016/j.conengprac.2016.07.008. http://dx.doi.org/10.1016/j.conengprac.2016.07.008
37. J. Guiochet, M. Machin, H. Waeselynck, Safety-critical advanced robots: A survey. Rob. Auton. Syst. **94**, 43–52 (2017). http://www.sciencedirect.com/science/article/pii/S0921889016300768
38. D. Hähnel, W. Burgard, G. Lakemeyer, GOLEX—bridging the gap between logic (GOLOG) and a real robot, in *KI Advances in Artificial Intelligence* (Springer, Berlin, 1998), pp. 165–176
39. R. Halder, J. Proença, N. Macedo, A. Santos, Formal verification of ros-based robotic applications using timed-automata, in *IEEE/ACM International FME Workshop on Formal Methods in Software Engineering (FormaliSE)* (2017). https://dblp.org/rec/conf/icse/HalderPMS17
40. M. Herrb, Pocolibs: POsix COmmunication LIbrary. Technical Report, LAAS-CNRS (1992). https://git.openrobots.org/projects/pocolibs/gollum/index
41. P.E. Hladik, Hippo. Technical Report, LAAS-CNRS (2020). https://redmine.laas.fr/projects/genom3-fiacre-template/gollum/hippo
42. J. Huang, C. Erdogan, Y. Zhang, B. Moore, Q. Luo, A. Sundaresan, G. Rosu, ROSRV: Runtime verification for robots, in *Runtime Verification* (Springer, Cham, 2014). http://link.springer.com/chapter/10.1007/978-3-319-11164-3_20
43. G. Infantes, M. Ghallab, F. Ingrand, Learning the behavior model of a robot. Auton. Rob. **30**, 1–21 (2010). https://homepages.laas.fr/felix/publis-pdf/arj10.pdf
44. F. Ingrand, M. Ghallab, Deliberation for autonomous robots: a survey. Artif. Intell. **247**, 10–44 (2017). https://doi.org/10.1016/j.artint.2014.11.003. http://dx.doi.org/10.1016/j.artint.2014.11.003
45. A. Kai, K. Hölldobler, B. Rumpe, A. Wortmann, Modeling robotics software architectures with modular model transformations. J. Softw. Eng. Rob. **8**(1), 3–16 (2017). https://doi.org/10.6092/JOSER. https://www.google.com/
46. J. Kober, J.A. Bagnell, J. Peters, Reinforcement learning in robotics: a survey. Int. J. Rob. Res. **32**, (2013). https://doi.org/10.1177/0278364913495721. http://ijr.sagepub.com/content/early/2013/08/22/0278364913495721.abstract
47. P. Koopman, M. Wagner, Challenges in autonomous vehicle testing and validation. SAE Int. J. Trans. Safety **4**(1), 15–24 (2016). https://doi.org/10.4271/2016-01-0128. http://papers.sae.org/2016-01-0128/
48. D. Kortenkamp, R.G. Simmons, Robotic systems architectures and programming, in *Handbook of Robotics*, ed. by B. Siciliano, O. Khatib (Springer, Berlin, 2008), pp. 187–206
49. H. Kress-Gazit, T. Wongpiromsarn, U. Topcu, Correct, reactive, high-level robot control. IEEE Rob. Autom. Mag. **18**(3), 65–74 (2011). https://doi.org/10.1109/MRA.2011.942116. http://ieeexplore.ieee.org/lpdocs/epic03/wrapper.htm?arnumber=6016593
50. C. Lesire, F. Pommereau, ASPiC: An acting system based on skill petri net composition, in *IEEE/RSJ International Conference on Intelligent Robots and Systems* (2018), pp. 1–7

51. H.J. Levesque, R. Reiter, Y. Lesperance, F. Lin, R.B. Scherl, GOLOG: A logic programming language for dynamic domains. J. Logic Program. **31**(1), 59–83 (1997). http://www. sciencedirect.com/science/article/pii/S0743106696001215

52. W. Li, A. Miyazawa, P. Ribeiro, A. Cavalcanti, J. Woodcock, J. Timmis, From formalised state machines to implementations of robotic controllers, in *Distributed Autonomous Robotic Systems* (Springer, Cham, 2018), pp. 1–14.

53. A. Lotz, A. Hamann, I. Lütkebohle, D. Stampfer, Modeling Non-Functional Application Domain Constraints for Component-Based Robotics Software Systems (2016). http://arxiv.org/abs/1601.02379

54. M. Luckcuck, M. Farrell, L. Dennis, C. Dixon, M. Fisher, Formal Specification and Verification of Autonomous Robotic Systems: A Survey (2018). http://arxiv.org/abs/1807.00048v1

55. A. Mallet, C. Pasteur, M. Herrb, S. Lemaignan, F. Ingrand, GenoM3: Building middleware-independent robotic components, in *IEEE International Conference on Robotics and Automation* (2010), pp. 4627–4632. https://doi.org/10.1109/ROBOT.2010.5509539. http://ieeexplore.ieee.org/xpls/abs_all.jsp?arnumber=5509539

56. W. Meng, J. Park, O. Sokolsky, S. Weirich, I. Lee, Verified ROS-based deployment of platform-independent control systems, in *NASA Formal Methods* (Springer International Publishing, Cham, 2015), pp. 248–262. https://doi.org/10.1007/978-3-319-17524-9_18. http://link.springer.com/10.1007/978-3-319-17524-9_18

57. A. Miyazawa, P. Ribeiro, W. Li, A. Cavalcanti, J. Timmis, Automatic property checking of robotic applications, in *IEEE/RSJ International Conference on Intelligent Robots and Systems* (2017). http://dblp.org/rec/conf/iros/Miyazawa0LCT17

58. C. Mühlbacher, S. Gspandl, M. Reip, G. Steinbauer, Improving dependability of industrial transport robots using model-based techniques, in *IEEE International Conference on Robotics and Automation* (2016), pp. 3133–3140. https://doi.org/10.1109/ICRA.2016.7487480. http://ieeexplore.ieee.org/xpls/abs_all.jsp?arnumber=7487480

59. A. Nordmann, N. Hochgeschwender, D. Wigand, S. Wrede, A survey on domain-specific modeling and languages in robotics. J. Softw. Eng. Rob. **7**(1), 1–25 (2016). https://scholar.google.com/

60. M. Quigley, B. Gerkey, K. Conley, J. Faust, T. Foote, J. Leibs, E. Berger, R. Wheeler, A.Y. Ng, ROS: an open-source Robot Operating System, in *IEEE International Conference on Robotics and Automation* (2009)

61. P. Ribeiro, A. Miyazawa, W. Li, A. Cavalcanti, J. Timmis, Modelling and Verification of Timed Robotic Controllers, in *International Conference on Integrated Formal Methods* (2017). http://dblp.org/rec/conf/ifm/0002MLCT17

62. K.Y. Rozier, Specification - The biggest bottleneck in formal methods and autonomy, in *Verified Software: Theories, Tools, and Experiments* (2016). https://doi.org/10.1007/978-3-319-48869-1. http://link.springer.com/chapter/10.1007/978-3-319-48869-1_2

63. Z. Saigol, Extending automotive certification processes to handle autonomous vehicles, in *RoboSoft: Software Engineering for Robotics* (Springer, Berlin, 2020)

64. C. Schlegel, Composition, separation of roles and model-driven approaches as enabler of a robotics software ecosystem, in *RoboSoft: Software Engineering for Robotics* (Springer, Berlin, 2020)

65. C. Schlegel, T. Hassler, A. Lotz, A. Steck, Robotic software systems: From code-driven to model-driven designs, in *International Conference on Advanced Robotics* (IEEE, Piscataway, 2009), pp. 1–8. http://ieeexplore.ieee.org/xpls/abs_all.jsp?arnumber=5174736

66. S.A. Seshia, D. Sadigh, S.S. Sastry, Towards Verified Artificial Intelligence (2016). http://arxiv.org/abs/1606.08514v3

67. R.G. Simmons, C. Pecheur, Automating model checking for autonomous systems, in *AAAI Spring Symposium on Real-Time Autonomous Systems* (2000)

68. D. Simon, R. Pissard-Gibollet, S. Arias, ORCCAD, a framework for safe robot control design and implementation, in *Control Architecture for Robots* (2006). https://hal.inria.fr/inria-00385258

69. D. Socci, P. Poplavko, S. Bensalem, M. Bozga, Modeling mixed-critical systems in real-time BIP, in *1st Workshop on Real-Time Mixed Criticality Systems* (2013). https://hal.archives-ouvertes.fr/hal-00867465/

70. A. Sorin, L. Morten, J. Kjeld, U.P. Schultz, Rule-based dynamic safety monitoring for mobile robots. J. Softw. Eng. Rob. **7**(1), 120–141 (2016). https://scholar.google.fr/

71. T. Sotiropoulos, H. Waeselynck, J. Guiochet, F. Ingrand, Can robot navigation bugs be found in simulation? An exploratory study, in *IEEE International Conference on Software Quality, Reliability and Security* (2017). https://dblp.org/rec/conf/qrs/SotiropoulosWGI17

72. H. Täubig, U. Frese, C. Hertzberg, C. Lüth, S. Mohr, E. Vorobev, D. Walter Guaranteeing functional safety: design for provability and computer-aided verification. Auton. Rob. **32**(3), 303–331 (2011). https://doi.org/10.1007/s10514-011-9271-y. http://www.springerlink.com/index/10.1007/s10514-011-9271-y

73. C.J. Tomlin, I. Mitchell, A.M. Bayen, M. Oishi, Computational techniques for the verification of hybrid systems. Proc. IEEE **91**(7), 986–1001 (2003). https://doi.org/10.1109/JPROC.2003.814621. http://ieeexplore.ieee.org/document/1215682/

74. V. Verma, A.K. Jónsson, C. Pasareanu, M. Iatauro, Universal executive and PLEXIL: engine and language for robust spacecraft control and operations, in *American Institute of Aeronautics and Astronautics Space, AIAA Space Conference* (2006). http://scholar.google.com/scholar?q=related:IpQ407u5_qsJ:scholar.google.com/&hl=en&num=20&as_sdt=0,5

75. F. Vicentini, M. Askarpour, M.G. Rossi, D. Mandrioli, Safety Assessment of Collaborative Robotics Through Automated Formal Verification. IEEE Trans. Rob. **36**(1), 42–61 (2020). https://doi.org/10.1109/TRO.2019.2937471. https://ieeexplore.ieee.org/document/8844289/

76. B.C. Williams, M.D. Ingham, Model-based programming of intelligent embedded systems and robotic space explorers. Proc IEEE Special Issue Model. Design Embedded Softw. **91**(1), 212–237 (2003)

77. K.W. Wong, H. Kress-Gazit, Robot operating system (ROS) introspective implementation of high-level task controllers. J. Softw. Eng. Rob. **8**(1), 1–13 (2017). https://doi.org/10.6092/JOSER. http://joser.unibg.it/index.php/joser/issue/view/9

78. J. Woodcock, P.G. Larsen, J. Bicarregui, J.S. Fitzgerald, Formal methods: Practice and experience. ACM Comput. Surveys **41**(4) (2009). https://dblp.org/rec/journals/csur/WoodcockLBF09

79. N. Yakymets, S. Dhouib, H. Jaber, A. Lanusse, Model-driven safety assessment of robotic systems, in *IEEE/RSJ International Conference on Intelligent Robots and Systems* (IEEE, Piscataway, 2013), pp. 1137–1142. https://doi.org/10.1109/IROS.2013.6696493. http://ieeexplore.ieee.org/xpl/articleDetails.jsp?tp=&arnumber=6696493&contentType=Conference+Publications

Chapter 9
RoboStar Technology: A Roboticist's Toolbox for Combined Proof, Simulation, and Testing

Ana Cavalcanti, Will Barnett, James Baxter, Gustavo Carvalho, Madiel Conserva Filho, Alvaro Miyazawa, Pedro Ribeiro, and Augusto Sampaio

Abstract Simulation is favored by roboticists to evaluate controller design and software. Often, state machines are drawn to convey overall ideas and used as a basis to program tool-specific simulations. The simulation code, written in general or proprietary programming languages, is, however, the only full account of the robotic system. Here, we present the RoboStar technology, a modern approach to design that supports automatic generation of simulation code guaranteed to be correct with respect to a design model, and complements simulation with model checking, theorem proving, and automatic test generation for simulation. Diagrammatic domain-specific, but tool-independent, notations for design and simulation use state machines, differential equations, and controlled English to specify behavior. We illustrate the RoboStar approach using an autonomous vehicle as an example.

1 Introduction

Advances in electronics and mechatronics are facilitating exciting applications of robotics. To realize their potential, however, we need to be able to ensure that robots do not fail in a way that can cause harm: a robot strong enough to help an elderly person out of a chair, for example, is strong enough to hurt that person.

Although many factors are involved in establishing the trustworthiness of a robotic system, software poses a key challenge for design and assurance. Full verification is beyond the state of the art due to the complexity of models and

A. Cavalcanti (✉) · W. Barnett · J. Baxter · A. Miyazawa · P. Ribeiro
University of York, York, UK
e-mail: Ana.Cavalcanti@york.ac.uk; wb689@york.ac.uk; James.Baxter@york.ac.uk; Alvaro.Miyazawa@york.ac.uk; Pedro.Ribeiro@york.ac.uk

G. Carvalho · M. C. Filho · A. Sampaio
Universidade Federal de Pernambuco, Recife, Brazil
e-mail: ghpc@cin.ufpe.br; mscf@cin.ufpe.br; acas@cin.ufpe.br

© Springer Nature Switzerland AG 2021
A. Cavalcanti et al. (eds.), *Software Engineering for Robotics*,
https://doi.org/10.1007/978-3-030-66494-7_9

249

properties. Lack of customized techniques and tools means that, despite the very modern outlook of the applications, the current practice of Software Engineering for Robotics is outdated.

What is routine in many engineering disciplines, that is, the use of models, tools, and techniques justified by mathematical principles, called formal methods, is becoming feasible for software developers. Main players in industry like Microsoft, Amazon, and Facebook have started using formal methods [78].

Several domain-specific languages for robotics are available in the literature [56], but, by far and large, their focus is support for programming and simulation. Modern verification techniques have not been widely explored, with a few notable exceptions [1, 25, 30, 38]; some are covered in this book, notably in Chaps. 4, 5, 7, 8, 11, 12, and 13. Applications of general-purpose formal techniques have shown the value that they can add to robotics. Due to lack of specialization and difficulties with automation, however, the cost involved and scalability achieved do not indicate a clear prospect of wide practical application.

The RoboStar framework for modeling, verification, simulation, and testing of mobile and autonomous robots uses three domain-specific languages: RoboChart [54], RoboSim [13], and RoboWorld. RoboChart includes a subset of UML-like state machines, a customized component model, and primitives to specify timed and probabilistic properties. RoboChart is an event-based notation for design; RoboSim is a matching cycle-based diagrammatic notation for simulation. RoboSim also includes block diagrams enriched to specify physical and dynamic behaviors of robotic platforms. RoboWorld uses controlled English to specify assumptions about the environment and the platform. It complements RoboChart.

RoboChart, RoboSim, and RoboWorld provide a solid foundation to deal with Software Engineering for Robotics. They are notations akin to those in widespread use but enriched to enable the use of modern design and verification techniques.

RoboChart, RoboSim, and RoboWorld can be used to generate mathematical models automatically. In the RoboStar approach, these models are hidden from practitioners, but can be used to prove properties of designs and simulations, consistency between them, and generate tests. The RoboStar testing approach is covered in Chap. 11. So far, we have had experience with the model checkers FDR [35] and PRISM [42], and the theorem prover Isabelle [55]. RoboStar work with probabilistic modeling and PRISM is the topic of Chap. 13.

RoboChart and RoboSim have an associated Eclipse-based tool, called RoboTool,[1] which supports graphical modeling, validation, and automatic generation of CSP [69] and reactive modules scripts, simulation code, and tests. It is integrated with FDR4 and PRISM, ARGoS [61], and CoppeliaSim [68]. A variety of plugins are available.

By ensuring that a RoboChart and RoboWorld design and a RoboSim simulation are consistent, the RoboStar framework guarantees that properties established by

---

[1] www.cs.york.ac.uk/robostar/robotool/.

analysis of the design are preserved in the simulation. So, problems revealed by simulation are design problems, not problems in the coding of the simulation itself.

RoboChart, RoboSim, and RoboWorld complement approaches that cater for a global view of the system architecture (like that in Chap. 3) by supporting modeling and verification of the components, covering interaction, time, and probabilistic properties. It also complements work on deployment of verified code.

In this chapter, we describe the RoboStar technology in detail and illustrate its application in modeling, verifying, and simulating control software for an autonomous vehicle. In the next section, we give an overview of the RoboStar technology, including the notations and techniques already available, and those that we envisage as part of our vision for Software Engineering for Robotics. Our running example is presented in Sect. 3. Section 4 presents a RoboChart model for its control software, and Sect. 5 presents a RoboSim model. Section 6 discusses environment modeling in RoboChart and RoboSim. Finally, Sect. 7 discusses related work, and Sect. 8 summarizes the RoboStar vision and agenda for future work.

2 RoboStar Vision

The RoboStar approach to Software Engineering for Robotics is presented in Fig. 9.1. For modeling designs, we envisage the combined use of RoboChart to model control software, and a controlled natural language, RoboWorld, under development to capture assumptions about the platform and environment. In the proposed workflow, a RoboChart model is the starting point, as indicated by the label (1) in Fig. 9.1. In the next section, we present a RoboChart model for our example in Sect. 4, where we also give more details about the RoboChart notation.

RoboTool supports the creation and editing of RoboChart diagrams and the automatic generation of CSP and PRISM models: label (2) in Fig. 9.1. RoboTool also provides a simple notation that uses controlled English, potentially mixed with CSP processes, to define assertions capturing properties of interest. These assertions can be checked using FDR4 or the PRISM model checker. We plan to enrich this language to cater for a readable account of properties specified in English or diagrammatically (e.g. using sequence diagrams).

It is possible, of course, to automatically generate other mathematical models for RoboChart. Currently, we are considering the integration of UPPAAL [6]. A crucial point, however, is that these models are consistent with each other. The definitive semantics of RoboChart is given by the CSP model, and our vision is one of integration of techniques and tools (model checkers and theorem provers) via the justification of soundness based on semantics. It would not be useful, if, for instance, deadlock checks carried out with different tools gave different results.

RoboTool generates two CSP models. The first gives an untimed semantics to RoboChart and ignores the time primitives. The definitive semantics uses the tock dialect of CSP, in which a special event *tock* marks the passage of discrete time. Probabilistic modeling is captured via a semantics given using reactive modules.

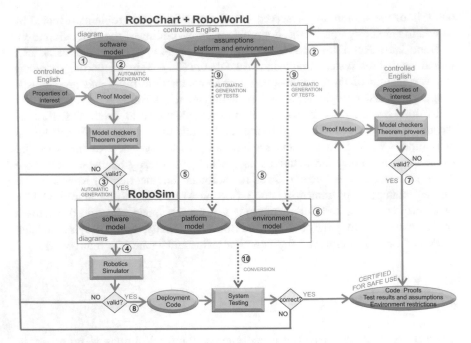

Fig. 9.1 Idealized workflow using RoboStar technology. Copyright University of York, printed with permission in this chapter.

Consistency between the probabilistic and timed semantics is ongoing work. An early approach that combines CSP and PRISM is available [24].

For theorem proving based on the CSP semantics, an initial approach is described in [27–29]. Automatic generation of Isabelle theories for proof is ongoing work. Use of Isabelle is our current route to address the issues of scalability that we can expect with the use of model checking. High levels of automation are still possible as evidenced by Adams and Clayton [2], given the use of proof models that are automatically generated.

In the RoboStar approach, initial proofs to check the models can validate them by establishing core properties, like deadlock and livelock freedom. Checks for the presence or absence of nondeterminism can also often reveal modeling problems. These checks are automatically generated by RoboTool.

If any of the properties of interest do not hold, the RoboChart model should be changed (unless, of course, further work reveals that the property is not actually relevant). For a property that does not hold, a model checker provides an example that illustrates the problem, and the relationship between such examples and the diagrams is simple. The examples can inform how the model should be changed, and the iterative process of debugging of a model is much cheaper than that of debugging a program. Automation of the generation of proof models makes this iterative process of validation of the RoboChart model using the properties of interest cheap. An animator would help, though, and it is part of our agenda for future work. Such a tool allows tracing the diagram based on a sequence of updates to variables, calls to operations, or occurrence of events indicated by a model checker.

Once we are convinced that enough validation has taken place, it then makes sense to consider simulation of the model. For cyber-physical systems, in general, and robotics systems, in particular, it is often the case that we do not have a full specification of behavior. Such a specification is normally highly dependent on the platform and environment. Simulations can, therefore, be very useful to validate the design. Moreover, simulations are core to current practice.

A simulation requires a cyclic account of the design model. This is what is provided by a RoboSim model, which can be automatically generated from a RoboChart model with guaranteed consistency: label (3) in Fig. 9.1.

For a simulation, however, we also need a physical and behavioral model of the platform and of the environment. RoboSim includes a block diagram notation to describe these models. A domain-specific language for robotics based on XML, namely, SDF (Scene Description Format),[2] provides inspiration and domain knowledge. RoboSim block diagrams, however, afford readability, modularity, and extensibility to models. Moreover, from such diagrams, it is possible to generate SDF documents for use in robotics simulators. This is ongoing work.

A RoboSim block diagram specifies a particular platform. Roboticists routinely write such descriptions, either using an XML-based notation like SDF or using tool-specific graphical or programming facilities. So, RoboSim block diagrams improve usability and do not require a significant change in the language used by practitioners.

For a design, however, a model for a particular platform or scenario is too specific. Instead, we need an account of operational requirements: assumptions that are made of the platform and environment, and ensure proper behavior of the robot. A proper account of such requirements is often neglected by practitioners. The RoboStar vision is, therefore, to provide support for their description based on the RoboSim block diagrams: label (5) in Fig. 9.1. The RoboStar approach is for the requirements to be captured in RoboWorld, a controlled natural language under development.

Just like for RoboChart, RoboTool can automatically generate CSP models for RoboSim: label (6) in Fig. 9.1. That model has a dual role: justify the soundness of an automatically generated RoboSim model with respect to an original RoboChart model, and support proof of properties of interest just like for RoboChart.

Soundness requires consideration of the design model given some assumptions. For example, a RoboChart state machine that is in a state with a transition triggered by an event obstacle, for instance, reacts instantaneously to the detection of such an obstacle (by the platform). A simulation (whether described by RoboSim or not), on the other hand, is a cyclic mechanism, where events are only observed and handled at sample times characterized by a cycle period. If an obstacle is detected in between

---

[2]sdformat.org.

sample times, it is ignored until the next sample time. So, consistent behavior only happens if we assume that events do not happen at all between sample times.

With these assumptions, proof of properties of a RoboSim simulation is not needed if the simulation is automatically generated. On the other hand, a RoboStar developer may well decide to write a RoboSim simulation directly, rather than start from a design model. In that case, proof of properties for RoboSim is useful.

Simulation may reveal problems, in which case, given the high level of automation, the right and cost-effective way to proceed is to update the RoboChart model, rerun proofs of properties, and regenerate the simulation. There is no need to deal with the (low-level) simulation code, and there is value in keeping models up to date.

If the simulation (eventually) suggests that the expected behavior is ensured by the design, a proof can add value by confirming that the property holds. Running simulations is a form of testing and cannot be used to guarantee properties. Proofs provide evidence of the quality of the design (label (7) in Fig. 9.1) in addition to the evidence provided by running the simulations. On the other hand, attempting to prove properties before checking behavior via simulation can lead to wasted effort, if the simulation can reveal that the property does not hold.

Simulation code for the control software can often be used as a basis for deployment: label (8) in Fig. 9.1. For deployment, system testing, considering both the target platform and environment, if possible, is necessary. This is essential, for example, to confirm that the operational requirements are satisfied, or that errors are not introduced in the (generation of the) deployment code.

The RoboStar approach to testing is the automatic generation of tests from RoboChart models, label (9) in Fig. 9.1, and the conversion of simulation tests to deployment tests, label (10) in Fig. 9.1. More information about testing from RoboChart models is given in Chap. 11; conversion is in our agenda for the future.

In a simulation, we control the whole robot and the environment. In deployment, since we test the controller on the actual hardware, we typically have less control over the software and only limited observability of its state. (It is possible to instrument the controller to record, for instance, values of variables and the execution path and to add a probe to control nondeterministic and probabilistic choices. Instrumentation can, however, be inappropriate if we are interested in timing or probabilities; we end up testing a program that is not quite the controller.) In deployment tests, we control just the initial state of the robot, often cannot control the entire environment, and have even weaker observational power. So, in each case, we have different notions of the test case. Our approach uses conversion to ensure traceability of tests. In this way, if a deployment test fails, and it corresponds to a simulation test that passed, we gather information about the system under test and the simulation.

If a deployment test fails, automation of the whole approach, including the automatic generation and traceability of tests, encourages update of models. Development and change efforts are concentrated on diagrammatic and controlled English artifacts, rather than low-level code. This increases productivity, lowering costs, and enables the production of evidence of quality as well as production of the code itself.

Next, we illustrate the RoboStar technology in Fig. 9.1 in the example of an autonomous vehicle. We focus on the mature components of the framework.

3 Autonomous Vehicle

Our case study has been described in [34]. It is a fully autonomous vehicle whose software has been developed by the UK Connected Places Catapult.[3] The pod has been developed as part of the LUTZ Pathfinder project: see Fig. 9.2a. Our focus is on the Basic Autonomous Control System (B-ACS), which is implemented using C++ and the ROS (Robot Operating System) middleware.

The B-ACS directs the pod around a predetermined route, specified by a sequence of latitude and longitude points. It ensures the pod follows the route and keeps within the speed limit for the current location. Reaction to obstacles, pre-emption, and mitigation of hazardous situations is carried out by a safety driver, who can override autonomous control by limiting the pod's speed or stopping the pod altogether. Figure 9.2b is an overview of the B-ACS inputs and outputs, corresponding to onboard sensors and actuators, extra sensors added to the pod, and interactions with the embedded controller (shown in the shaded area).

The use of ROS strongly influences the structure of the controller software. It is composed of many modules executed concurrently as an individual process known as a ROS node. The ROS nodes typically communicate with each other using asynchronous messages via a publish/subscribe mechanism provided by ROS.

The behavior of the pod's controller is determined by the function of each ROS node and the messages communicated among them. We describe all these nodes and their relationship in [5]. The original documentation identifies three groups of nodes called b-acs, lutz, and data logging. They are characterized in Table 9.1.

The nodes for data logging are for evaluation; they are not central to the application and not considered here. More information about them is available at [5].

Figure 9.3 depicts the nodes of the b-acs group and the messages communicated between them (via the publish/subscribe ROS mechanism). (The nodes of Fig. 9.3 represent each ROS node and the edges represent the messages that the nodes communicate.) Groups of ROS nodes are indicated in the figure by a shaded node with a dashed outline. A description summarizing the behavior of the b-acs nodes we model in the next section can be found in Table 9.2. A full list is in [5].

---

[3]Formerly, the Transport Systems Catapult.

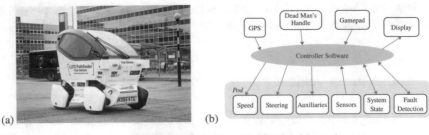

(a)   (b)

GPS	The current latitude and longitude of the pod.
Dead man's handle	Status of the safety driver's speed limiting control.
Gamepad	Manual commands for pod control.
Display	Instructions to the user of the autonomous pod system.
Pod speed	The speed the pod currently maintains.
Pod steering	The steering angle the pod currently maintains.
Pod auxiliaries	The state of the pod's peripheral devices, such as indicators and horn.
Pod sensors	The state of the pod's various internal sensors, such as bump sensors or seat occupancy.
Pod system state	Indication of the status of the pod embedded controller: drive disabled, manual-drive enabled, or autonomous-drive enabled.
Pod fault detection	A seed-key response to the pod's embedded controller to enable fault detection.

Fig. 9.2 The autonomous pod, and its inputs and outputs. Copyright University of York, printed with permission in this chapter.

Table 9.1 Uses of the ROS nodes, message definitions, and classes in each group of nodes of the pod software controller

Group of nodes	Description
b-acs	Generation and processing of demand messages that guide the pod around a predetermined path. The messages specify the speed and steer values required for the pod to follow the path, based on its current location. Processing involves geofencing, enforcing speed limits, and validity checks, for example
Lutz	Interfacing with the pod to receive and manage its state, translating demands into control messages, carrying out the fault detection protocol, and generating user instructions to guide the safety driver
Data logging	Recording sensor data for later evaluation

4 RoboChart Model

This section presents a RoboChart model for the pod controller. Section 4.1 outlines the overall structure of the model. The following sections present the robotic platform and the controllers. Section 4.2 covers the RoboChart module and its robotic platform, and Sect. 4.3 covers the controller for the group of b-acs nodes and their state machines. The data types used in the RoboChart model reflect those used in the source code; they are all defined in [5]. Section 4.4 discusses verification.

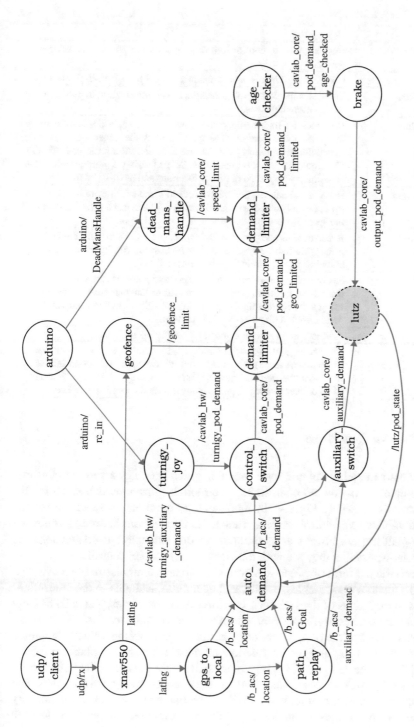

Fig. 9.3 The ROS nodes of the b_acs group, adapted from transport systems catapult specifications. Copyright University of York, printed with permission in this chapter.

Table 9.2 b_acs ROS nodes

Node	Description
turnigy_joy	Converts remote control into messages for the pod and its peripheral devices.
dead_mans_handle	Converts dead man's handle messages to publish speed-limit messages.
gps_to_local	Translates the pod's global location sensor information to publish local 2D location messages.
path_replay	Publishes an intermediary goal toward the route's destination using the pod's local location. The goal contains the target local location, curvature, maximum speed, and duration describing the path to take. The route to the destination is defined using a configuration file containing a sequence of global locations and auxiliary state.
auto_demand	Receives a goal, the pod's local location, and speed from the pod state to create and publish a demand message. The demand message contains the specific steer and speed values to meet the goal.
control_switch	Receives multiple pod demand inputs and publishes only the highest-priority non-abdicating input. The highest-priority input can abdicate, allowing lower-priority inputs to be published.
geofence	Receives the global location of the pod and publishes the speed limit for the current location. The speed limit for rectangular areas is configured using a file containing a sequence of two pairs of latitude and longitude points, each with an associated speed limit.
demand_limiter	Adjusts the speed of a demand based on a speed limit.
age_checker	Checks the age of a demand; demand messages that are older than a defined time have their speed reduced. The speed reduction increases the older the demand message is, until the speed is zero. The age-checked demand with appropriately adjusted speed is published.

4.1 Overall Structure

A RoboChart model is defined by a module, which specifies a robotic platform, and one or more (software) controllers. The module AutonomousVehicle for the pod system is shown in Fig. 9.4. In this example, there are, besides the robotic platform Vehicle, two controllers, B-acs and Lutz. Connections between a controller and the platform are always asynchronous. In our example, all connections are asynchronous, whether they are with the platform or between controllers.

The robotic platform of a RoboChart model captures a representation of its sensors and actuators via variables, events, and operations available to the controllers. Changes to the values of the variables, occurrences of the events, and calls to these operations define the observable interactions between the robot control software and its environment. For the pod, the robotic platform represents the Vehicle, including all of the extra sensors in Fig. 9.2. Section 4.2 defines the robotic platform for the pod and details the analysis of the inputs and outputs for the controller software.

Because RoboChart controllers can describe concurrent behavior, they can represent either the individual ROS nodes of the pod or higher-level functionality provided by the groups of nodes in Table 9.1. Representing each node as a

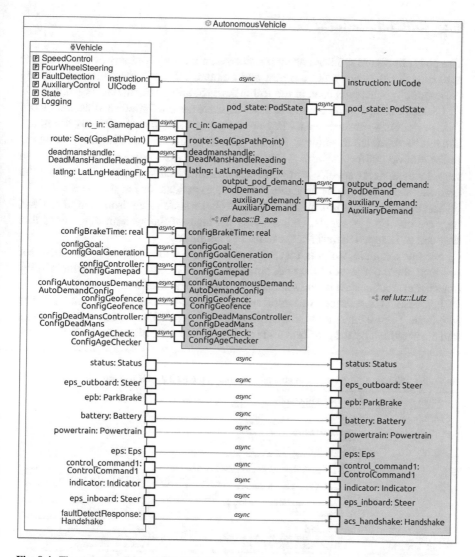

Fig. 9.4 The autonomous pod RoboChart module. Copyright University of York, printed with permission in this chapter.

controller would mean that there are more controllers in the module, making it more difficult to understand. Representing functionally related groups of ROS nodes as controllers emphasizes the coupling between related areas of functionality. This means, however, that ROS nodes are represented as RoboChart machines, and so their behavior is predominately sequential. For our example, this is not an issue because the nodes of the pod are sequential. In addition, in general, a node that has a parallel implementation can be modeled by a group of machines inside a controller.

The controller for the b-acs group that determines the autonomous behavior of the pod is in Sect. 4.3. The rest of the model is discussed in [5].

4.2 Robotic Platform

The pod sensors are modeled as inputs and the actuators as outputs of the controller software; Fig. 9.2 shows these inputs and outputs. Table 9.3 lists the names of the corresponding elements used in the RoboChart robotic platform (see Fig. 9.4).

Besides the inputs and outputs of Fig. 9.2, there are configuration files that record ROS parameters for the nodes. They are stored in a server accessible by the nodes and can be modified at runtime. So, the parameters are a form of input represented in the robotic platform. For example, configDeadMansController communicates a record whose fields specify minimum and maximum ranges for analog inputs; the DeadMansHandle node uses this parameter to validate its inputs.

The inputs provide information about a robot's state or environment; therefore, they are modeled as events. Outputs alter the state of the system; this means that they can be mapped to variables, events, or operations.

For our example, outputs that significantly affect the state of the system, for instance, the movement of the pod, are modeled as operations. Because the display does not affect the state of the system, it is modeled as an event.

For ease of reference, variable, events, and operations can be grouped into interfaces. Figure 9.5 shows the interfaces of our model. They group the operations and are provided by the platform and required by the controllers, as described next.

Table 9.3 Mapping from the pod inputs and outputs to the RoboChart robotic platform

System input	Name in the model	System output	Name in the model
Sensors	status		
	eps_outboard		
	epb		
	battery		
	powertrain	System output	Name in the model
	eps	Speed	setSpeed
	indicator		setParkingBrake
	epsInboard	Steering	setFrontSteering
Fault detection	faultDetectResponse		setRearSteering
System state	control_command1	Auxiliaries	setAuxiliaries
GPS	latlng	System state	requestState
Dead man's handle	deadmanshandle	Fault detection	sendHandshakeResponse
Gamepad	rc_in	Display	display
Configuration files	configGoal		
	configController		
	configAutonomousDemand		
	configGeofence		
	configDeadMansController		

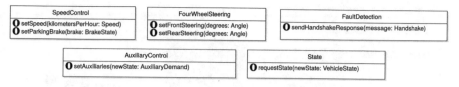

Fig. 9.5 The interfaces. Copyright University of York, printed with permission in this chapter.

4.3   B_acs Controller

To define the B_acs controller, the behavior of the nodes in Fig. 9.3 has to be captured. Some of them are modeled via the abstraction provided by the robotic platform. The UDP/client node provides connectivity between the several devices that comprise the robotic platform and are abstracted away by the definition of the platform as a single component. The xnav550 and arduino nodes translate low-level sensor data into ROS messages captured as input events in the platform: location, safetyDriverInput, and remoteControl. All other nodes contribute to the functionality of the controller and so are modeled as state machines.

Figure 9.6 depicts the B_acs controller: its machines and connections. The messages used for communication between nodes are captured by events of their corresponding machines. Their connections correspond to ROS topics, which are used to communicate the messages via the publish and subscribe ROS mechanism. So, overall, the connections mirror the edges in Fig. 9.3. Given the nature of the publish and subscribe mechanism, the connections are all asynchronous.

The structure of the messages is modeled using RoboChart data types and fields. For example, the LatLngHeadingFix message consists of three doubles representing a latitude, a longitude, and a heading; this can be represented using a RoboChart data type with three fields of type real. These message types are used to define the types of the events that represent the communications via these messages.

The states of the RoboChart state machines are determined by a control flow analysis of the source code for the corresponding ROS nodes. As a small example, we show in Listing 1 a pseudocode that is representative of the implementation of the dead_mans_handle node. The corresponding state machine is shown in Fig. 9.7.

The method deadMansHandleNode() (line 7) is the node's constructor; it gets configuration information from the ROS parameter server (lines 8 and 9) and subscribes and publishes to the topics used in the node (lines 11 and 12). In the state machine DeadMansHandle, the check for availability of new parameter values, carried out by the calls to getParam, is captured by a communication via a configuration event. The subscribed and published topics are represented by the events deadmanshandle and speed_limit. The type ConfigDeadMans of configuration is a record with three fields: two corresponding to the variables rangeMin and rangeMax (lines 1 and 2) and a third Boolean field retrieved to indicate whether a new configuration is available. The types of deadmanshandle and speed_limit are primitive (basic) types corresponding to those in the code.

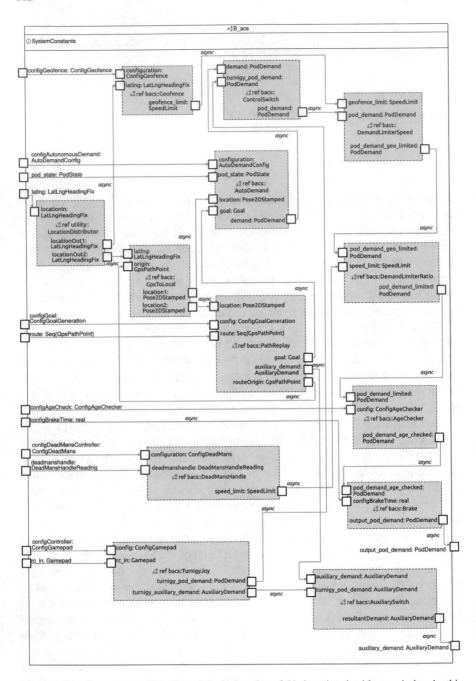

Fig. 9.6 The B_acs controller. Copyright University of York, printed with permission in this chapter.

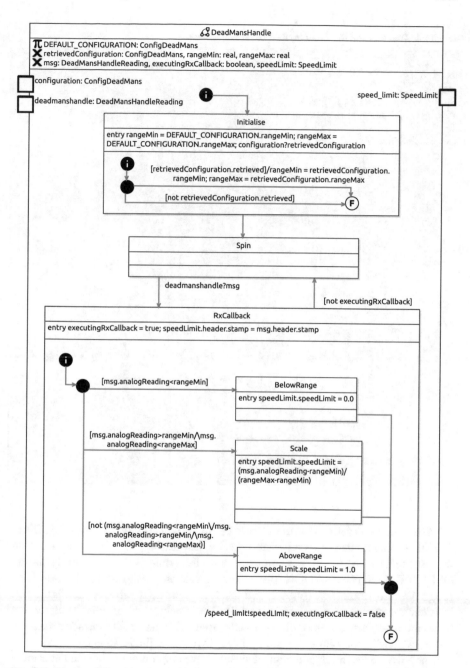

Fig. 9.7 The DeadMansHandle machine. Copyright University of York, printed with permission in this chapter.

Listing 1 Pseudo code for the ROS node dead_mans_handle

```
1  rangeMin  100
2  rangeMax  500
3
4  sub      // Instance of ros :: Subscriber  class
5  pubSpeedLimit    // Instance  of ros :: Publisher  class
6
7  deadMansHandleNode () {
8       nodeParameters . getParam ("rangeMin", rangeMin);
9       nodeParameters . getParam ("rangeMax", rangeMax);
10
11      sub = node . subscribe ("arduino / deadMansHandle", 10, rxCallback
         , this )
12      pubSpeedLimit = node . advertise ("cavlab_core / speed_limit", 10)
13 }
14
15 rxCallback ( deadMansHandleReading & msg ) {
16      speedLimit     // Instance  of SpeedLimit class
17
18      speedLimit . header . stamp = msg . header . stamp ;
19
20      if (msg . analogReading < rangeMin ) {
21           speedLimit . speedLimit = 0.0;
22      } else if (msg . analogReading > rangeMin && msg . analogReading
         < rangeMax ) {
23           speedLimit . speedLimit = (msg . analogReading − rangeMin ) /
         (rangeMax − rangeMin );
24      } else {
25           speedLimit . speedLimit = 1.0;
26      }
27
28      pubSpeedLimit . publish (speedLimit)
29 }
```

When the deadMansHandleNode() node constructor is executing (lines 7–13), the node can be considered to be initializing. So, the first state in the RoboChart model is a composite Initialise state. On entry to Initialise, DEFAULT_CONFIGURATION parameter values are assigned to the variables rangeMin and rangeMax (lines 1 and 2). Next, an input retrievedConfiguration is taken via configuration (lines 8 and 9). The machine in Initialise captures the behavior of getParam. If the call returns true, that is, retrievedConfiguration.retrieve holds, rangeMin and rangeMax are updated. Otherwise, the input is ignored.

The ROS nodes in the pod control system have either a periodic or an aperiodic control flow. In both cases, after the initialization, the spin() method is invoked. It simply calls the ROS ros_spin() method, which blocks handling asynchronously received messages on subscribed topics by calling the corresponding callback methods defined by the node. So, the callback methods define the node's behavior. In Listing 1, the callback method rxCallback is in lines 15–29.

In the RoboChart model, from the Initialise state, a transition (without a label) moves immediately to the state Spin. This captures the behavior as the node waits for a message msg. When it arrives (via the event deadmanshandle), a transition leads to a (composite) state RxCallback that models the callback method.

A Boolean variable executingRxCallback is used to ensure that the state RxCallback is not left until the behavior corresponding to the execution of the method is finished. So, upon entry of RxCallback, this variable is set to true, and the only transition out of RxCallback has a guard that requires it to be false.

In the machine in RxCallback, the initial junction leads to a junction that corresponds to the if-else structure in rxCallback (lines 20–26). The guard in each transition coming out of the junction matches those in lines 20, 22, and 24 (which has an implicit condition). Each target state has entry actions that match the assignments in the code (lines 21, 23, and 25). From each state, there is an immediate transition to a junction with a single transition to a final state. The action of that transition communicates the output via the event speed_limit, corresponding to the publish statement in line 28 of Listing 1. The transition action also updates executingRxCallback so that a transition out of RxCallback leads back to Spin.

Other machines are presented in [5]. We now briefly discuss their verification.

4.4 Verification

The RoboChart model of the autonomous vehicle enables both core and user-specified timed and untimed properties of the controller software to be verified automatically. The core properties that are automatically generated include deadlock and livelock freedom, determinism, termination, and reachability. Other properties of interest can be, for the moment, defined as CSP processes that are verified by refinement checking against the calculated semantics.

RoboTool supports the specification of the core properties to be verified utilizing an assertion language that uses controlled English [53]. For example, to check the core property that the DeadMansHandle machine is deterministic, the corresponding statement written using the assertion language is as follows.

assertion DMH_1:
    bacs::DeadMansHandle::DeadMansHandle **is deterministic**

The assertion keywords are indicated in boldface and specify the property to be checked. DMH_1 is a user-defined label for easy identification of the property. The next part of the **assertion** statement is a fully qualified name of a component (module, controller, or machine) from the RoboChart model. Figures 9.8 and 9.9 show RoboTool and a report generated in the verification of the DeadMansHandle machine.

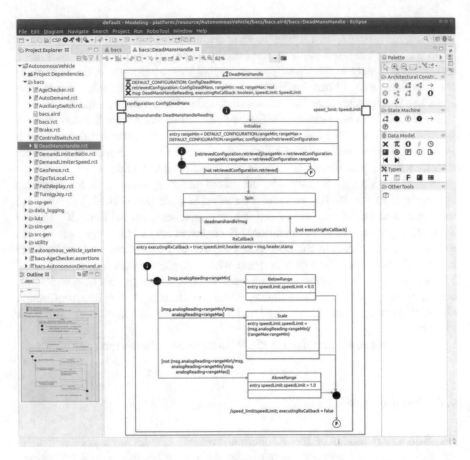

Fig. 9.8 The DeadMansHandle machine in RoboTool. Copyright University of York, printed with permission in this chapter.

For the DeadMansHandle machine, an important property, which we name DMH_INOUT, is that for every input received from deadmanshandle, an output is generated via speed_limit; this can be expressed as shown below in CSP.

The expression defines a CSP process *OFEI* (Output For Every Input) that offers a choice of events corresponding to the three events of the DeadMansHandle machine. (Fully qualified names have been truncated for readability.)

$$OFEI= \quad configuration.in?currentConfiguration \longrightarrow OFEI$$
$$\square$$
$$speed\_limit.out?speedLimit \longrightarrow OFEI$$
$$\square$$
$$deadmanshandle.in?msg \longrightarrow speed\_limit.out?speedLimit \longrightarrow OFEI$$

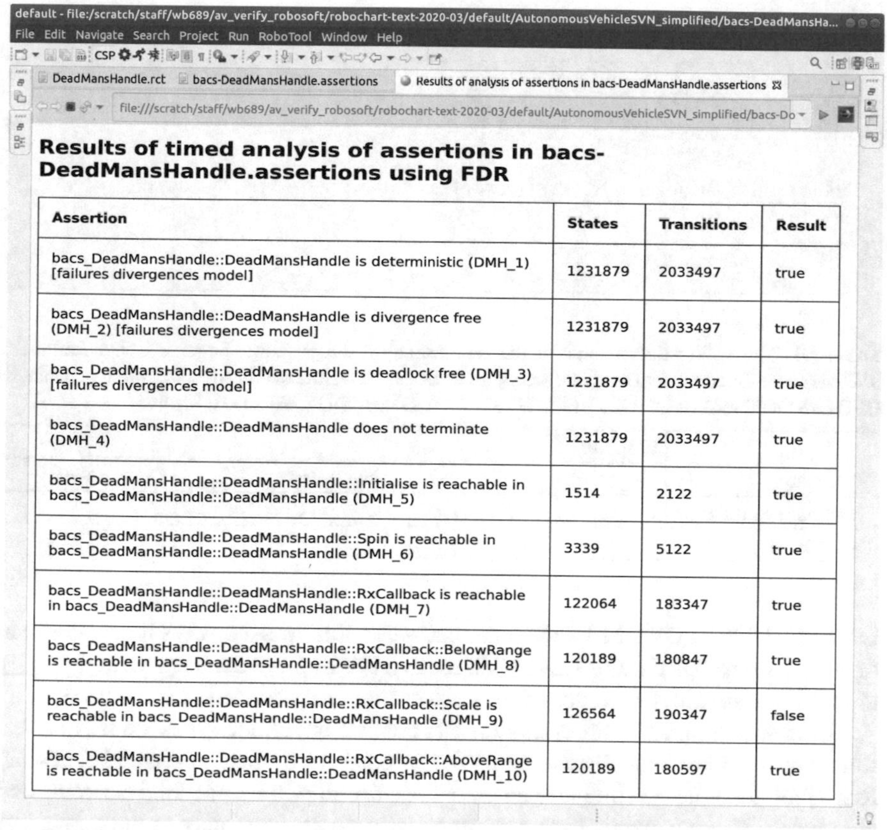

Results of timed analysis of assertions in bacs-DeadMansHandle.assertions using FDR

Assertion	States	Transitions	Result
bacs_DeadMansHandle::DeadMansHandle is deterministic (DMH_1) [failures divergences model]	1231879	2033497	true
bacs_DeadMansHandle::DeadMansHandle is divergence free (DMH_2) [failures divergences model]	1231879	2033497	true
bacs_DeadMansHandle::DeadMansHandle is deadlock free (DMH_3) [failures divergences model]	1231879	2033497	true
bacs_DeadMansHandle::DeadMansHandle does not terminate (DMH_4)	1231879	2033497	true
bacs_DeadMansHandle::DeadMansHandle::Initialise is reachable in bacs_DeadMansHandle::DeadMansHandle (DMH_5)	1514	2122	true
bacs_DeadMansHandle::DeadMansHandle::Spin is reachable in bacs_DeadMansHandle::DeadMansHandle (DMH_6)	3339	5122	true
bacs_DeadMansHandle::DeadMansHandle::RxCallback is reachable in bacs_DeadMansHandle::DeadMansHandle (DMH_7)	122064	183347	true
bacs_DeadMansHandle::DeadMansHandle::RxCallback::BelowRange is reachable in bacs_DeadMansHandle::DeadMansHandle (DMH_8)	120189	180847	true
bacs_DeadMansHandle::DeadMansHandle::RxCallback::Scale is reachable in bacs_DeadMansHandle::DeadMansHandle (DMH_9)	126564	190347	false
bacs_DeadMansHandle::DeadMansHandle::RxCallback::AboveRange is reachable in bacs_DeadMansHandle::DeadMansHandle (DMH_10)	120189	180597	true

Fig. 9.9 The DeadMansHandle machine verification in RoboTool. Copyright University of York, printed with permission in this chapter.

The first two choices consider occurrences of the events *configuration*, which is an *in*put, and *speed_limit*, an *out*put. In these cases, *OFEI* recurses, placing no constraints on the behavior of the machine being verified. The final choice is an *in*put via *deadmanshandle*, which is immediately followed by a *speed_limit* *out*put event. This describes the desired property DMH_INOUT shown below.

The assertion language can be used to specify refinement checks against the defined CSP expressions and instances of RoboChart components. The **assertion** DMH_INOUT specifies a refinement check and the **model** to use for verification; in our example, we consider just the **traces** of the process.

assertion DMH_INOUT:
  bacs::DeadMansHandle::DeadMansHandle **refines** OFEI **in the traces model**

Similar processes and assertions can be used to verify that the value of the output speed limit must always be less than or equal to a given maximum (assertion

Listing 2 DeadMansHandle instantiations

```
1 — generate real not
2 nametype core_real = { −2..2}
3
4 — generate
5 — const_bacs_DeadMansHandle_DeadMansHandle_DEFAULT_CONFIGURATION
    not
6 const_bacs_DeadMansHandle_DeadMansHandle_DEFAULT_CONFIGURATION =
    (−1, 1, false)
```

Table 9.4 The verification results for the DeadMansHandle machine. Legend: ① Deterministic, ② Divergence freedom, ③ Deadlock freedom, ④ Does not terminate, ⑤ All states are reachable, ⑥ DMH_INOUT, ⑦DMH_OUT_BELOW_MAX, ⑧ DMH_OUT_ABOVE_MIN

State machine	Property								Note
	①	②	③	④	⑤	⑥	⑦	⑧	
DeadMansHandle	✓	✓	✓	✓	✓	✓	✓	✓	
DeadMansHandle	✓	✓	✓	✓	✓	✓	✓	✓	Timed

DMH_OUT_BELOW_MAX) and the out speed limit must always be greater than or equal to zero (DMH_OUT_ABOVE_MIN). For that, we define processes similar to OFEI, but restrict the values of the outputs that are produced to the valid sets.

For abstraction, RoboChart models can contain undefined types, constants, and functions that must be defined to verify the properties specified. For model checking, RoboTool generates preliminary instantiations for all of the undefined elements and core types. These instantiations need to be tailored appropriately to the domain of the system, noting that a large cardinality of definitions of sets leads to models that are complex and require significant resources to verify. The instantiation file used for verification of the DeadMansHandle state machine is shown in Listing 2. To prevent RoboTool from overwriting customized instantiations, not is appended to the end of a description to suppress regeneration of the corresponding definition.

Types have been represented by minimal sets, for example, core_real (line 2) ranges from −2 to 2. These values can be used to represent data domains of the DeadMansHandle machine; for example, we can have 2 to represent values above the maximum range, 1 for the maximum, 0 for values in range, and so on. The FDR model checker does not support real numbers, so, for verification, we need to use integers and may need to make approximations in the model. To the best of our knowledge, there are no model checkers for FDR that deal with real numbers.

The results obtained in verifying properties for the aperiodic DeadMansHandle machine are summarized in Table 9.4. Further verification, associated with the generation of a simulation, is discussed in the next section.

5 Simulation

As well as verification, a RoboChart model can also be used as a basis to develop a simulation. It is, of course, possible to develop a simulation from scratch, as it is usually the case. Even in this scenario, there is still value derived from using a RoboChart model for guidance because it is precisely described in an organized notation. As already mentioned, with the use of RoboTool, it can be guaranteed that the model is valid (well typed, all operations are declared, all states are connected, and so on) and core properties can be checked automatically. So, a RoboChart model is a high-quality starting point for further work on coding.

In addition, it is possible to check whether the RoboChart model can be accurately described by a simulation at all. A simulation is an iterative mechanism; in each cycle, the inputs are read, the data is processed, the outputs are produced, and then time is advanced. So, developing a simulation corresponding to a RoboChart model requires scheduling the processing in the state machines in cycles. This may not be possible if, for example, the RoboChart model requires at some point two urgent calls to the same operation. It is not possible to call the same operation twice in a simulation cycle, and, so, if both calls are urgent, that timed behavior cannot be reproduced in a simulation. Such a RoboChart model needs to be revisited.

An automated schedulability check can reveal such a problem. It consists of verifying whether, in the presence of the assumptions normally made when writing a simulation, the RoboChart model does not deadlock. The assumptions relate the events, variables, and operations of the robotic platform of a RoboChart model to the inputs and outputs read in a simulation cycle. They also enforce the restriction on operation calls. Absence of deadlock in the presence of these assumptions ensures that there is no behavior of the RoboChart model that cannot be implemented in the simulation paradigm that reflects the assumptions.

For our example, DeadMansHandle state machine, this check passes. It can be carried out automatically by RoboTool using the assertion below.

simulation DMHSim **of**
  bacs::DeadMansHandle::DeadMansHandle { **cycleDef** cycle == 1 }
  assertion Schedulable: DMHSim **is schedulable**

The **simulation** clause defines a simulation specification from a RoboChart component. In this example, we have a specification DMHSim based on DeadMansHandle. The specification requires the definition of a cycle, which is given in the **cycleDef** clause and embeds the simulation assumptions mentioned above. The **cycleDef** clause specifies the value of a variable cycle; in our example, it is 1. The actual **assertion**, called Schedulable in the example, requires that DMHSim **is schedulable**. Given that the pod model is constructed to reflect a (cyclic) implementation, it is not a surprise that the machines are schedulable in simulations.

If the RoboChart model is schedulable, we can generate simulations automatically. For constrained RoboChart models, it is possible to generate a C++ simulation for ARGoS. The gap between event-based control flow embedded in a design state machine and cycle-based control flow of a simulation, however, is very large. To bridge this gap and produce an artifact that describes the cycle-based mechanism faithfully, we have developed RoboSim. This is also a diagrammatic notation, with support for modeling, validation, and verification in RoboTool.

In what follows, we present two aspects of RoboSim. In Sect. 5.1, we describe a RoboSim module corresponding to the RoboChart module in Fig. 9.4. This is an account of the simulation of the control software; it is called a d-model (for data model). We focus our discussion on the simulation of the state machine DeadMansHandle. In Sect. 5.2, we describe the RoboSim support to describe physical models, called p-models, and illustrate our ideas using a simplified version of the pod vehicle. Finally, Sect. 5.3 discusses our approach to generate simulation code from RoboSim models for use with robotics simulators.

5.1 RoboSim: d-model

RoboSim has the same structure of modules, controllers, and machines as RoboChart. For our example, we use the same module and controller definitions in RoboChart to define the RoboSim simulation, except only that the RoboSim versions define a value for the length of the simulation cycle. For the machines, however, different models give a cycle-based account of the behavior. We present in Fig. 9.10 a RoboSim machine corresponding to DeadMansHandle in Fig. 9.7. The RoboSim machine has the same name but defines the cycle in a cycleDef clause next to the name.

In a RoboSim machine, there is only one event (in the RoboChart sense) called exec, available without declaration. It controls the cyclic flow of the simulation. The processing phase of the simulation is defined by the machine, using the inputs, and the event exec to indicate when processing is finished and outputs can be provided.

In each cycle, inputs are read to determine whether the corresponding events have happened and, if so, the values that are provided as input. In the definition of the machine, if needed, this information is available. For our example, the inputs are deadmanshandle and configuration, declared as events in the RoboChart model, but used in RoboSim differently. Rather than as a trigger, we use $deadmanshandle?msg as a Boolean in a guard, which is true if the input deadmanshandle event has happened. In this case, the input value is recorded in the local variable msg.

The transition whose guard includes $deadmanshandle?msg is part of the translation of the transition from Spin to RxCallback in the RoboChart model. In the first cycle, like in the RoboChart model, the machine starts in Initialise and proceeds to Spin, and from there to a junction. In the junction, there is a decision characterized by the guards of the outgoing transitions. If $deadmanshandle?msg is true, the simulation machine moves to the state RxCallback, as in the RoboChart machine.

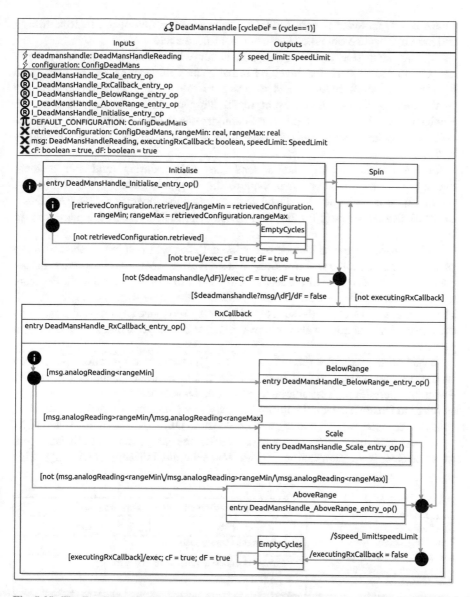

Fig. 9.10 The DeadMansHandle RoboSim machine. Copyright University of York, printed with permission in this chapter.

The guard uses also a local Boolean variable dF initialized to true. Our model-transformation strategy for generating a RoboSim model from a RoboChart model uses such variables to record whether an associated input event, in this case, deadmanshandle, has been referenced. It ensures that only one reference is possible in each cycle. This is to match the structure of the RoboChart model, where multiple references to an event denote different occurrences of that event. In RoboSim, a different occurrence can take place only in a future cycle.

Another transition from the junction is for when deadmanshandle does not happen. In RoboChart, the machine remains in Spin waiting to be interrupted by the event. In RoboSim, exec is raised to allow the simulation to progress to the next cycle (when the event might happen). In this extra transition, all Boolean variables (dF and cF, for $configuration) are reset to true. No variable is associated with the reference to $deadmanshandle, so no variable is updated.

In the translation, all state actions (i.e. entry, during, and exit actions) become calls to operations that capture the original action in the RoboChart model. In our example, for instance, we have the operation DeadMansHandle_Initialise_entry_op() for the entry action of Initialise. Its definition is shown in Fig. 9.11.

This transformation to include operation calls is part of an initial normalization phase in the model transformation. It removes or reduces complexity in the potentially rich structure of the RoboChart model to simplify translation. For state actions, normalization ensures that they all have the same form: an operation call. In the actions themselves, the structure defined by sequence (;) and conditionals (if-then-else), for example, is also removed in favor of the exclusive use of states and junctions.

Based on the variables (including constants), events, and operations referenced in the original action, an interface is defined and required in the corresponding operation definition. The interface ReqVars_DeadMansHandle_Initialise_entry required in DeadMansHandle_Initialise_entry_op() collects retrieveConfiguration, rangeMin, rangeMax, cF, and dF, and the constant DEFAULT_CONFIGURATION.

Each primitive action (assignment or input) in the entry action of Initialise occurs as an isolated transition action in DeadMansHandle_Initialise_entry_op(). This

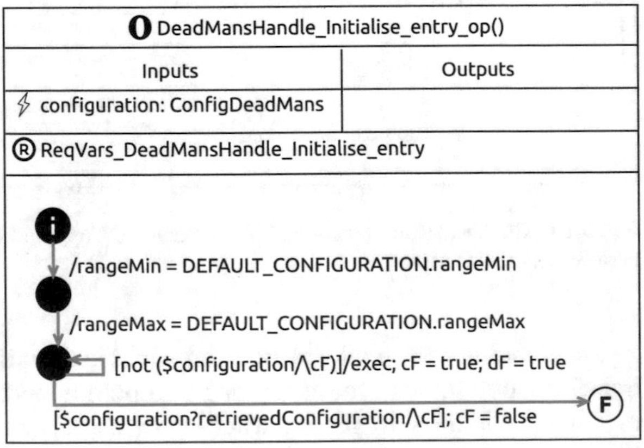

Fig. 9.11 The DeadMansHandle_Initialise_entry_op() RoboSim operation. Copyright University of York, printed with permission in this chapter.

results from the normalization. In the case of the input, similar to what is done in DeadMansHandle, an extra transition with event exec allows the cycles to proceed.

The structures of the machines for Initialise and RxCallback are very similar to those of the original RoboChart machines. Of note is the fact that, during processing, outputs are defined, but become visible to other machines, controllers, and the platform only when exec occurs. So, in the RoboChart machine for RxCallback, the output via speed_limit is the trigger of the transition to the final state. In the RoboSim machine, the output occurs as a transition action, $speed_limit!speedLimit. The use of $speed_Limit, instead of simply speed_limit, indicates that this is an output that is only actually visible when the processing phase of the simulation terminates.

The final states of the machines for Initialise and RxCallback are replaced with a new state EmptyCycles. This is because, in general, the final state of a machine of a composite state needs to explicitly raise the exec event to allow the cycles to proceed. So, instead of the special final state, the translation uses a new state with an extra transition that has an exec action. This transition is guarded by the negation of the guards of the transitions that leave that composite state.

In the example, there is only one outgoing transition from Initialise and one from RxCallback. In the case of Initialise, the negation of its outgoing transition is just false; in the case of RxCallback, it is executingRxCallback. In both cases, the extra transitions are never taken because their guards never hold. For RxCallback, as soon as its machine finishes, the outgoing transition is enabled because of the action executingRxCallback = false, and the state machine DeadMansHandle exits RxCallback. In general, however, this is not the case.

Generating code for a simulation from a RoboSim module is much more direct than from a RoboChart module. In the case of RoboSim, cyclic behavior is already identified. If the RoboSim model is verified against, or generated automatically, from a RoboChart model, we can be certain that its properties are preserved (given the assumptions characteristic of the cyclic control flow of a simulation). For example, we know that the outputs produced are as specified by the RoboChart model for the inputs provided, and time budgets and deadlines are preserved.

For a simulation, however, as well as an account of the control software, we also need a model of the physical robot. This is discussed in the next section.

5.2 RoboSim: p-model

RoboSim has a block-diagram notation that allows us to describe robotic platforms by characterizing their physical properties and behaviors using systems of differential-algebraic equations. Figure 9.12 presents a RoboSim specification of a (simplified) physical model (p-model) for the pod in our running example.

A diagram for a p-model defines blocks to represent links (i.e. rigid bodies), joints, sensors, and actuators, as well as blocks to represent some of their properties. The diagram defines a tree structure that specifies a containment relationship

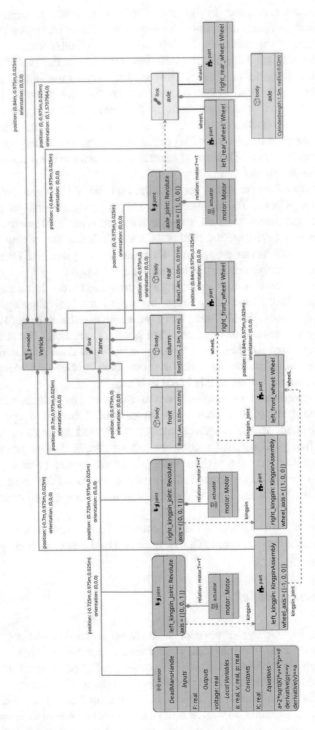

Fig. 9.12 The pod p-model. Copyright University of York, printed with permission in this chapter.

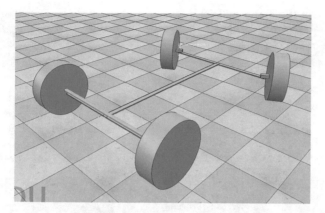

Fig. 9.13 The pod in CoppeliaSim. Copyright University of York, printed with permission in this chapter.

between elements. The root of the tree represents the physical component as a whole. Its children are blocks representing links (or parts) of that physical component. Links may contain junctions, links and junctions may contain sensors and actuators, and so on. In Fig. 9.12, the tree represents a physical model for the pod called Vehicle.

This simplified model for the vehicle omits its body (just for conciseness). What is modeled are the frame and the wheels, as shown in Fig. 9.13. This p-model contains several links. A link called frame has three bodies, namely, front, column, and rear. These are the bars that define the H shape in Fig. 9.13. Since frame is a link, its multiple bodies are pieces of a single rigid component.

The containment association is represented by connections between blocks with a closed lozenge on the side of the containing block. They can be annotated with the position coordinates (x, y, and z) and orientation (roll, pitch, and yaw) of the contained block. These are defined with respect to the frame of reference of the containing block. For the p-model block, by convention, the position and orientation are (0,0,0) and (0,0,0). In Fig. 9.12, the front and rear bars are positioned by identifying their y coordinates as 0.975 m to the front and to the back (−0.975 m) of the column. They have the same orientation as the vehicle.

The wheels are also links of the Vehicle. In Fig. 9.12, they are left_front_wheel, right_front_wheel, left_rear_wheel, and right_rear_wheel, defined as parts. Each part is an instance of a p-model defined separately by another block diagram. In this example, the p-model is Wheel; it is shown in Fig. 9.14. Each Wheel has a link wheelL with a body wheelB defined as a Cylinder.

The left_rear_wheel and right_rear_wheel are connected by an axle, whose body is a Cylinder. The connection to the wheels is fixed: represented by solid lines.

Two more parts, left_kingpin and right_kingpin, are instances of a King-pinAssembly that can be used to turn a wheel. As shown in Fig. 9.14, this is a Box with a Revolute joint kingpin_joint, whose axis is the wheel_axis defined

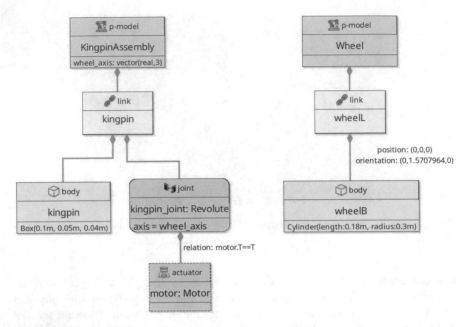

Fig. 9.14 The physical models for the parts used in the pod. Copyright University of York, printed with permission in this chapter.

as a parameter to KingpinAssembly. In the definitions of left_rear_wheel and right_rear_wheel, the values of this parameter are specified for the wheel associated with the kingpin.

Every joint is contained in a link and has a flexible connection, identified by a dashed arrow, to another link. The joint kingpin_joint of a KingpinAssembly is contained in its link. Its flexible connection is specified when defining a part. In Fig. 9.12, the flexible connections are to the wheels. Connections to and from a part are annotated with the elements of the part that are being connected. In the case of both left_kingpin and right_kingpin, the kingpin_joint is flexibly connected to the link wheelL (of left_front_wheel and of right_front_wheel).

The frame has three Revolute joints: left_kingpin_joint and right_kingpin_joint, flexibly connected to the links in left_kingpin and right_kingpin to turn the front wheels, and axle_joint, flexibly connected to the axle for the rear wheels. The behavior of a Revolute joint, and others, is defined as part of a library, using a system of differential-algebraic equations. Sensors and actuators can also have their behavior defined in this way, and the library includes a number of such definitions. In our example, all joints contain a motor, an actuator defined in a library.

As an example, we show a sensor DeadMansHandle contained in the frame. Its input is the force F applied to the handle, and its output is a voltage. The equations at the bottom of the DeadMansHandle block use local variables a, v, and p, to record the acceleration, velocity, and position of the handle, and the spring constant K. When the driver releases the handle, it goes back to the initial position.

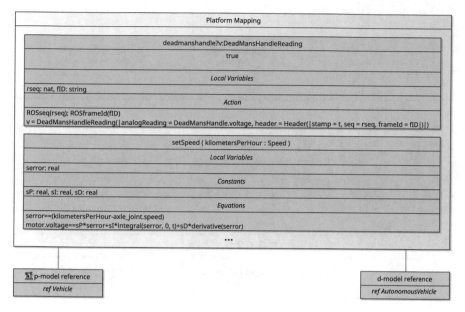

Fig. 9.15 The platform mapping for the pod. Copyright University of York, printed with permission in this chapter.

A more realistic physical model for the pod includes a description of its body and properties such as its weight and shape, which have an effect on its motion. Moreover, the simple steering mechanism based solely on kingpin joints is not what is used in the pod. It uses two sets of rack-and-pinion steering mechanisms, although only the front set is used by the control software. The functionality of the simplified steering component presented here is all that is needed for the control software.

To give an account of the behavior of the robot, considering both its control software, defined by a d-model, and its physical platform, defined by a p-model, we need to combine these models. Their connection is via the variables, events, and operations defined in the d-model for the platform. These elements identify the visible behavior of the software. So, to make the connection, we need to define them using elements of the p-model: outputs of sensors and inputs of actuators.

This definition is provided via a platform mapping. For our example, a sketch is presented in Fig. 9.15. The mapping box connects a d-model, in our example, AutonomousVehicle, to a p-model, here, Vehicle. In the box, we define each variable, event, or operation of the connected d-model. In Fig. 9.15, we present just two such definitions: for the event deadmanshandle and for operation setSpeed.

For events, a condition specifies when they happen. In our example, the condition for deadmanshandle is just true, indicating that this event is always available. The Action defines the communicated value v, using two Local Variables rseq and fID to record elements of a ROS header defined by built-in platform software operations called ROSseq and ROSframeId. These variables are used to define the record of

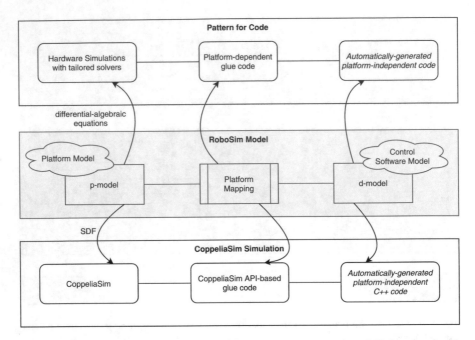

Fig. 9.16 The simulation approach for RoboSim. Copyright University of York, printed with permission in this chapter.

type DeadMansHandleReading that is assigned to v. It records the voltage output by the sensor DeadMansHandle in the p-model, and, in the header, the time t in which deadmanshandle occurs and the message is communicated. The operation setSpeed is defined by Equations that define a PID that reaches the speed kilometersPerHour.

To validate a p-model, and to execute the simulation as a whole, we need code for a robotics simulator. The generation of such code is discussed next.

5.3   Simulation Code

Figure 9.16 summarizes our approach to simulation of RoboSim models. As already mentioned, a RoboSim model (light-gray box in Fig. 9.16) is composed of three distinct components. The *p-model* is a block diagram that describes the physical platform in terms of links, joints, sensors, actuators, and their equations. The *d-model* is a RoboSim module that specifies the control software in terms of variables, events, and operations of the platform. The *platform mapping* describes how they are interpreted in terms of variables (continuous flows) of the physical model.

Accordingly, a simulation derived from a RoboSim model consists of three components, matching the structure of the RoboSim model; see Fig. 9.16. In general, the p-model needs to be used to produce a system of differential-algebraic equations that is passed to an off-the-shelf or tailor-built solver to simulate the continuous behaviors of the system. The d-model needs to be used to automatically generate platform-independent code. Finally, platform mapping needs to be used to generate solver-dependent interface code that bridges the gap between the control software and platform simulations, passing data back and forth between them.

This general pattern can be adopted to develop a simulation from scratch, but can also be instantiated for an existing simulator to streamline the simulation process. Our approach to generating simulations relies on existing simulators. We have experience with ARGoS and CoppeliaSim. Such robotics simulators embed domain knowledge, such as multi-body physics notions like links and joints. They also provide the means to simulate their behaviors efficiently through the use of various physics engines.

Figure 9.16 describes the instantiation of our pattern for use with CoppeliaSim. The d-model is used in the same way, with the generated code suitable for any simulator that adopts the used programming language. For CoppeliaSim, we use C++. The p-model does not need to be used to provide the equations for simulation. Abstractions such as link and joints are available in CoppeliaSim and can be imported via SDF, which can be used as an input to various robotics simulators. Finally, platform mapping is used to produce an interface that implements the platform's variables, events, and operations used by d-model in terms of the API of CoppeliaSim.

With a high level of automation, fixing problems found during simulation costs much less. Tests for use with the simulation are discussed in Chap. 11.

6 Environment Modeling

Another core component of a robotic system is the environment in which it operates. For RoboSim, as a simulation language, we need a notation to characterize a particular scenario (or a collection of specific scenarios). We envisage the use of a block diagram, like for a p-model definition, to specify an e-model, that is, an environment model. Like for a p-model, such a diagram defines the physical elements of the scenario and their behavior. An environment mapping needs to describe how elements of the scenario are perceived and affected by the sensors and actuators.

For a RoboChart model, the possibility of defining only a particular scenario is too restrictive. In a design model, we need, instead, to identify the assumptions about the environment and the robotic platform that need to be satisfied to ensure the proper behavior of the robotic system. The assumptions are the operational conditions.

When documented, if at all, these assumptions are commonly expressed in natural language. In the RoboStar technology, these assumptions, which abstract general properties of any valid environment and platform, are also written in natural language, but using controlled English. For natural-language processing (NLP), we need to manage the trade-off between unconstrained text and automation capabilities.

Some NLP techniques, such as the one described in [14], do not restrict the text and, thus, can be considered to deal with fully natural languages. In general, they are built on artificial intelligence techniques, and rely on a large corpus of sentences to train the underlying models to be capable of processing new unseen text. This approach is not suitable for RoboStar, since we do not have a large corpus of environment and platform assumptions; many times, they are not properly documented and left as part of the roboticists tacit knowledge.

At the other end of the spectrum of NLP techniques, we have very constrained languages, such as that in [21]. They require loss in naturalness by considering fragments of formal definitions and programming concepts. The imposed structure, however, favors text processing automation even without a corpus of examples.

In the middle of this spectrum, we have approaches that seek for a compromise between naturalness and constrained writing [11, 46, 77]; we seek this compromise. We have devised RoboWorld, a controlled natural language (CNL) for the specification of environment and platform assumptions with a precise semantics. From the controlled English, we can automatically generate CSP scripts of models that support the consideration of these assumptions within the RoboStar technology.

RoboWorld is defined using the Grammatical Framework (GF) [65], a special-purpose functional programming language for developing grammars. It supports the complexities found in different natural languages, such as word inflections and agreement between elements of a sentence. In Sect. 6.1 we comment on the syntax of RoboWorld, followed by an overview of its semantics in Sect. 6.2.

6.1   *RoboWorld Syntax*

RoboWorld is defined with abstract and concrete grammars. The abstract grammar defines the types of assumptions that can be described in the language. It can be seen as a metamodel of the supported controlled English. Differently, the concrete grammar relates the metamodel with actual English sentences.

The grammar establish that a RoboWorld specification defines assumptions about the world, including the environment and the platform, and mapping information. The mapping explains how the world influences and is influenced by the values of the variables, events, and operations of the platform of a RoboChart module. The concept is similar to that of a platform mapping described in Sect. 5.2.

Here, we illustrate the syntax of RoboWorld via an example presented in Fig. 9.17. It specifies some assumptions and part of the mapping for the autonomous pod. The environment assumptions highlight that the arena is two-dimensional, the ground is flat, and we have obstacles in some locations of the arena.

To exemplify a mapping definition, we show how the input event deadmanshandle can be defined. The English description defines when the event occurs, and the value communicated. The assumption is slightly more abstract than the definition in

```
## World assumptions ##
The arena is two-dimensional.
The gradient of the ground is 0.
There are entities called obstacle.
Some of the locations contain an obstacle.

## Mapping definitions ##
** Output event definitions **
** Input event definitions **
The event deadmanshandle is always available, and it communicates a record
whose field analogReading records the voltage obtained from the
DeadMansHandle, and whose field header is a record whose field stamp
records the time the event occurs.
** Operation definitions **
When the operation setSpeed(x) is called, the speed of the robot is set
to x km/h.
** Variable definitions **
```

Fig. 9.17 RoboWorld assumptions for the pod. Copyright University of York, printed with permission in this chapter.

the platform mapping in Fig. 9.15, since it does not impose any restriction on values of the header that are not relevant for the software. The RoboChart module for the pod reflects in many ways the fact that we have a ROS application, so it is to be expected that features of ROS are assumed (like the format of messages).

In general, we can be more abstract if it is convenient. For example, if an application has an input event obstacle, the RoboWorld mapping can specify "The event obstacle happens when the robot is less than one meter from a location in which there is an object". In this case, there is no reference to a particular sensor, and a variety of technologies can be used to satisfy such an assumption.

In Fig. 9.17, the definition of the operation setSpeed explains how it affects the robotic platform. Here, we have another example of a very abstract definition, where the use of motors needed to achieve the required speed is not mentioned. This is in direct contrast with the platform mapping in Fig. 9.15.

As already said, RoboWorld has a precise formal semantics given in CSP. As we explain in the next section, in the semantics, there are concepts that are application independent, whereas others are derived directly from the controlled English specification, such as that presented in Fig. 9.17.

6.2   RoboWorld Semantics

A CSP process that captures a RoboWorld specification consists of two components, shown in Fig. 9.18: a process that captures the world assumptions and another for the mapping. The process that captures the whole robotic system behavior also includes the CSP definition for the RoboChart module.

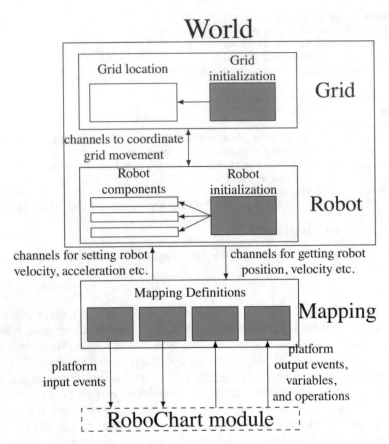

Fig. 9.18 The structure of the RoboWorld CSP semantics. Copyright University of York, printed with permission in this chapter.

CSP processes communicate via channels. In the case of a process for a RoboChart module, there are channels to represent each of the variables, events, and operations of the robotic platform. In the system process in Fig. 9.18, these channels are used for communication with the mapping process. A process for a RoboChart module, which defines control software, does not communicate directly with the world process.

The mapping process captures how the variables, events, and operations of the control software affect and are affected by the world. To specify that, the mapping needs to set and get information about the world. So, the mapping process communicates with the world process via get and set channels for properties like position, velocity, and acceleration of the robot, for example. The world process describes the layout and behavior of objects in the world.

The parts of the CSP process that are application dependent and generated from the RoboWorld description are indicated in Fig. 9.18 by gray shading. Roughly speaking, as probably expected, the world process is generated from the world assumptions, and the mapping process is generated from the mapping definitions.

The world process has two components: the grid and the robot processes. The grid process models a 2D or 3D arena as a grid of locations. With that, we can represent the position of each of the entities in the environment, including the robot itself. The world assumptions define the initial positions and properties of each of the entities on the grid, in terms of application independent location processes that manage the movement and placing of entities on the grid.

The robot process has several components for managing the robot's dynamic behavior, such as movement and carrying of objects. These components are application independent. For example, the processes that manage the robot's movement compute its path, but the robot is moved by communication with the grid process. The world assumptions are used to generate an initialization process that defines the movement components required, and specifies initial values for the robot's position and velocity.

The mapping includes a process for each mapping definition, specifying how the variables, events, and operations of the module affect and are affected by the behavior of the robot. For an input event, information about the robot, such as its position, is obtained from the world process and the event generated when the conditions for its occurrence specified in the mapping definition are met. For an output event or operation, the mapping process gets information from the module and communicates with the world to get and set values, such as the robot's velocity and acceleration, as specified in the mapping definition. The mapping-definition processes are combined in parallel so that each input and output is handled separately.

We automate the translation of RoboWorld descriptions to CSP using the GF Java API. We use GF to parse the input text, and Java code to traverse the AST and generate the CSP script. RoboWorld embeds primitive notions of 2D and 3D arenas, regions, areas, objects (with a physical body), entities (like gas and light), and so on. More examples are required to enrich RoboWorld further.

7 Related Work

Early on, existing mathematical techniques [20, 38] were applied to robotics. Model-checking techniques are available for many general-purpose languages. The goal of the RoboStar technology, however, is customization to produce simple domain-specific languages for practitioners [10, 18, 60], with tool support for graphical modeling, and optimizations in the semantics and verification that do not apply in general. Nordmann et al. [56] suggest that domain-specific languages for robotics like RoboChart and RoboSim are becoming popular. Our work is distinctive in its use of mathematical models for verification.

There are several general languages for architectural and behavioral modelling: SysML [58], AADL [23], Focus [9], and others. For SysML, a comprehensive semantics in a CSP-like language is available [43]. The RoboSim block diagrams used to specify physical models are based on SysML block diagrams.

UML (and its derivatives) has been used in various application domains, including safety-critical systems. There are many formalizations of UML, but, in general, covering subsets of UML. There are tailored semantic domains [8] and applications of existing techniques: graph transformations [41], CSP [16, 66], and others.

The AutoFocus [74] approach caters for the whole development process, from informal textual specification to code. This tool chain is similar to RoboTool in its goals. On the other hand, where AutoFocus targets embedded software with behavior defined by automata or functions, RoboTool focuses on robotic applications with behavior defined by state machines. Verification in AutoFocus uses theorem proving with Isabelle/HOL; similar goals are explored in [27] for RoboTool. Semi-automatic model transformation encodes properties into temporal logic; the transformation generates a refinement of the original model, rather than encoding its semantics. So the properties of the generated model can be slightly different. AutoFocus also provides facilities for code generation.

In this book, other domain-specific approaches to modeling and verification of robotic systems are presented. In Chap. 1, we have an approach to deal with the challenges of creating product lines for robotic systems that uses a few domain-specific languages. RoboChart and RoboSim can be useful in that context to specify functionality, physical elements such as sensors, and (non-functional) time requirements. They would complement, rather than replace, the use of the domain-specific notations that deal with feature modeling and can support use of verification.

Model-based and component-based development is at the heart of the very ambitious RobMoSys framework presented in Chap. 2. That effort proposes a modeling approach for development of robotic software based on loosely connected components. It puts these forward as a basis for the collaborative construction of a base of reusable resources developed and used by a variety of stakeholders. For description of behavior, they use data sheets, an abstraction mechanism. RoboChart (or RoboSim) could be used in conjunction with data sheets to provide a layer of formality while maintaining abstraction, something that cannot be achieved with code. A challenge would be to map the concepts in RoboChart and RoboSim related to the abstraction of the robotic platform and of the environment, since these are not present in RobMoSys.

Chapter 8 also reports on a very successful approach based on a domain-specific language called GenoM. Like RoboStar notations, GenoM covers architectural design, concurrency, control of events, and verification by translation to existing formal notations and tools. GenoM is an executable language (potentially including C code). RoboChart, on the other hand, is a self-contained modeling language supporting various levels of abstraction, but indeed requiring extra modeling effort from users.

SafeRobots [63] is a general component-based framework in which components have a data-flow architecture. OCL is adopted for definition of properties, but specification of behavior is via code from libraries rather than state machines.

MontiArcAutomaton [67] comprises an ADL based on components and connectors that allows extension with component-behavior modeling languages. There is support for use and integration of multiple modeling languages and code generators and for heterogeneous target platforms. RoboChart, as a language based on components and connectors, could be integrated with this setting.

FlexBE [70] is a behavior engine for ROS that enables human operators to specify and observe a robot's behavior and intervene at runtime by pausing or modifying it. Behaviors are specified by hierarchical state machines with actions implemented in Python. Similar, but more abstract, models can be developed in RoboChart for verification using shared variables and multiple state machines. FlexBE's tool does not support formal verification. Thus, our approach is complementary.

MissionLab [19] supports end users in specifying behavior as mission plans in military applications. A wizard allows the definition of tasks, environment, the possibility of presence of enemies, and other parameters. Behavior is defined using simple state machines. Verification is not mentioned, but usability studies indicate ease of use. Such studies for RoboChart and RoboSim are not available yet.

SPECTRA allows modeling of behavior and environment assumptions using patterns [48] of LTL with efficient synthesis algorithms [7]. This requires discrete data type abstractions [50]. Time constraints cannot be directly specified, and so the model needs to account for the target cyclic paradigm. Evidence [49] suggests that modeling of realistic environments and traceability are challenging.

Mauve [36] supports component-based models with interfaces defined by constants, operations, and ports, but not shared variables like RoboChart and RoboSim. Behavior can be defined just by code or simple textual state machines. Specifications, however, can use a contract language based on temporal logic and observation points of the code or machine. Code generation is for Orocos [73] platforms, with an optimized WCET analysis used to ensure schedulability. Time properties are derived from this analysis, rather than specified like in RoboChart.

The work in [25] is for an adaptive architecture; the verification enables identification of optimal configurations based on various proof techniques, including model checking. Verification of behavioral properties, however, is not the focus.

Orccad [20, 38] is a notation for modeling, simulation, and programming, with verification (of timed properties) based on the translation of models into formal languages like for RoboChart and RoboSim. Orccad models are formed from tasks defined by control laws, combined by procedures defined by reactive programs. The combined use of RoboChart with control laws is addressed in [12], and that approach, based on modern co-simulation standards [26], can be used for RoboSim.

The RoboChart time primitives are inspired by timed automata and Timed CSP. Timed automata use synchronous continuous-time clocks, and properties expressed in temporal logic can be checked using the model checker UPPAAL. RoboChart, in contrast, provides abstractions specific for robotic applications and has a semantics for refinement. Comparable UPPAAL models require additional states, interleaved automata, and state invariants. Ongoing work is exploring a RoboChart and a RoboSim semantics using timed automata for property verification.

A real-time extension of UML statecharts called Hierarchical Timed Automata (HTA) is proposed in [15]. Roughly speaking, HTA are timed automata with hierarchy and history, but no operations. In [15], HTA are translated to timed automata for use with UPPAAL. Some of the restrictions on UML are similar to those of RoboChart, but some impose severe constraints on data, guards, and use of events. On the positive side, the target UPPAAL timed automata remain decidable.

UML [37] has a simple notion of time and little support to model timed properties. On the other hand, UML-MARTE [72] is a profile with support for logical, discrete, and continuous time using clocks. Clock constraints may be specified using CSSL [47], and a constraint solver [17] can find solutions for deployment. Specification of deadlines and time budgets is through sequence and time diagrams. While it is possible to define budgets for a particular behavior, it is not possible to define timed constraints in terms of transitions and states. Limited support for UML-MARTE is available in the freely available Papyrus tool [32], an Eclipse plugin for UML.

UML-RT, an extension to UML, focuses on the architectural description of systems using the notions of capsules, ports, and protocols. Capsules encapsulate state machines, while communication between capsules is via ports and defined by protocols. A timing protocol can act as a timer by raising timeouts [71], but it is not obvious how deadlines can be specified directly on UML-RT state machines.

A CSP-based semantics for UML-RT is defined in [64], but it does not cover time. An extension to UML-RT [3] has a timed semantics defined using CSP+T [79], an extension of CSP that supports the timing of events. CSP+T is the inspiration for the work in [3], where annotations are added to record the occurrence time of events and constrain the occurrence time of subsequent events. Some RoboChart and RoboSim time primitives are similar, but we have a richer set of primitives.

General-purpose simulation frameworks such as Simulink and Stateflow [51, 52], 20-sim,[4] and Modelica [31] are in widespread use. They can be used in robotics, but roboticists often describe state machines using an informal notation [59, 62, 75] before writing optimized code (in C or C++, for instance) for a simulator tailored for robotics. When there are complex control laws involved, the general simulators are useful. This is, however, not the case in many applications, and the flexibility of code-based simulations enables the development of more efficient simulations.

Compared to Stateflow, which is a statechart simulation language, RoboSim is, on one hand, much more restrictive, but, on the other hand, the cycle of a RoboSim model can be more flexibly defined. The occurrence of events or the structure of the machine does not implicitly define the behavior in each cycle. Moreover, in Stateflow, in each cycle, the machine is potentially executed several times, once for each event that has occurred. RoboSim adopts the approach of reactive simulators, where the machine is executed just once, when all events are normally considered. We expect, however, that it is possible to define a pattern for Stateflow models to allow their verification following the RoboSim approach.

---

[4] www.20sim.com/.

Robotics simulators vary in their coding language. Webots [57] and CoppeliaSim (previously called V-REP) [68] provide (different) graphical interfaces. Webots adopts a human-readable customized notation; in CoppeliaSim, several general languages are available. ARGoS and Enki (home.gna.org/enki/) are programmed using different C++ libraries. The Microsoft Robotics Developer Studio (www.microsoft.com/robotics) has environments and platforms that can be programmed in VPL or C#. Player/Stage [33] provides a device server, and clients can be programmed using popular languages. MASON [45] and BREVE [39] adopt the agent paradigm; BREVE adopts a custom language or Python, and MASON, Java.

None of these simulators adopts a diagrammatic notation like RoboSim to specify simulation code. Moreover, there is no portability between them. The RoboStar vision is that a RoboSim model can be used for automatic generation of code for such simulators. We have illustrated the results for CoppeliaSim.

RobotML [18] is a domain-specific language for robotics based on UML. It has support for automatic generation of platform-independent code, but reasoning about non-functional properties is envisaged although not available yet.

In the same vein, rFSM [40] is a domain-specific language for simulation and deployment but does not have a formal semantics. There is no support for analysis of models, either in isolation or in relation to designs, like we have for RoboSim.

RoboFlow [4] is a programming language with operational semantics. This formal semantics provides a clear way to define sound tools, but there is no support for reasoning about RoboFlow models in relation to designs.

ArmarX and Rafcon are programming languages for robotics based on state machines, but without formal semantics [10, 76]. Some of their restrictions, like the absence of inter-level transitions, are similar to those of RoboChart and RoboSim and ultimately can facilitate the provisioning of reasoning facilities.

In summary, what is distinctive about the RoboStar technology are small and controlled domain-specific languages. The architectural pattern that they embed can guide roboticists in developing models; the same is not true of open and general languages. The restrictions of RoboChart and RoboSim simplify their semantics and facilitate verification. Beyond support for verifying desirable properties of individual models, we have a conformance notion for a simulation with respect to a more abstract design model. More than new notations, the RoboStar technology provides a modeling and verification approach for simulation of robotic applications that can be useful in the context of all notations based on the state machines mentioned above.

A comprehensive survey on formal specification and verification in robotics [22, 44] highlights model checking as the most prominent verification approach in the literature. As illustrated here, RoboChart supports verification by model checking. Our long-term plan, however, is the use of theorem proving to deal with larger models. Ongoing and future work will explore the combination of verification approaches.

8 Final Considerations and Future Work

Current practice in robotics is normally based on standard state machines [10, 18, 60, 76], without formal semantics, to specify the robot controller only. The state machine that gives an abstract account of the robot controller guides the development of a simulation, but no rigorous connection between them is established. For implementation in a robotic platform, ad hoc adjustments are normally required to cater for the reality gap between the simulated and the actual environment. Numerous iterations of (re)development and testing, tool dependency, and low-level programming are prevalent, with an impact on cost, maintainability, and reliability.

RoboStar technology addresses the issues of principled modeling, verification, sound simulation generation, and testing. With domain-specific languages, and support for automatic generation of artifacts, and verification, it enables significant advances in the practice of Software Engineering for Robotics.

Reactive robotics simulators do not normally generate code specifically for deployment. Instead, simulation code is often reused after changes, because the API for simulation and for deployment is different, and simulation code is based on a cyclic executive. (This is a simple programming pattern for single-processor architectures; it cannot easily cope with multiple processors, heterogeneous architectures, or mixed criticality.) There is potential loss of properties observed via simulation: because of the possibility of changes introducing errors, and of the reality gap.

In future work, we will develop a domain-specific language for modeling deployment (layered) architectures and code, and a library of architectures. For each of them, we will define how to derive code automatically from a RoboChart or RoboSim model. Automation will ensure preservation of properties. The code will use multiple processors and have components with differentiated levels of guarantee: hard results for high-criticality and probabilistic guarantees for low-criticality components. Fault models will justify the adequacy of the approach to fault tolerance. Monitors will enable update of the deployment, simulation, and design models.

RoboWorld provides the basis for further work on identification of additional environment concepts of relevance to particular areas of application and categories of robotic platforms. In Chap. 7, the authors quite rightly state that a mathematical model cannot capture the physical world. Our approach with RoboWorld is to capture the assumptions that are necessary to prove properties of interest. These assumptions are operational conditions that, currently, are, at best, left implicit.

CorteX, described in Chap. 10, provides support for principled programming. It is a middleware designed to deal with the maintainability challenges faced by large-scale long-running applications, typical of those in the nuclear industry. Code generation for CorteX from RoboChart or RoboSim models is a very interesting avenue for future work. CorteX is equipped with validation support based on testing, and complementarity with RoboStar technology is promising.

Our vision is a twenty-first-century toolbox for robot-controller developers. In this toolbox, a developer can find unambiguous diagrammatic notations to specify models for the environment, the robotic platform, and the controller. For commonly used environments and robotic platforms, the toolbox includes a range of ready-made models. Because these models are precise, there is no scope for misunderstanding, and, most importantly, the toolbox includes techniques for desirable properties of the models: deadlock freedom, speed limits, and so on.

Since the technique for validation that robot controller developers favor nowadays is simulation, in the twenty-first-century toolbox, there are tools for automatic generation of these simulations. The ingenuity of the developer is now focused on the optimization of the simulation and of the associated deployed code. Because the languages used for simulation and programming are high level, the results are tool independent and can be deployed in a variety of robotic platforms.

With the twenty-first-century toolbox, the costly cycles of iterations of design and testing, with problems found very late, even just at deployment time, are reduced. Moreover, the developer can demonstrate that the controller produced satisfies essential properties. Software for mobile and autonomous robots is cheaper and more reliable.

Acknowledgments The B-ACS work was done as part of the CAVlab project in 2017–2018. The team involved includes Dave Barnett, Servando German Serrano, Ujjar Bhandari, Nastaran Shatti, and Alan Peters. Zeyn Saigol is proposing and developing the B-ACS work as a suitable autonomy verification case study. All members of the RoboStar group (www.cs.york.ac.uk/robostar/) have contributed directly or indirectly to the vision described here. Our work is funded by the Royal Academy of Engineering under Grant No. CiET1718/45, and by the UK EPSRC (Engineering and Physical Sciences Research Council) under Grant Nos. EP/M025756/1 and EP/R025479/1.

References

1. T. Abdellatif, S. Bensalem, J. Combaz, L. deSilva, F. Ingrand, Rigorous design of robot software: a formal component-based approach. Robot. Autonom. Syst. **60**(12), 1563–1578 (2012)
2. M.M. Adams, P.B. Clayton, Cost-effective formal verification for control systems, in *ICFEM 2005: Formal Methods and Software Engineering*, ed. by K. Lau, R. Banach. Lecture Notes in Computer Science, vol. 3785 (Springer, Berlin, 2005), pp. 465–479
3. K.B. Akhlaki, M.I.C. Tunon, J.A.H. Terriza, L.E.M. Morales, A methodological approach to the formal specification of real-time systems by transformation of UML-RT design models. Sci. Comput. Program. **65**(1), 41–56 (2007)
4. S. Alexandrova, Z. Tatlock, M. Cakmak, Roboflow: a flow-based visual programming language for mobile manipulation tasks, in *IEEE International Conference on Robotics and Automation* (2015), pp. 5537–5544
5. W. Barnett, Architectural data modelling for robotic applications. Technical report (2019)
6. G. Behrmann, A. David, K.G. Larsen, J. Hakansson, P. Petterson, W. Yi, M. Hendriks, UPPAAL 4.0, in *3rd International Conference on the Quantitative Evaluation of Systems* (IEEE Computer Society, Washington, 2006), pp. 125–126
7. R. Bloem, B. Jobstmann, N. Piterman, A. Pnueli, Y. Sa'ar, Synthesis of reactive(1) designs. J. Comput. Syst. Sci. **78**(3), 911–938 (2012). In Commemoration of Amir Pnueli

8. M. Broy, M.V. Cengarle, B. Rumpe, Semantics of UML - towards a system model for UML: The state machine model. Technical Report TUM-I0711, Institut für Informatik, Technische Universität München (2007)
9. M. Broy, K. Stølen, *Specification and Development of Interactive Systems: Focus on Streams, Interfaces, and Refinement* (Springer, Berlin, 2001)
10. S.G. Brunner, F. Steinmetz, R. Belder, A. Domel, Rafcon: a graphical tool for engineering complex, robotic tasks, in *IEEE/RSJ International Conference on Intelligent Robots and Systems* (2016), pp. 3283–3290
11. G. Carvalho, A.L.C. Cavalcanti, A.C.A. Sampaio, Modelling timed reactive systems from natural-language requirements. Formal Aspects Comput. **28**(5), 725–765 (2016)
12. A.L.C. Cavalcanti, A. Miyazawa, R. Payne, J. Woodcock, Sound simulation and co-simulation for robotics, in *Present and Ulterior Software Engineering*, ed. by M. Mazzara, B. Meyer (Springer, Berlin, 2017), pp. 173–194
13. A.L.C. Cavalcanti, A.C.A. Sampaio, A. Miyazawa, P. Ribeiro, M. Conserva Filho, A. Didier, W. Li, J. Timmis, Verified simulation for robotics. Sci. Comput. Program. **174**, 1–37 (2019)
14. D. Chen, C. Manning, A fast and accurate dependency parser using neural networks, in *Conference on Empirical Methods in Natural Language Processing* (Association for Computational Linguistics, Stroudsburg, 2014), pp. 740–750
15. A. David, M.O. Möller, W. Yi, Formal verification of UML statecharts with real-time extensions, in *Fundamental Approaches to Software Engineering*, ed. by R.-D. Kutsche, H. Weber (Springer, Berlin, 2002), pp. 218–232
16. J. Davies, C. Crichton, Concurrency and refinement in the unified modeling language. Formal Aspects Comput. **15**(2–3), 118–145 (2003)
17. J. DeAntoni, F. Mallet, TimeSquare: treat your models with logical time, in *Objects, Models, Components, Patterns* (Springer, Berlin, 2012), pp. 34–41
18. S. Dhouib, S. Kchir, S. Stinckwich, T. Ziadi, M. Ziane, RobotML, a domain-specific language to design, simulate and deploy robotic applications, in *Simulation, Modeling, and Programming for Autonomous Robots* (Springer, Berlin, 2012), pp. 149–160
19. Y. Endo, D.C. MacKenzie, R.C. Arkin, Usability evaluation of high-level user assistance for robot mission specification. IEEE Trans. Syst. Man Cybern. C (Appl. Rev.) **34**(2), 168–180 (2004)
20. B. Espiau, K. Kapellos, M. Jourdan, *Formal Verification in Robotics: Why and How?* (Springer London, 1996), pp. 225–236
21. M. Esser, P. Struss, Obtaining models for test generation from natural-language like functional specifications, in *International Workshop on Principles of Diagnosis* (2007), pp. 75–82
22. M. Farrell, M. Luckcuck, M. Fisher, Robotics and integrated formal methods: necessity meets opportunity, in *Integrated Formal Methods*, ed. by C.A. Furia, K. Winter. Lecture Notes in Computer Science, vol. 11023 (Springer, Berlin, 2018), pp. 161–171
23. P.H. Feiler, D.P. Gluch, *Model-Based Engineering with AADL: An Introduction to the SAE Architecture Analysis & Design Language* (Addison-Wesley, Boston, 2012)
24. M.S. Conserva Filho, R. Marinho, A.C. Mota, J.C.P. Woodcock, Analysing robochart with probabilities, in *Formal Methods: Foundations and Applications*, ed. by T. Massoni, M.R. Mousavi (Springer, Berlin, 2018), pp. 198–214
25. F. Fleurey, A. Solberg, A domain specific modeling language supporting specification, simulation and execution of dynamic adaptive systems, in *12th International Conference on Model Driven Engineering Languages and Systems* (Springer, Berlin, 2009), pp. 606–621
26. FMI development group. Functional mock-up interface for model exchange and co-simulation, 2.0 (2014). https://www.fmi-standard.org
27. S. Foster, J. Baxter, A.L.C. Cavalcanti, A. Miyazawa, J.C.P. Woodcock, Automating verification of state machines with reactive designs and isabelle/UTP, in *Formal Aspects of Component Software*, ed. by K. Bae, P.C. Ölveczky (Springer, Cham, 2018), pp. 137–155
28. S. Foster, A.L.C. Cavalcanti, S. Canham, J.C.P. Woodcock, F. Zeyda, Unifying theories of reactive design contracts. Theor. Comput. Sci. **802**, 105–140 (2020)

29. S. Foster, Y. Nemouchi, C. O'Halloran, K. Stephenson, N. Tudor, Formal model-based assurance cases in Isabelle/SACM: an autonomous underwater vehicle case study, in *8th International Conference on Formal Methods in Software Engineering* (ACM, New York, 2020)

30. M. Foughali, B. Berthomieu, S. Dal Zilio, F. Ingrand, A. Mallet, Model checking real-time properties on the functional layer of autonomous robots, in *Formal Methods and Software Engineering*, ed. by K. Ogata, M. Lawford, S. Liu (Springer, Berlin, 2016), pp. 383–399

31. P. Fritzson, *Principles of Object-Oriented Modeling and Simulation with Modelica 2.1* (Wiley-IEEE Press, Hoboken, 2004)

32. S. Gérard, C. Dumoulin, P. Tessier, B. Selic, Papyrus: a UML2 tool for domain-specific language modeling, in *Model-Based Engineering of Embedded Real-Time Systems: International Dagstuhl Workshop, Dagstuhl Castle, Germany, November 4-9, 2007. Revised Selected Papers*, chap. 19 (Springer, Berlin, 2010), pp. 361–368

33. B. Gerkey, R.T. Vaughan, H. Andrew, The player/stage project: tools for multi-robot and distributed sensor systems, in *11th International Conference on Advanced Robotics* (2003), pp. 317–323

34. S. German, A. Peters, D. Barnett, U. Bhandari, N. Shatti, Connected and autonomous vehicles laboratory (CAVLab) - an accessible facility for development and integration of CAV technologies, in *ITS World Congress* (2018)

35. T. Gibson-Robinson, P. Armstrong, A. Boulgakov, A.W. Roscoe, FDR3 - a modern refinement checker for CSP, in Tools and Algorithms for the Construction and Analysis of Systems (Springer, Berlin, 2014), pp. 187–201

36. N. Gobillot, C. Lesire, D. Doose, A modeling framework for software architecture specification and validation, in *Simulation, Modeling, and Programming for Autonomous Robots* ed. by D. Brugali, J.F. Broenink, T. Kroeger, B.A. MacDonald (Springer, Berlin, 2014), pp. 303–314

37. Object Management Group. OMG Unified Modeling Language (2015). https://www.omg.org/spec/UML/2.5/About-UML/

38. K. Kapellos, D. Simon, M. Jourdant, B. Espiau, Task level specification and formal verification of robotics control systems: state of the art and case study. Int. J. Syst. Sci. **30**(11), 1227–1245 (1999)

39. J. Klein, BREVE: a 3D environment for the simulation of decentralized systems and artificial life, in *8th International Conference on Artificial Life* (The MIT Press, Cambridge, 2003), pp. 329–334

40. M. Klotzbucher, H. Bruyninckx, Coordinating robotic tasks and systems with rFSM statecharts. J. Softw. Eng. Robot. **2**(13), 28–56 (2012)

41. S. Kuske, M. Gogolla, R. Kollmann, H.-J. Kreowski, An integrated semantics for UML class, object and state diagrams based on graph transformation, in *Integrated Formal Methods*, ed. by M. Butler, L. Petre, K. SereKaisa. Lecture Notes in Computer Science, vol. 2335 (Springer, Berlin, 2002), pp. 11–28

42. M. Kwiatkowska, G. Norman, D. Parker, Probabilistic symbolic model checking with PRISM: a hybrid approach. Int. J. Softw. Tools Technol. Transf. **6**(2), 128–142 (2004)

43. L. Lima, A. Miyazawa, A.L.C. Cavalcanti, M. Cornélio, J. Iyoda, A.C.A. Sampaio, R. Hains, A. Larkham, V. Lewis, An integrated semantics for reasoning about SysML design models using refinement. Softw. Syst. Model. **16**(3), 1–28 (2017)

44. M. Luckcuck, M. Farrell, L.A. Dennis, C. Dixon, M. Fisher, Formal specification and verification of autonomous robotic systems: a survey. *CoRR*, abs/1807.00048 (2018)

45. S. Luke, C. Cioffi-Revilla, L. Panait, K. Sullivan, G. Balan, Mason: a multiagent simulation environment. Simulation **81**(7), 517–527 (2005)

46. B. Luteberget, J.J. Camilleri, C. Johansen, G. Schneider, Participatory verification of railway infrastructure by representing regulations in RailCNL, in *Software Engineering and Formal Methods*, ed. by A. Cimatti, M. Sirjani (Springer, Berlin, 2017), pp. 87–103

47. F. Mallet, Clock constraint specification language: specifying clock constraints with UML/-MARTE. Innov. Syst. Softw. Eng. **4**(3), 309–314 (2008)

48. S. Maoz, J.O. Ringert, GR(1) synthesis for LTL specification patterns, in *10th Joint Meeting on Foundations of Software Engineering, ESEC/FSE 2015* (Association for Computing Machinery, New York, 2015), pp. 96–106
49. S. Maoz, J.O. Ringert, Synthesizing a lego forklift controller in GR(1): a case study, in *4th Workshop on Synthesis* (2015)
50. S. Maoz, J.O. Ringert, On the software engineering challenges of applying reactive synthesis to robotics, in *1st International Workshop on Robotics Software Engineering* (Association for Computing Machinery, New York, 2018), pp. 17–22
51. The MathWorks, Inc., Simulink. www.mathworks.com/products/simulink
52. The MathWorks, Inc., *Stateflow and Stateflow Coder 7 User's Guide*. www.mathworks.com/products
53. A. Miyazawa, P. Ribeiro, A.L.C. Cavalcanti, W. Li, J. Timmis, J.C.P. Woodcock, RoboChart and robotool: modelling, verification and simulation for robotics. Technical report, University of York, Department of Computer Science, York (2020). www.cs.york.ac.uk/circus/RoboCalc/robosim/robosim-reference.pdf
54. A. Miyazawa, P. Ribeiro, W. Li, A.L.C. Cavalcanti, J. Timmis, J.C.P. Woodcock, RoboChart: modelling and verification of the functional behaviour of robotic applications. Softw. Syst. Model. **18**(5), 3097–3149 (2019)
55. T. Nipkow, M. Wenzel, L.C. Paulson, *Isabelle/HOL: A Proof Assistant for Higher-Order Logic* (Springer, Berlin, 2002)
56. A. Nordmann, N. Hochgeschwender, D. Wigand, S. Wrede, A survey on domain-specific modeling and languages in robotics. J. Softw. Eng. Robot. **7**(1), 75–99 (2016)
57. M. Olivier, Webots^{TM}: professional mobile robot simulation. Int. J. Adv. Robot. Syst. **1**(1), 39–42 (2004)
58. OMG, OMG systems modeling language (OMG SysML), version 1.3 (2012)
59. H.W. Park, A. Ramezani, J.W. Grizzle, A finite-state machine for accommodating unexpected large ground-height variations in bipedal robot walking. IEEE Trans. Robot. **29**(2), 331–345 (2013)
60. I. Pembeci, H. Nilsson, G. Hager, Functional reactive robotics: an exercise in principled integration of domain-specific languages, in *4th ACM SIGPLAN International Conference on Principles and Practice of Declarative Programming* (ACM, New York, 2002), pp. 168–179
61. C. Pinciroli, V. Trianni, R. O'Grady, G. Pini, A. Brutschy, M. Brambilla, N. Mathews, E. Ferrante, G. Di Caro, F. Ducatelle, M. Birattari, L.M. Gambardella, M. Dorigo, ARGoS: a modular, parallel, multi-engine simulator for multi-robot systems. Swarm Intell. **6**(4), 271–295 (2012)
62. C.A. Rabbath, A finite-state machine for collaborative airlift with a formation of unmanned air vehicles. J. Intell. Robot. Syst. **70**(1), 233–253 (2013)
63. A. Ramaswamy, B. Monsuez, A. Tapus, Saferobots: a model-driven framework for developing robotic systems, in *2014 IEEE/RSJ International Conference on Intelligent Robots and Systems* (2014), pp. 1517–1524
64. R. Ramos, A.C.A. Sampaio, A.C. Mota, A semantics for iUML-RT active classes via mapping into Circus, in *Formal Methods for Open Object-Based Distributed Systems, Lecture Notes in Computer Science*, vol. 3535 (2005), pp. 99–114
65. Aarne Ranta, *Grammatical Framework: Programming with Multilingual Grammars* (CSLI Publications, Stanford, 2011)
66. H. Rasch, H. Wehrheim, Checking consistency in UML diagrams: classes and state machines, in *Formal Methods for Open Object-Based Distributed Systems*, ed. by E. Najm, U. Nestmann, P. Stevens. Lecture Notes in Computer Science, vol. 2884 (Springer, Berlin, 2003), pp. 229–243
67. J.O. Ringert, A. Roth, B. Rumpe, A. Wortmann, Code generator composition for model-driven engineering of robotics component & connector systems. J. Softw. Eng. Robot. **6**(1), 33–57 (2015)

68. E. Rohmer, S.P.N. Singh, M. Freese, V-rep: a versatile and scalable robot simulation framework, in *IEEE International Conference on Intelligent Robots and Systems*, vol. 1 (IEEE, Piscataway, 2013), pp. 1321–1326
69. A.W. Roscoe, *Understanding Concurrent Systems*. Texts in Computer Science (Springer, Berlin, 2011)
70. P. Schillinger, S. Kohlbrecher, O. von Stryk, Human-robot collaborative high-level control with application to rescue robotics, in *IEEE International Conference on Robotics and Automation* (2016), pp. 2796–2802
71. B. Selic, Using UML for modeling complex real-time systems, in *Languages, Compilers, and Tools for Embedded Systems*, ed. by F. Mueller, A. Bestavros. Lecture Notes in Computer Science, vol. 1474 (Springer, Berlin, 1998), pp. 250–260
72. B. Selic, S. Grard, *Modeling and Analysis of Real-Time and Embedded Systems with UML and MARTE: Developing Cyber-Physical Systems* (Morgan Kaufmann, Burlington, 2013)
73. P. Soetens, H. Bruyninckx, Realtime hybrid task-based control for robots and machine tools, in *2005 IEEE International Conference on Robotics and Automation* (2005), pp. 259–264
74. M. Spichkova, F. Hölzl, D. Trachtenherz, Verified system development with the autofocus tool chain. in *Workshop on Formal Methods in the Development of Software* (2012)
75. T. Tomic, K. Schmid, P. Lutz, A. Domel, M. Kassecker, E. Mair, I.L. Grixa, F. Ruess, M. Suppa, D. Burschka, Toward a fully autonomous UAV: research platform for indoor and outdoor urban search and rescue. IEEE Robot. Autom. Mag. **19**(3), 46–56 (2012)
76. M. Wachter, S. Ottenhaus, M. Krohnert, N. Vahrenkamp, T. Asfour, The ArmarX statechart concept: graphical programing of robot behavior. Front. Robot. AI **3**, 33 (2016)
77. C. Wang, F. Pastore, A. Goknil, L. Briand, Z. Iqbal, Automatic generation of system test cases from use case specifications, in *International Symposium on Software Testing and Analysis* (Association for Computing Machinery, New York, 2015), pp. 385–396
78. J.C.P. Woodcock, P.G. Larsen, J. Bicarregui, J.S. Fitzgerald, Formal methods: practice and experience. ACM Comput. Surv. **41**(4), 19 (2009)
79. J.J. Zic, Time-constrained buffer specifications in CSP + T and timed CSP. ACM Trans. Program. Lang. Syst. **16**(6), 1661–1674 (1994)

Chapter 10
CorteX: A Software Framework for Interoperable, Plug-and-Play, Distributed, Robotic Systems of Systems

Ipek Caliskanelli, Matthew Goodliffe, Craig Whiffin, Michail Xymitoulias, Edward Whittaker, Swapnil Verma, Craig Hickman, Chen Minghao, and Robert Skilton

Abstract Robots are being used for an increasing number and range of tasks in the fields of nuclear energy, mining, petrochemical processing, and sub-sea. This is resulting in ever more complex robotics installations being deployed, maintained, and extended over long periods of time. Additionally, the unstructured, experimental, or unknown operational conditions frequently result in new or changing system requirements, meaning extension and adaptation is necessary. Whilst existing frameworks allow for robust integration of complex robotic systems, they are not compatible with highly efficient maintenance and extension in the face of changing requirements and obsolescence issues over decades-long periods. We present CorteX that attempts to solve the long-term maintainability and extensibility issues encountered in such scenarios through the use of a standardised, self-describing data representations and associated communications protocols. Progress in developing and testing the CorteX framework, as well as an overview of current and planned deployments, will be presented.

1 Introduction

Systems that will be used to maintain and inspect future fusion powerplants (e.g. ITER [2] and DEMO [17]) are expected to integrate hundreds of systems from multiple suppliers with a lifetime of several decades, over which requirements evolve and obsolescence management is required. There are significant challenges

I. Caliskanelli · M. Goodliffe · C. Whiffin · M. Xymitoulias · E. Whittaker · S. Verma · C. Hickman · C. Minghao · R. Skilton (✉)
Remote Applications in Challenging Environments (RACE), Culham Science Centre, Abingdon, UK
e-mail: ipek.caliskanelli@ukaea.uk; matthew.goodliffe@ukaea.uk; craig.whiffin@ukaea.uk; michail.xymitoulias@ukaea.uk; edward.whittaker@ukaea.uk; swapnil.verma@ukaea.uk; craig.hickman@ukaea.uk; chen.minghao@ukaea.uk; robert.skilton@ukaea.uk

© Springer Nature Switzerland AG 2021
A. Cavalcanti et al. (eds.), *Software Engineering for Robotics*,
https://doi.org/10.1007/978-3-030-66494-7_10

associated with the integration of such large systems from multiple suppliers, each operating using bespoke interfaces and having their training requirements.

Nuclear robotics in complex facilities, comprising hundreds of interoperating systems, will become increasingly commonplace as safety and productivity requirements drive facility design and operation. In order to be efficient and operate for years or decades without significant downtime, future systems will be dependent on automation and the sharing of information. Within current nuclear robotics applications, there are fundamental limitations regarding long-term maintainability, extensibility, dependability, reliability, security, and compatibility with regulatory requirements. Existing networked and cloud-based robotics lack interoperability, heterogeneity, security, multi-robot management, common infrastructure design, quality-of-service (QoS), and standardisation. There is a need for a new robotic framework, specifically tailored for such applications, that can adapt to new challenges, new user requirements, and new advances in technology. RACE is working towards providing a future-proof communications and control framework, capable of meeting these requirements, essential for future large-scale nuclear robotic facilities.

Development has been focused around five key design principles, namely, reusability, extensibility, modularity, standardisation, and integrated user interfaces. To guarantee reusability, control solutions must be implemented in a generic fashion in order to be agnostic of their application. These control solutions must also be implemented in such a way as to allow them to be extended by later applications, which may require additional functionality. Modularity ensures the ability to replace components of a control system with newer or alternative counterparts, providing the ability to adapt to changing requirements. Interoperability is essential for large-scale integration applications and can only be achieved through the use of agreed standards. However, these standards should not restrict the capability of a platform and therefore need to be customisable whilst maintaining backwards compatibility. If these four principles are followed, it is then possible to create standard integrated user interfaces for all systems, providing users and developers with a standardised experience.

CorteX is built around a self-describing protocol that contains both the data in the system and supporting metadata. The structure of data within the system is similar to that found in other robot control system platforms, such as ROS [41]. The key difference is that CorteX communication messages are standardised; there is only one message type. In the proposed interoperable solution, data belonging to a particular component is complimented with a combined architecture and inheritance type structure of the component, allowing for interpretation and discovery of previously unknown systems. The implementation of this solution takes the form of a core library containing standardised structures for storing the data and metadata. Additionally, a selection of supporting libraries has been developed to allow communication between instances of these data structures to create a distributed system.

The proposed solution delivers a long-term maintainable and extensible robotic framework, including an interoperable communication standard, control methods applicable to current robotic technologies, and validation routines to test the stability of the developed platform.

The chapter is organised as follows. Section 2 provides background information on the sectoral requirements of the nuclear industry and reviews the related academic work and available market products in the areas of robotic middleware and control platforms. Our specific problem definition is presented in Sect. 3. Section 4 covers future-proofing, interoperable implementation of the proposed framework. Section 5 describes the evaluation techniques and analysis of the experimental results. The chapter is closed with software infrastructure quality and provides information on software maintainability in Sect. 6 and the main conclusion of this study is presented in Sect. 7.

2 Background and Related Work

2.1 Software-Engineering Requirements in the Nuclear Industry

High-energy physics research devices are often large and complex environments that are hazardous to humans due to temperature, radiation, and toxic materials. Joint European Torus (JET) [30] experimental fusion reactor is the only active one in Europe, whereas a few others including the ITER [2], the European Spallation Source (ESS) [12], and DEMO [17] are currently under construction across the world.

Systems that will operate future fusion powerplants are expected to integrate hundreds of systems from multiple suppliers with multiple bespoke interfaces and training requirements. The lifespans of these reactors are expected to exceed 30 years, where maintenance and reconfiguration requirements evolve and obsolescence management is required. Such operations are typically conducted using tele-robotic devices and remotely controlled tools and equipment. Tele-robotic systems used for the remote handling and maintenance themselves require frequent reconfiguration to adapt to the evolving needs of the facility and maintenance operations (i.e. tasks required to perform maintenance activities). Furthermore, operations carried out in these facilities are often time, safety, or mission critical, causing an additional burden on the control systems in terms of high-fidelity, real-time performance to successfully meet the time, safety, or mission goals.

In the scientific context, the sectoral requirements and challenges presented above can be translated into software requirements for long-term maintainability, extensibility, high fidelity, and interoperability. However, the existing control system architectures used for remote handling and maintenance are highly coupled, often monolithic, systems, which fail to adapt to the changing nature of hazardous nuclear environments and often become unmanageable due to ad hoc, bespoke adaptations causing them to become prohibitively expensive to rectify. An extensive literature review is presented in the next section on the existing market products and control systems explaining why these systems are not well suited for nuclear applications.

The most important criteria for future-proofing long-lived nuclear applications are long-term maintainability, extensibility, and interoperability. The design principles for CorteX development have been mostly focused on these three key factors. Nevertheless, performance measures including fidelity and real-time performance are also important for time or mission-critical nuclear applications. First, we define functional requirements that affect performance measures, namely, fidelity and real-time character. Later, we follow defining three non-functional design principles in this section.

In generic terms, *fidelity* is the extent to which the appearance and behaviour of a system/simulation reflect the appearance/behaviour of the real world [14, 43, 45]. The concept of fidelity has two distinctive types: physical and functional fidelity. Physical fidelity refers to the degree of similarity between the equipment, materials, displays, and controls used in the operational environment and those available in the simulation (e.g. EtherCAT, TCP, serial). Functional fidelity refers to how the processes are implemented (e.g. how information requirements are mapped onto response requirements).

Real-time performance of a system can be defined as the system's reaction time to apply environmental changes [8]. Real-time systems guarantee a response within a defined period, and missing a deadline has varied effects depending on the constraints on the system. Real-time systems are divided into two categories based on their constraints: hard and soft deadlines [8]. Most nuclear applications contain a safety or mission-critical (sub-)system (i.e. hard constrained) where a missed deadline is greater damage than any correct or timely computation. Although the solution provided in the chapter is not a real-time system, going forward further research and development is required to make the provided solution real time given that most nuclear systems contain safety or mission criticality.

Maintainability is a concept that suggests how easily the software can evolve and change over time [38]. Measuring maintainability is not a straightforward task; however, cohesion, coupling, and granularity are measurable metrics that collectively provide information about the level of maintainability of the software. Cohesion and coupling are attributes of software that summarise the degree of connectivity or interdependence among and within subsystems. *Cohesion* refers to the strength of association of elements within a system [23, 34]. Gui and Scott defined cohesion as the extent to which the functions performed by a subsystem are related [21]. If a subcomponent is responsible for a number of unrelated functions, then the functionality has been poorly distributed to subcomponents. Hence, high cohesion is a characteristic of a well-designed subcomponent. *Coupling* is a measure of independence among modules in a computer program [27]. Coupling can also be described as inter-relatedness among components [23]. High cohesion and loose coupling are tactics to enhance modifiability of a complex system. As Bass et al. so concisely described it: "high coupling is an enemy of modifiability" [3]. *Granularity* is associated with the size or the complexity of system components [20, 23]. As the granularity size of a module increases, the probability of interchangeability decreases. On the other hand, as the granularity size of a module decreases, the maintenance of the system gets more complex. Low coupling between modules and high cohesion of a fine-grained module are desired properties in the modular architecture design [20].

Extensibility is a design principle that provides for future growth. Extensibility is a measure of the ability to extend a system and the level of effort required (i.e. cost, development time, development effort) to implement the extension [32]. Extensions can be through the addition of new functionality or modification of existing functionality. The principle provides for enhancements and increases in software consistency, through reusing system components where possible, without impairing existing system functions. Literature in the field of software engineering collectively suggests a strong relationship between extensibility and highly cohesive, loosely coupled, highly granular software.

Interoperability is defined as the ability of two or more software components to cooperate despite differences in language, interface, and execution platform [47]. In the context of software design, modularity or modular software components/programs help improve software reliability, allow multiple uses of common designs and programs, make it easier to modify programs (i.e. improve modifiability), and support extensibility.

Component-based software engineering is a field that focuses on creation and maintenance of software at a lower cost with increased stability through the reuse of approved components in flexible software architecture [22]. We believe the key to achieving long-term maintainability and extensibility whilst developing modular, reusable, interoperable architecture that holds high-fidelity, deterministic functional characteristics is through developing a component-based software solution. Using modern object-orientated design paradigms, encapsulation, low coupling between components, and high cohesion of fine-grained components will help us to achieve the desired functional and non-functional sectoral requirements that long-lived nuclear facilities require.

2.2 Related Work

There are several existing middleware, communication frameworks, and control systems in the market able to perform some of the tasks required for remote handling of robotic systems and that could be considered for use in the nuclear sector. Whilst existing frameworks allow for robust integration of complex robotic systems, they are not compatible with highly efficient maintenance and continue extension in the face of changing requirements and obsolescence issues over decades-long periods.

Simulators play a critical role in robotics research as these tools facilitate rapid and efficient testing of new concepts, strategies, and algorithms [33]. The Player/Stage project [19] began in 1999 as one of the first distributed multi-robot platforms and has widely been used among the research community. Player [18] is a robotic device server allowing offline development of robot control algorithms and is also capable of interfacing with robotic hardware. Although Player does not provide a high level of fidelity, or determinism, it has been well accepted as an open-source, general-purpose, multi-lingual robotic development and deployment platform. Stage is a configurable, lightweight 2D robot simulator capable of supporting large multi-robot simulations. Given that scalability is one of the most important aspects of the studies involving multi-robot research, Player/Stage provides a balance between fidelity and abstractions for its users. The 3D dynamics simulator Gazebo [33] was also developed within the Player/Stage Project at first as one of the components; it later became independent. Gazebo is integrated with the Open Dynamics Engine (ODE) [49], Bullet [10], and a few other high-performance physics engines, and therefore is capable of simulating rigid body dynamics. To facilitate fast and complex visualisation, Gazebo chose OpenGL and GLUT (OpenGL Utility Toolkit) [31] as its default visualisation tools for 3D rendering. Today, Gazebo continues to gain popularity not only in the robotic research community but also in industry, and it can simulate complex, real-world scenarios with high-quality graphics.

Robot Operating System (ROS) [41] came after the Player/Stage Project as a follow-up project with the intention of implementing a more modular, tool-based, reusable system. ROS is an open-source, multi-lingual platform, primarily used within the academic community. Over the years, it has gained popularity and has also been accepted in industry, for non-critical applications where time, mission, safety criticality, and QoS are not required. ROS is a powerful tool. It provides a structured communications middleware layer which is designed around commonly used sensory data (e.g. images, inertial measurements, GPS, odometry). Although the structured messages promote modular, reusable software, ROS messages do not cope well with the continuously evolving nature of software, causing compatibility issues. The highly coupled solutions created in ROS create issues for lasting maintainability and extensibility, crucially important factors for large-scale industrial systems. Integration of ROS components is fairly easy for small-scale projects, but they are not practical solutions for large-scale engineering problems due to the efforts required for integration and modification when the system configuration changes (i.e. not extensible easily).

The second generation of Robot Operating System, ROS2 [37], provides deterministic real-time performance in addition to the existing ROS features. Proprietary ROS message formats are converted into Distributed Data Service (DDS) [46] participants and packages, thus providing a high-performing, reliable communication backbone which helps to achieve determinism at the communication layer. ROS2 is backwards compatible with ROS via message converters. Although ROS2 has resolved the reliability, timeliness, determinism, and high-fidelity issues ROS previously contained, it has not resolved the maintainability and limited reusability issues for large-scale engineering problems, as there is no change in message structures.

The evolution of Player/Stage to ROS2 took over 20 years, with thousands of contributors to the open-source platform developing extraordinary, state-of-the-art robotics research on it. Player/Stage, ROS, and ROS2 frameworks are important milestones in today's robotic community and technologies. These three platforms

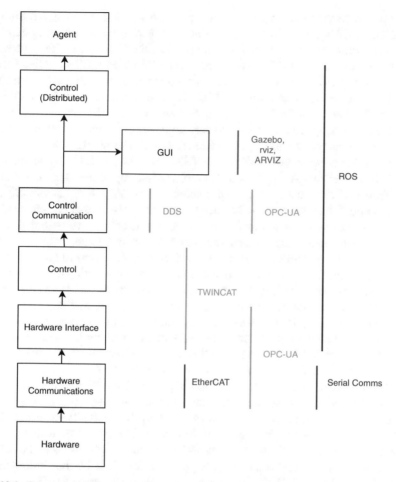

Fig. 10.1 Comparison between market products on control system stack.

are some of the shining examples of visionary research, leading to the development of the ROS2 platform for wide industrial use. Despite all the useful tools ROS and ROS2 contain (e.g. rviz, relaxed_ik, Gazebo), the structured message types offer limited capacity to support interoperable, future-proof, modular architecture required/desired in long lived nuclear facilities.

Figure 10.1 illustrates the modules of a control stack on the left of the image and lists the associated products capable of carrying out the set of tasks required in each module on the right. Hardware is placed at the bottom of the stack, and the applied control principles (therefore the modules drawn) become more abstract going further away from the hardware. Hardware communications are the protocols that interact with the hardware directly. Fieldbus protocols (e.g. EtherCAT, Modbus, PROFIBUS, Control Area Network (CAN) bus, serial communications) that are standardised as IEC 61158 for industrial use are listed in this category. The hardware interface module represents the fieldbus network (e.g. TwinCAT, Modbus, PROFINET, CAN open, OPC-UA). TwinCAT contains an operating system that hosts the control systems. It also has a built-in EtherCAT master that interfaces with the fieldbus network. Modbus can run over TCP, UDP, or RS485. OPC Unified Architecture (OPC-UA) [25] is a machine-to-machine communication protocol used in industrial automation under IEC 62541 specification. OPC-UA TSN [6] is a specific version of OPC-UA that implements IEEE 802.1 standards for time-sensitive networking (TSN). These add deterministic behaviour to standard Ethernet and therefore guarantee quality-of-service (QoS). Unlike regular OPC-UA, which is a client-server application, OPC-UA TSN introduces a publish-and-subscribe (Pub/Sub) model for OPC-UA. The Control module represents particular control algorithms and tools used on a local machine. TwinCAT, ROS, and ROS2 have control ability integrated and can be used to operate hardware that is connected to a local machine running the control algorithm. In order to achieve a distributed control system, the information from a local machine has to be distributed over a network.

The Control Communications module refers to middleware (e.g. DDS, OPC-UA, MQTT, ZeroC ICE) that can be used to drive the information out from a local machine to networked devices. The data-centric Pub/Sub protocol Data Distribution Service (DDS) OpenSplice [5] offers highly dynamic, timely, reliable QoS. Device-centric OPC-UA [25] standardises the communication of acquired process data, alarm and event records, historical data, and batch data to multi-vendor enterprise systems. The standardised communication process allows users to organise data and the semantics in a structured manner, which makes OPC-UA an interoperable platform unique for multi-vendor, industrial systems. To ensure interoperability and increase reusability, standardised but extensible base message types are provided by the OPC-UA Foundation. The Message Queuing Telemetry Transport (MQTT) protocol [1] provides a lightweight and low-bandwidth approach which is more suitable for resource-constraint Internet-of-Things (IoT) applications and machine-to-machine communications and is orthogonal to OPC-UA, but not interoperable like OPC-UA. ZeroC ICE [24] provides a remote procedure call (RPC) protocol that can use either TCP/IP or UDP as an underlying transport. Similar to DDS, MQTT, and OPC-UA, ZeroC ICE is also a client-server application. Although the asynchronous, event-driven nature of ZeroC ICE makes it unsuitable for real-time applications where QoS and durability are key, the same characteristics help improve scalability. Its neatly packaged combination of a protobuf-like compact IDL, an MQTT-like architecture, some handy utility executables to run brokers, autodiscovery features, and APIs in half a dozen languages make ZeroC ICE a popular middleware choice for non-real-time applications. Createc Robotics has been developing Iris [26], an open platform for deployment, sensing, and control of robotics applications. Iris combines 3D native visualisation, a growing suite of ready-to-use robotics applications and system administration tools for application deployment. As a platform, Iris intends to introduce an open standard designed to enable high-level interoperability of robotics and telepresence system modules. Iris is not a framework or middleware in itself, but aims to provide an abstraction layer to a growing number of middleware such as ZeroC ICE and ROS.

Table 10.1 compares and categorises existing frameworks and middleware products in terms of source openness, development language, distributed or centralised control, deterministic real-time characteristic, interface to hardware features, existing graphical user interface (GUI), scalability, extensibility, interoperability, security, physics engine, and control system capabilities. Source openness encourages transparency, correctness, and repeatability and is extremely beneficial in academia. The openness of Player/Stage, ROS, and ROS2 to the robotics community is invaluable in this regard. Within the nuclear industry, open-source applications are not preferred due to the perceived security risks that arise. OPC-UA and ROS2 have built-in security authentication, whereas DDS and IRIS are compatible with secure protocols. Single points of failure should be avoided at all cost. Distributed and scalable architectures are desired for the nuclear industry. Modularity encourages reusability; a key to achieving maintainable and extensible software is through highly granular, loosely coupled, and highly cohesive design. Although ROS, ROS2, and IRIS are modular and encourage reusability, integration efforts to put in these frameworks suggest that they are not easily extensible. This results in lasting maintainability issues which are far from being ideal for long-lived nuclear applications.

Gazebo is a 3D dynamics simulator, whereas rviz is a visualisation tool. CorteX comes with several integrated GUIs, and example screens are illustrated in Sect. 4.6. Orocos [7], Open Motion Planning Library (OMPL) [50], OpenRave [13], and Robotics Library (RL) [42] are robotic calculation libraries providing kinematics, motion planning including 3D paths and trajectory estimations, vector translations, etc. Orocos and ROS were integrated to combine the deterministic real-time aspect of Orocos with large ROS control paradigms.

The Experimental Physics and Industrial Control System (EPICS) [11] was designed by Los Alamos National Laboratory, and ITER has identified EPICS as a standard operating framework for all of ITER control systems [9, 44]. EPICS is a distributed process control system built on a software communication bus. As such, it provides brook-less communications where computer processes run EPICS databases that represent system units. EPICS databases hold records of functional algorithms used for the system. EPICS is capable of running on RTOS and provides deterministic real-time performance with fairly high fidelity. ITER has identified EPICS as its control system framework due to EPICS' wide range of functionality, rapid development, and modifiability and extensibility characteristics. As such, ITER's choice of control system clearly shows the importance of maintainability, extensibility, and high fidelity for nuclear operations.

ASEBA [35, 36] is an event-driven, distributed, lightweight simulation platform for mobile robotics. CLARAty [53] is developed by JPL and has a multi-layer abstraction model to ensure interoperability. The control architecture promotes reusable components and modularity through using layers of abstraction, thus assuring extensibility. Lower levels of abstraction focus on integration of devices, motors, and processors, whereas high levels of abstraction integrate the lower-level abstractions to provide higher levels of functionality such as manipulation, navigation, trajectory planning, etc. From this perspective, CLARAty aims to achieve maintainable and extensible frameworks, where CorteX provides timeliness and determinism, which CLARAty lacks.

Table 10.1 Comparison between control systems and middleware frameworks

Frameworks	Open source	Dev. language	Distri-buted	Real time	Hardware intf.	GUI	Scalability	Extensibility	Inter-operability	Security	Physics engine	Control system
DDS [46]	Yes	C++	Yes	Yes	No	No	High	High	High	Compatible	No	No
OPC-UA [25]	Yes	C,C++,Java,Python	Yes	Yes	Yes	Yes	High	Medium	High	Built-in	No	Limited
ZeroC ICE [24]	Yes	C++,C#,Java	Yes	Yes	No	Yes	High	Low	High	High	No	No
Player [18]	Yes	C,C++	Yes	Yes	Yes	Compatible	Medium	Low	Low	Low	No	Internal/limited
Stage [19]	Yes	C,C++	No	No	No	Yes	High	Low	None	Low	No	External
ROS [41]	Yes	C++,Python,Lisp	Yes	No	Yes	No	Medium	Medium	Low	Low	No	Internal
ROS2 [37]	Yes	C++	Yes	Yes	Yes	No	Medium	Medium	Low	Built-in	No	Internal
Gazebo [33]	Yes	C,C++	No	No	No	Yes	Low	Low	Low	Low	Yes	External
Hop [48]	Yes	Bigloo+JavaScript	Yes	Yes	Yes	Yes	High	High	Medium	Medium	No	Internal/limited
MSRS [29]	No	C#	Yes	No	Yes	Yes	Medium	Low	No	Low	Yes	No
ARIA [4]	Yes	C++	No	No	Yes	Yes	High	High	No	Low	No	Internal/limited
ASEBA [35, 36]	Yes	C	Yes	No	Yes	No	Medium	Low	No	Low	No	Internal/limited
Carmen[51, 52]	Yes	C	Yes	No	Yes	Yes	Low	Low	No	Low	No	Internal/limited
CLARAty [53]	Yes	C++	Yes	No	Yes	No	Medium	Medium	High	Low	No	Internal/limited
CoolBOT [15, 16]	Yes	C++	Yes	Yes	Yes	No	Medium	Medium	Low	Low	No	Internal/limited
Orocos [7]	Yes	C++	Yes	Yes	Yes	Yes	Low	Low	Low	Low	No	Internal
MOOS [40]	Yes	C++	No	No	No	No	Low	Medium	None	No	No	Internal/limited
Iris	No	C++,C#	Yes	External	External	Yes	High	High	High	External	External	External
CorteX	No	C++	Yes	Yes	Yes	Yes	High	High	High	Compatible	Compatible	Internal

RACE UKAEA is a UK government-funded lab, pioneering technologies on industrial problems, providing state-of-the-art solutions to niche problems in the nuclear industry. CorteX attempts to solve the main problems associated with interoperable, plug-and-play, distributed robotic systems of systems, at least from a data/communications perspective. CorteX is designed from the ground up to work as a decentralised, distributed control system compatible with Pub/Sub, service-oriented application. Although DDS is used to distribute information across the CorteX network, the middleware agnostic nature of CorteX allows it to interface to ZeroC ICE, OPC-UA, or ROS. This means that unlike ROS or custom solutions using plain TCP or other transport layers, there's no central "CorteX server" to set up, configure, and act as a potential single point of failure.

3 Problem Statement

The problems facing the development of a control system software, or in this case a framework, that is capable of supporting an application such as nuclear remote handling can be grouped into four main categories: interoperability, maintainability, extensibility, and performance.

Interoperability In order for a control system to support the various bespoke interfaces that are required when integrating hardware from a range of suppliers, whilst concurrently reducing integration efforts, a high level of interoperability is required.

The framework should have an architecture and supporting middleware that aim to abstract application-specific data and function, in order to allow as much of the system to be application agnostic as possible. Part of this abstraction is the standardisation of interfaces, which can be applied both at the architectural level and within the middleware.

A standard interface within the middleware will allow the various components of the system to be interoperable at the fundamental level, i.e. discovery and exploration of data and functionality within the system.

As much standardisation as possible within the architecture, i.e. in the contents of data and functionality, will maximise interoperability between the components of the system and allow them to be reused and replaced. This also increases maintainability.

Maintainability The lifespan of robotics applications in the nuclear sector, and their supporting control systems, is usually in terms of years and decades. This creates a significant challenge supporting a list of requirements that will evolve as the tasks the control system is to perform change over time. The introduction of new technologies required to support these tasks and avoid obsolescence requires the control system software to be easily modifiable whilst ensuring the effort required to do so remains low.

A high level of modularity in the design and structure enforced by the framework architecture will ensure components of the system can be swapped out whilst minimising the impact on the rest of the system. As previously mentioned, this is also strongly coupled with the standardisation of the interfaces between components so as to abstract as much of the control solution from the hardware as possible.

The modular nature of a system with this structure also allows testing at the component level. This is advantageous when replacing or altering components as usually little to no modification is required to re-run the test with the new changes. This means changes can be made with a higher level of confidence as the test ensures the same level of functionality is maintained.

Extensibility A change in requirements often requires an increase in functionality. Where possible, the additional functionality should be added without altering the previous components of the system. Therefore, any errors introduced are unlikely to be at the expense of the original functionality and result in a higher level of reliability. The ability to extend the capabilities of a system whilst minimising the components affected can be achieved with a highly cohesive, loosely coupled, high granularity architecture.

Performance Whilst the previously described features are beneficial to control systems for nuclear applications, they cannot come at the expense of performance. Robotic systems must be reliable, predictable, and responsive in order to meet their critical system requirements.

To ensure performance is maintained, the proposed control system software is evaluated against the following metrics:

- *Loop Cycle Duration*—the time between two consecutive tick signals
- *Loop Cycle Jitter*—the deviation in the loop cycle duration from a set frequency
- *Loop Cycle Duty Cycle*—the percentage of the loop cycle duration spent executing the job
- *Overflow Count*—the number of ticks missed because the previous job has not been completed
- *Command Latency*—the time between a simplex sending a command to another and the command being executed

In the next section, we describe our proposed solution, CorteX, in detail and explain how we address these challenges.

4 CorteX Design

CorteX attempts to solve the main problems associated with interoperable, plug-and-play, distributed robotic systems of systems, at least from a data and communications perspective.

CorteX could be thought of as:

- A standardised graphical data representation for robotic systems that can be
  modelled as directed graphs
- A method for communicating this representation
- A software framework that implements the above
- Additional software tools to add functionality

To achieve a highly modular, loosely coupled, highly granular system infrastructure,
CorteX implements a building blocks methodology. The required functionality
is provided by bringing together plug-and-play building blocks. CorteX consists
of several modules, namely, CorteX Core, CorteX CS, CorteX Toolkit, CorteX
Explorer, VirteX, and many other potential future extensions.

CorteX applies an inherently concurrent architectural design through the use of
object-orientated design methodologies and implements the imperative paradigm
within the single methods only (i.e. control system functions) and for main
executables. The concept of distributed objects and actor model design methodology
is used. CorteX Core and CorteX CS, the communication backbone and the control
system module, are set up to work with real-time constraints, where possible. The
main network middleware layer (i.e. DDS) is chosen specifically for its real-time
capabilities (DDS is able to operate as a real-time publish-subscribe protocol).

In the rest of this section, we describe the modules of CorteX in detail. We will
start by exploring the standardised interface in Sect. 4.1 followed by an explanation
on how we achieve interoperability within CorteX using the standardised interface
in Sect. 4.2. Section 4.3 will describe CorteX Core that is the data, and the control
system module CorteX CS will be described in Sect. 4.5. This section will end with
several examples on CorteX's human-machine interfaces in Sect. 4.6.

4.1   Standard Interface

A system in the CorteX environment can be represented with a graph consisting of
nodes and edges, similar to the graph presented in Fig. 10.6. Every node in a CorteX
environment is called a *simplex* and every edge denotes information flow between
simplexes. Each simplex uses the same format for internal data representation and
has the same external interface, as shown in Fig. 10.2. This data representation can
be used as part of a communications protocol to allow distributed components of a
single control system to exchange data without prior knowledge of each other. This
means a CorteX control system can grow to incorporate new hardware and control
features without modifying other distributed components.

The term *simplex* is used to describe units of data within CorteX, facilitating
functionality within the control system, including: defining data, describing input
and output relationships, calling functionality, representing (a part of) a system, controlling an active element (i.e. controller), or communicating with other simplexes
within the system.

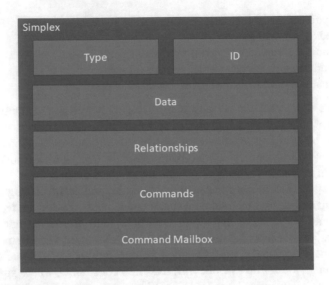

Fig. 10.2 Simplex.

A typical *simplex* $s_m \in S$, a self-contained unit of information, is the tuple: $s_m =< type_m, id_m, data_m, rlsp_m, cmd_m, cmdbox_m >$

- $type_m$: Set based to the ontological structure.
- id_m: Unique identifier, including path.
- $data_m$: Either an int, float, bool, or string. May be a single element or an array. Auto-set by the architecture based on the ontological type.
- $rlsp_m$: A link to another *simplex*, in order to obtain information or call a command. It may be a single element or an array. This is automatically generated and set based on the ontological type.
- cmd_m: A function which can be called in order to perform a task or change behaviour. Commands are declared with a list of request parameters, which are sent with the call, and response parameters, which are returned upon completion.
- $cmdbox_m$: A standardised interface to exchange commands.

Simplexes are common building blocks in a CorteX environment, at a level of granularity associated with minimal reusable units. Each simplex has a type, and in the next section, we describe their types in detail.

4.2 Interoperablility

The basic concept behind the self-describing, distributed data model is built upon simplexes with types that are associated with a software ontology and morphology techniques. We have used software ontologies to provide semantic meaning to robotic and control system components and have implemented morphological rules to associate syntax with the components represented in the system. In the remainder of this section, we describe applied ontology and morphology techniques using a case study to illustrate the principles described.

Software ontologies have been around since the early 1990s and applied broadly in multiple disciplines. Software ontologies aim to build the structure of information of a certain domain, to share a common understanding of the information available in that specific domain. Computer agents can extract and aggregate information shared within a structure to make the best use of the presented information in an ontology. Ontologies, by creating a structure for information, help increase common understanding in the following ways:

- To enable reuse of domain knowledge
- To make domain assumptions explicit
- To separate domain knowledge from operational knowledge
- To analyse domain knowledge

CorteX makes use of an ontology to ensure common and consistent use of domain knowledge and to make domain assumptions explicit. CorteX shares the domain-specific structure (the robotic and control system ontology) with every agent that runs CorteX, before execution. At runtime, distributed CorteX agents can make explicit assumptions using the knowledge represented in the ontology, thus ensuring the consistent reuse of distributed domain knowledge among agents. Figure 10.3 illustrates some elements of an example CorteX ontology relating to robotic and control systems.

In Fig. 10.3, the *Manipulator* type model is split into four categories. In addition to traditional serial and parallel manipulator categorisation, the ontology presented in this chapter also considers gripper and one degree-of-freedom (DOF) categories under the *Manipulator* subgroup. *Kuka LBR* is placed under the *Serial Manipulator* category. Using the ontology built for the type inheritance model structure, the CorteX system is capable of discovering that the *Kuka LBR IIWA* type also inherits the *Serial Manipulator* interface, which can be used instead, if needed. Careful examination can detect that *Robotiq 2-Finger Gripper* is classified under the *1DOF Manipulator* and listed under the *Gripper* type model. This is not a mistake; one can use the same technique to operate a 1DOF manipulator and a Robotiq 2-Finger Gripper. Therefore, we decide to allow multiple occurrences in this ontology-based knowledge structure. The effects of multiple inheritance and polymorphism on the ontology and interoperability are out of the scope of this chapter and will be analysed in the future separately.

The *Concept* type in this case study represents pure data modules, and they are used as inputs and outputs for processing simplexes. CorteX applies standardised (but extensible) data blocks and assures standardised data exchange between blocks. The standardised information exchange is guaranteed by the standardised simplex interface. In Fig. 10.3, Concept data types are categorised into three example categories: axis, Cartesian, and digital concept type models.

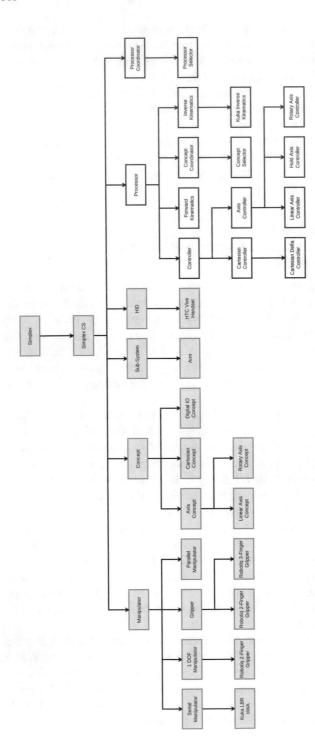

Fig. 10.3 Robotic and control system ontology—type model. Within our implementation, type modules fall into two main categories: descriptive shown in blue and active shown in white. Descriptive modules, as the name implies, are there to describe the state of the system, either currently (which we refer to as an observation) or as we wish it to be (which we refer to as a modification). In terms of hardware, an observation contains data read *from* the hardware, and a modification would contain data to be written *back* to the hardware as a demand. Active modules, on the other hand, are used in full CorteX control systems (CorteX CS) in which functionality has been added, in order for the modules to manipulate the data during its loop cycle task. In a system where the functionality was provided by some other means, for example, a ROS implementation, the CorteX system would likely be comprised of only descriptive modules. Values from ROS would then be injected into the simplex data, in order for it to be communicated. Although the "role" of these two categories of modules differs, this difference is purely implicit and is not reflected in their structure or contents.

The ability to use a simplex of a given type at various levels of its inheritance hierarchy allows for standardisation but also extension, in the form of polymorphism. This is one of the advantages of using an ontological type system.

The second feature of the CorteX type model is that it does not require prior knowledge. The type model is distributed at runtime and can be completely customised for the particular application, rather than using predetermined types. If two systems are required to interoperate, then the types within their type models must not contradict each other, but the overall models do not need to be identical.

Consider the following example: Agent A declares an Axis type, and it contains position, velocity, and acceleration data. Agent B declares an Axis type but states that it contains position, velocity, and torque data. These two agents are not interoperable as they do not agree on the definition of the Axis type.

However, now consider this example: Agent A declares the Axis type as before. Agent B declares an Axis type with a definition that matches Agent A, but *also* declares another type that inherits from Axis, states it contains torque data, and calls it AxisWithTorque. These two are now interoperable as they do not conflict in their definitions of any types, but Agent B can extend the Axis type to include their required data. Not only does this mean that Agent B can add their data to the system, but Agent A can also use the simplex of type AxisWithTorque with any interface designed for a simplex of type Axis as one inherits from the other, and therefore must contain all the required information from the Axis type.

Arm is listed under the *Sub-System* category, whereas *HTC Vive Headset* is placed under *Human Interaction Device (HID)*. One can think that UAV and UGV categories are entirely missing, or ask why different brands of headsets and robotic arms are not represented. Figure 10.3 does not illustrate the full ontology CorteX implements; the figure only contains the required knowledge to represent the case study presented towards the end of this subsection, shown in Fig. 10.6. We suggest the reader get in touch with the authors to find out details of the full ontology.

Inverse and forward kinematic modules, different types of controllers, and concept (data) coordinators are all categorised under the *Processor* type model, whereas *Processor Coordinator* has associated only with the *Processor Selector* type model in the provided hierarchy. Fundamentally, kinematics modules are responsible for the motion of objects and the forces required to provide the motion.

Morphology is the study of form and structure. In linguistics, it generally refers to the study of form and structure of words. In the fields of computer science, computational linguistics have been used to analyse complex words to define their component parts or to analyse grammatical information.

Within the CorteX framework, we use ontology to build a common structure of the domain-specific information, to distribute and reuse to make explicit assumptions. Therefore, a robotics and control system ontology is used to provide *semantic* meaning to components of robotic and control system elements, represented by simplexes. On the other hand, morphology is used to provide a set of rules to enforce the *syntactic* meaning and a binding structure to the component types represented in the ontology to achieve operational success among the distributed components,

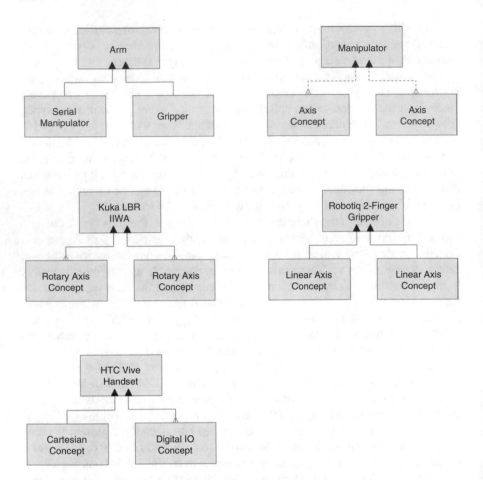

Fig. 10.4 Robotic and control system morphology—morphological rules.

and allow explicit assumptions to be made with regard to the structure of a given system, to facilitate interoperability.

The types that are provided in the ontology not only define functionality and structure but also data represented and external interfaces. Relationships in the context of components presented in the ontology define connections to other components. Types can define not only the relationships a component must have (to be considered of that particular type), but also how many (minimum, maximum, or absolute) components must be related to it, and define a particular type that the related components must be. The result of these relationship rules is that a system develops a particular *morphology*, which is consistent between all systems using common types. These morphologies tend to fall into one of two distinct groups: structural, as shown in Fig. 10.4, and behavioural, as shown in Fig. 10.5.

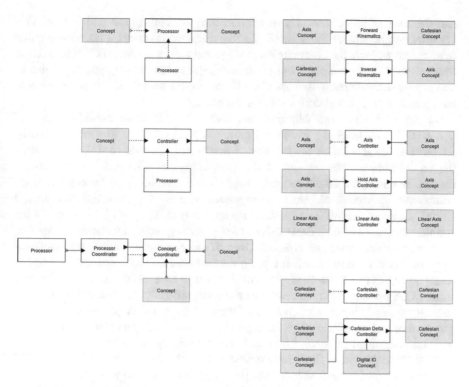

Fig. 10.5 Robotic and control system morphology—morphological rules.

Structural morphologies are either used to describe how the physical counterparts of descriptive simplexes are connected to one another in reality or to compose several granular descriptive simplexes under grouping nodes. Some structural morphologies can be seen in Fig. 10.4. The *Arm* (Fig. 10.4, top left) is an example of how relationships can be used to describe physical assemblies. In this example an *Arm* is physically comprised of a *Serial Manipulator* and a *Gripper*. This is achieved using two relationship rules: the first is an *Arm* must have an input relationship of type *Serial Manipulator*, and the second is an *Arm* must have an input relationship of type *Gripper*. This means that an *Arm* component cannot be added to a CorteX system without a *Serial Manipulator* component and a *Gripper* component and the relationships configured to connect them. This makes CorteX systems and their component structure highly discoverable and navigable, as these structures can be recognised and explored.

The *HTC Vive Handset* (Fig. 10.4, bottom left) is an example of composition via relationships. In this case, the *HTC Vive Handset* is described by combining (via relationships) a single *Cartesian Concept* and one or more (signified by the many-to-one symbol) *Digital IO Concepts*. These components are used to describe an HTC Vive Handset's position in 3D space (which is provided by a tracking system) and the state of its buttons. These relationships are both inputs of the *HTC Vive Handset* component as the state of the individual components must be evaluated first, before we have the complete knowledge that makes up the *HTC Vive Handset*. Similar to the *Arm* example, these relationship rules allow us to create recognisable and expected structures within the CorteX component model, which can be used to contextualise groups and derive semantic meaning.

In Fig. 10.4, the *Arm*, *Manipulator*, *Kuka LBR*, *Robotiq 2-Finger Gripper*, and *HTC VIVE Handset* behavioural morphological rules are illustrated visually. Arrowheads in the boxes denote the ownership of the rule; boxes that contain the arrowheads are the owners of the morphological rules and are responsible for implementing the associated rule. Behavioural morphologies are created when combining descriptive and active components and are therefore common in full CorteX control systems. The *Processor* example (Fig. 10.5, top left) is one of the simplest examples of how descriptive and active components can be used together. This example uses the basic types of *Concept* and *Processor* and has no functional purpose, but is used to establish a template that specific types can follow (similar to a purely abstract class in C++). The *Processor* type defines three relationship rules: (1) A processor *may* have one or more (signified by the dotted line and one-to-many symbol) input relationships of type *Concept*. (2) A processor *may* have one or more input relationships of type *Processor*. (3) A processor *must* have at least one (signified by a solid line and one-to-many symbol) output relationships of type *Concept*. These three relationship rules exist for behavioural reasons and to ensure the processor is able to function. The first rule is to provide input to the processor. This may not be required if the process is "open loop" and is therefore optional. The second rule is designed to allow for sub-processes that may need to be completed first for the processor to function. This may also not be required and is therefore also optional. The final rule is to provide an output for the processor. This is essential, as without an output, the processor is not making a contribution to the state of the system and is therefore redundant. In the given examples, the *Arm* type model has two components: a serial manipulator and a gripper. In CorteX, this translates as a robotic component and is considered as a robotic arm (see Fig. 10.4) if it consists of a serial manipulator (e.g. Kuka LBR) and a gripper (e.g. Robotiq 2-Finger gripper). A serial manipulator on the other hand would input and output multiple axis data information that implements positional move. Therefore, in Fig. 10.4 the *Serial Manipulator* type model is visualised with axis concepts. *Kuka LBR* is a 7DOF serial manipulator that contains rotary joints only and is therefore represented with rotary axis concepts. *Robotiq 2-Finger Gripper* is a 1DOF linear axis slider and is visualised with linear axis concepts. An HTC Vive Handset can output Cartesian position data and digital data, separately. Therefore, in the figure *HTC Vive Handset* type model is visualised as a model containing *Cartesian* and *Digital IO* concepts.

Figure 10.5 illustrates some of the morphological rules of control system components. Morphology rules for the *Processor* type model and its sub-categories, including variations of different *Controller* and kinematic type models, are presented as well as for the *Processor Coordinator*.

The *Forward Kinematics* example (Fig. 10.5, top right) is a less abstract example of the *Processor* morphology. In this case, the generic *Concept* types have been inherited by more concrete *Axis Concept* and *Cartesian Concept* types. Note that this change still satisfies the rules established by the base *Processor* type, as both *Axis Concept* and *Cartesian Concept* inherit from the *Concept* type (see Fig. 10.3). As those familiar with control theory will be aware, the purpose of forward kinematics is to derive the tip position of a manipulator using its joint positions. This functionality is partly evident in the morphology, as the *Forward Kinematics* simplex takes one-or-many *Axis Concept* as input (in this case the joint positions) and produces an output of a single *Cartesian Concept* (in this case the tip position). Similarly to the structural morphologies described earlier, these behavioural morphologies help create consistent, discoverable, and navigable structures within the CorteX simplex model.

RACE, within the Robotics and AI in Nuclear Hub (RAIN) project, has been working on a set of complex research problems to find state-of-the-art engineering solutions for decommissioning gloveboxes for the nuclear industry. The tele-operated dual-arm platform, shown in Fig. 10.7, has been used as a development and deployment platform for decommissioning tasks. The CorteX control system designed for one of the tele-operated Kuka LBRs is illustrated in Fig. 10.6 and provided in this section as a case study to explore the ontology and morphology principles described above. Figure 10.6 illustrates an HTC Vive Handset tele-operated Kuka LBR and its working principle. When the operator clicks on the digital output button and starts moving the HTC Vive Handset, the Kuka LBR mimics the move of the handset and moves accordingly. When the operator is not moving the handset, Kuka LBR is being kept on hold using the *Hold Axis Controller*. The *Processor Selector* picks one of the controllers based on the registered delta of the move (i.e. if there is move) and the activity on the *Digital IO Concept* (i.e. if the operator passed the button). *Robotiq 2-Finger Gripper* is operated independent of the manipulator and shown separately.

In summary, to achieve interoperability, CorteX applies a self-describing model of the system. We decided to use the following concepts to achieve a modular, self-describing, and distributed CorteX control system:

- One standard building block type.
- Standardised (but extensible) data blocks.
- Standardised data exchange between blocks.
- These standard structures of the ontology must be distributed.

The common building block types are called simplexes. Each common building type, simplex, has a type, which is the information described in the ontology. In the next section, we will describe simplexes in detail. A small part of the ontology tailored for the case study has been illustrated in Fig. 10.3. Standardised but extensible data blocks are the data type model represented under the *Concept* sub-category in the ontology. Standardised data exchange between blocks is enforced and implemented through morphological rules. Some of the rules used in the case study are presented in Figs. 10.4 and 10.5. In our current study, we implement the

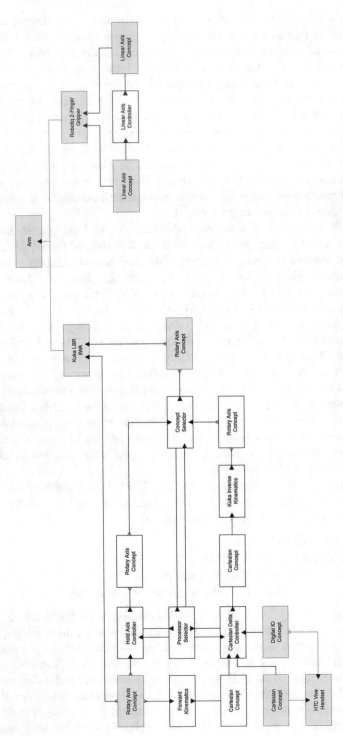

Fig. 10.6 Case study: system built using the building blocks represented in the ontology and assembled according to the morphological rules.

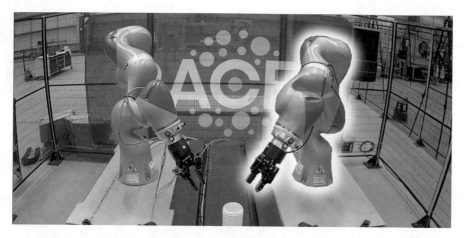

Fig. 10.7 Tele-operated dual-arm used in the RAIN project and case study.

ontology and morphology in the XML format and distribute it across computer agents that run CorteX to enable reuse of the domain knowledge.

4.3   Core Architecture

CorteX Core is the communication component of CorteX. The CorteX method for communicating data (simplexes) is kept in data tables. The creation of these data tables is initiated by a validation process. Figure 10.8 shows the fundamentals of the Core architecture using a UML diagram. Ontology, morphology, and the type model that are explained in the previous section are validated in the CorteX Core against a given use case. For example, the case study presented in Fig. 10.7 and the associated types of each simplex are validated against the ontology and the morphological rules.

This process is performed at the initialisation stage. The initialisation process achieves success if, and only if, all the simplexes can compile correctly, thus requiring construction of data, relationships, commands, and command parameters to match with the associated simplex type and morphological rules. If the provided application by the user contains simplexes that do not match the ontological structure and morphological rules, the CorteX system does not compile (Fig. 10.8).

Figure 10.9 gives an abstract of the CorteX Core components and illustrates the information flow between these components step by step. Simplexes update the system and type tables in CorteX Core, illustrated with step 1.1. Domain information is parsed through the XML file, and tables are updated in steps 2.1 and 2.2, respectively. The type model (Fig. 10.3) and the system model (Fig. 10.6) inform system tables in step 3.3. For example, when a simplex type, $type_m$, is

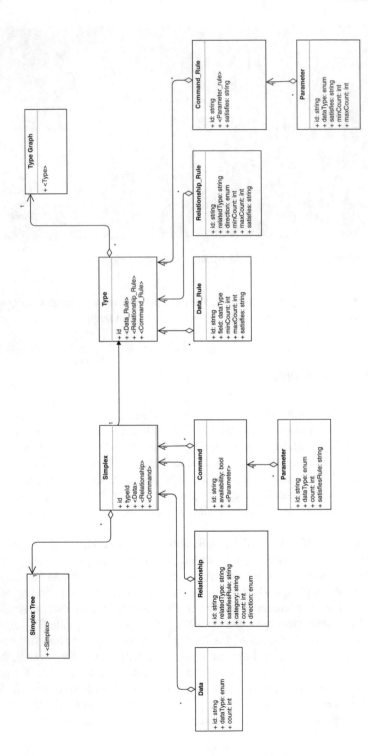

Fig. 10.8 Core architecture.

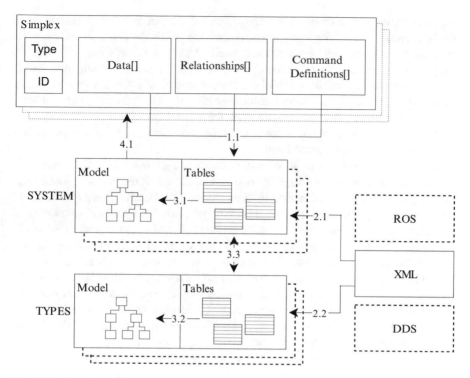

Fig. 10.9 Core data tables and interfaces.

set to Kuka LBR, as in step 1.1, CorteX auto-generates a *Rotary Axis Concept* in
the simplex data field, $data_m$ (step 4.1). The knowledge that a 7DOF Kuka LBR
contains *Rotary Axis Concept*s is embedded into the ontology, illustrated in steps
2.2 and 3.2, using the morphological rules shown in steps 2.1 and 3.1.

4.4   Distributing CorteX Data

CorteX agents (i.e. at least one cluster of simplexes running on a single PC) can
publish their current states to a shared network via the real-time publish-subscribe
protocol. This protocol is based on a standard that is maintained by the Object
Management Group, and is more commonly known as the Data Distribution Service
(DDS). CorteX abstracts the DDS communications and allows users to write new
transport methods using an API to access the data inside the CorteX Core. CorteX
also abstracts this interface and allows its users to use an API to construct simplex
models in CorteX either programmatically or as a result of a parsed interface of their
choosing.

DDS allows a DDS "participant" to publish tables of data in a broadcast
message to all other DDS "participants". DDS participants can filter these broadcast

messages and process them as required. This means that as long as two or more CorteX agents are on the same network and are using the DDS interface, their contents are automatically transmitted to each other. New CorteX agents (Clusters) may also be discovered at runtime, so when new CorteX participants "come online", they will appear in the DDS interface automatically. CorteX provides fully distributed data transmission and communication via DDS, in (almost) real-time manner. Therefore, unlike MQTT or ROS, there is no need for a central server, as each individual CorteX agent is able to broadcast messages to all other agents in a fully decentralised, distributed manner.

CorteX has interfaces to ROS, which make use of the ROS API; therefore, CorteX can read and write to ROS nodes. CorteX converts Interface Definition Language (IDL) files that describe ROS messages into CorteX system model-based structure, which is then accessible and can be processed in a CorteX environment. CorteX currently is only able to read ROS messages and does not translate CorteX commands into ROS service calls, although this is a planned future development.

In summary, CorteX Core is responsible for encoding and extracting system information, making use of the *semantic* meaning of the domain-specific information shared in XML format using ontologies and handling interoperable communications between systems. CorteX CS carries the functionality of the control system to meet the functional and performance requirements of the robotic control application. CorteX CS is discussed in the next section.

4.5   Control System Architecture

An obvious extension of the CorteX Core paradigm is the introduction of functionality. In addition to simply describing a system, by augmenting the ability to manipulate the data within each simplex, we can produce a system capable of control.

Building on the architecture described in Sect. 4.3, a full CorteX control system (CS) extends the original simplex by adding a *task* ($task_m$) resulting in the following *simplexCS* tuple:

$$scs_m = < type_m, id_m, data_m, rlsp_m, cmd_m, cmdbox_m, task_m >$$

The definitions of $type_m$, id_m, $data_m$, $rlsp_m$, cmd_m, and $cmdbox_m$ are provided in Sect. 4.3 and used accordingly in simplexCS; therefore, we only describe $task_m$ in this section.

Each simplex task is broken up into four main parts. The first is the processing of any commands. These commands may have been sent from a simplex task executed previously in the current job, from a simplex task executed in the previous loop cycle job, or from another distributed agent of the control system. These commands may request to change the behaviour of a simplex, or change the value of a piece of data, or start and stop the functionality of the simplex entirely (Fig. 10.10).

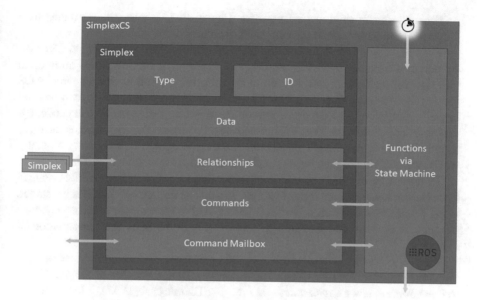

Fig. 10.10 SimplexCS.

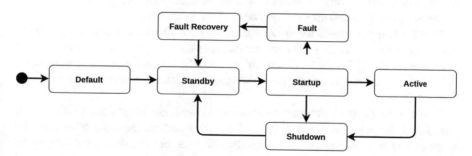

Fig. 10.11 SimplexCS state machine.

The final three parts of the simplex task $(task_m)$ are executed as follows: first, a *loopStart* function which is executed every loop cycle regardless of state; second, a state-specific function based on the current state of the state machine (e.g. *stateStandby* or *stateActive*); and finally, a *loopEnd* function which is executed every loop cycle regardless of state. The various states of the *simplexCS* state machine and its transitions are shown in Fig. 10.11.

The fundamental properties of a control system, including *Loop Cycle, Task, Job, Tick signal*, are defined in Sect. 3. The control system *loop cycle* is intended to be run periodically with minimised jitter, and features are available to halt or send warnings if this is breached. A CorteX environment may contain multiple CorteX agents (pc running CorteX), and each CorteX agent should at least have one cluster.

Each *ClusterCS* contains *simplexCS*s. Figure 10.12 shows the attributes of a *ClusterCS* and its interactions in the UML format.

The *Loop Cycle Scheduler* constructs the execution sequence of simplexCS tasks. As part of a running CorteX agent, it is also responsible for generating an internal *tick signal*. The *Thread Pool* is used to optimise the execution of each simplexCS loop cycle, by using parallel threads. Threads, configured by the user, can be used to pull tasks from the loop cycle scheduler, whilst still obeying any dependency-driven order (often in the form of Directed Acyclic Graph (DAG)), and execute them in parallel where possible. This ensures tasks are executed quickly and efficiently whilst still maintaining the dependencies as a result of data flow requirements.

To provide a big picture, our implementation of a CorteX-compatible control system framework (which is one example way of creating a CorteX CS) is illustrated using a sequence diagram in Figs. 10.13 and 10.14. Please note that the sequence diagram is divided into two due to its size. Figure 10.14 is the continuation of Fig. 10.13.

When a CorteX environment is created at runtime, the executable (main.cpp) constructs an instance of a *ClusterCS*, a *Simplex Tree*, a number of *SimplexCS*s, a *Trigger Source*, and a *Communication Interface* (in this case, DDS). The executable then calls the populate() function on the simplexCS instances (which constructs the various data, relationships, and command items. i.e. a simplex) before adding them to the Simplex Tree instance via the addSimplex() function.

The next step is to add the collection of *simplexCS*s to the *ClusterCS* using the addSimplexTree() function. The executable sets the *Trigger Source* by calling the setTriggerSource() function upon itself, passing the TriggerSource as a parameter. With the system model constructed, the main.cpp calls the start() function of the *ClusterCS*.

The *ClusterCS* then executes its constructLoopCycle() function, which creates the *Scheduler* and *Threadpool*. With these constructed, the *Scheduler* must be populated with each *simplexCS*'s task so that they can be used during the loop cycle. This is done via a call of the *Scheduler*'s generateTasks() function, which iterates through each *simplexCS* and retrieves its task via the getLoopCycleTask() function. Finally, with the contents of the *simplexCS*s established, the System Model and Type Model can be communicated via a write() call on the *Communication Interface(s)*.

The *ClusterCS* is now ready to begin the loop cycle. At this point, the *Cluster* creates a thread on which to run the loop cycle, due to it being a blocking function. The first call is to the waitForTick() function of the *TriggerSource*, which blocks until a trigger is received. In the case of the default internal clock, this is at a regular interval set by a configured frequency. Upon receiving a trigger, the function continues to the following steps:

1. Read any updates from the Communications Interface(s) via the read() function. This will update any information required from simplexes running on another agent.

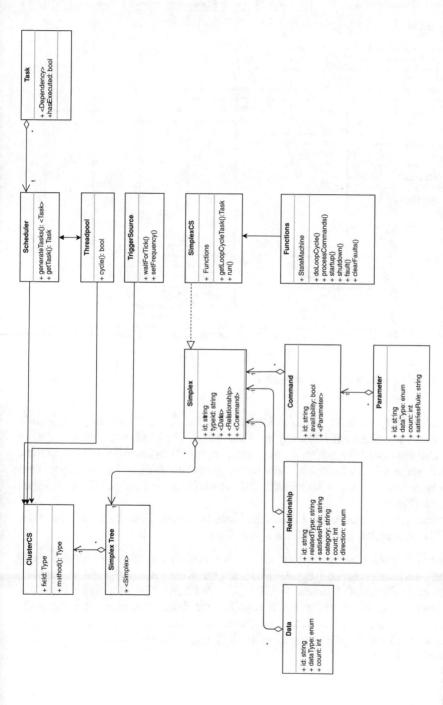

Fig. 10.12 CorteX CS UML diagram.

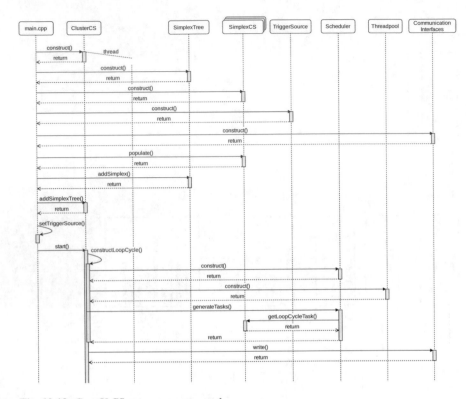

Fig. 10.13 CorteX CS run sequence—part 1.

2. Call cycle() on the *Threadpool*. Execution of the cycle() function is an initiation of several actions carried out by the *Threadpool*. The first is to get the list of tasks from the *Scheduler*, which returns each simplexCS task and the order they must be executed in, as derived by input and output relationships. The *Threadpool* then calls executeTask() on simplexCS instances so that each simplexCS executes its task ($Task_m$).
3. Write any updates to the Communications Interface(s) via the write() function. This will update any subscribed agents.

ClusterCS calls heartbeatFunctions() on the main executable to signal activity at the end of each loop cycle. When a user terminates the main executable, main.cpp calls stop() function on ClusterCS and breaks the loop cycle loop. The destruction() function on ClusterCS initialises the destruction phase and ClusterCS calls destruction() on Scheduler and Threadpool. The main executable then calls destruction() on *Simplex Tree, simplexCS, Trigger Source,* and *Communications Interface(s).*

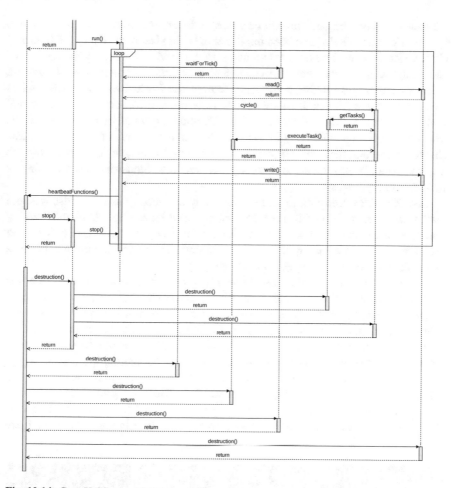

Fig. 10.14 CorteX CS run sequence—part 2.

4.6   CorteX Explorer

The CorteX Explorer is a feature of CorteX that collects all the graphical user interfaces (GUI) and human-machine interfaces (HMI) under the same umbrella. It is an essential module of a control system that is designed to be used in the nuclear industry, where the majority of people in operations teams are likely to consist of non-technical operations engineers. CorteX GUIs and HMIs are fundamental to guarantee high operational success. Therefore, there is a great need for highly reliable, responsive, and efficient visualisation and reporting features.

To provide the required high level of reliability, responsiveness, and efficiency, an event-driven methodology is used for the Explorer, where the front-end interaction with the back-end (i.e. pulling for updates) is on a time-based loop. The loop frequency is parametrised and can be changed depending on the user requirement. Custom-designed GUIs have been implemented using freely available QT libraries. Cortex visualises system parameters using a tabular view to display the content of each simplex in a CorteX environment to help the operator in monitoring the system activities. In addition to the monitoring features, the CorteX Explorer also renders the content of the (selected) simplexes and provides control screens. Some CorteX HMIs are provided below as an example. Performance evaluation such as reliability, responsiveness, and efficiency of the CorteX Explorer is out of the scope of this chapter; therefore, there are no results presented in this chapter on these. Through this approach, it is possible to automate adherence to industrial and accessibility standards and guidelines.

CorteX's self-describing protocol allows any connecting agent to discover and read the system model and its contents without prior knowledge. This functionality also extends to the CorteX Explorer GUI allowing an operator, user, or a developer to explore the contents of any simplex in the system. To provide an intuitive interface to do this, we have developed the *Base View*.

Figure 10.15 illustrates the Base View. It displays the list of simplexes in a tree down the left side of the screen in much the same way as a file browser. Upon selecting a simplex in the tree, the right side of the screen is populated with the simplex's contents. Data, relationships, and command definitions are shown in a tab view along with each item's value, path, and availability, respectively. Arrays are shown in collapsible rows (as shown) to avoid over-populating the screen. In the case of a simplexCS being selected, the simplex's state is shown in the top left, and command buttons to *Startup*, *Shutdown*, and *Clear Faults* are provided.

As any value within a CorteX system can be read and updated over time, it is also possible to plot any numerical value on a graph. Within the Base View, there

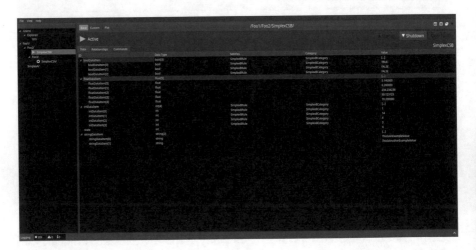

Fig. 10.15 CorteX explorer: base view.

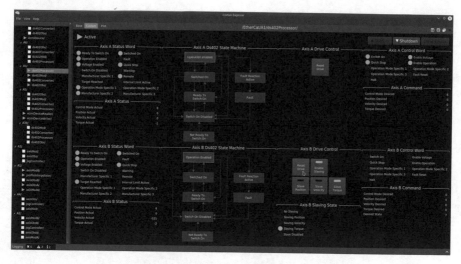

Fig. 10.16 CorteX explorer: Dual DS402 processor view.

is the option to select any number of numerical data items within CorteX and place them onto a graph to be plotted over time. This can be very useful when debugging or watching for certain signals within a system.

The CorteX Explorer GUI also allows users to develop custom views for certain simplex types or create views for entire systems comprised of both static and dynamic elements, examples of which are given in Fig. 10.16.

Figure 10.16 shows a view used to observe and control a dual-axis EtherCAT DS402 (motor drive) device. The view is split into two halves (top and bottom) to display both axes: Axis A and Axis B, respectively. Although only the *DS402 Processor* simplex is selected in the tree on the left, the view not only uses data from that particular simplex but also uses its relationships to pull data from both the input *DS402 Observation* simplex and the output *DS402 Modification* simplex. The left side of each axis view shows the current state of the axis from the *DS402 Observation*, the centre drive control area shows state values and control buttons for the *DS402 Processor*, and the right side shows the demand state of the drive from the *DS402 Modification*. This shows how using the standard processor morphology (see Fig. 10.5) we can design views around these structures and pull data from multiple simplexes to give clear contextualised information.

Similar to Fig. 10.16, Fig. 10.17 shows a view that encompasses data from multiple simplexes. This is the view for the *Jog Controller* simplex, but again pulls data from a related input *Axis Concept* and output *Axis Concept*. The current axis values from the input *Axis Concept* are shown on the graph and in the displays on the top row. The operator can enter demand values and change the operating mode of the *Jog Controller* using the bottom row controls. Demand values from the output *Axis Concept* are also displayed on the graph as targets.

Fig. 10.17 CorteX explorer: jog controller view.

Figure 10.18 demonstrates several capabilities of the CorteX GUI framework. First, the view is split into two halves. The left side shows a representation of the physical TARM manipulator, posed to show not only the current position (in white) but also the target position (in green). This pulls data from a number of *Axis Concepts*, both observations and modifications, to pose the TARM image. This side of the view is specific to the TARM manipulator as the robot joints are preloaded.

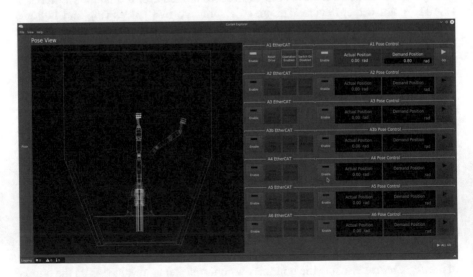

Fig. 10.18 CorteX explorer: TARM pose view.

However, the right side of the view is generic and can be used for any multi-axis manipulator. This selection of controls is generated dynamically by searching through the simplex model for any *Axis Controllers* and creating a set of controls for each type of controller—in this case, they are all *Axis Pose Controllers*. You will also notice that to the left of each Pose Control area is an EtherCAT control area. The controls in this area were also auto-generated, using the morphology to discover the device *Processor* related to each *Axis Controller* and then generating a view for the specific device processor type—in this case, an *EtherCAT DS402 Processor*. Notice how the EtherCAT control area discovers the number of axes each processor is controlling, producing two status displays for axes A1, A2, A3B, A5, and A6, but only one for A3 and A4.

CorteX Explorer is the manifestation of CorteX's discoverability (via the use of the morphology, the type model) and the self-describing nature (through the use of standardised simplex interface) that provides auto-generating, dynamic, contextual, and universal GUIs for exploring any and all CorteX systems.

5 Performance Evaluation

In the previous section, the *CorteX* framework is described in detail. In this section, we discuss performance evaluation and the configuration parameters used to evaluate *CorteX*. The section starts with a typical CorteX system definition in Sect. 5.1, followed by a performance analysis of *CorteX* in terms of memory allocation and real-time characteristics, in Sects. 5.2 and 5.3, respectively.

5.1 A Typical CorteX System

There are 50–70 simplexes in an average CorteX control system. Figure 10.7 which is a typical robotic arm that can be represented with 44 simplexes (each rotary axis has 7 joints and therefore is equal to 7 simplexes) as shown in Fig. 10.6 to illustrate the ontology and morphology concept. The communication (e.g. Fieldbus protocol) and diagnostic functionality also use an additional 20 simplexes. Appropriately, between a third and a half of the entire system consists of Concepts. Referring back to Fig. 10.6 again, 30 out of 44 simplexes are Concept types. This means a minimum of a third of the simplexes in the CorteX system do not send commands, but do contain numerous pieces of data and some relationships. An average Concept type simplex contains 5–10 pieces of data; some Cartesian Concepts used for the inverse kinematics or arms contain up to 20 data item each. Therefore, the performance analysis presented in this chapter does not represent a typical control system; the examinations on CorteX are performed for stress-testing purposes.

5.2 Memory Footprint

Dynamic memory allocation is permitted within CorteX simplex functions but should be avoided if real-time performance is required (see Sect. 5.3), due to how long it can take. Within the internal CorteX Core and CS infrastructure, dynamic memory allocation is avoided wherever possible to produce a fixed memory footprint. Figure 10.19 illustrates the allocated memory for 2 min for the increased number of simplexes. For this test, we start analysing the allocated memory for 100 empty simplexes where there are no data, relationships, or commands defined. We increase the number of simplexes by 100 each time and log the memory change over the total runtime. Figure 10.19 provides evidence for zero memory allocation by the CorteX framework over a runtime of 2 min. The same experimental data is also used to show the linear increase of the allocated memory for an increased number of empty simplexes in Fig. 10.20. These two figures hint at the scalability of the platform, without providing further information.

Figure 10.21 illustrates memory allocation for an increased number of data items over a fixed number of simplexes. We gradually increase the number of data items by 10 per simplex over the 25 simplexes and record the memory allocation. These 25 simplexes do not have any relationship or command values; they only contain data items. As such, Fig. 10.21 shows the linear increase in allocated memory as the data items increase for 25 simplexes, again hinting at the ability of the platform to scale linearly.

Figure 10.22 shows memory allocation for an increased number of relationship items over a fixed number of simplexes. We increase the number of relationship items by 10 per simplex over a total of 25 simplexes and record the memory allocation. These 25 simplexes do not have any data or command values; they only contain relationship items. Figure 10.23 presents the allocated memory for the increased number of commands over a fixed number of simplexes, increasing the

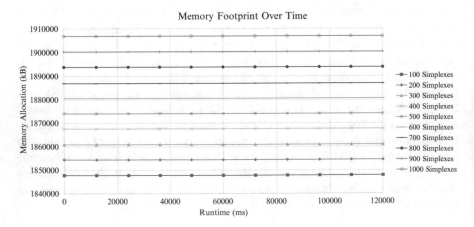

Fig. 10.19 Memory footprint over time for the increased number of simplexes.

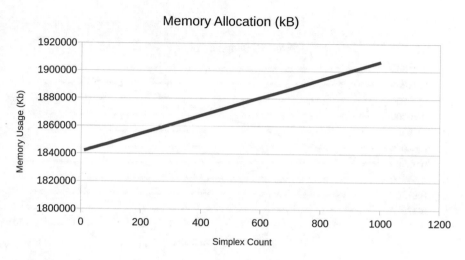

Fig. 10.20 *CorteX* memory profile over the increased number of simplexes.

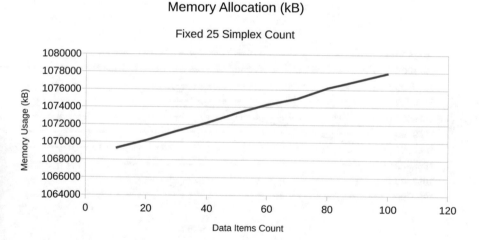

Fig. 10.21 *CorteX* memory profile over the increased number of data.

number of defined commands by 10 per simplex over 25 simplexes and recording the memory allocation. In this set, 25 simplexes do not contain any data or relationship items; they only contain command items. As such, Figs. 10.22 and 10.23 show a linear increase in the allocated memory for relationship and command items, hinting at the system's ability to scale linearly.

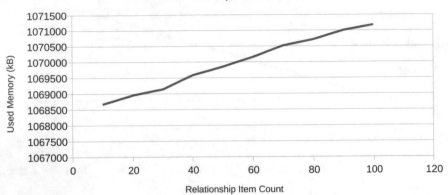

Fig. 10.22 *CorteX* memory profile over the increased number of relationships.

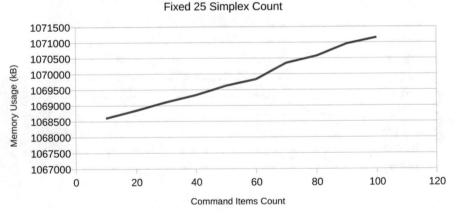

Fig. 10.23 *CorteX* memory profile over the increased number of commands sent.

5.3 Real-Time Characterisation

The importance of timeliness, QoS, and high fidelity within control systems for the nuclear industry was previously explained in Sect. 2. The examinations of CorteX in this section use a loop cycle frequency of 1 kHz and a sample size of 5000 loop cycles. The simplexes also use a parallel dependency structure, meaning all simplex tasks are capable of executing concurrently. However, running all simplex tasks on

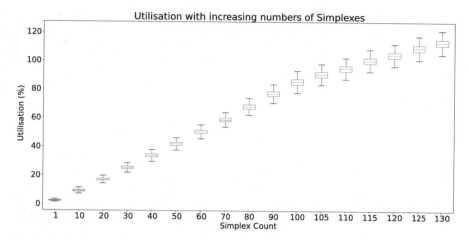

Fig. 10.24 Duty cycle over the increased number of simplexes each sending one command per loop cycle.

their thread would require heavy context switching, which is not an efficient method for parallel systems. For this reason, the tasks are distributed by the *Scheduler* among four reserved threads owned by the *Threadpool*, each running on one core of the CPU of the CI machine.

Figure 10.24 presents how system utilisation is affected as the number of simplexes is increased. As we increase the system workload by inserting more simplexes, the utilisation percentage increases. As each simplex task has an equal duration for completion (in this case), the percentage of time between ticks spent performing simplex tasks increases linearly with the number of simplexes in the system. In this particular test, the CorteX control system achieves almost 100% utilisation with 105 simplexes in the system.

For the next test, we create dependency pairs. Each pair has a sender and a recipient simplex; commands are sent from the sender and received by the recipient. The read() and write() functions are called on both simplexes, which are known to be more time consuming than processing empty tasks and no commands. Figure 10.25 illustrates the interval between task executions on the same experimental data. As it can be seen in the figure, Cortex runs with 1 kHz when the simplex count is less than 105 or more. When there are 110 or more simplexes in the system, task executions can no longer be complete in one loop cycle period. This causes tasks to overflow into the next loop cycle. Given that each loop cycle has only one command to execute per loop cycle, the interval between the task executions should be close to loop cycle duration when jitter is low in a system. Figure 10.25 shows that a CorteX system can run with 1 kHz and execute commands of 105 simplexes within 1 ms interval, whereas as we introduce more simplexes into the system, the loop cycle frequency drops down to 500 Hz and therefore the interval between task executions jumps to 2 ms.

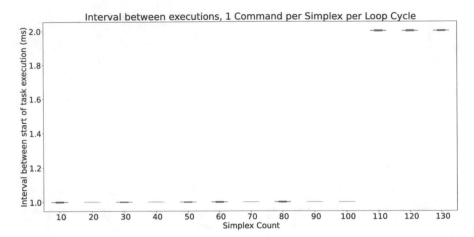

Fig. 10.25 Duration between task executions over the increased number of simplexes each sending one command per loop cycle.

Deviation in the duration between task execution hints at a level of latency and jitter in a system. Figs. 10.26 and 10.27 illustrate the difference between a timely and an overflown CorteX system. A CorteX system, running at 1 kHz with 105 simplexes without overflowing, has a minimal deviation in the interval between the start of task executions. This provides suggestions on the determinism of a system. As shown in Fig. 10.26, a deviation of a maximum of 3 microseconds illustrates

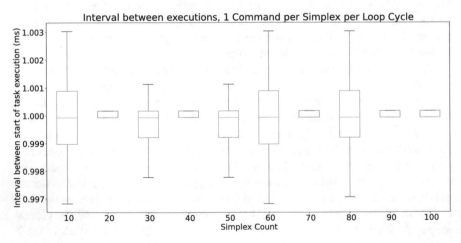

Fig. 10.26 Deviation in duration between task executions over the increased number of simplexes up to 100, each sending 1 command per loop cycle.

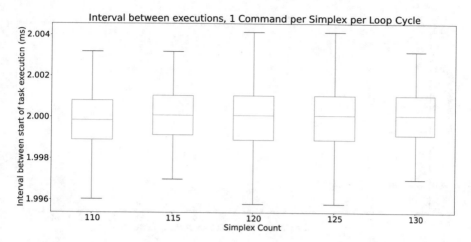

Fig. 10.27 Deviation in duration between task executions over the increased number of simplexes between 110 and 130, each sending 1 command per loop cycle.

how timely the CorteX system is. The noise introduced into the CorteX system is minimal.

Figures 10.28 and 10.29 analyse the latency in CorteX. For this set of experiments, we introduce 10 simplexes in a CorteX system and increase the number of commands each simplex sends. A fixed number of simplexes, with increased workloads (commands and tasks), increases the latency in the system. The deviation of latency increases as the workload of the system increases and at the maximum the deviation gets to $14\,upmu$ without an overflown system running at 1 kHz. The system starts to overflow when there are more than 10 commands per simplex per

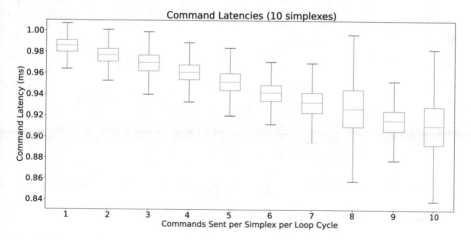

Fig. 10.28 Latency over the increased number of commands per loop cycle.

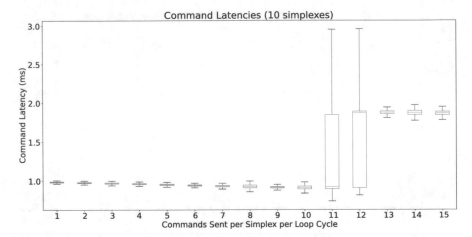

Fig. 10.29 An overflow analysis of latency over the increased number of commands per loop cycle.

loop cycle. When the system overflows, it outputs in 500 Hz and latency increases to 2 ms and above. The deviation is reduced by 2 ms when there are 13 or more commands per loop cycle.

The experimental results presented in Fig. 10.30 were performed on a fixed number of simplexes each sending one command per loop cycle. The analysis presented shows the effects of changing loop cycle frequency over latency in Fig. 10.30.

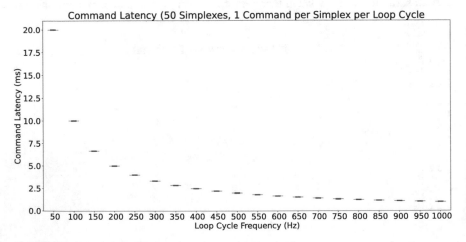

Fig. 10.30 Latency over the increased loop cycle frequency.

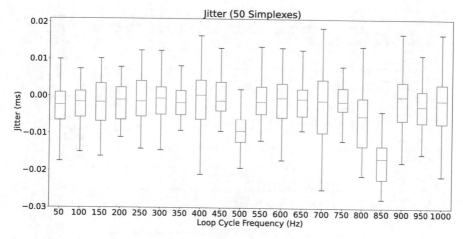

Fig. 10.31 Jitter over the increased loop cycle frequency.

To report jitter of a CorteX system, we use 50 empty simplexes. Figure 10.31 shows the jitter in the CorteX system when there is no processing or communication happening, which means there is no injected noise due to execution or communication. This experiment is performed to provide a baseline for the increase in loop cycle frequency and to improve our perspective on any other noise that the system could have contained. According to Fig. 10.31, deviation in jitter varies between −0.02 and 0.01 for loop cycle frequency less than 400 Hz. Jitter increases and fluctuates more as the loop cycle frequency increases.

6 Software Infrastructure Quality and Maintainability of the High-Performing System

The quality of the software is paramount in achieving the high performance required for control systems in long-lived nuclear facilities. In this section we describe maintainability of the *CorteX* codebase over the applied life-cycle management procedures. This section is followed by the unit testing that is used to assure the correctness of the required functionalities in Sect. 6.2. In Sect. 6.3 we explain the additional optimisation applied on the CorteX codebase to increase the efficiency of the software.

6.1 Life-Cycle Management

Component-based software engineering and life-cycle management techniques are practised during *CorteX* development. Modern, maintainable software infrastruc-

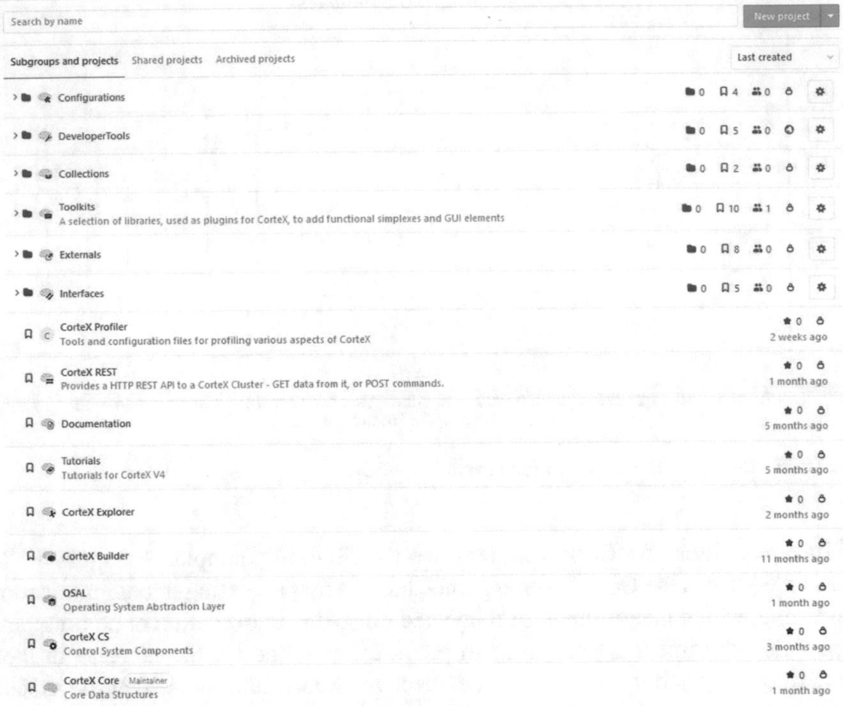

Fig. 10.32 Life-cycle and process management.

tures help us in achieving the required high performance (high fidelity, timeliness, real-time characteristics) over a long period of time. Therefore, the importance of component-based software engineering for long-lived nuclear facilities is discussed in Sect. 2. Some of the main components of *CorteX*, namely, CorteX Core, CorteX CS, and CorteX Explorer, are described in Sect. 4 in detail. CorteX Profiler is the built-in evaluation module within the *CorteX* framework that analyses the system performance as a whole. The results presented in Sect. 5 are obtained using CorteX Profiler.

Continuous Integration (CI), Continuous Delivery, and Continuous Deployment are commonly used processes within modern software life-cycle management, especially for large-scale engineering solutions. *CorteX* is compliant with sustainable development through continuous delivery routines and continuous deployment procedures. As part of the commit process, *CorteX* CI runners perform several automated tests on the modified codebase to check that alterations have not introduced any undesired change in functionality or performance (Fig. 10.32).

CorteX Core

CorteX CS

	dev-unstable	dev	master	production
Status	pipeline passed	pipeline passed	pipeline passed	pipeline passed
Coverage	coverage 91.10%	coverage 91.20%	coverage unknown	coverage unknown

Fig. 10.33 Percentage of unit-tested code coverage over the CI pipelines.

6.2   Unit Testing

Each *CorteX* repository is unit tested individually, so any errors can be more easily traced to a particular component. These tests are automated for the CI runner and the entirety of the code library is tested module by module. These are combined with integration tests that then assemble various modules together and test their overall functionality and integration. As shown in Fig. 10.33 dev-unstable and dev pipelines reach over 85% tested code coverage for both CorteX Core and CorteX CS modules. The hardware specifications of the CI runner are a 4-core (2 real, 2 virtual) Intel(R) Core(TM) i3-4170 CPU with a 3.70GHz processor and 4GB memory running Debian 9.5 (stretch) OS with the 4.16.8 kernel, RT patched. Soon, we intend to perform the test routines on RTOS-like operating systems, which have shown to have greater RT performance [28].

6.3   Optimising the CorteX Codebase

The Valgrind tool suite [39] provides a number of debugging and profiling tools that assist with bug fixing and optimising software. Callgrind is a profiling tool within the Valgrind suite that records the call history among functions in a program's run as a call-graph. By default, the collected data consists of the number of instructions executed, their relationship to source code, the caller/callee relationship between functions, and the number of each call. Optionally, cache simulation and/or branch prediction can produce further information about the runtime behaviour of the codebase. Figure 10.34 shows a call-graph generated by Callgrind applied to

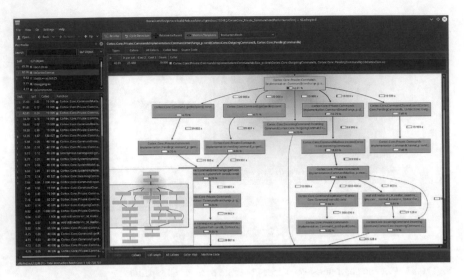

Fig. 10.34 Valgrind's Callgrind profiling *CorteX*.

Simplex Commands and the Command Mailbox, to optimise implementation of command exchange in *CorteX*. Analysing the codebase function by function, we detect bottlenecks that can limit performance over case studies and make necessary changes to increase code efficiency. This reflects positively on the performance of CorteX systems and helps us understand the difference between the newly written code versus the excising code from a traceability point of view and informs us on the improvement areas of the entire codebase.

7 Conclusion

This chapter has proposed a novel software framework for interoperable, plug-and-play, distributed robotic systems of systems. We developed CorteX to tackle the implementation of control systems for robotic devices in complex, long-lived nuclear facilities.

CorteX attempts to solve the main problems associated with interoperability and extensibility using self-describing data representation. Standardised but extensible data interfaces are developed to provide interoperability, whilst *semantic* meaning is self-described by the components through typing, where the types are associated with a software ontology for robotic and control system components. To aid with structural interpretation in data exchange between these interfaces, software morphologies are implemented and used to provide *syntactic* meaning. The robotic and control system knowledge structure is distributed across the CorteX agents before runtime.

Encapsulation combined with low coupling between components and high cohesion of fine-grained components, along with the use of standardised interfaces, helps in achieving modularity and testability. Quality and maintainability required from a software platform are achieved by modern life-cycle management processes and effective component-based development techniques. Unit-tested components and a high level of code coverage of CorteX are illustrated in Sect. 6 as part of the software quality control.

An extensive analysis of CorteX memory profiling is provided in Sect. 5.2. The results illustrated that CorteX is a lightweight framework. In addition to this, CorteX runs with a constant memory footprint, which translates as no memory leakage. Scalability, which is crucial to achieving extensibility, is briefly illustrated as part of the memory tests with 1000 simplexes.

Timeliness and fidelity are important features of nuclear applications. Real-time characterisation of CorteX is shown in Sect. 5.3. Although CorteX is not a deterministic system, the deviation in latency, jitter, and loop cycle duration is less than $40\,upmu$ whilst the CorteX loop cycle is running at 1 kHz. Stress testing on the increased number of commands sent per simplex and the increased number of simplexes, each executing one command per loop cycle, identifies that error and noise in the system is minimal. Based on the real-time characterisation and the applied software quality management, we believe CorteX promises to deliver the needed control system solutions for long-lived nuclear facilities.

Future work will focus on the evaluation of CorteX on large-scale, distributed clusters to inspect scalability and fidelity even further. Analysis of interoperability will be part of these tests. In addition to these points, further analysis is required in the fields of metric-based software engineering to characterise the low coupling, high cohesion, and fine-granular nature of CorteX. We believe this will be key to achieving long-term maintainability in software. Furthermore, performance improvements on CorteX that will lead to increased determinism will be carried out. We envision further investigation on appropriation of task executions and methods to improve different approaches in overflow compensation to take place.

Acknowledgements CorteX is an intellectual property of the UKAEA. This work is partly supported by the UK Engineering & Physical Sciences Research Council (EPSRC) Grant No. EP/R026084/1.

References

1. A. Banks, R. Gupta, MQTT version 3.1.1. OASIS Standard, 29:89 (2014)
2. V. Barabash, The ITER International Team, A. Peacock, S. Fabritsiev, G. Kalinin, S. Zinkle, A. Rowcliffe, J.-W. Rensman, A.A. Tavassoli, P. Marmy, Materials challenges for ITER–current status and future activities. J. Nucl. Mater. **367**, 21–32 (2007)
3. L. Bass, P. Clements, R. Kazman, *Software Architecture in Practice* (Addison-Wesley, Boston, 2003)

4. E.T. Bekele, U. Lahiri, A.R. Swanson, J.A. Crittendon, Z.E. Warren, N. Sarkar, A step towards developing adaptive robot-mediated intervention architecture (ARIA) for children with autism. IEEE Trans. Neural. Syst. Rehabil. Eng. **21**(2), 289–299 (2013)
5. P. Bellavista, A. Corradi, L. Foschini, A. Pernafini, Data distribution service (DDS): a performance comparison of opensplice and RTI implementations, in *2013 IEEE Symposium on Computers and Communications (ISCC)* (IEEE, Piscataway, 2013), pp. 000377–000383
6. D. Bruckner, M.-P. Stănică, R. Blair, S. Schriegel, S. Kehrer, M. Seewald, T. Sauter, An introduction to OPC UA TSN for industrial communication systems. Proc. IEEE **107**(6), 1121–1131 (2019)
7. H. Bruyninckx, Open robot control software: the OROCOS project, in *IEEE International Conference on Robotics and Automation (Cat. No. 01CH37164) Proceedings 2001 ICRA*, vol. 3 (IEEE, Piscataway, 2001), pp. 2523–2528
8. A. Burns, A.J. Wellings, *Real-Time Systems and Programming Languages: Ada 95, Real-Time Java, and Real-Time POSIX* (Pearson Education, London, 2001)
9. P.F. Carvalho, B. Santos, B. Goncalves, B.B. Carvalho, J. Sousa, A.P. Rodrigues, A.J.N. Batista, M. Correia, Á. Combo, C.M.B.A. Correia, et al., EPICS device support module as ATCA system manager for the ITER fast plant system controller. Fusion Eng. Design **88**(6–8), 1117–1121 (2013)
10. E. Coumans, Bullet physics engine. Open Source Softw. **1**(3), 84 (2010). http://bulletphysics.org
11. L.R. Dalesio, A.J. Kozubal, M.R. Kraimer, EPICS architecture. Technical report, Los Alamos National Lab., NM (1991)
12. C. Darve, M. Eshraqi, M. Lindroos, D. McGinnis, S. Molloy, P. Bosland, S. Bousson, The ESS superconducting linear accelerator. MOP004, SRF2013, Paris (2013), p. 168
13. R. Diankov, J. Kuffner, OpenRAVE: a planning architecture for autonomous robotics 79. Tech. Report, CMU-RI-TR-08-34, Robotics Institute, Carnegie Mellon University (2008)
14. P. Dieckmann, D. Gaba, M. Rall, Deepening the theoretical foundations of patient simulation as social practice. Simul. Healthcare **2**(3), 183–193 (2007)
15. A.C. Domínguez-Brito, CoolBOT: a component-oriented programming framework for robotics (2003)
16. A.C. Domínguez-Brito, F.J. Santana-Jorge, S. Santana-De-La-Fe, J.M. Martínez-García, J. Cabrera-Gámez, J.D. Hernández-Sosa, J. Isern-González, E. Fernández-Perdomo, CoolBOT: an open source distributed component based programming framework for robotics, in *International Symposium on Distributed Computing and Artificial Intelligence* (Springer, Berlin, 2011), pp. 369–376
17. G. Federici, C. Bachmann, L. Barucca, W. Biel, L. Boccaccini, R. Brown, C. Bustreo, S. Ciattaglia, F. Cismondi, M. Coleman, et al., Demo design activity in Europe: progress and updates. Fusion Eng. Design **136**, 729–741 (2018)
18. B.P. Gerkey, R.T. Vaughan, K. Stoy, A. Howard, G.S. Sukhatme, M.J. Mataric, Most valuable player: a robot device server for distributed control, in *Proceedings 2001 IEEE/RSJ International Conference on Intelligent Robots and Systems. Expanding the Societal Role of Robotics in the the Next Millennium (Cat. No. 01CH37180)*, vol. 3 (IEEE, Piscataway, 2001), pp. 1226–1231
19. B. Gerkey, R.T. Vaughan, A. Howard, The player/stage project: tools for multi-robot and distributed sensor systems, in *Proceedings of the 11th International Conference on Advanced Robotics*, vol. 1 (2003), pp. 317–323
20. M. Ghasemi, S.M. Sharafi, A. Arman, Towards an analytical approach to measure modularity in software architecture design. JSW **10**(4), 465–479 (2015)
21. G. Gui, P.D. Scott, Measuring software component reusability by coupling and cohesion metrics. J. Comput. **4**(9), 797–805 (2009)

22. W. Hasselbring, Component-based software engineering, in *Handbook of Software Engineering and Knowledge Engineering: Volume II: Emerging Technologies* (World Scientific, Singapore, 2002), pp. 289–305
23. W. Hasselbring, Software architecture: past, present, future, in *The Essence of Software Engineering* (Springer, Cham, 2018), pp. 169–184
24. M. Henning, M. Spruiell, Distributed programming with ICE. ZeroC Inc. Revision 3:97 (2003)
25. R. Henßen, M. Schleipen, Interoperability between OPC UA and AutomationML. Proc. Cirp **25**, 297–304 (2014)
26. http://www.createcrobotics.com. Iris
27. IEEE Standards Association et al. ISO/IEC/IEEE 24765: 2010 systems and software engineering-vocabulary Institute of Electrical and Electronics Engineers, Inc. (2010)
28. B. Ip, Performance analysis of VxWorks and RTLinux. Languages of Embedded Systems Department of Computer Science (2001)
29. J. Jackson, Microsoft robotics studio: a technical introduction. IEEE Robot. Autom. Mag. **14**(4), 82–87 (2007)
30. Jet Team, Fusion energy production from a deuterium-tritium plasma in the JET tokamak. Nucl. Fusion **32**(2), 187 (1992)
31. M.J. Kilgard, The OpenGL utility toolkit (GLUT) programming interface (1996)
32. J. Kim, S. Kang, J. Ahn, S. Lee, EMSA: extensibility metric for software architecture. Int. J. Softw. Eng. Knowl. Eng. **28**(03), 371–405 (2018)
33. N. Koenig, A. Howard, Design and use paradigms for Gazebo, an open-source multi-robot simulator, in *2004 IEEE/RSJ International Conference on Intelligent Robots and Systems (IROS)(IEEE Cat. No. 04CH37566)*, vol. 3 (IEEE, Piscataway, 2004), pp. 2149–2154
34. H. Leung, Z. Fan, *Handbook of Software Engineering and Knowledge Engineering* (World Scientific, Singapore, 2002)
35. S. Magnenat, V. Longchamp, F. Mondada, ASEBA, an event-based middleware for distributed robot control, in *Workshops and Tutorials CD IEEE/RSJ 2007 International Conference on Intelligent Robots and Systems* (IEEE Press, Piscataway, 2007)
36. S. Magnenat, P. Rétornaz, M. Bonani, V. Longchamp, F. Mondada, ASEBA: a modular architecture for event-based control of complex robots. IEEE/ASME Trans. Mechatron. **16**(2), 321–329 (2010)
37. Y. Maruyama, S. Kato, T. Azumi, Exploring the performance of ROS2, in *Proceedings of the 13th International Conference on Embedded Software* (2016), pp. 1–10
38. D. Morris, *Concise Encyclopedia of Software Engineering*, vol. 1 (Elsevier, Amsterdam, 2013)
39. N. Nethercote, R. Walsh, J. Fitzhardinge, Building workload characterization tools with Valgrind, in *2006 IEEE International Symposium on Workload Characterization* (IEEE, Piscataway, 2006), p. 2
40. P.M. Newman, MOOS-mission orientated operating suite, OUEL Report, Department of Engineering Science, University of Oxford (2008)
41. M. Quigley, K. Conley, B. Gerkey, J. Faust, T. Foote, J. Leibs, R. Wheeler, A.Y. Ng, ROS: an open-source robot operating system, in *ICRA Workshop on Open Source Software*, Kobe, vol. 3 (2009), p. 5
42. M. Rickert, A. Gaschler, Robotics library: an object-oriented approach to robot applications, in *2017 IEEE/RSJ International Conference on Intelligent Robots and Systems (IROS)* (IEEE, Piscataway, 2017), pp. 733–740
43. D.T. Ross, J.B. Goodenough, C.A. Irvine, Software engineering: process, principles, and goals. Computer **8**(5), 17–27 (1975)
44. D. Sanz, M. Ruiz, R. Castro, J. Vega, M. Afif, M. Monroe, S. Simrock, T. Debelle, R. Marawar, B. Glass, Advanced data acquisition system implementation for the ITER neutron diagnostic use case using EPICS and FlexRIO technology on a PXIe platform. IEEE Trans. Nucl. Sci. **63**(2), 1063–1069 (2016)
45. M.W. Scerbo, S. Dawson, High fidelity, high performance? Simul. Healthc. **2**(4), 224–230 (2007)

46. J.M. Schlesselman, G. Pardo-Castellote, B. Farabaugh, OMG data-distribution service (DDS): architectural update, in *IEEE MILCOM 2004. Military Communications Conference, 2004.*, vol. 2 (IEEE, Piscataway, 2004), pp. 961–967
47. R. W. Schwanke, An intelligent tool for re-engineering software modularity, in *Proceedings of the 13th International Conference on Software Engineering* (IEEE Computer Society Press, Piscataway, 1991), pp. 83–92
48. M. Serrano, E. Gallesio, F. Loitsch, Hop: a language for programming the web 2. 0, in *OOPSLA Companion* (2006), pp. 975–985
49. R. Smith, Open dynamics engine (2005)
50. I.A. Sucan, M. Moll, L.E. Kavraki, The open motion planning library. IEEE Robot. Autom. Mag. **19**(4), 72–82 (2012)
51. S. Thrun, Robotic mapping: A survey (2002). http://citeseerx.ist.psu.edu/viewdoc/download? doi=10.1.1.319.3077&rep=rep1&type=pdf
52. S. Thrun, D. Fox, W. Burgard, F. Dellaert, Robust Monte Carlo localization for mobile robots. Artif. Intell. **128**(1), 99–141 (2001)
53. R. Volpe, I. Nesnas, T. Estlin, D. Mutz, R. Petras, H. Das, The CLARAty architecture for robotic autonomy, in *2001 IEEE Aerospace Conference Proceedings (Cat. No. 01TH8542)*, vol. 1 (IEEE, Piscataway, 2001), pp. 1–121

Chapter 11
Mutation Testing for RoboChart

Robert M. Hierons, Maciej Gazda, Pablo Gómez-Abajo, Raluca Lefticaru, and Mercedes G. Merayo

Abstract This chapter describes a test-generation approach that takes as input a model S of the expected behavior of a robotic system and seeds faults into S, leading to a set of mutants of S. Given a mutant M of S, we check whether M is a valid implementation of S, and, if it is not, we find a test case that demonstrates this: a test case that reveals the seeded fault. In order to automate this approach, we used the Wodel tool to seed faults and a combination of two tools, RoboTool and FDR, to generate tests that detect the seeded faults. The result is an overall test-generation technique that can be automated and that derives test cases that are guaranteed to find certain faults.

1 Introduction

It is well known that there is a growing use of robotic systems and there is significant interest in the use of robotic systems in critical areas (Chap. 4). There is thus a need to develop techniques for validating such robotic systems before they are used. There are a number of techniques that can be used in validation, including formal verification and testing. Testing is the focus of this chapter, but it can be

R. M. Hierons (✉) · M. Gazda
The University of Sheffield, Sheffield, UK
e-mail: r.hierons@sheffield.ac.uk; m.gazda@sheffield.ac.uk

P. Gómez-Abajo
Universidad Autónoma de Madrid, Madrid, Spain
e-mail: Pablo.GomezA@uam.es

R. Lefticaru
University of Bradford, Bradford, UK
e-mail: r.lefticaru@bradford.ac.uk

M. G. Merayo
Universidad Complutense de Madrid, Madrid, Spain
e-mail: mgmerayo@fdi.ucm.es

© Springer Nature Switzerland AG 2021
A. Cavalcanti et al. (eds.), *Software Engineering for Robotics*,
https://doi.org/10.1007/978-3-030-66494-7_11

complemented, for example, by formal verification techniques that find certain classes of faults (see Chap. 5).

Testing involves executing the *system under test (SUT)*, possibly within a simulation, based on the given *test cases*. A *test case* defines the inputs that the SUT should receive and any other elements of the setup (such as properties of the environment in which the SUT will run). A test case will often also include a *test oracle*, which is a method for determining whether the behavior observed in a test run is allowed. Unfortunately, however, the number of possible test cases is usually vast (often infinite) and so it is only possible to use a very small proportion of these. In testing, we therefore have the problem of choosing a set of test cases (a *test suite*) to use.

Ideally, we would like to use a test suite that finds any faults that are present in the SUT, but we normally do not know whether the SUT is faulty and, if it is, what faults are present. As a result, there is a need to devise *test-generation techniques*, and such techniques typically aim to achieve a particular objective, with the objective often being to cover parts of a model or the code [2]. However, it is unclear how coverage relates to test suite effectiveness, except that a lack of coverage will usually indicate that certain types of faults cannot be found by the test suite (those in the parts of the SUT not covered).

In contrast to coverage, the intention of *mutation testing* is to provide some guarantees regarding test suite effectiveness. The original work, by DeMillo et al. [17], proposed mutation testing as an approach to assess how good a test suite T is likely to be when testing a program P. The approach involved generating a set of *mutants* $M_1, \ldots, M_n$ of P, by using mutation operators to make small changes to P and then determining how many of the mutants are distinguished from P (*killed* by T). Ideally, mutation operators represent the types of mistakes that programmers make.

The essential idea is that a "good" test suite is one that tends to distinguish between correct and faulty programs and that the proportion of mutants killed (the *mutation score*) is thus a way in which one can assess the quality of a test suite. More recent work has used models and specifications, with researchers having applied mutation testing to, for example, statecharts [23, 74], finite state machines [21], Petri nets [22] and CSP [73].

Although the original purpose of mutation testing was to assess test suite quality, it can also be used to drive test-generation: test-generation can be based on a process in which we try to produce test cases that kill the mutants. The first such approach, by DeMillo and Offutt [16], represented test-generation as a constraint satisfaction problem. The use of constraint programming to automate test-generation has since received much more attention, including for the testing of robotic systems (Chap. 4). A number of other approaches have been used, with these applying solution mechanisms such as metaheuristic search [24], dynamic symbolic execution [61], and a combination of the two [31]. For a review, see the work of Souza et al. [72].

In this chapter, we describe an approach developed for RoboChart [51] (Chap. 9) models. Test-generation is based on mutation, and so we start with a RoboChart model S that provides a specification and generates mutants of this model. A mutant need not represent a fault since it is possible, for example, that M and S differ syntactically but that M is a correct implementation of S. As a result, for each mutant M of S, we determine whether M represents a fault. If M does represent a fault (is not a correct implementation of S), then we produce a test case t that demonstrates this.

The result of this process is a set of test cases that is guaranteed to find the faults seeded during the mutation process. The use of seeded faults has previously been justified by the *competent programmer hypothesis*, which states that competent programmers make small faults that are similar to mutants [17]. Although one cannot expect this to generally hold, there is some empirical evidence to suggest that test suites that are good at killing mutants are also good at finding real faults [4]. As we explain in Sect. 3, empirical studies show that mutation testing can be effective.

This chapter describes the overall mutation testing approach, as well as reviewing relevant literature. It also explains how this approach has been implemented through three tools.

In this work, we used the Wodel tool [27, 28] to generate mutants of a RoboChart model, the key advantage of Wodel being that it can be used to mutate models in any language that has a metamodel. For automated test-generation, we use the fact that RoboChart has been given a semantics in CSP [66], a process algebra that has an associated formal semantics and a model checker (FDR [25]). The existence of such a CSP semantics is crucial since it makes it possible to formally analyze the semantics of a model. In particular, given a RoboChart model S that describes how the system should behave, and a mutant M of S generated by Wodel, we can use RoboTool to automatically generate CSP models $C(S)$ and $C(M)$ that provide the semantics of S and M, respectively. We then use the FDR model checker to determine whether $C(M)$ is a correct implementation of $C(S)$, with FDR returning a counter-example if it is not (and a counter-example can be found in a reasonable amount of time). Such a counter-example is a trace (sequence of events) that is a trace of M and is not a trace of S. As noted by Cavalcanti et al. [8], a counter-example defines a test case that can find the seeded fault. The approach described in this chapter is based on the work of Cavalcanti et al. [8], (Chap. 9) extending this with additional mutation operators and a more substantial review of related work.

This chapter is structured as follows. First, in Sect. 2 we provide a brief introduction to RoboChart and Sect. 3 reviews previous work on mutation testing. In Sect. 4, we then describe the use of Wodel to mutate RoboChart models. The mutation operators are given in Sect. 5, where we also explain how they were implemented in Wodel. Automated test-generation is described in Sect. 6. Section 7 describes an initial experimental study. Finally, in Sect. 8 we draw conclusions.

2 RoboChart

This section provides a short overview of RoboChart [51]; more details are provided by Cavalcanti et al. (Chap. 9). RoboChart is a state-based language that has been developed for a specific domain: robotics. A RoboChart model contains a number of elements including a module, which specifies the robotic platform. The module, for example, uses events (inputs and outputs) and variables to model sensors and actuators.

A RoboChart model will also describe one or more *controllers* that form the software. The controllers are described using an approach similar to UML statecharts, the intention being to make RoboChart accessible to a relatively wide range of developers. The software communicates with its environment via the robotic platform and so inputs correspond to the events of the platform. Outputs can alter the state of the system and so an output can be represented as an event, a change in the value of a variable or an operation. Variables, events and operations can be grouped to form interfaces. The models of the controllers are the focus of the work described in this chapter; we are not interested in potential faults in the platform.

We now briefly describe a model of an autonomous vacuum robot for solar panels [11], which can be found online.[1] The robot performs a two-stage cleaning process to remove the dust that sticks to the solar panels. It aims to follow an efficient navigation pattern on the inclined surfaces of the solar panels and regularly checks its battery. When needed, it finishes the current cleaning cycle and returns to its docking station for charging.

The RoboChart model of the robot consists of a module that connects the platform with the only controller, namely, PathPlanningController, and 12 interfaces, such as those for cleaning operations (vacuum, brush), for moving operations or commands (move_forward, turn, stop), for sensor events (battery_level, charging, ultrasonic) or for movement events (acc_l, acc_r – to receive input from the left and right accelerometers, speed_l, speed_r, angle).

The controller model contains nine state machines. One of them, the Path-PlanningSM model, is depicted in Fig. 11.1. The PathPlanningSM model makes navigational decisions based on sensor input, and then these are communicated to other state machines, like MidLevelSM. This machine is responsible for converting the movement commands received from PathPlanningSM into voltages to send to the left and right motors of the robot. Based on the PathPlanningSM commands, operations such as brush and vacuum are handled by another state machine, CleanSM.

---

[1] https://robostar.cs.york.ac.uk/case_studies/RoboVacuum/.

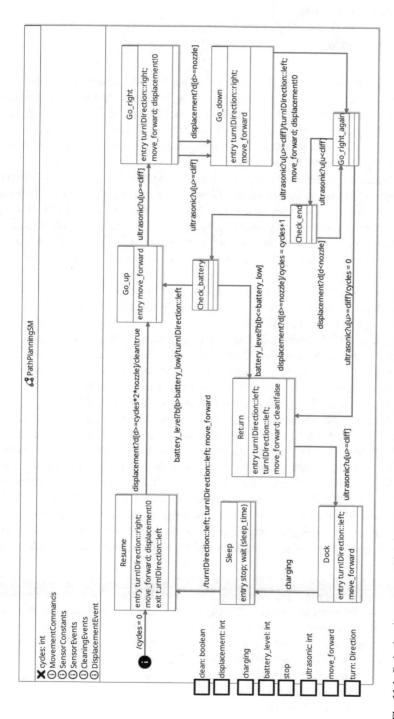

Fig. 11.1 Path planning state machine for the solar panel vacuum cleaner robot [11]. ©2020 University of York, reprinted with permission.

3 Mutation Testing

3.1 *Introduction*

Mutation testing is an approach that was developed with the aim of assessing the quality of a test suite, that is, to estimate how effective the test suite is at finding faults. To do this, mutation testing techniques have traditionally introduced small syntactic changes into a program to generate faulty versions called *mutants*. After a set of mutants has been produced from a program, the test suite is applied to both the original program and the mutants.

If the behavior of a mutant is different from that exhibited by the original program when given a test case t, then the mutant is said to have been *killed* by t. If none of the test cases kills a mutant M, then M is *alive*. If a test suite does not kill all of the mutants, then the test suite may be extended with new test cases to kill the mutants that are alive. The underlying motivation for mutation testing is the belief that a test suite that is good at distinguishing a program from its mutants is likely to be good at discovering faults.

In early work, mutation testing was based on source code and so mutation operators have been proposed for a range of languages, including Fortran [36], Ada [58], C [1, 33, 69], Java [5, 35, 46], C++ [14, 40], SQL [76], WS-BPEL [20] and JavaScript [50]. However, mutation testing approaches have also been devised for specification languages. Examples of this include finite state machines [21], statecharts [23, 74], Petri nets [22] and CSP [73]. Mutation testing has been applied to a range of systems, including access control policies [53], web services [42], simulation platforms [7], security issues [44] and Android apps [43].

There is empirical evidence to suggest that mutation testing can be effective. For example, Chekam et al. [9] found mutation testing to be more effective than two standard white-box coverage criteria, statement and branch coverage, in the sense that the experiments found a correlation between the mutation score of a test suite and its fault detecting ability, while no such correlation was found for statement and branch coverage. Ramler et al. [65] report on the use of mutation testing in the context of a safety-critical system. They report that a test suite that had previously been produced for this system failed to kill a number of non-equivalent mutants. This led to new test cases being devised and these test cases found two additional faults.

A wide range of mutation testing tools have been developed to support mutation testing for different programming languages. Initially, these tools were designed to deal with the mutation testing techniques proposed in the academic environment [12, 18]. However, since the early 2000s many tools have been developed that have helped to turn mutation testing into a practical testing technique for a range of notations such as C [34], Java [47, 68], SQL [75] and UML [39].

The rest of this section is structured as follows. It starts, in Sect. 3.2, by describing core concepts. Section 3.3 then reviews some previous uses of mutation testing and Sect. 3.4 discusses a number of practical limitations.

3.2   Core Concepts

Mutants are produced by injecting small changes into an entity, often a program. These changes are defined by *mutation operators* that are developed to simulate common faults. An example of a small change is a relational mutant operator, which could replace $<$ with $\leq$. The use of such syntactically small changes has been justified by the *competent programmer assumption*, which states that competent programmers make syntactically small faults [17]. As previously noted, there is some empirical evidence that demonstrates that test suites that are effective in finding such mutants tend also to be effective in finding real faults [4]. A study of software repositories found that most bug fixes were syntactically simple, involving only three or four tokens being changed [29]. This provides some support for the competent programmer assumption, but the authors also found that most bug fixes were unlike standard mutation operators, suggesting a need to devise new mutation operators.

The application of the mutation operators to a program generates a set of *mutants*. If a mutant M is generated by applying one mutation operator, and only applying this once, then M is said to be a *first-order mutant*. If M was generated by two or more applications of mutation operators, then M is a *higher-order mutant*. Most work has only concerned first-order mutants. This has been justified by the *coupling effect* [56], which states that "Complex mutants are coupled to simple mutants in such a way that a test data set that detects all simple mutants in a program will detect a large percentage of the complex mutants". Early empirical studies provided evidence for the coupling effect [55, 56], although it is unclear whether the results generalize to, for example, specification languages.

Test suite effectiveness is typically assessed by computing the *mutation score*, which is the proportion of mutants that are killed by the test suite. Unfortunately, there may be *equivalent mutants*, where a mutant M of an entity P is an equivalent mutant if it is semantically equivalent to P: for every input x, M and P can produce the same set of outputs in response to x. An equivalent mutant is obtained, for example, by mutating if (x>1) to if (x>=1) at a point in a program where x cannot take on the value 1. Clearly, no test case can kill an equivalent mutant, and so ideally these mutants are not included when computing the mutation score: we are interested in the proportion of non-equivalent mutants that are killed. However, equivalence is undecidable unless significant restrictions are imposed and so it may not be possible to compute this score (see Sect. 3.4).

Mutation can also produce what are called duplicated and redundant mutants. A *duplicated* mutant is one that is not equivalent to the original program, and so can be killed, but it is equivalent to another mutant. A *redundant* mutant is semantically different from the original program and also other mutants, but it is killed when other mutants are killed because it is *subsumed* by them [3]. That is, if we have a set $\mathcal{M}$ of mutants, then $M \in \mathcal{M}$ is redundant if every test suite that kills all the mutants in $\mathcal{M} \setminus \{M\}$ also kills M. The presence of duplicated and redundant mutants can also affect the mutation score. For example, if there are many duplicates of a mutant

M, then we are effectively giving a higher weighting to (the semantics of) M when we compute the mutation score. As a result, ideally one eliminates redundant and duplicated mutants before computing the mutation score.

3.3   Uses of Mutation Testing

One of the main applications of mutation testing is to evaluate testing techniques by estimating their effectiveness [17]. The essential idea is that a test suite T_1 is superior to another test suite T_2 if T_1 has a higher mutation score than T_2. As a result, one test technique is superior to another if the first tends to produce test suites that have a higher mutation score.

Mutation testing has also been used for other objectives. As previously mentioned, one use of mutation testing is to drive test-generation [16, 24, 31, 61, 62, 72]. In principle, the idea here is simple: test-generation involves producing a test suite that kills all of the non-equivalent mutants. Thus, if some non-equivalent mutants are alive, then we try to extend the test suite with test cases that kill these mutants. Again, decidability issues arise (see Sect. 3.4).

Mutation testing has also been used in test case prioritization. Test case prioritization techniques order the execution of test cases based on some criterion. The prioritization goal of the *fault exposing potential* [67] approach is to order test cases so that faults are likely to be detected as early as possible. This technique considers whether a test case executes a potentially faulty statement and the probability that it reveals a fault in that statement, using mutation analysis to estimate this probability. Basically, given a program P, a test suite and a set of mutants, the test cases are executed against the mutants, and for each statement s of P that has been mutated, the fault exposing potential of a test case t on s is calculated as the ratio of mutants of s killed by t to the total number of mutants of s. Then, each test case t is associated with a value that corresponds to the sum, over all statements in P, of the obtained values for t. The test cases are sorted in descending order of these values.

The *additional fault exposing potential* also takes into account, in the calculation of the values, the *confidence* in a statement, that is, the probability that the statement is correct. The application of a test case, which executes a statement and does not reveal a fault, increases the level of confidence.

Another approach to prioritization [45] presents two models, a statistics-based model and a probability-based model. The former uses the number of mutants killed by a test case as the basis for estimating the fault detection capability of the test case. The latter considers the distribution of mutants over statements, that is, the number of ways in which each statement can be mutated. Both of them focus on software evolution, that is, the differences in source code between two different versions of a program. The idea is that test cases that have the potential to reveal faults in the latter version should be executed early. The approach therefore generates mutants only on statements in the difference between the versions. Then, the capacity of each test case (to detect faults) is calculated and the test cases are ordered in descending order of the obtained values.

Finally, recent work has proposed a new prioritization objective based on the diversity-aware mutation adequacy criterion [70]. The issue addressed is that the mutation adequacy criterion does not consider the diversity of killed mutants, only the number of killed mutants. The diversity-aware mutation adequacy criterion computes the number of mutants distinguished from *other* mutants rather than the number of mutants distinguished from the original program.

The application of diversity-aware mutation adequacy has been empirically compared with traditional mutation adequacy (i.e. killing all mutants as early as possible) in several settings [71]. The results show that in some cases, the combination of the two mutation-based objectives is more effective than considering one objective at a time.

Mutation testing has also been used as the basis for fault localization: the problem of trying to determine where a fault is located. The first mutation-based fault localization approach corresponded to the Metallaxis method [59, 60]. The underlying idea was extended in the MUSE method [52]. In fault localization, it is normal to have the results of the execution of a number of test cases, with the SUT having failed some test cases (*failing tests*). Metallaxis is based on the idea that if a mutant is mostly killed by failing tests, then this provides evidence that the statement mutated is a faulty program location. The extension, MUSE, is based on the mutation of statements and how this changes the number of failing test cases. The assumption is that, on the one hand, the mutation of faulty statements will make more failed test cases pass, when compared to the mutation of correct statements. On the other hand, the mutation of correct statements will make more passed test cases fail, when compared to mutating faulty statements. These two considerations are used as the basis of the proposed metric for fault localization.

3.4 Practical Limitations

This section reviews some the main factors that can make it difficult to use mutation testing and how they have been addressed in the past. For a more complete overview, see the recent systematic literature review on this topic [63].

The first issue is that there are often many mutants and these have to be generated, compiled and tested. This makes it difficult to scale mutation testing to larger programs, models or specifications. Two main approaches address this problem: either reduce the number of mutants or reduce the time taken to compile or test mutants.

There are at least two approaches that reduce compilation or execution time without reducing the number of mutants. One approach uses parallel compilation and execution: it is possible to compile or execute the mutants on multiple machines and then bring together the results of testing (see, e.g. [38]). Another approach uses one program that captures a set of mutants, with this program being called a *mutant schema*. A mutant schema contains a meta-mutant that can be parametrized to capture any one of a set of mutants. To see how this can be done, let us suppose that there is only one way of mutating a statement s of a program P, and this mutates s to form s'. Then, in forming the mutant schema from P, s can be replaced by a conditional statement that leads to a copy of s being executed for certain parameter values, while s' is executed for the other possible parameter values. Such transformations can be applied so that for every mutant M of P, there is a parameter value under which the meta-mutant behaves like M. Such an approach will tend to reduce storage requirements, since a single (traditional) mutant M includes copies of all statements not mutated in addition to the mutated statement. It also reduces compilation time, since only one compilation is required.

The problem of reducing the number of mutants has been addressed by many researchers. The simplest technique proposes to use a randomly generated sample of the possible mutants [30, 77]. Another proposal uses a relatively small number of second-order [49, 64] or higher-order [33] mutants rather than a large number of first-order mutants [48, 49, 64]. A third approach, called *selective mutation*, focuses on operator-based mutant selection and uses only a subset of the mutation operators, with the choice of operators being based on the results of experiments. Offutt et al. [57] introduced selective mutation, finding that it was possible to reduce the number of mutants by a factor of four or above while losing little in effectiveness: test suites that killed the mutants generated by a selection of mutation operators also killed over 98% of mutants generated using the complete set of mutation operators. This early work was based on FORTRAN, but selective mutation approaches have since been developed for a number of other languages such as C [54], C++ [15] and Java [19] and also for concurrent code [26]. Some researchers have proposed combinations of both operator-based selection and random sampling techniques [77, 78]. Finally, there is work that aims to seed a relatively small number of faults that model misunderstandings of the language being used rather than small syntactic mistakes, with this approach being called *semantic mutation testing* [10].

A second practical issue concerns the potential presence of equivalent and duplicated mutants. As noted previously, ideally we should find and eliminate such mutants if we are interested in using the mutation score to assess the quality of a test suite. If we base test-generation on finding test cases to kill mutants, then the presence of equivalent or duplicated mutants can lead to wasted effort, with equivalent mutants being particularly problematic since a significant amount of effort may be expended in trying to find test cases to kill them.

The identification of equivalent and duplicated mutants is undecidable [6]. As a result, one cannot expect a technique, for detecting equivalent and redundant mutants, to be entirely automated unless restrictions are imposed, such as the use of a language for which equivalence is decidable. As a result, the elimination of such mutants typically involves a manual element, which is error-prone and time-consuming.

Several approaches have been proposed to tackle this problem [48]. For example, Trivial Compiler Optimization [37] identifies equivalent mutants by checking whether the compiled object code of the mutant and the compiled object code of the original program are identical. Experiments performed on both C and Java showed that this approach has value, with the results indicating that the number of mutants can be reduced by 11% for Java and 28% for C. Although equivalence is undecidable for programming languages, it is decidable for some less expressive languages such as regular languages, with this increasing the degree of automation that is possible. It is also important to note that even when equivalence is undecidable, there may be solvers that can decide equivalence for a significant number of cases.

4 Mutation Testing and Wodel

Wodel is a domain-specific language (DSL) that has been designed to make it possible to define mutation operators for models. Wodel was developed by the Miso team[2] and was introduced recently [27, 28]. Wodel is provided as an Eclipse plug-in that has been made available.[3]

The Wodel tool includes a complete IDE with some useful development facilities, such as code completion, syntax highlighting and an automated compilation of Wodel programs into Java code. The Wodel DSL is largely domain-independent, i.e. it can be used with models of arbitrary domains as long as they are described using metamodels. Wodel programs must refer to a domain metamodel to check type conformance and to ensure that the generated mutants are valid.

Figure 11.2 shows the modular and component-based architecture of the Wodel environment. The workflow is as follows. First, the user must specify the domain concepts using an Ecore metamodel (label 1). Next, the user defines the desired mutation operators and their execution details, such as the number of times they are going to be applied or the execution order. This specification is the Wodel program. The Wodel execution generates mutant models from the seed models (label 4 in Fig. 11.2), as well as a registry of the applied mutation operators and the corresponding affected objects (label 5 in Fig. 11.2).

The generated mutants conform to the given domain metamodel (label 3). Additionally, the user can use the seed models synthesizer provided by the Wodel tool to automatically generate example models from the given Wodel program (label 7). The Wodel tool generates a set of mutation footprints, which are values of metrics regarding, for example, the coverage of the metamodel and models used, and this can help the user identify missing mutation operator (label 8).

Wodel provides nine mutation primitives to select, create, remove, clone and retype objects and create, remove and modify their features. There are several approaches to selecting candidate objects. For example, they can be randomly selected, selection could be based on a required property or selection might be based on object type. The Wodel tool has been used in diverse domains such as Finite

---

[2] http://www.miso.es/.
[3] gomezabajo.github.io/Wodel/.

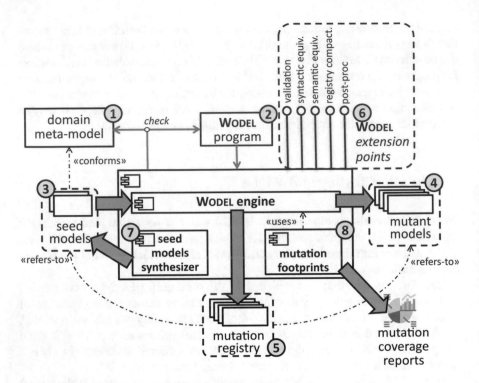

Fig. 11.2 Architecture of the Wodel development environment.

Automata, Probabilistic Automata, UML class diagrams, Security Policies, BPEL and BPMN.

Listing 1 shows a Wodel program applied to the RoboChart domain metamodel. Lines 1–3 define the configuration of the Wodel program. Line 1 of Listing 1 states that three mutants will be generated in the "out/" folder from the seed models stored in the "models/" folder. Next, line 2 defines the domain metamodel, and line 3 ends the configuration with an optional description in natural language that can help identify the Wodel program.

The presented mutation operator (line 8 of Listing 1) retypes an EntryAction object as a DuringAction object. This mutation operator works as follows: first, an object of the EntryAction type is randomly selected as a source; second, a new object of the DuringAction type is created, and all the compatible features of the source object are copied into it; and finally, the source object is deleted. Therefore, this mutation primitive has the effect that we have retyped the source object as a new object of the target type.

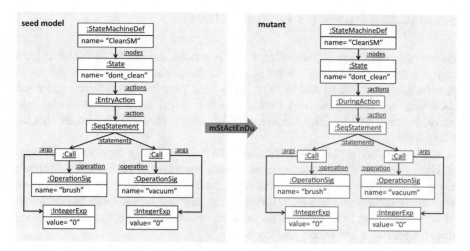

Fig. 11.3 Example of the application of the mStActEnDu mutation operator to a RoboVacuum model.

```
 1   generate 3 mutants in "out/" from "models/"
 2   metamodel "http://RoboChart.com"
 3   description "Simple Wodel program"
 4
 5   with blocks {
 6     mStActEnDu "Retypes an EntryAction object as a DuringAction object"
 7     {
 8        retype one EntryAction as DuringAction
 9     }
10   }
```

Listing 1 A simple Wodel program

Figure 11.3 shows an example of the application of the mutation operator of Listing 1 to a RoboChart model of a solar panel vacuum cleaner robot described in Sect. 2. The left side of the figure shows an excerpt from the seed model and the right side presents the resulting mutant. The objects and references affected by the mutation operator are shown in red.

Mutation operators might generate invalid mutants: mutants that do not satisfy the metamodel. In order to address this, we translated the RoboChart well-formedness conditions into Object Constraint Language (OCL). Wodel eliminates any mutant that does not satisfy this OCL condition.

The next section presents a set of mutation operators that have been derived for the RoboChart domain metamodel.

5 Mutation Operators for RoboChart

As explained in Chap. 10, RoboChart [51] is a notation born from the need for a rigorous development approach for robotic models, enhanced with simulation and verification tools. RoboTool[4] is a set of Eclipse plug-ins that supports graphical modeling, validation and model checking. For model checking, the RoboChart model is translated to CSP, for use of FDR, or the probabilistic model checker PRISM [41].

The RoboChart notation is distinctive in its features, which support architectural modeling as well as timed constructs in state machines. The RoboChart models are specified in packages that contain types, interfaces, modules, robotic platforms, controllers and state machines. The RoboChart core language permits the definition of connections, events, operations, variables and complex statements, including timed expressions and timed statements.

As state machines are frequently used to model the behavior of robotic systems, initial work on mutation testing for RoboChart focused mainly on defining, in Wodel, mutation operators for state machines [8]. The nine mutation operators initially defined for RoboChart models are given in Table 11.1. These mutation operators were originally given by Cavalcanti et al. [8], where they were employed in a small case study involving a drone. The model considered [8] contained only a small subset of the RoboChart language, which in fact is much richer.

In this section, we propose additional mutation operators for RoboChart and classify them based on the language elements covered. We also illustrate them on a more complex model, the model of an autonomous vacuum robot for solar panels [11], described in Sect. 2 and shown in Fig. 11.1.

Although Wodel offers several mutation primitives to *create*, *remove* and *modify* objects and their features, we mainly focus on using primitives for *deletion* and *modification*, rather than new object *creation*. There are two main reasons for this choice: (a) as the RoboChart language is quite rich, complex objects can be defined, and the *creation* of new complex structures via mutation is more likely to lead to invalid models; (b) this choice reduces the number of mutants generated.

The choice to exclude creation mutation operators was also motivated by empirical evidence [13] that shows that, for program code, mutation testing can be effective even if we only use statement deletion: it was found that test suites that killed these mutants also killed a high percentage of other mutants. The main practical benefit was that far fewer mutants were obtained when only using statement deletion, with the number of mutants being linear in terms of the program size. This motivated us to search for similar deletion operators for RoboChart models, although we supplemented these with modification operators.

---

[4]https://robostar.cs.york.ac.uk/robotool/.

Table 11.1 Mutation operators previously defined and corresponding Wodel blocks. The first column gives the name of the mutation operator and the second gives the Wodel code used to implement it

Mutation	Wodel blocks
mStActEnDu	1**retype one** EntryAction **as** DuringAction 2// modifies a state by changing an entry action into a during action
mStActEnEx	1**retype one** EntryAction **as** ExitAction 2// modifies entry into exit action
mTransSource	1tr = **select one** Transition **where** {^**source** <> **one** Initial} 2**modify target** ^**source from** tr **to other** State 3// changes the start state of a transition, except the **one from** the initial junction
mTransTarget	1**modify target** ^**target from one** Transition **to other** State 2// changes the ending state of a transition
mTransTrigger	1interf = **select one** Interface **where** {events <> **null**} 2ev = **select one** Event **in** interf→events 3tg = **create** Trigger **with** {event = ev} 4**modify one** Transition **with** {trigger = tg} 5// modifies a transition by replacing its trigger **with** another event
rSeqStatement	1ss = **select one** SeqStatement 2**remove one** Statement **from** ss→statements [1..5] 3**remove all** SeqStatement **where** {statements = **null**} 4// randomly deletes 1−4 statements **from** a sequence **and all** empty sequences
rState	1st = **select one** State 2**remove all** Transition **where** {^**source** = st} 3**remove all** Transition **where** {^**target** = st} 4**remove** st 5// removes a non−initial state **and all** transitions **from or to** that state
rTran	1**remove one** Transition **where** {^**source** <> **one** Initial} 2// deletes **one** transition, except the **one from** initial junction
rTranAction	1tr = **select one** Transition **where** {action <> **null**} g 2**remove one** Call **from** tr→action 3**remove one** SendEvent **from** tr→action 4**remove one** Action **from** tr 5// removes the action associated **with** a transition (call **or** send event)

The following gives some new mutation operators for RoboChart and groups them by the language elements they cover. The list is not exhaustive: we show a few examples for each category, providing the corresponding Wodel blocks.

Mutations for Types
The RoboChart notation has primitive types, enumerations and record types. For example, the orientation of the robot is represented using an Enumeration with the values up, right, down, left, while the position can be given by a triple-axis (X, Y, Z) using a RecordType element. Some simple mutations are: rElemEnumeration, removes one Literal from an Enumeration object; mElemEnumeration, modifies one Literal from an Enumeration; and rFieldRecordType, removes one Field from a RecordType.

```
1    rElemEnumeration "Removes one literal from an enumeration"
2    {
3      e = select one Enumeration
4      remove one Literal from e→literals
5    }
6
7    mElemEnumeration "Modifies one element from an enumeration"
8    {
9      e = select one Enumeration
10     l = select one Literal in e→literals
11     modify l with {name= random−string(3, 6)}
12   }
```

Listing 2 Wodel mutation operators for types

Listing 2 presents the Wodel code for two of these mutation operators: rElemEnumeration and mElemEnumeration. Consider first the Wodel code for rElemEnumeration. Here, the Wodel tool randomly chooses (selects) an Enumeration object from the model, randomly chooses a literal and deletes this literal. To see how this works, let us suppose that the Enumeration object chosen is the object Direction with the possible values up, right, down, left. The operator might remove the last element (left) from the list and then replace all its occurrences (references in the model) with one of the remaining items in the list predecessor (e.g. down).

Now consider the mutation operator mElemEnumeration from Listing 2. The first step randomly chooses an Enumeration, and then it chooses an element of the enumeration and replaces it with a randomly chosen value. For example, down might be replaced by the string "Phgx".

Mutations for Expressions

The RoboChart metamodel supports expressions of different types: primitive ones, like IntegerExp and BooleanExp; arithmetic expressions such as Plus, Minus, Mult, Div and Modulus; comparisons Equals, Different, GreaterThan, GreaterOrEqual, LessThan and LessOrEqual; logical expressions Not, And, Or, Implies, Iff, Forall and Exists; and many others. Some of the syntax elements mentioned extend the abstract class BinaryExpression, and a straightforward mutation operator that applies to all of them is swapBinaryExpression.

```
1    mIntegerExp "Modifies the value of an integer expression"
2    {
3      exp = select one IntegerExp
4      modify exp with {value = random−int(1, 10)}
5    }
6
7    swapBinaryExpression "Changes the left and right side of a binary expression"
8    {
9      modify one BinaryExpression with {swapref(left, right)}
10   }
11
12   mRelationalOperator "Retypes one relational operator as another one"
13   {
14     retype one [LessThan, LessOrEqual, GreaterThan, GreaterOrEqual, Different, Equals]
15       as [LessThan, LessOrEqual, GreaterThan, GreaterOrEqual, Different, Equals]
16   }
```

Listing 3 Wodel mutation operators for expressions

We now explain how the Wodel code in Listing 3 operates. First, consider the mIntegerExp mutation operator. This randomly selects an integer expression and assigns to it a random value between 1 and 10. For example, if the RoboChart model contained an expression such as x > 1, then this could be changed to an expression such as x >5.

The swapBinaryExpression operator starts by randomly selecting a binary expression in the RoboChart model and swaps its left-hand and right-hand terms. For example, it would replace u >= cliff with cliff >= u. Naturally, this operator might be expected to generate relatively large numbers of equivalent mutants since some binary operators are commutative.

Finally, the mRelationalOperator swaps one relational operator for another. This starts by randomly selecting an instance, in the model, of a relational operator from the given list of operators. The Wodel code then replaced this instance of the relational operator by another relational operator, keeping the same left-hand side and right-hand side. For example, if the selected relational operator was GreaterThan in the expression b > battery_low, then the mutant might be generated by replacing this with b >= battery_low.

As noted, the mutants generated can sometimes be equivalent. Section 6 describes how FDR can be used to eliminate equivalent mutants.

Mutations for Actions and Statements
The RoboChart notation includes an action language with concrete statements used to define actions in states and transitions. A State can have entry, during and exit actions, executed in particular phases of its life cycle. Transitions can be triggered by an event, guarded by a condition and contain an action that is executed when the transition is taken.

```
1   rAssignment "Removes one assignment "
2   {
3     remove one Assignment
4   }
5
6   rSendEvent "Removes a SendEvent statement"
7   {
8     remove one SendEvent
9   }
10
11  rSeqStatement "Removes a sequence of statements"
12  {
13    remove one SeqStatement
14  }
```

Listing 4 Wodel mutation operators for actions and statements

A few examples of mutation operators for actions and statements are given in Listing 4; these are all remove operators. In the first one, Wodel randomly selects an Assignment object from the RoboChart model and deletes it. Similarly, rSendEvent deletes an object of type SendEvent, and rSeqStatement deletes an entire block: a sequence of statements that can have different types (Call, SendEvent, Wait, etc.).

Mutations for Timed Primitives

Timed primitives are used in expressions, statements and transitions. Some examples of mutation operators for timed primitives are given in Listing 5. The rWait mutation operator removes a Wait statement that could occur in the action statements associated with a state or a transition.

The mutation operators rStateClockExp and rStateClockExp2 relate to state clock expressions. A StateClockExpression counts the elapsed time since entry in a state S. Such an expression is permitted only in transition guards and may only occur in comparisons where the other branch is a constant. Typically, the StateClockExp occurs in the left-hand side of a GreaterOrEqual expression but it need not. rStateClockExp and rStateClockExp2 delete a state clock expression, along with the associated condition. The result is a mutant that has a transition with the time expression removed.

```
1    rWait "Removes one wait statement "
2    {
3      remove one Wait
4    }
5
6    rStateClockExp "Removes a state clock expression from a GreaterOrEqual guard"
7    {
8      sce = select one StateClockExp
9      goe = select one GreaterOrEqual where {left = sce}
10     remove sce
11     remove goe
12   }
13
14   rStateClockExp2 "Removes a state clock expression from any binary expression"
15   {
16     sce = select one StateClockExp
17     be = select one BinaryExpression where {left = sce or right = sce}
18     remove sce
19     remove be
20   }
```

Listing 5 Wodel mutation operators for timed primitives

Mutations for State Machines

A RoboChart model can include a number of state machines, which are similar to UML state machines, but stripped of features deemed non-essential for robotics and augmented with the possibility of specifying timing properties and probabilities. The mutation operators could tackle different aspects, such as states or transitions, and their distinctive features such as triggers, conditions (or guards), actions, start and end deadlines and probabilities.

Some examples are provided in Listing 6. For example, in order to remove a state, the Wodel block rState first randomly chooses a state st, deletes all the transitions connected to st and then removes st. If the mutation operator were to only contain the statement remove one State, then this would cause the deletion of the state and consequently the appearance of hanging transitions, with a null destination or source.

The mTransTrigger operator first randomly chooses an event ev from a randomly chosen Interface (in RoboChart events are defined in interfaces, hence the requirement events <> null). It then creates a new object tg of type Trigger, based on this event. It then modifies a randomly chosen transition by adding or updating its trigger to refer to the new object tg.

The last block rCondTrans modifies a randomly chosen transition tr (that has a guard or condition) by deleting a condition.

```
1   rState "Removes one state and all transitions connected to it "
2   {
3     sl = select one State
4     remove all Transition where {^source = st}
5     remove all Transition where {^target = st}
6     remove st
7   }
8
9   mTransTrigger "Modifies a transition by replacing its trigger with another event"
10  {
11    interf = select one Interface where {events <> null}
12    ev = select one Event in interf→events
13    tg = create Trigger with {event = ev}
14    modify one Transition with {trigger = tg}
15  }
16
17  rCondTrans "Removes a guard (transition condition)"
18  {
19    tr = select one Transition where {condition <> null}
20    remove one Expression from tr→condition
21  }
```

Listing 6 Wodel mutation operators for state machines

Mutations for Modules, Controllers and Other Elements

An RCPackage (or RoboChart package) can include declarations of types, interfaces, modules, robotic platforms, controllers and state machines. An Interface groups variable lists, operations and events. A Module contains a number of elements that can be connected, namely, platforms, controllers and state machines. The Connections occur between source (from) and target (to) nodes, and in a module they establish the relationship between a platform and its controllers. They can be asynchronous and bidirectional, as indicated by the Boolean attributes async and bidirec. A module gives a complete account of a robotic system by associating their robotic platforms with particular controllers to specify behavior.

A few mutation operators that can be applied are given in Listing 7.

```
1   mConnectionAsyn "Modifies the asynchronous boolean attribute of the connection"
2   {
3     modify one Connection with {reverse(async)}
4   }
5
6   mConnectionBidirec "Modifies the bidirectional attribute of the connection"
7   {
8     modify one Connection with {reverse(bidirec)}
9   }
10
11  rStaMachController "Removes a state machine from a controller and all its connections"
12  {
13    con = select one Controller
14    stm = select one StateMachine in con→machines
15    remove all Connection where {^from = stm}
16    remove all Connection where {^to = stm}
```

```
17        remove stm
18    }
19
20    rEventController "Removes an event from a controller and all its connections"
21    {
22        con = select one Controller
23        ev = select one Event in con→events
24        remove all Connection where {efrom = ev}
25        remove all Connection where {eto = ev}
26        remove ev
27    }
28
29    rPostCondFunction "Removes a postcondition of a function"
30    {
31        f = select one Function where {postconditions <> null}
32        remove one Expression from f→postconditions
33    }
```

Listing 7 Wodel mutation operators for modules, controllers and other objects

The first two mutation blocks in Listing 7 follow the same pattern: first, a Connection is randomly selected, and then a Boolean attribute of this Connection (async or bidirect) is modified using the reverse keyword, i.e. true value is changed to false and vice versa.

Similar to rState, the rStaMachineController mutation operator randomly chooses a state machine (that is found in a randomly chosen Controller), deletes first all the Connections to it and then removes it. Following the same template, the rEventController operator randomly chooses a Controller to mutate, and then it chooses an Event defined in its event list, deletes all the Connections having efrom (event from) or eto (event to) relation with this event and finally removes the event. Removing the object this way avoids the introduction of objects that reference null objects. Finally, rPostCondFunction first selects a Function object that has a postcondition component that is not null and then removes an element (Expression) from the list of postconditions.

6 Automated Test-Generation

We have seen that Wodel can be used to generate mutants of a RoboChart model. Each mutant that is not a valid implementation of the original model represents a potential fault that the SUT might have. In this section, we explain how we automated the process of generating test cases that kill these mutants; such test cases find the seeded (potential) faults and so will detect corresponding faults in the SUT.

Let us suppose that S is a RoboChart model and M is a mutant of S produced by Wodel. RoboTool can map a RoboChart model S to a CSP model $C(S)$ that provides the semantics of S and also map M to a CSP model $C(M)$ that provides the semantics of the mutant M. We pass $C(S)$ and $C(M)$ to the FDR model checker and ask whether $C(M)$ is a valid implementation (a refinement) of $C(S)$.

FDR implements several notions of refinement and for this work we use *traces refinement*. So, $C(M)$ is a refinement of $C(S)$ if and only if every sequence of events (*trace*) of $C(M)$ is also a trace of $C(S)$. Observe that this is not an equivalence relation since the implementation can exhibit fewer traces. Note that this means that we are not concerned with equivalent mutants and instead are interested in what might be called *refining mutants*. Specifically, a mutant M is a refining mutant, given specification S, if every trace of $C(M)$ is also a trace of $C(S)$. This does not imply that $C(M)$ and $C(S)$ have the same sets of traces.

Once FDR has been used, there are three possible cases:

1. FDR finds that $C(M)$ is a refinement of $C(S)$. In this case, $C(M)$ does not represent a fault and so the mutant M can be discarded.
2. FDR finds that $C(M)$ is not a refinement of $C(S)$. In this case, FDR returns a counter-example, which is a trace of $C(M)$ (and so also of M) that is not a trace of $C(S)$ (and so also is not a trace of S). This defines a test case that kills M.
3. FDR runs out of resources, failing to prove refinement and also not returning a counter-example.

The overall approach therefore uses three tools to provide automation: Wodel to generate mutants, RoboTool to map the original model and the mutants to CSP and FDR to generate test cases from the CSP models. The use of FDR has the potential to eliminate the equivalent (or refining) mutant problem.

Naturally, automation is not provided in the third case above; the experience to date is that this case is rare with small specifications but this is likely to change when more complex models are used. If the third case were to be more common, then potential solutions might revolve around model simplification: separately testing different aspects of the SUT and/or using more abstract models. Alternatively, test-generation might be expressed as either an optimization problem, to which one can apply metaheuristic search [32], or as a Constraint Satisfaction Programming problem, potentially drawing on the approach described by Gotleib et al. (Chap. 4). We intend to explore this issue further by carrying out a range of experiments. The following section gives the results of some very initial experiments.

7   Experimental Evaluation

This section reports the results of initial experiments used to evaluate the proposed mutation operators. We used three RoboChart models for the experiments: RoboVacuum (described previously), Drone and ForagingRobot. Further details regarding these models can be found on the RoboStar website.[5] The aim of the experiments was to assess the following:

---

[5] https://robostar.cs.york.ac.uk/.

Table 11.2 The
experimental subjects

Model	Number of states	Number of transitions
Drone	4	5
ForagingRobot	10	28
RoboVacuum	36	75

1. How many mutants are generated by each operator? This concerns both whether an operator is applicable (whether any mutants are produced) and also the scalability of the operators. An operator that leads to many mutants may be useful but suggests the need for techniques that either choose a subset of the mutants or prioritize them.
2. Whether the mutants tended to be valid. The issue here is that mutation operators can generate models that are not valid RoboChart models. Such mutants are of no value and their generation consumes resources.
3. Whether FDR was able to decide whether (valid) mutants represent faults and, if it was, what proportion led to the generation of test sequences. This essentially assesses the scalability of FDR and, when FDR was able to process mutants, the proportion of valid mutants that were not refining mutants and so could contribute to test-generation (refining mutants are of no value in test-generation).

Table 11.2 gives properties of the experimental subjects. One is relatively small (Drone), with the other two (ForagingRobot and RoboVacuum) being more complex models. The mutation operators used in the experiments are those described earlier. For completeness, the mutation operators are listed in Table 11.3.

We used Wodel to apply each mutation operator to each model. One of the parameters used with Wodel is the number of mutants that should be produced, with this value essentially providing an upper bound. This parameter was set to 100 in order to limit the time taken by the experiments. Table 11.4 shows the number of mutants produced in the experiments. Since we set the maximum number of mutants to 100, if 100 mutants were produced for a particular operator and model, then it is possible that more mutants would have been produced if a larger bound had been used. As one might expect, there is significant variation in the number of mutants produced, with many operators only producing a few mutants for each model. The operators mTransSource and mTransTarget, which change the source and target states of a transition, produced relatively large numbers of mutants for each model. This is to be expected since, for models with n states and t transitions, the number of such mutants is of $O(nt)$.

Once mutants had been generated, the next stage was to determine which were valid RoboChart models. This was achieved by importing the mutants into RoboTool, which carries out two validity checks. First, RoboTool checks whether the model satisfies the RoboChart metamodel. If the mutant does satisfy the metamodel, then RoboTool then checks whether the mutant satisfies the RoboChart

Table 11.3 Mutation operators

Class of operator	Mutation operator
	mStActEnDu
	mStActEnEx
Mutations for types	rElemEnumeration
	mElemEnumeration
Mutations for expressions	mIntegerExp
	swapBinaryExpression
	mRelationalOperator
Mutations for actions and statements	rAssignment
	rSendEvent
	rSeqStatement
Mutations for timed primitives	rWait
	rStateClockExp
	rStateClockExp2
Mutations for state machines	rState
	mTransSource
	mTransTarget
	mTransTrigger
	rTran
	rTranAction
	rCondTrans
Mutations for modules, controllers and other elements	mConnectionAsyn
	mConnectionBidirec
	rStaMachController
	rEventController
	rPostCond

well-formedness conditions. Table 11.5 provides the results of these checks, giving numbers for the individual models and also the total numbers. In this table, a triple X, Y, Z represents the situation in which the corresponding mutation operator produced X mutants, a total of Y of these satisfied the metamodel and Z of these were well-formed. Only the first of the three values is given in the case where no mutants were produced.

As can be seen, there was significant variation in the results. For example, there are several operators, such as mElemEnumeration and mIntegerExp, where all mutants generated were valid. In contrast there are three operators (rElemEnumeration, rState and rEventController) where no mutants were valid. It may be that the mutation operators that returned relatively few valid mutants will not be useful in practice, since only valid mutants can for the basis of test-generation. However, currently it is difficult to draw strong conclusions since there is potential for a mutation operator to lead to only a small number of valid mutants but for these mutants to lead to particularly useful tests. In addition, there is a need for experiments with

Table 11.4 Number of mutants

Mutation operator	RoboVacuum	Drone	ForagingRobot
mStActEnDu	20	2	6
mStActEnEx	20	2	6
rElemEnumeration	6	0	0
mElemEnumeration	100	0	0
mIntegerExp	100	9	91
swapBinaryExpression	99	0	20
mRelationalOperator	100	0	100
rAssignment	80	0	7
rSendEvent	54	4	0
rSeqStatement	100	18	3
rWait	5	3	1
rStateClockExp	6	0	0
rStateClockExp2	6	0	0
rState	36	4	10
mTransSource	100	12	100
mTransTarget	100	15	100
mTransTrigger	100	18	54
rTran	75	4	18
rTranAction	3	2	1
rCondTrans	41	0	7
mConnectionAsyn	29	10	8
mConnectionBidirec	29	10	8
rStaMachController	13	1	3
rEventController	2	1	4
rPostCond	15	0	3

additional examples. There is also a need for additional experiments that explore how mutants relate to real faults and what the mutation score tells us about the expected effectiveness of a test suite.

Interestingly, there is noticeable variation in results between the models. As one might expect, the larger models tended to lead to more mutants. However, even if one focuses only on the two larger models (RoboVacuum and ForagingRobot), there are some noticeable differences. Some of these differences are simply the result of the features used in defining the models. For example, the two operators that mutate clock expressions only produced mutants for RoboVacuum, and there were far more rSeqStatement mutants for RoboVacuum (100) than for ForagingRobot (3). Other differences do not have such a clear explanation. For example, all of the 29 mConnectionBidirc mutants for RoboVacuum were valid but all of the 8 mConnectionBidirc mutants for ForagingRobot were invalid. There is a need for additional experiments, with new case studies, to look for general patterns.

Table 11.5 Mutant numbers for each model and in total, giving the number of mutants generated, then numbers that satisfy the metamodel and finally numbers that are also well-formed

Mutation operator	RoboVacuum	Drone	ForagingRobot	Total
mStActEnDu	20, 20, 20	2, 2, 2	6, 6, 3	30, 30, 27
mStActEnEx	20, 20, 19	2, 2, 2	6, 6, 6	30, 30, 29
rElemEnumeration	6, 0, 0	0	0	6, 0, 0
mElemEnumeration	100, 100, 100	0	0	100, 100, 100
mIntegerExp	100, 100, 100	9, 9, 9	91, 91, 91	226, 226, 226
swapBinaryExpression	99, 87, 87	0	20, 20, 20	121, 109, 109
mRelationalOperator	100, 42, 41	0	100, 38, 37	220, 85, 82
rAssignment	80, 56, 56	0	7, 4, 4	87, 60, 60
rSendEvent	54, 40, 40	4, 3, 3	0	58, 43, 43
rSeqStatement	100, 39, 39	18, 10, 10	3, 0, 0	124, 49, 49
rWait	5, 1, 1	3, 1, 1	1, 0, 0	10, 2, 2
rStateClockExp	6, 6, 6	0	0	6, 6, 6
rStateClockExp2	6, 6, 6	0	0	6, 6, 6
rState	36, 0, 0	4, 0, 0	10, 0, 0	52, 0, 0
mTransSource	100, 19, 19	12, 12, 12	100, 42, 37	214, 75, 70
mTransTarget	100, 15, 15	15, 15, 15	100, 34, 34	218, 67, 67
mTransTrigger	100, 24, 23	18, 18, 6	54, 0, 0	175, 45, 31
rTran	75, 75, 74	4, 4, 4	18, 18, 3	99, 99, 83
rTranAction	3, 3, 3	2, 2, 2	1, 1, 1	6, 6, 6
rCondTrans	41, 41, 41	0	7, 7, 7	49, 49, 49
mConnectionAsyn	29, 29, 28	10, 10, 5	8, 8, 4	49, 49, 38
mConnectionBidirec	29, 29, 29	10, 10, 0	8, 8, 0	49, 49, 29
rStaMachController	13, 0, 0	1, 0, 0	3, 1, 1	18, 1, 1
rEventController	2, 0, 0	1, 0, 0	4, 0, 0	7, 0, 0
rPostCond	15, 15, 15	0	3, 3, 3	18, 18, 18

We also used FDR to explore the nature of the mutants and, specifically, how many refine the specification and how many do not (are faulty) and so lead to the generation of test cases. As a result of practical constraints, we only ran these experiments with the Drone example, although it should be noted that preliminary tests with a larger model (RoboVacuum) suggested that more work is required in order to make FDR scale. The results of this experiment can be found in Table 11.6. It was found that the majority of mutants represented faults and so led to the generation of test cases. Further experiments might explore how the mutants relate. For example, there may be cases where test cases generated using one mutation operator tend to kill the (non-refining) mutants generated by another.

Table 11.6 Refining and incorrect mutant numbers for the Drone. Also it includes numbers of mutants that satisfy the metamodel and numbers that are also well-formed

Mutation operator	No mutants	Metamodel	Well-formed	Refinements	Faulty mutants
mStActEnDu	2	2	2	0	2
mStActEnEx	2	2	2	1	1
rElemEnumeration	0	0	0	0	0
mElemEnumeration	0	0	0	0	0
mIntegerExp	9	9	9	0	9
swapBinaryExpression	0	0	0	0	0
mRelationalOperator	0	0	0	0	0
rAssignment	0	0	0	0	0
rSendEvent	4	3	3	0	3
rSeqStatement	18	10	10	1	9
rWait removes	3	1	1	1	0
rStateClockExp	0	0	0	0	0
rStateClockExp2	0	0	0	0	0
rState	4	0	0	0	0
mTransSource	12	12	12	6	6
mTransTarget	15	15	15	0	15
mTransTrigger	18	18	6	0	6
rTran	4	4	4	4	0
rTranAction	2	2	2	0	2
rCondTrans	0	0	0	0	0
mConnectionAsyn	10	10	5	5	0
mConnectionBidirec	10	10	0	0	0
rStaMachController	1	0	0	0	0
rEventController	1	0	0	0	0
rPostCond	0	0	0	0	0
total	115	98	71	18	53

8 Conclusions

The increasing importance of robotic systems means that we need cost-effective testing techniques that can be used by practitioners. Such techniques should be automated as much as possible and ideally should also provide guarantees regarding test effectiveness.

This chapter has described a mutation testing approach to drive automated test-generation. The approach bases test-generation on a RoboChart model S of how the system should behave, with mutants of S being generated through fault seeding. Mutants represent possible faulty implementations, and the generation of mutants has been implemented in the Wodel tool, with this providing flexibility and the potential to work with other formalisms and languages.

Given a mutant M of S, we use RoboTool to map S and M to CSP models $C(S)$ and $C(M)$ respectively, with these CSP models providing semantics of the corresponding RoboChart models. The final step involves using FDR to check whether $C(M)$ is a valid implementation of $C(S)$; if this is the case, then we can discard M (it does not model a fault), and otherwise, FDR might return a counter-example that defines a test case. For larger models, FDR might not have sufficient resources to complete the comparison, and so there is a need to explore alternatives that scale further.

The overall approach has a number of benefits, in addition to automation. The test cases have guaranteed fault detection capability (they find the faults seeded in mutation). In addition, flexibility is provided by the potential to add new mutation operators. Where FDR scales, it can be used to eliminate mutants that are correct implementations (the refining mutant problem).

In addition, the use of mutation at a high level of abstraction should mean that the number of mutants is relatively small, when compared to approaches that mutate the code. As a result, we hope that the approach will scale much better than traditional code-based mutation.

There are several lines of future work. First, there would be value in exploring the nature of actual faults in robotic systems, and such a study might lead to new mutation operators being defined. Second, we need additional case studies and experiments. Such experiments might explore, for example, the nature of the mutation operators and how they relate. The initial focus of the work was on relatively simple state machines, and there is a need to include additional features such as probability and models that have a mixture of discrete and continuous values (hybrid systems). Finally, we will require alternative test-generation techniques for cases where FDR is unsuitable.

Acknowledgements This work has been supported by EPSRC grant EP/R025134/2 RoboTest: Systematic Model-Based Testing and Simulation of Mobile Autonomous Robots, Spanish MINECO-FEDER grant FAME RTI2018-093608-B-C31 and the Comunidad de Madrid project FORTE-CM S2018/TCS-4314.

References

1. H. Agrawal, R.A. DeMillo, B. Hathaway, W. Hsu, W. Hsu, E.W. Krauser, R.J. Martin, A.P. Mathur, E. Spafford, Design of mutant operators for the C programming language. Technical report, Purdue University, 1989
2. P. Ammann, J. Offutt, *Introduction to Software Testing* (Cambridge University Press, Cambridge, 2008)
3. P. Ammann, M.E. Delamaro, J. Offutt, Establishing theoretical minimal sets of mutants, in *Verification and Validation 2014 IEEE Seventh International Conference on Software Testing* (2014), pp. 21–30 ISSN: 2159–4848

4. J.H. Andrews, L.C. Briand, Y. Labiche, Is mutation an appropriate tool for testing experiments? in *27th International Conference on Software Engineering (ICSE 2005), 15–21 May 2005, St. Louis, Missouri*, ed. by G.-C. Roman, W.G. Griswold, B. Nuseibeh (ACM, New York, 2005), pp. 402–411

5. J.S. Bradbury, J.R. Cordy, J. Dingel, Mutation operators for concurrent Java (J2SE 5.0), in *Second Workshop on Mutation Analysis (Mutation 2006 - ISSRE Workshops 2006)* (2006), pp. 83–92

6. T.A. Budd, D. Angluin, Two notions of correctness and their relation to testing. Acta Inform. **18**(1), 31–45 (1982)

7. P.C. Cañizares, A. Núñez, M.G. Merayo, Mutomvo: mutation testing framework for simulated cloud and HPC environments. J. Syst. Softw. **143**, 187–207 (2018)

8. A. Cavalcanti, J. Baxter, R.M. Hierons, R. Lefticaru, Testing robots using CSP, in *Proceedings of Tests and Proofs - 13th International Conference, TAP 2019, Held as Part of the Third World Congress on Formal Methods 2019, Porto, 9–11 Oct 2019*, ed. by D. Beyer, C. Keller. Lecture Notes in Computer Science, vol. 11823 (Springer, Berlin, 2019), pp. 21–38

9. T.T. Chekam, M. Papadakis, Y.L. Traon, M. Harman, An empirical study on mutation, statement and branch coverage fault revelation that avoids the unreliable clean program assumption, in *Proceedings of the 39th International Conference on Software Engineering, ICSE 2017, Buenos Aires, 20–28 May 2017*, ed. by S. Uchitel, A. Orso, M.P. Robillard (IEEE/ACM, New York, 2017), pp. 597–608

10. J.A. Clark, H. Dan, R.M. Hierons, Semantic mutation testing. Sci. Comput. Program. **78**(4), 345–363 (2013)

11. B. Darolti, Software engineering for robotics: an autonomous robotic vacuum cleaner for solar panels. Master's thesis, University of York, 2019

12. M.E. Delamaro, J.C. Maldonado, Proteum-a tool for the assessment of test adequacy for C programs, in *Proceedings of the Conference on Performability in Computing Systems (PCS'96)*, New Brunswick, NJ, July 1996, pp. 79–95

13. M.E. Delamaro, J. Offutt, P. Ammann, Designing deletion mutation operators, in *Verification and Validation 2014 IEEE Seventh International Conference on Software Testing* (2014), pp. 11–20

14. P. Delgado-Pérez, I. Medina-Bulo, J.J. Domínguez-Jiménez, A García-Domínguez, F. Palomo-Lozano, Class mutation operators for C++ object-oriented systems. Ann. Télécomm. **70**(3–4), 137–148 (2015)

15. P. Delgado-Pérez, S. Segura, I. Medina-Bulo, Assessment of C++ object-oriented mutation operators: a selective mutation approach. Softw. Test. Verif. Reliab. **27**(4–5), e1630 (2017)

16. R.A. DeMillo, A.J. Offutt, Constraint-based automatic test data generation. IEEE Trans. Softw. Eng. **17**(9), 900–910 (1991)

17. R.A. DeMillo, R.J. Lipton, F.G. Sayward, Hints on test data selection: help for the practicing programmer. Computer **11**(4), 34–41 (1978)

18. R.A. DeMillo, D.S. Guindi, W.M. McCracken, A.J. Offutt, K.N. King, An extended overview of the Mothra software testing environment, in *[1988] Proceedings. Second Workshop on Software Testing, Verification, and Analysis*, July 1988, pp. 142–151

19. L. Deng, J. Offutt, N. Li, Empirical evaluation of the statement deletion mutation operator, in *Sixth IEEE International Conference on Software Testing, Verification and Validation, ICST 2013, Luxembourg, 18–22 March 2013* (IEEE Computer Society, Washington, 2013), pp. 84–93

20. A. Estero-Botaro, F. Palomo-Lozano, I. Medina-Bulo, Quantitative evaluation of mutation operators for WS-BPEL compositions, in *Proceedings of the 3rd International Conference on Software Testing, Verification, and Validation Workshops (ICSTW'10)* (2010), pp. 142–150

21. S.C.P.F. Fabbri, M.E. Delamaro, J.C. Maldonado, P.C. Masiero, Mutation analysis testing for finite state machines, in *Proceedings of the 5th International Symposium on Software Reliability Engineering, Monterey, CA, 6–9 Nov 1994*, pp. 220–229

22. S.C.P.F. Fabbri, J.C. Maldonado, P.C. Masiero, M.E. Delamaro, W.E. Wong, Mutation testing applied to validate specifications based on Petri Nets, in *Proceedings of the IFIP TC6 8th International Conference on Formal Description Techniques VIII* (1995), pp. 329–337

23. S.C.P.F. Fabbri, J.C. Maldonado, T. Sugeta, P.C. Masiero, Mutation testing applied to validate specifications based on Statecharts, in *Proceedings of the 10th International Symposium on Software Reliability Engineering (ISSRE'99), Boca Raton, FL*, 1–4 Nov 1999, pp. 210

24. G. Fraser, A. Zeller, Mutation-driven generation of unit tests and oracles. IEEE Trans. Softw. Eng. **38**(2), 278–292 (2012)

25. T. Gibson-Robinson, P. Armstrong, A. Boulgakov, A.W. Roscoe, FDR3 - a modern refinement checker for CSP, in *Tools and Algorithms for the Construction and Analysis of Systems* (2014), pp. 187–201

26. M. Gligoric, L. Zhang, C. Pereira, G. Pokam, Selective mutation testing for concurrent code, in *Proceedings of the 2013 International Symposium on Software Testing and Analysis, ISSTA 2013* (Association for Computing Machinery, New York, 2013), pp. 224–234

27. P. Gómez-Abajo, E. Guerra, J. de Lara, A domain-specific language for model mutation and its application to the automated generation of exercises. Comput. Lang. Syst. Struct. **49**, 152–173 (2017)

28. P. Gómez-Abajo, E. Guerra, J. de Lara, M.G. Merayo, A tool for domain-independent model mutation. Sci. Comput. Program. **163**, 85–92 (2018)

29. R. Gopinath, C. Jensen, A. Groce, Mutations: How close are they to real faults? in *25th IEEE International Symposium on Software Reliability Engineering, ISSRE 2014, Naples, 3–6 Nov 2014* (IEEE Computer Society, Washington, 2014), pp. 189–200

30. R. Gopinath, A. Alipour, I. Ahmed, C. Jensen, A. Groce, How hard does mutation analysis have to be, anyway? in *2015 IEEE 26th International Symposium on Software Reliability Engineering (ISSRE)* (2015)

31. M. Harman, Y. Jia, W.B. Langdon, Strong higher order mutation-based test data generation, in *Proceedings of the 19th ACM SIGSOFT symposium and the 13th European conference on Foundations of software engineering (ESEC/FSE'11)* (Association for Computing Machinery, New York, 2011), pp. 212–222

32. M. Harman, S.A. Mansouri, Y. Zhang, Search-based software engineering: trends, techniques and applications. ACM Comput. Surv. **45**(1), 11:1–11:61 (2012)

33. Y. Jia, M. Harman, Constructing subtle faults using higher order mutation testing, in *Eighth IEEE International Working Conference on Source Code Analysis and Manipulation (SCAM 2008), Beijing*, 28–29 Sept 2008, pp. 249–258

34. Y. Jia, M. Harman, Milu: a customizable, runtime-optimized higher order mutation testing tool for the full C language, in *Proceedings of the 3rd Testing: Academic and Industrial Conference Practice and Research Techniques (TAIC PART'08), Windsor*, 29–31 Aug 2008, pp. 94–98

35. S.-W. Kim, J.A. Clark, J.A. McDermid, Investigating the effectiveness of object-oriented testing strategies using the mutation method. Softw. Test. Verif. Reliab. **11**(3), 207–225 (2001)

36. K.N. King, A.J. Offutt, A Fortran language system for mutation-based software testing. Softw. Pract. Exper. **21**(7), 685–718 (1991)

37. M. Kintis, M. Papadakis, Y. Jia, N. Malevris, Y. Le Traon, M. Harman, Detecting trivial mutant equivalences via compiler optimisations. IEEE Trans. Softw. Eng. **44**(4), 308–333 (2018)

38. E.W. Krauser, A.P. Mathur, V. Rego, High performance software testing on SIMD machines. IEEE Trans. Softw. Eng. **17**(5), 403–423 (1991)

39. W. Krenn, R. Schlick, S. Tiran, B. Aichernig, E. Jobstl, H. Brandl, MoMut::UML model-based mutation testing for UML, in *2015 IEEE 8th International Conference on Software Testing, Verification and Validation (ICST)*, April 2015, pp. 1–8

40. M. Kusano, C. Wang, Ccmutator: a mutation generator for concurrency constructs in multithreaded C/C++ applications, in *Proceedings of the 28th IEEE/ACM International Conference on Automated Software Engineering, ASE'13* (IEEE Press, Piscataway, 2013), pp. 722–725

41. M.Z. Kwiatkowska, G. Norman, D. Parker, PRISM 4.0: Verification of probabilistic real-time systems, in *Proceedings of Computer Aided Verification - 23rd International Conference, CAV 2011, Snowbird, UT, USA, July 14–20, 2011*, ed. by G. Gopalakrishnan, S. Qadeer. Lecture Notes in Computer Science, vol. 6806 (Springer, Berlin, 2011), pp. 585–591

42. S.C. Lee, J. Offutt, Generating test cases for XML-based web component interactions using mutation analysis, in *12th International Symposium on Software Reliability Engineering (ISSRE 2001), Hong Kong*, 27–30 Nov 2001, pp. 200–209

43. M. Linares-Vásquez, G. Bavota, M. Tufano, K. Moran, M. Di Penta, C. Vendome, C. Bernal-Cárdenas, D. Poshyvanyk, Enabling mutation testing for Android apps, in *Proceedings of the 2017 11th Joint Meeting on Foundations of Software Engineering, ESEC/FSE 2017* (Association for Computing Machinery, New York, 2017), pp. 233–244

44. T. Loise, X. Devroey, G. Perrouin, M. Papadakis, P. Heymans, Towards security-aware mutation testing, in *Proceedings of the 10th International Conference on Software Testing, Verification and Validation Workshops (ICSTW'17)* (2017), pp. 97–102

45. Y. Lou, D. Hao, L. Zhang, Mutation-based test-case prioritization in software evolution, in *2015 IEEE 26th International Symposium on Software Reliability Engineering (ISSRE)* (2015), pp. 46–57

46. Y.-S. Ma, Y.R. Kwon, J. Offutt, Inter-class mutation operators for Java, in *13th International Symposium on Software Reliability Engineering (ISSRE 2002), Annapolis, MD*, 12–15 Nov 2002, pp. 352–366

47. Y.-S. Ma, A.J. Offutt, Y.-R. Kwon, Mujava: a mutation system for Java, in *Proceedings of the 28th international Conference on Software Engineering (ICSE '06)*, Shanghai, 20–28 May 2006, pp. 827–830

48. L. Madeyski, W. Orzeszyna, R. Torkar, M. Józala, Overcoming the equivalent mutant problem: a systematic literature review and a comparative experiment of second order mutation. IEEE Trans. Softw. Eng. **40**(1), 23–42 (2014)

49. P.R. Mateo, M.P. Usaola, J.L.F. Alemán, Validating second-order mutation at system level. IEEE Trans. Softw. Eng. **39**(4), 570–587 (2013)

50. S. Mirshokraie, A. Mesbah, K. Pattabiraman, Guided mutation testing for JavaScript Web applications. IEEE Trans. Softw. Eng. **41**(5), 429–444 (2015)

51. A. Miyazawa, P. Ribeiro, W. Li, A. Cavalcanti, J. Timmis, J. Woodcock, RoboChart: modelling and verification of the functional behaviour of robotic applications. Softw. Syst. Model. **18**(5), 3097–3149 (2019)

52. S. Moon, Y. Kim, M. Kim, S. Yoo, Ask the mutants: mutating faulty programs for fault localization, in *Verification and Validation 2014 IEEE Seventh International Conference on Software Testing* (2014), pp. 153–162

53. T. Mouelhi, Y. Le Traon, B. Baudry, Mutation analysis for security tests qualification, in *Testing: Academic and Industrial Conference Practice and Research Techniques - MUTATION (TAICPART-MUTATION 2007)*, Sept 2007, pp. 233–242

54. A.S. Namin, J.H. Andrews, D.J. Murdoch, Sufficient mutation operators for measuring test effectiveness, in *Proceedings of the 30th International Conference on Software Engineering, ICSE '08* (Association for Computing Machinery, New York, 2008), pp. 351–360

55. A. Offutt, The coupling effect: fact or fiction. ACM SIGSOFT Softw. Eng. Notes **14**(8), 131–140 (1989)

56. J. Offutt, Investigations of the software testing coupling effect. ACM Trans. Softw. Eng. Methodol. **1**(1), 5–20 (1992)

57. A.J. Offutt, A. Lee, G. Rothermel, R.H. Untch, C. Zapf, An experimental determination of sufficient mutant operators. ACM Trans. Softw. Eng. Methodol. **5**(2), 99–118 (1996)

58. A.J. Offutt, J. Voas, J. Payne, Mutation operators for Ada. Technical report, Information and Software Systems Engineering, George Mason University, 1996

59. M. Papadakis, Y. Le Traon, Using mutants to locate "Unknown" faults, in *Verification and Validation 2012 IEEE Fifth International Conference on Software Testing* (2012), pp. 691–700

60. M. Papadakis, Y. Le Traon, Metallaxis-FL: mutation-based fault localization. Softw. Test. Verif. Reliab. **25**(5–7), 605–628 (2015)

61. M. Papadakis, N. Malevris, Automatic mutation test case generation via dynamic symbolic execution, in *2010 IEEE 21st International Symposium on Software Reliability Engineering* (2010), pp. 121–130

62. M. Papadakis, N. Malevris, Mutation based test case generation via a path selection strategy. Inf. Softw. Technol. **54**(9), 915–932 (2012)

63. A.V. Pizzoleto, F.C. Ferrari, J. Offutt, L. Fernandes, M. Ribeiro, A systematic literature review of techniques and metrics to reduce the cost of mutation testing. J. Syst. Softw. **157**, 110388 (2019)

64. M. Polo, M. Piattini, I. García-Rodríguez, Decreasing the cost of mutation testing with second-order mutants. Softw. Test. Verif. Reliab. **19**(2), 111–131 (2009)

65. R. Ramler, T. Wetzlmaier, C. Klammer, An empirical study on the application of mutation testing for a safety-critical industrial software system, in *Proceedings of the Symposium on Applied Computing, SAC 2017, Marrakech, April 3–7, 2017*, ed. by A. Seffah, B. Penzenstadler, C. Alves, X. Peng (ACM, New York, 2017), pp. 1401–1408

66. A.W. Roscoe, *Understanding Concurrent Systems*. Texts in Computer Science (Springer, Berlin, 2011)

67. G. Rothermel, R.H. Untch, C. Chu, M.J. Harrold, Prioritizing test cases for regression testing. IEEE Trans. Softw. Eng. **27**(10), 929–948 (2001)

68. D. Schuler, A. Zeller, Javalanche: efficient mutation testing for Java, in *Proceedings of the 7th joint meeting of the European Software Engineering Conference and the International Symposium on Foundations of Software Engineering, Amsterdam*, 24–28 Aug 2009, pp. 297–298

69. H. Shahriar, M. Zulkernine, Mutation-based testing of buffer overflow vulnerabilities, in *Proceedings of the 32nd Annual IEEE International Computer Software and Applications Conference, COMPSAC 2008, Turku*, 28 July–1 Aug 2008, pp. 979–984

70. D. Shin, S. Yoo, D.-H. Bae, A theoretical and empirical study of diversity-aware mutation adequacy criterion. IEEE Trans. Softw. Eng. **44**(10), 914–931 (2018)

71. D. Shin, S. Yoo, M. Papadakis, D.-H. Bae, Empirical evaluation of mutation-based test case prioritization techniques. Softw. Test. Verif. Reliab. **29**(1–2), e1695 (2019)

72. F.C. Souza, M. Papadakis, V. Durelli, M. Delamaro, Test data generation techniques for mutation testing: a systematic mapping (2014)

73. T. Srivatanakul, J.A. Clark, S. Stepney, F. Polack, Challenging formal specifications by mutation: a CSP security example, in *Proceedings of the 10th Asia-Pacific Software Engineering Conference (APSEC'03)* (2003), pp. 340–350

74. M. Trakhtenbrot, New mutations for evaluation of specification and implementation levels of adequacy in testing of Statecharts models, in *Proceedings of the 3rd Workshop on Mutation Analysis (MUTATION'07), Windsor*, 10–14 Sept 2007, pp. 151–160

75. J. Tuya, M.J. Suarez Cabal, C. de la Riva, SQLmutation: a tool to generate mutants of SQL database queries, in *Proceedings of the 2nd Workshop on Mutation Analysis (MUTATION'06), Raleigh, NC*, Nov 2006, p. 1

76. J. Tuya, M.J. Suarez Cabal, C. de la Riva, Mutating database queries. Inf. Softw. Technol. **49**(4), 398–417 (2007)

77. L. Zhang, S.-S. Hou, J.-J. Hu, T. Xie, H. Mei, Is operator-based mutant selection superior to random mutant selection? in *Proceedings of the 32nd ACM/IEEE International Conference on Software Engineering - Volume 1, ICSE '10* (Association for Computing Machinery, New York, 2010), pp. 435–444

78. L. Zhang, M. Gligoric, D. Marinov, S. Khurshid, Operator-based and random mutant selection: better together, in *2013 28th IEEE/ACM International Conference on Automated Software Engineering (ASE)* (2013), pp. 92–102

Chapter 12
Languages for Specifying Missions of Robotic Applications

Swaib Dragule, Sergio García Gonzalo, Thorsten Berger,
and Patrizio Pelliccione

Abstract Robot application development is gaining increasing attention both from
the research and industry communities. Robots are complex cyber-physical and
safety-critical systems with various dimensions of heterogeneity and variability.
They often integrate modules conceived by developers with different backgrounds.
Programming robotic applications typically requires programming and mathematical or robotic expertise from end-users. In the near future, multipurpose robots will
be used in the tasks of everyday life in environments such as our houses, hotels,
airports or museums. It would then be necessary to democratize the specification
of missions that robots should accomplish. In other words, the specification of
missions of robotic applications should be performed via easy-to-use and accessible
ways, and, at the same time, the specification should be accurate, unambiguous, and
precise. This chapter presents domain-specific languages (DSLs) for robot mission
specification, among others, profiling them as internal or external and also giving an
overview of their tooling support. The types of robots supported by the respective
languages and tools are mostly service mobile robots, including ground and flying
types.

S. Dragule
Department of Computer Science and Engineering, Chalmers University of Gothenburg,
Gothenburg, Sweden

Department of Computer Science, Makerere University, Kampala, Uganda
e-mail: dragule@chalmers.se

S. G. Gonzalo · T. Berger
Department of Computer Science and Engineering, Chalmers University of Gothenburg,
Gothenburg, Sweden
e-mail: sergio.garcia@gu.se; thorsten.berger@chalmers.se

P. Pelliccione (✉)
Gran Sasso Science Institute (GSSI), L'Aquila, Italy

Chalmers University of Technology, Gothenburg, Sweden
e-mail: patrizio.pelliccione@gssi.it

© Springer Nature Switzerland AG 2021
A. Cavalcanti et al. (eds.), *Software Engineering for Robotics*,
https://doi.org/10.1007/978-3-030-66494-7_12

1   Introduction

Inexpensive and reliable robot hardware—including ground robots, multicopters, and robotic arms—is becoming widely available, according to the H2020 Robotics Multi-Annual Roadmap (MAR).[1] As such, robots will soon be deployed in a large variety of contexts, leading to the presence of robots in everyday life activities in many domains, including manufacturing, healthcare, agriculture, civil, and logistics.

Robots are complex safety-critical cyber-physical systems that are subject to various dimensions of heterogeneity and variability [15, 39].

In addition, the robotics domain is divided into a large variety of sub-domains, including vertical ones (e.g. drivers, planning, navigation) and horizontal ones (e.g. defence, healthcare, logistics), with a vast amount of variability [7, 36], further complicating robotics software engineering. Due to this heterogeneity, a robot typically integrates modules conceived by developers with different backgrounds. For instance, electrical engineers design the robot's hardware, control engineers develop planning and control algorithms, and software engineers architect and quality-assure the software system. Coordinating the integration of all these modules from developers with different backgrounds is one of the major challenges that characterize the domain of robotics [15, 39]. Further challenges comprise identifying stable requirements, defining abstract models to cope with hardware and software heterogeneity, seamlessly transitioning from prototype testing and debugging to real systems, and deploying robotic applications in real-world environments.

A core activity when engineering robotics software is defining and implementing the robot's behavior. Specifically, in addition to building and integrating modules that define the lower-level behavior, the overall behavior of robots needs to be defined. This behavior, often called a *mission*, coordinates the lower-level behaviors that are typically defined in modules realizing the different skills. While this coordination has traditionally been implemented in plain code [47], this will not be feasible in the near future when multipurpose robots will be used in our houses, hotels, hospitals, and so on to accomplish tasks of the everyday life. For these reasons, the use of dedicated (domain-specific) languages is becoming increasingly popular [32, 81, 91]. These languages target end-users without robotic, ICT, or mathematical expertise and allow them to conveniently command and control robots. This trend is also expressed by the MAR roadmap, given the increasing involvement of robots in our society, especially service robots (i.e. robots that perform useful tasks for humans excluding industrial automation applications.[2]) In fact, the MAR roadmap describes DSLs [54, 81, 93], together with model-driven engineering [8, 14, 48, 83, 85], as core technologies required to achieve a separation of roles in the robotics domain while also improving, among others, modularity and system integration.

---

[1] https://eu-robotics.net/sparc/upload/about/files/H2020-Robotics-Multi-Annual-Roadmap-ICT-2016.pdf.

[2] https://www.iso.org/standard/55890.html.

The specification of mission ranges from (1) very intricate and difficult-to-use [3, 46] logical languages, such as Linear-Temporal Logic or Computation Tree Logic [30, 42, 64, 66, 96], whose instances are directly fed into planners; via (2) common notations for specifying behavior, such as Petri nets [94, 97] and state machines [12, 55, 88], which require low-level and step-by-step descriptions of missions; to (3) robotics-specific DSLs tailored to the robot at hand [25, 26, 35, 41, 69, 82], which often allow a more high-level mission specification.

This chapter contributes to the state of the art in mission specifications for robots. We present an overview of programming languages for robotic applications and respective IDEs (integrated development environments) in Sect. 2. Thereafter, we present DSLs for mission specification in Sect. 3, including internal and external DSLs, together with their tooling. In Sect. 4, we discuss how robots are usable in everyday life, with specific reference to the PROMISE tool for specifying missions for multi-robot applications. We put PROMISE into practice by describing a real mission with PROMISE we realized, together with the rest of the robotic software, including a multi-layer architecture. We conclude and discuss areas for future work in Sect. 6.

2   Programming Languages and IDEs for Robotic Applications

The software of a robotic application can be conceptually organized into two main parts: (1) the software controlling the various modules (written once and embedded into the robot) and (2) the software that permits the specification and execution of the mission (potentially changing from mission to mission, especially for multipurpose robots). Traditionally, these two parts are mixed for robots capable of doing specific tasks, where the mission specification only involves setting some parameters that are specific for the environment in which the mission will be executed. In this section, we briefly describe programming languages (Sect. 2.1) and IDEs (Sect. 2.2) used in robotics.

2.1   Programming Languages for Robotic Applications

Many different languages are used for the development of mobile robotic applications. Starting from the lowest level of abstraction, hardware-description languages (e.g. Verilog or VHDL) are mainly used by electronic engineers to "program" the low-level electronics of robots [70]. Hardware-description languages are commonly used to program field-programmable gate arrays [72], which are devices that make it possible to develop electronic hardware without having to produce a silicon chip.

At the level of microcontrollers, a widely used option is Arduino[3] [52]. It is an open-source electronics platform that consists of a board with assembled sensors (and potentially actuators) that can be controlled using specific software. Software for Arduino-based applications may be developed using an open-source IDE,[4] which supports the languages C and C++, applying a wrapper around programs written in these languages and using special rules of code structuring. The hardware manufacturers typically also provide proprietary software, such as RAPID[5] technical reference manual from ABB and KRL[6] reference guide from Kuka.

More powerful machines in terms of computation—including single-board computer solutions such as Raspberry Pi—support Ubuntu distributions, and, therefore, the Robot Operating System (ROS) [59]. ROS [75] is an open-source middleware offering a framework for structured communication among various robotic components using a peer-to-peer connection. ROS currently runs on Unix-based platforms, and the software for ROS is primarily tested on Ubuntu. Therefore, a typical setup for a roboticist includes a certain version of Ubuntu[7] with a certain distribution of ROS.[8]

Most packages and libraries of ROS are developed using either C++ or Python so those languages are the most commonly used. However, ROS's communication system is language-agnostic, which enables several languages such as C++, Python, Octave, Java, and LISP to be used depending on the user's proficiency. ROS also offers modularized tool-based microkernel design to aggregate various tools performing specific tasks such as navigating source code tree, get and set configuration parameters, and visualize the peer-to-peer connection topology, among others [44, 58].

ROS has evolved with a number of distributions, supporting more than 20 robotic systems,[9] including drones, arm robots, humanoids, and wheeled mobile-base robots. Among the robot-agnostic middleware, ROS is considered the de facto standard for robot application development [39], officially supporting more than 140 robots (including ground mobile robots, drones, cars, and humanoids) [28]. Examples of repositories from robotics companies that support the integration of ROS are the one from Kuka[10] or from Aldebaran and Softbank Robotics.[11]

---

[3] https://www.arduino.cc.

[4] https://arduino.en.softonic.com.

[5] https://library.e.abb.com/public/688894b98123f87bc1257cc50044e809/Technical%20reference%20manual_RAPID_3HAC16581-1_revJ_en.pdf.

[6] http://robot.zaab.org/wp-content/uploads/2014/04/KRL-Reference-Guide-v4_1.pdf.

[7] https://wiki.ubuntu.com/Releases.

[8] https://wiki.ros.org/Distributions.

[9] http://wiki.ros.org/Distributions.

[10] https://wiki.ros.org/kuka.

[11] https://wiki.ros.org/Aldebaran.

MATLAB (and its open-source relatives, such as Octave) is a popular option among engineers for analysing data and developing control systems. It has also been used for robotics software development [87], and there even exists a robotics-dedicated toolbox.[12] The toolbox contains tools that support functionalities ranging from producing advanced graphs to implementing control systems.

Machine learning is another technique applied in the context of robotics, as is being used in decision making and image recognition. Machine-learning models are first trained using platforms such as TensorFlow or PyTorch and then implemented as ROS nodes [57]. These training platforms provide dedicated APIs, and they are commonly Python or C++-based. Finally, image processing is a key functionality in robotics, and the most used library in this domain is OpenCV[13] [19], written in C++. Its primary interface is written in and uses C++, but there are bindings for Python, Java, and MATLAB.

2.2 IDEs for Developing Robotic Applications

IDEs aid software engineering by providing editing, compilation, interpretation, debugging, and related automation facilities. They often come with version-control, refactoring, visual-programming, and multi-language support. The usage of IDEs improves efficiency in software development and makes it less error-prone.

Working with general IDEs, such as Eclipse or Qt Creator, appears to be the most popular option among roboticists, despite the existence of a few free robotic-centered IDEs. For many IDEs, there are instructions for configuring toward robotics. For instance, the ROS community provides configurations for several IDEs, including Eclipse, Netbeans, KDevelop, Emacs, and RoboWare studio, a variant of Microsoft Visual Studio.

Eclipse, in particular with its tooling for model-driven software engineering (e.g. Eclipse Modeling Framework), has been used to realize DSLs and respective environments for building robotics applications in a model-driven way. For instance, Arias et al. [1] offer a complete robotics toolset upon Eclipse to support the engineering from design to code generation, called the ORCCAD model.

Similar to general-purpose IDEs, Robotics IDEs offer facilities for robotics software engineering, including code editors, robotics libraries, build tools, and quality-assurance tools (i.e. debuggers, test environments, and simulators). As opposed to general IDEs, robotics IDEs primarily target building robotic applications, without support for other domains.

Table 12.1 summarizes all the IDEs with details on target users, languages supported, and features that go beyond a general IDE. To illustrate one of the IDEs, Fig. 12.1 shows a screenshot of the Robot Mesh Studio. The user interface

---

[12]https://www.mathworks.com/products/robotics.html.

[13]https://opencv.org.

Table 12.1 List of dedicated robotic IDEs

Name	IDE details
RobotC [77, 80]	C-based educational environment providing two notations, RobotC Graphical, e.g. Fig. 12.5, and RobotC Natural Language, e.g. Listing 4
Robot Mesh Studio [76]	IDE for programming educational robots from Arduino, PICAXE, Parallax, and Raspberry Pi microcontrollers. It offers two graphical DSLs: Flowol, a flowchart-based language, and a Blockly-based language. Textual languages: C++, Python
VEX Coding Studio [18, 92]	A robot vendor's environment for programming educational robot kits. The IDE offers Scratch-based syntax (VEXcode Blocks) and a text-based syntax (VEXcode Text)
PICAXE [50, 71]	For programming educational PICAXE microcontroller-based robots. It offers the PICAXE language in three syntaxes: PICAXE BASIC—textual, PICAXE Blockly—graphical, and PICAXE Flowchart syntax
ROS Development Studio [79]	An online IDE with ready-to-use tools, such as simulators and AI-based libraries. The ROS Development Studio supports all robots compatible with ROS and a variety of languages, such as C++, Python, Java, MATLAB, and Lisp
Microsoft Robotic Developer Studio (MRDS) [45]	Microsoft product for hobbyist, academic, and commercial robot application developers. The IDE supports programming robot applications in Microsoft's Visual Programming Language (MVPL) and C#
MATLAB and Simulink [73]	IDE offers hardware-agnostic robot control for Arduino and Raspberry Pi microcontrollers that can be connected to ROS and ROS2. Code from a variety of embedded hardware, such as Field Programmable Gate Arrays (FPGAs), Programmable Logic Controllers (PLCs), and Graphics Processing Units (GPUs), can be generated to various target languages including C/C++, VHDL/Verilog, Structured Text, the PLC language, and Compute Unified Device Architecture (CUDA) language
Webots [67]	An open-source, online IDE simulator that supports a number of robots and a range of languages such as C, C++, Python, Java, MATLAB, and ROS-supported languages C, C++, Python, Java, and MATLAB
Robot Task Commander (RTC) [43]	The IDE is meant for automated task planning for robot(s) using one or more computing devices over a network. It supports humanoid robots programmed using Python scripting language and RTC visual programming language
The SmartMDSD Toolchain [86]	IDE for developing robot systems by providing building blocks that can be used for composing new systems from existing components. The IDE applies modeling techniques using tools such as Xtext, Xtend, and Sirius from Eclipse
BRICS Integrated Development Environment (BRIDE) [10]	IDE for developing editors in robotics based on model-driven engineering principles. BRIDE incorporates the OROCOS and ROS frameworks. The ROS version offers features such as graphical modeling of ROS nodes, code generation in C++ or Python, and generation of launch files

Table 12.1 Continued.

Name	IDE details
Universal Robotic Body Interface (URBI) [4]	Open-source IDE for programming robot controls, using client-server architecture. The server manages low-level hardware controls for sensors, camera, and speakers, and the client sends high-level behavior commands like "walk" to the server. Languages supported include C++, Urbiscript scripting language, MATLAB, Java, and Lisp
TeamBots [5, 58]	A Java-based environment for developing and executing control systems on teams of robots and on simulation using the application TBSim. The IDE provides a set of applications and packages for multi-agent mobile robots
Pyro [11]	An educational IDE that abstracts low-level details, making it suitable for students learning to program robots using the C++, Java, and Python. Pyro wraps Player/Stage and ARIA, for easy access to its users
CopeliaSim (VREP) [23]	A multi-robot IDE, which uses distributed control architecture to model objects through embedded script, a plugin, a ROS or BlueZero node, a remote API client, or a custom solution. The IDE supports programming using C/C++, Python, Java, Lua, MATLAB, or Octave

is separated into three main panes. Pane A shows a description of the current mission—rich text entered by the developer to describe and illustrate the mission (here, the visual recognition and lifting of an object by the robot). Pane B shows the actual mission expressed in an external DSL (with Blockly syntax) provided by the IDE, or alternatively the generated textual code. Pane C shows help text, or alternatively the interactive debugger or an overview on the current robot configuration.

3 Robot Mission Specification

As robots become an integral part of the everyday life, we need better ways to instruct robots on the tasks they should accomplish. Mission specification is a process that relies on a strategy and mechanism that determines the steps a robot takes when performing a given task [24, 27, 35, 90].

The specification of a robot mission is influenced by the range of tasks the robot can execute, the end-user of the robot, the number of robots involved, the physical environment in which the mission will be executed, and the programming languages provided by the robot manufacturer. Robots performing a specific task are normally pre-programmed by manufacturers, while those with the ability to do a number of tasks require frequent change of what they do depending on the need at a given time—calling for flexible ways of specifying missions.

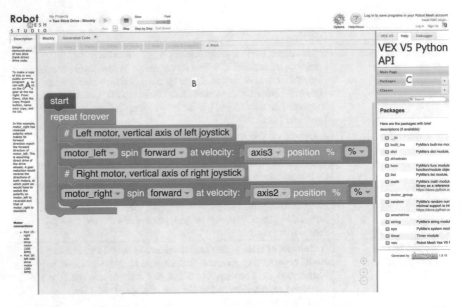

Fig. 12.1 Screenshot of the Robot Mesh Studio IDE [76] with three panes. Pane A provides a rich-text description of the mission, Pane B shows the actual mission expressed in a DSL with Blockly syntax, and Pane C shows help text or alternatively the debugger or an overview on the current robot configuration.

Any mission specified using a DSL should be easily understood by experts in that domain—e.g. logistics, commerce, and health. DSLs are recognized for their ability to abstract low-level details of robotic implementations and allowing users to specify their concerns from higher levels by using common terms in the domain. This abstraction further enhances effective communication of concepts with the domain experts. Due to these reasons, DSLs have been studied and proposed for mission specification by the community [25, 26, 35, 41, 69, 82].

DSLs typically work based on the underlying formalisms such as state machines, flowcharts, and behavior trees [21, 22, 40]. We assume that the reader has already some knowledge on state machines and flowcharts. Before presenting the selected DSLs, we give a brief introduction to behavior trees, which are less widely known.

A behavior tree is a hierarchical model in which the nodes of the tree are tasks to be executed [21, 22, 40]. Behavior trees emphasize modularity, coupled with two-way control transfer using function calls, unlike one-way (transitions) in finite-state machines. The modular character in behavior trees makes the reuse of behavior primitives feasible. Behavior trees have been applied in computer science, robotics, control systems, and video games.[14] Behavior trees consist of control-flow nodes (namely, *Parallel*, *Fallback*, *Decorator*, and *Sequence*) and executor nodes (i.e.

---

[14]http://wiki.ros.org/behavior_tree.

Table 12.2 List of DSLs, their notation used, and their styles (internal or external DSL)

Name of DSL	Notation	Style
Choregraphe	Visual	External
NaoText	Textual	External
Microsoft Visual Programming Language	Visual	External
EasyC	Textual, Visual	External
SMACH	Textual	Internal
Open Roberta	Visual	External
FLYAQ	Visual	External
Aseba	Textual, Visual	External
LEGO Mindstorms EV3	Visual	External
MissionLab	Visual	External
CABSL	Textual	Internal
BehaviorTree.CPP	Textual	External
ROS Behavior Tree	Textual	Internal
Unreal Engine 4 Behavior Trees	Textual	External
PROMISE	Textual, Visual	External

Action and *Condition*). An action node executes a task and returns success or failure, while the condition node tests if a certain condition is met.

In the following subsections, we describe a selection of internal and external DSLs for mission specification together with examples. Internal DSLs are extensions of a general-purpose (i.e. programming) language—often called host language. An external DSL is a language with independent syntax, semantics, and other related language resources and designed with notation and abstractions suitable to the user domain. Table 12.2 shows an overview of these DSLs with the notations supported (visual or textual) and style (internal or external DSL).

3.1 Internal DSLs

Internal DSLs follow the host language's syntax, and their execution is limited to the host language's infrastructure. They provide features specific to given end-user domains, such as robotics engineering, which simplify specification of domain user's concerns. In each of the following internal DSLs, we look into the host language, the developing organization (company), its semantics (compiled/interpreted), features specific to the internal DSL, and the end-user domain the language is targeting.

3.1.1 ROS Behavior Tree

ROS Behavior Tree [21] is an open-source C++ library for creating behavior trees. The DSL's aim is to be used by expert robot developers, who are conversant with the ROS framework and C++ or Python languages. Listing 1 shows sample code for creating a behavior tree.[15] It consists of header files and demonstrates how the action node and the condition node are executed in the behavior tree.

```
1  #include<actions/actiontestnode.h>
2  #include<conditions/conditiontestnode.h>
3  #include<behaviortree.h>
4  #include<iostream>
5  int main(int argc, char **argv){
6    ros::init(argc, argv, "BehaviorTree ");
7    try{
8      int Tick Period_millisecond s = 1000;
9      BT::ActionTestNode *action1 = new BT::ActionTestNode ("Action1");
10     BT::ConditionTestNode *condition1 = new BT::ConditionTestNode ("Condition1");
11     action1−>set_time(5);
12     BT::SequenceNodeWithMemory* sequence1= new BT::SequenceNodeWithMemory("seq1");
13
14     condition1−>set_boolean_value(true);
15     sequence1−>AddChild(condition1);
16     sequence1−>AddChild(action1);
17     Execute(sequence1, Tick Period_milliseconds);
18   } catch (BT::BehaviorTreeException& Exception){
19     std::cout<<Exception.what()<<std::endl;
20   }
21   return 0;
22 }
```

Listing 1 Creation of a new behavior tree using the ROS Behavior Tree DSL

Selector nodes are used to find and execute the first child that does not fail. A selector node immediately returns success or running when one of its children returns success or running. Sequence nodes are used to find and execute the first child that has not yet succeeded. A sequence node returns failure or running when one of its children returns failure or running. The parallel node ticks its children in parallel and returns success if $M \leq N$ children return success, it returns failure if $N - M + 1$ children return failure, and it returns running otherwise. The decorator node manipulates the return status of its child according to the policy defined by the user. Decorator Retry retries the execution of a node if this fails, and Decorator Negation inverts the Success/Failure outcome.

---

[15]https://github.com/miccol/ROS-Behavior-Tree.

3.1.2   SMACH

SMACH [12] is a non-commercial application programming interface written in Python, based on hierarchical concurrent state machines. It allows executions to be controlled by a higher-level task-planning system.

The library enables a quick way to create robust robot missions with maintainable and modular code. The DSL provides integration with ROS for developing robot applications using state machines. The actionlib library in SMACH provides an interface for tasks such as moving the base to a target location, performing a laser scan and returning the resulting point cloud, and detecting the handle of a door. SMACH Viewer is a graphical interface that shows a hierarchy of state machines, transitions between states, active states, and data passed between states. Once a state machine for a given mission is created, it is executed in the ROS environment.

Listing 2 demonstrates how to create a state machine, adding states to the state machine. In the state execution (line 10), "event" depicts the condition to execute outcome1 if true, outcome2 otherwise.

```
1  #!/usr/bin/env python
2  import rospy
3  import smach
4  # creating a state
5  class Foo(smach.State):
6      def __init__(self, outcomes=['outcome1', 'outcome2']):
7  # Your state initialization goes here
8      def execute(self, userdata):
9  # Your state execution goes here
10         if event:
11             return 'outcome1'
12         else:
13             return 'outcome2'
14 # Adding states
15  sm = smach.StateMachine(outcomes=['outcome4','outcome5'])
16    with sm:
17       smach.StateMachine.add('FOO', Foo(),
18                    transitions={'outcome1':'BAR',
19                         'outcome2':'outcome4'})
20       smach.StateMachine.add('BAR', Bar(),
21                    transitions={'outcome2':'FOO'}) \
```

Listing 2   Creation of a state and adding the state to a state machine in SMACH

3.1.3   C-Based Agent Behavior Specification Language

The C-based agent behavior specification language (CABSL) [78] enables the description of robot behaviors as a hierarchy of finite state machines. The control program executes behaviors based on the acquired sensor data, which maps the sensor data to actions the robot executes. In a state when an action is taken, either the state generates an output or calls another state machine. Otherwise, there is a transition from one state to another state.

An active graph in CABSL is a tree consisting of a set of state machines being executed. Each state machine can call any other state machine. The language is implemented and compiled in C++.[16] CABSL does not provide the functionality of replacing the behavior on the fly in case the acquired sensor data requires a change in behavior.

The DSL's textual notation makes it suitable for developers with experience in using the C language. It has reportedly been used for the NAO robot.

3.2 External DSLs

External DSLs have no dependence on the resources of another language. In profiling the external DSLs, the following information is considered: the developing organization (company), its semantics (compiled/interpreted), notation, features specific to the external DSL, type of robots the DSL supports, and the domain the language is targeting.

3.2.1 NaoText

NaoText [41] is an external DSL developed by the research group QualiTune. The DSL is a role-based language for specifying collaborative missions for NAO robots using a textual notation. NaoText uses CPSTextInterpreter, which runs on the Java runtime environment using Maven to manage dependencies.[17]

The code below shows the declaration of a pass action in a soccer game between NAO robots.[18] Some of the domain terms used in specifying the mission in Listing 3 include striker, ballpossesor, and ballseeker.

```
1  activate for {       // (1) player selection
2      BallPossessor p;
3      BallSeeker s;
4  } when {          // (2) condition
5      ((p.robotInVision(s)) and
6      !(p as Striker).isGoalShotPossible());
7  }
8  with bindings {     // (3) role binding
9      p + Sender;    //   bind Sender role
10     s + Receiver;  //   bind Receiver role
11     s − BallSeeker; //  unbind BallSeeker role
12  } with settings {  // (4) evaluation time settings
```

---

[16]https://github.com/bhuman/CABSL.

[17]https://github.com/max-leuthaeuser/CPSTextInterpreter.

[18]http://www.qualitune.org/?page_id=453.

```
13    interval 500;   // check every 500ms
14    after 1000;    // start after 1s
15    continuously true;
16  }
```

Listing 3 A snippet of a mission for pass action in a soccer game between NAO robots

3.2.2 EasyC

EasyC is a commercial product of Intelitek that provides a graphical notation for programming VEX robots. The DSL auto-generates C code from missions specified using the drag and drop graphical editor. Experienced C programmers can seamlessly switch to a fully text-based development environment. This DSL has been enriched with robotic abstractions such as robot driving—Drive, Turn, Stop, or Drive for Time.

EasyC uses a graphical interface on top of Intelitek's own C library,[19] which was custom made for the VEX Cortex and IQ robot controllers. Figure 12.2 is a screenshot of the DSL showing abstractions of the language concepts and a sample mission specified.

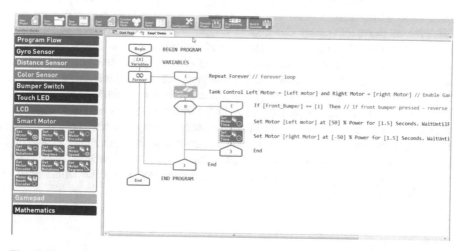

Fig. 12.2 A screenshot showing EasyC DSL: the first frame presents the language feature abstractions and the second frame is the active mission being specified.

---

[19] https://www.slideserve.com/tova/april-27-2006-programming-with-easyc-and-wpilib.

3.2.3 BehaviorTree.CPP

BehaviorTree.CPP [29] is a C++ library for creating behavior trees. It is developed by a research group at the Eurecat Technology Centre. The library provides a flexible framework to easily specify robot mission as behavior trees that can be loaded at runtime for execution. The nodes of the tree are either actions the robot can execute or conditions to be fulfilled before an action is taken.

The BehaviorTree.CPP DSL provides mechanisms to monitor, log, and debug the execution of a tree. The behavior trees for robot missions are executed using the C++ language runtime environment. Groot[20] provides a graphical editor for the C++ library to create and edit behavior trees. The primitives in Groot, built-in nodes, or custom nodes can be dragged and dropped to build a required behavior tree. Domain terms and expressions such as DetectObject, Grasp, GetMapLocation, and MoveTo have been used in the DSL for mobile robots with the ability to move, recognize, and grasp objects.

3.2.4 Unreal Engine 4 Behavior Trees

The Unreal Engine 4 (UE4) Behavior Tree [6] is a commercial DSL developed and maintained by Epic Games, Inc. Behavior trees define the Unreal AI agent's processor, which makes decisions and executes various branches based on the outcome of those decisions. The Unreal Engine implements behavior trees using the Blackboard tool which acts as the "brain" of the AI character and stores key values that the behavior tree uses to make its decisions. A behavior tree task is an action the AI character can perform, for instance, move to a location or rotate to face an object.[21] Some examples of domain expressions are SetMovementSpeed, LookStraightAhead, and RapidMoveTo. The DSL has been used for simulation characters in video games, representing humans, helicopters, and vehicles. The Unreal Engine uses the Unreal scripting languages with a graphical editor for creating UE4 behavior trees, related blackboards for the behavior trees, and tasks—i.e. actions. The scripting languages are compiled using the UnrealScript Compiler.[22]

3.2.5 Choregraphe

Choregraphe [68, 74] is a commercial DSL produced and maintained by SoftBank Robotics for programming Aldebaran robots such as NAO. The language aids users

---

[20]https://github.com/BehaviorTree/Groot.

[21]https://docs.unrealengine.com/en-US/Engine/ArtificialIntelligence/BehaviorTrees/BehaviorTreesOverview/index.html.

[22]https://docs.unrealengine.com/udk/Three/UnrealScriptReference.html.

to create animations, behaviors, and dialogues for the NAO humanoid robot—meant for experimentation and research. Choregraphe also offers simulation support for the NAO robot. The graphical DSL provides a flowchart-like interface in which end-users specify missions by connecting boxes to construct a behavior for the robot.[23]
  Boxes are pre-programmed libraries, which abstract mission primitives. Some of the mission primitives include Play Sound, Set Speech Lang, and Speech Reco.

3.2.6   Microsoft Visual Programming Language

The Microsoft Visual Programming Language (MVPL) [45] is a DSL in Microsoft Robot Development Studio used for programming robotic applications based on the idea of boxes and arrows. The language concepts (activities) are represented by boxes, while the arrows connect the boxes to build a program.[24] The MVPL data flow diagram consists of a connected sequence of activities represented as blocks with inputs and outputs that can be connected to other activity blocks. Activities can represent data flow control or processing functions, or user-defined activities, which the user creates in MVPL.

3.2.7   Open Roberta

Open Roberta [33, 51, 53] is a web-based educational DSL developed by the Fraunhofer Institute, which offers free use for individuals, but commercial use for institutional use. The Blockly-based DSL can be used to program a variety of robots: Lego Mindstorms EV3 and NXT, Calliope mini, micro:bit, Bot'n Roll, NAO, and BOB3. This DSL provides a rich set of behavior abstractions and primitives, which are mainly categorized into actions (drive, turn, steer, show, play, say), sensors (touch, ultrasonic, colour, infrared, temperature, gyro, timer), control (program control flows), logic (comparisons, AND, OR, Boolean), math (constants and arithmetic operators), text, colour, and variables. The specifics of these abstractions vary according to the robot for which the mission is specified. This DSL has a considerable potential in harnessing end-user programming, since it is a drag and drop graphical language with syntactic and semantic editor services. The DSL can either be run on the cloud or installed on a local server. Open Roberta generates code in Python, Java, JavaScript, and C/C++ depending on the target robot.

---

[23] http://doc.aldebaran.com/1-14/software/choregraphe/interface.html.

[24] https://acodez.in/microsoft-robotics-developer-studio/.

Fig. 12.3 Specifying a mission for a drone to hover in given space while avoiding no-fly zones [25].

3.2.8   FLYAQ

FLYAQ [13, 24, 27, 31] is an open-source research prototype developed and maintained by a team of researchers that provides an extensible DSL for specifying missions for a variety of robots, including quadrotors. The monitoring mission language (MML) allows specification of mission context such as obstacles, flight path (i.e. starting point, action points, ending point), and no-fly zones on a live map. The executable code is automatically generated to be executed by a robot or a swarm of robots as shown in Fig. 12.3. The DSL is suitable for missions such as surveillance, public order management, and agriculture. The concrete syntax (i.e. the map) used for specifying the mission context makes the language reachable for end-users. The FLYAQ virtual machine provides a ready-to-run version for the end-user.

3.2.9   Aseba

Aseba [62, 89] is a DSL with variants of visual and text syntaxes, created by Mobsya under Creative Commons Attribution-ShareAlike 3.0 License. The language syntax variants are the visual programming language (VPL), a Blockly-based language, a Scratch-based language, and the Aseba textual language. VPL provides events and action-based programming in which, for a given event, there is a corresponding action. These events are triggered by data from sensor readings. Examples of events are press button, obstacle detector, ground detector, robot tapped, and hand clap. In

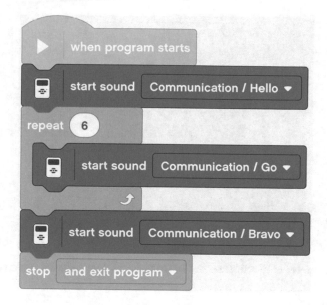

Fig. 12.4 Specifying a loop in LEGO Mindstorms EV3.

turn, examples of actions are set motor speed, set top or bottom colour, and play music. The same language concepts can be programmed using the other DSLs of Aseba. Some common behaviors[25] associated with the Thymio robot are friendly (follow hand and react to another Thymio robot), explore (avoid obstacles and stop when the ground is dark), fearful (goes away when approached and scream when cornered), attentive (changes colour and moves depending on the number of claps detected), investigator (follows a black track), and obedience (reacts to button and remote control).

3.2.10 LEGO Mindstorms EV3

The LEGO Mindstorms EV3 builder [16, 60] makes it possible to create robots that can do a number of things such as walk, talk, or drive. The graphical DSL provides a rich set of language constructs categorized into action, flow, sensor, data, and advanced blocks. For instance, the action blocks include move steering block, display block, and sound block, which can be used for specifying a mission by kids learning how to program. The DSL is a visual language with blocks connected to form missions. Figure 12.4 shows a mission specification in which the robot says "Hello" once, then "Go" six times, and then "Bravo" once. The sound blocks are used for creating the respective sounds while the loop is used for repeating the "Go" sound. Each block is an icon of the function it executes.

---

[25]https://www.thymio.org/basic-behaviours/.

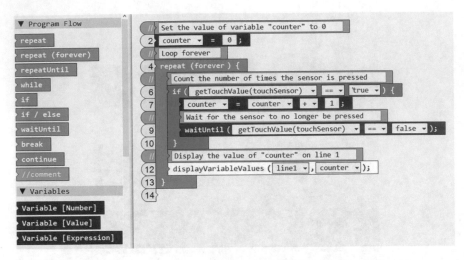

Fig. 12.5 Screenshot of RobotC Graphical DSL's click-and-drag command blocks [77].

3.2.11 MissionLab

MissionLab [2, 90] was created by the Mobile Robot Laboratory at Georgia Tech and is a research prototype DSL that facilitates mission specification through a state-machine-based visual language. The DSL uses assemblage and temporal sequencing constructs to create a temporal chain of behaviors as a mission. The assemblage construct defines behavior primitives and coordination mechanisms. During mission specification, the assemblage is instantiated. The temporal sequencing creates states with perceptual triggers to enable transitions between states. MissionLab provides a graphical editor-based configuration description language (CDL) to specify multi-agent missions. Missions can be executed on a simulator or on the following wheeled robots used for smaller commercial applications: ATRV-Jr, Urban Robot, AmigoBot, Pioneer AT, and Nomad 150 & 200.

3.2.12 RobotC

Figure 12.5 shows a sample program demonstrating how RobotC graphical language is used to write a robot program that counts and displays the number of times a button is pressed. The language primitives are in the form of blocks, which users drag and drop to build a program, making it easy for novice programmers to write robot missions. Listing 4 illustrates the same program in textual version of RobotC Natural language.[26] The expressions preserve the C language syntax, while the natural language makes it easy for novice programmers to comprehend the programs.

---

[26]http://www.robotc.net/NaturalLanguage/.

```
1    task main(){
2      int counter = 0; //set the value of variable "counter" to zero
3      while(true) { //loop forever
4      //count the number of times the sensor is pressed
5        if(getTouchValuetouchSensor) == true){
6          counter = counter +1;
7          waitUntil(getTouchValue(touchSensor) == false); // wait for the sensor to no longer be
           pressed
8        }
9        displayVariableValue(line1, counter); //display the value of "counter" on line 1
10     }
11   }
```

Listing 4 Snippet of a program in RobotC Natural Language

4 Making Robots Usable in the Everyday Life

Mobile robots are increasingly used in everyday life to autonomously realize missions such as exploring rooms, delivering goods, or following certain paths for surveillance. The current robotic market is asking for a radical shift in the development of robotic applications where mission specification is performed by end-users that are not highly qualified and specialized in robotics or ICT. To this end, in the context of the Co4Robots EU H2020 project,[27] we developed our contribution in two steps.

- First (Sect. 4.1), with the aim of understanding the missions that are currently expressed in practice, we surveyed the state of the art and formulated and formalized a catalogue of 22 mission specification patterns for mobile robots. We also provide tooling for instantiating, composing, and compiling the patterns to create mission specifications [65, 66].
- Second (Sect. 4.2), using specification patterns as the main building blocks, we proposed a DSL that enables non-technical users to specify missions for a team of autonomous robots in a user-friendly and effective way [35, 37].[28]

4.1 Mission Specification Patterns

The proposed patterns provide solutions for recurrent mission specification problems for service robots, and they focus on robot movement and on how robots perform actions within their environment. The first step for creating the catalogue

---

[27] http://www.co4robots.eu.

[28] PROMISE webpage: https://sites.google.com/view/promise-dsl/home.

of patterns was the collection and analysis of 245 natural language mission requirements systematically retrieved from the robotics literature. From these requirements,
we identified recurrent mission specification problems to which we provided
solutions and organized them as patterns. The patterns provide a formally defined
vocabulary that supports robotics developers in defining mission requirements in an
unambiguous way.

The patterns provide a formal and precise description of what robots should
do in terms of movements and actions, and therefore, relying on the usage of the
pattern catalogue as a common vocabulary makes it possible to mitigate ambiguity
in natural language formulations. Moreover, the patterns also provide validated
mission specifications for recurrent mission requirements, facilitating the creation
of correct mission specifications.

A pattern is described in terms of a structured English formulation, its usage
intent, known uses, relationships to other patterns, and, most importantly, a template
mission specification in temporal logics. Since the patterns do not contain an explicit
time or probability, the temporal logics used are LTL and CTL. This catalogue might
be extended in many directions, e.g. by considering explicit time, probability, cost,
utility, and other aspects. Patterns, while keeping their roots in a formal language,
can be used by non-experts as well.

To further support developers in designing missions, we have implemented the
tool PsALM (Pattern bAsed Mission specifier). PsALM allows the user (1) to
specify a mission requirement through a structured English grammar, which uses
patterns as the basic building blocks and operators that enable composition of the
patterns into complex missions, and (2) automatically generate specifications from
mission requirements. PsALM also enables the composition of patterns toward
the specification of complex missions by the conjunction or disjunction of the
patterns [65].

We thoroughly validated the patterns [66]. We evaluated the benefits of using
our patterns for designing missions by collecting 441 mission requirements in
natural language: 436 obtained from robotics development environments used by
practitioners and 5 defined in collaboration with 2 well-known robotics companies.
Further information about the theoretical aspects might be found in [66], and about
the tool in [65], while details about the specification of each pattern might be found
in the website.[29]

4.2 PROMISE

In order to support the specification of more complex missions with respect
to those that can be specified using the specification patterns, and in order to
enable the specification of missions for multiple robots, we proposed a domain-

---

[29]Specification Patterns for Robotic Missions webpage: http://roboticpatterns.com.

specific language called PROMISE. PROMISE considers the mission specification patterns as atomic tasks that can be executed by robots and proposes sophisticated composition operators for describing complex and multi-robot missions. These operators are inspired by behavior trees [20, 49] in their style and notation. The DSL is integrated into a framework,[30] which allows the seamless specification and execution of a mission. The framework contains:

1. The realization of the language using Eclipse and two plugins for language workbench, namely, Xtext[31] and Sirius.[32] In this way, mission specification can be performed through textual and graphical interfaces, which are synchronized.
2. A compiler implemented using Xtend[33] for mission code generation.
3. An interpreter, which parses the mission code and gives the low-level commands to each robot accordingly.

While the DSL support is provided by a standalone tool and can be integrated within a variety of frameworks, the current implementation has been integrated with a software platform [34] that provides a set of functionalities, including motion control, collision avoidance, image recognition, SLAM, and planning. This software platform has been implemented in ROS.

Our DSL has been successfully validated through experimentation with both simulation and real robots. Footage of the validation through experimentation we have conducted can be found on the dedicated website. The experimentation led to a demonstration of several missions to the Co4Robots consortium, which triggered important feedback. For instance, an industrial partner from the Bosch Center for Artificial Intelligence suggested that practitioners from their logistics facilities would appreciate a response from the tool stating a natural English description of the mission that had been specified. An example of such a description is provided in Sect. 5. We targeted specific robots during the experimentation; however, PROMISE is intended to be robot-agnostic, so it could be integrated with any robot by modifying the interpreter with the interfaces required for the new robot. The experimentation enabled us to validate PROMISE from the point of view of expressiveness by measuring the ability to write missions defined by practitioners, as we will detail in Sect. 5.

Our language and its framework implementation have been also validated in terms of usability, by measuring the ability of potential end-users in using the DSL for specifying missions. To this end, we conducted two user studies, where participants were instructed before the study and then received a set of tasks to be fulfilled within a given time frame. After the tasks' completion, the participants were asked to submit their results and to fill in a questionnaire. The first of the studies was conducted at the University of L'Aquila as an exploratory validation, which

---

[30] https://github.com/SergioGarG/PROMISE_implementation.

[31] https://www.eclipse.org/Xtext/.

[32] https://www.eclipse.org/sirius/.

[33] https://www.eclipse.org/xtend/.

triggered important refinements in PROMISE, especially in its implementation. Examples of refinements are the inclusion of a wizard to help the users in the first steps of mission specification—e.g. defining the number of robots and locations.

The second user study was designed to understand the elements of PROMISE that could be perceived as error-prone by the participants and to measure how confident the participants were of their provided solutions. During this study, the participants had to specify missions using PROMISE from textual descriptions within a time frame of 30 min. Furthermore, the participants were requested to validate their solutions through experimentation using a ROS and Gazebo-based setup in a provided laptop. All the participants were able to correctly specify their missions within the given time frame and to validate the results of two thirds of their missions through simulation. Based on the responses to the questionnaire, the perception from the users was positive toward the language and its implementation not considered error-prone. We collected qualitative data from the questionnaire using open-ended questions, which also triggered refinements in the language and its implementation. Some of the responses to those open-ended questions remain as future lines of work, as, for example, enhancing the feedback offered to the user during mission specification.

Further information regarding PROMISE and the validation procedure we followed during its development might be found in our previous study [35], in a tool paper [37], and in the DSL's dedicated webpage.

5 Putting PROMISE into Practice

In the previous section, we introduced the methods and mechanisms we developed to make robots usable in the everyday life in a descriptive way. In the following, we present an example of a mission and its specification using PROMISE together with a comprehensive discussion of the context in which it has been defined. This example originates from our work in the Co4Robots project,[34] which aimed at developing a full functioning robot that integrates several robotic skills that we have developed, including navigation, self-localization, and planning. Its focus is on robotic applications realized on top of robotic platforms provided by our industrial partners, including a TIAGo robot[35] and an ITA robot,[36] both in real life and simulation. To test our developments, when we could not directly access any of these robots, we used an economic and easy-to-use robot, the Turtlebot 2.[37] It does not provide a wide range of functionalities, but allows easy prototyping, while testing

---

[34] http://www.co4robots.eu.

[35] http://pal-robotics.com/robots/tiago/.

[36] https://www.bosch-presse.de/pressportal/de/en/current-examples-of-robotics-research-102528.html.

[37] https://www.turtlebot.com/turtlebot2.

recognition and navigation skills before deployment to the production-level robots TIAGo or ITA.

Our example scenario is inspired by a mission proposed for the 2018 edition of the well-known robotics competition RoboCup@Home. We replicated and made available in PROMISE's repository several missions proposed in the rules of this RoboCup@Home'18 [63]. Concretely, we use here the restaurant simulation scenario as an example. In this scenario, two robots collaborate to help clients in a simulated restaurant at the same time. The robots are required to ask the customers for their order and deliver drinks or snacks provided by a barman (i.e. the human *operator*), while people walk around. Both robots must work in parallel.

For our project, we have used Python as the development programming language since it is one of the most common languages used in robotics together with C++ [39], as discussed in Sect. 2. It is also well-supported by the Robot Operating System (ROS) [75] middleware. Many libraries, such as for testing or developing dedicated skills, are also written in Python. As anticipated above, we make use of ROS since it is the most widespread middleware and it is used by the Turtlebot 2 and TIAGo.

Next, also influenced by our middleware choice, we have designed a three-layered software architecture for the software, because it supports the separation of concerns among processes with different layers of abstraction [34]. We have also opted to adhere to a component-based approach, mostly because ROS enforces the component-based software development with its clustering of software modules into packages and nodes. If properly performed, the step of designing and adhering to a software architecture simplifies the later integration of robotic skills while promoting their documentation. As a mainstream IDE, we used Eclipse. In Sect. 2.2 we present and discuss popular IDEs that support users in developing robotics software, distinguishing between mainstream IDEs, such as Eclipse, which are extensible via plugins for various robotics aspects, and dedicated robotics IDEs.

Figure 12.6 shows the representation of the restaurant scenario using the graphical syntax of PROMISE. The image has been edited with circled numbers to label nodes and ease the explanation of the mission. In turn, Fig. 12.7 shows the textual representation of the same mission. This figure also contains circled numbers, which label the same nodes, and therefore supports the reader while linking the graphical-textual mapping.

Running Example: Mission Defined Using PROMISE

The root of the mission specification, i.e. the operator *parallel*, is identified by the node ① and specifies that *robot1* and *robot2* must perform their missions in parallel. A robot is assigned to each branch associated with this operator, as indicated with labels in the edges between ① and ② in Fig. 12.6, and with the name of the assigned robot (i.e. *robot1* and *robot2*) in Fig. 12.7. Since the mission for *robot2* (㉒) is a replica of the one for *robot1*, we only show the latter for the sake of conciseness.

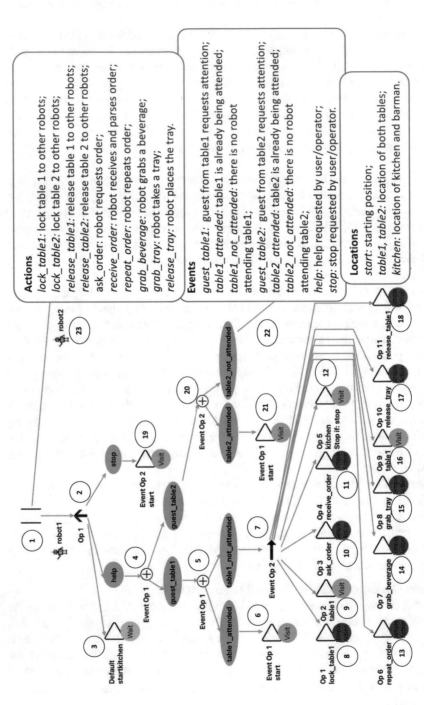

Fig. 12.6 Running example specified with the graphical syntax of PROMISE.

```
operators{ parallel{
    robot1 (    1
        eventHandler (  2
            default ( delegate ( Wait locations start , kitchen ) ) 3
            except help (
                condition ( 4
                    if guest_table1 (
                        condition ( 5
                            if table1_attended ( delegate ( Visit locations start ) ) 6
                            if table1_not_attended (
                                sequence (  7
                                    delegate (  SimpleAction actions lock_table1 ) , 8
                                    delegate ( Visit locations table1 ) , 9
                                    delegate ( SimpleAction actions ask_order ) , 10
                                    delegate ( SimpleAction actions receive_order ) , 11
                                    delegate ( Visit locations kitchen stoppingEvents stop) 12
                                    delegate ( SimpleAction actions repeat_order ) , 13
                                    delegate ( SimpleAction actions grab_beverage ) , 14
                                    delegate ( SimpleAction actions grab_tray ) , 15
                                    delegate ( Visit locations table1 ) , 16
                                    delegate ( SimpleAction actions release_tray ) , 17
                                    delegate ( SimpleAction actions release_table1 )))) 18
                    ) if guest_table2 (
                        condition ( 20
                            if table2_attended ( delegate ( Visit locations start ) ) 21
                            if table2_not_attended ( 22
                                sequence ( □
                            except stop ( delegate ( Visit locations start )))) , 19
    robot2 ( 23
```

Fig. 12.7 Running example specified with the graphical syntax of PROMISE.

The operator labeled with ② is the *eventHandler*—more information regarding PROMISE's operators is available in [35]. It has a default behavior; in our example, it forces the robot to wait in location *start* (③). This behavior is paused when one of the events that are assigned to the *eventHandler* is detected by the robot. The default robot's behavior (③) is resumed whenever any of the behaviors triggered by an event is finished (either succeeding or failing).

Each event is assigned to a child of the *eventHandler* (as represented in Fig. 12.6) as gray circles and invoked in Fig. 12.7 by the keyword except. If the event "help" is detected, the first operator *condition* (④) is executed. This operator evaluates its associated events in order, and if they hold, it triggers the behaviors associated with them. In this case, the operator *condition* evaluates whether the request of help comes from *table*1 or *table*2.

In case "guest_table1" holds (i.e. the request of help comes from *table*1), another operator *condition* (⑤) is executed. This operator evaluates whether this table is already being attended to by another robot ("table1_attended") and, in this case, makes the robot return to the starting position *start*. This behavior is encoded by the instantiation of an operator *delegate* with a task *Visit* ((⑥).

The next operator *condition* (⑤) evaluates "table1_not_attended," and, if it holds, the execution of an operator *sequence* (⑦) is triggered. This operator executes in sequence a set of operators. Concretely, the sequence of operators starts with ⑧, which "locks" *table*1 from the rest of the robotic team by forwarding a message (in this case, other robots will recognize it with the event "table1_attended'). The robot will then move to *table*1 (⑨), ask the order ((⑩), and receive and parse it ((⑪)). The robot will then move to *kitchen* ((⑫)) to interact with the barman (i.e. the human operator). Note that this specific task can be stopped by the user or human operator by means of the event "stop" (see Fig. 12.6). Once the robot has reached location *kitchen*, it will repeat the order to the barman ((13)), after which the robot will grab beverages ((14)) and a tray with the ordered snacks ((15)). The robot will then return to *table*1 ((16)) with the order, where it will place the tray ((17)). The sequence of tasks finishes with the robot "releasing" the table for other robots, in a similar way as to how it locked it.

The operator *condition* (20) is a replica of (5)—see the conditions in the textual representation in Fig. 12.7 "if guest_table1" and "if guest_table2"—and, therefore, we do not show its whole graphical representation for the sake of conciseness.

As suggested by an industrial partner after a demonstration to the Co4Robots consortium, PROMISE prompts a natural English description of the mission once specified and saved. An excerpt from the description of the example introduced in this section is as follows.

> *Robot robot1 does by default wait in location start, and if event help occurs, it will, if event guest_table1 holds, and if event table1_attended holds, visit (without any specific order) location(s) start. If event table1_not_attended holds, it will perform action lock_table1 and then visit (without any specific order) location(s) table1 and then perform action ask_order, and then perform action receive_order, and then visit (without any specific order) location(s) kitchen, and then perform action repeat_order, and then perform action gra_beverage, and then perform action grab_tray, and then visit (without any specific order) location(s) table1 and then perform action release_tray and then perform action release_table1.*

The mission of the example was modeled through mission specification from a natural English description, in this case, from the rules of the RoboCup@Home'18 [63]. Once the mission was modeled, we proceeded to validate it through experimentation in an iterative way. The first step we took was simulation,[38] for which we used simulated models of the facilities and robotic models provided by the industrial partners of Co4Robots for Gazebo [56]. Once the simulation was performed and the mission specification validated, we proceeded with validation in real life. As explained above, we purchased a Turtlebot2 for experimentation. We validated the same restaurant scenario with the Turtlebot in the facilities of the University of Gothenburg.[39] The last step was to conduct a demonstration in the presence of the project consortium at the facilities of PAL Robotics, for which we used a TIAGo

---

[38] https://www.youtube.com/watch?v=F3BnIEPB8Sk.

[39] https://www.youtube.com/watch?v=Qr9FqzSrZuk.

robot.[40] Through this process, we demonstrated the ability of PROMISE to specify complex missions from textual descriptions. We also demonstrated its capability to operate with different robots by accordingly modifying the interpreter of its framework—see Sect. 4.2.

We invite the interested reader to learn more about the validation procedures we followed during the development of PROMISE in our published studies [35, 38] and on its dedicated website.

6 Discussion and Perspectives for Future Research

As discussed in this chapter, in the last years there have been many contributions from the research community to propose domain-specific languages for mission specification [13, 35], the description of missions in natural language [61], and visual and end-user-oriented mission environments [9, 17, 69, 95].

The approaches surveyed here greatly contribute to the field; however, the mission specification-problem still requires solutions able to make robots usable in everyday life for accomplishing complex missions. In the following, we highlight the limitations of current approaches and we devise perspectives for future research. As stated also in the Multi-Annual Roadmap (MAR) For Robotics in Europe [84], in order to reduce costs and establish a vibrant component market, there is a need for instruments for supporting mission *reuse* and diversification, as well as coping with the *variability* of conditions of application scenarios occurring in real environments. This is also testified by our findings during our collaboration with practitioners in the robotic domain: the complexity does not reside in commanding a robot with a set of tasks but in making the robotic application robust enough to be able to cope with the variability that characterizes the real environments in which the robots are required to operate, especially those that involve humans [39].

To the best of our knowledge, few approaches try to address the reusability and variability envisioned by the MAR. PROMISE and the specification patterns are greatly contributing; however, there are some aspects that should be investigated in the future. In the following, we devise important research directions, which we identified based on our collaboration with companies in the Co4Robots project and additional collaborations in the healthcare domain. Specifically, we believe that the main research directions go in the following directions:

- *Reusability*: the DSLs we will develop for enabling end users to specify missions will make use of libraries of tasks and skills. They will also integrate with libraries produced by other projects and initiatives, like RobMoSys.[41]

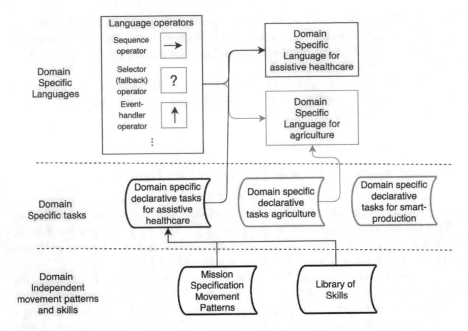

Fig. 12.8 Mission specification.

- *Variability of the real world*: the DSL will be conceived to enable the specification of the variability of conditions of complex real-world scenarios.
- *Fleet specification of a mission*: the end-user that will specify the mission does not need to assign tasks to specific robots, but the mission specification will represent the "needs" of the end-user and robots will be automatically assigned and potentially re-assigned during the mission execution, according to the capabilities of robots and various quality parameters.
- *Human-robot collaboration*: the mission specification will include also humans, with two different roles, namely, *operators*, able to perform actions needed to successfully accomplish the mission, and *patients*, which will require actions from robots.

In order to support what we believe we might need, various libraries of pre-defined solution schemes that can be reused, instantiated, and composed by means of properly defined operators need to be implemented. As shown in Fig. 12.8[42] we envision three different types of libraries organized on two levels, one being application-domain-independent (specific for service robots) and the other one being domain-specific, e.g. assistive healthcare, agriculture, or smart production.

---

[42]We use the same terminology in "Architectural Pattern for Task-Plot Coordination" of the EU H2020 RobMoSys project: https://robmosys.eu/wiki/general_principles:architectural_patterns:robotic_behavior.

- *Mission specification movement patterns* are pre-defined solutions concerning movements of robots and provide the bridge between a mission requirement expressed in structured English (a subset of English with a well-defined semantics) and a formulation in temporal logic. An initial result in this direction consists in the specification patterns described in Sect. 4.1.
- *Library of skills* contains the implementation of the modules for enabling the robot to do specific actions, like grasp object with constraint, low dexterity, soft grasping, image recognition, gesture recognition, and so on, that are compliant with the RobMoSys platform.
- *Domain-specific declarative tasks* are recurrent combinations of mission specification patterns and skills used to define declarative tasks for domain-specific operations. For instance, in the assistive healthcare domain, a declarative task can be "welcome" and would require patterns for movements and various skills such as human recognition, speech recognition, etc. The tasks are declarative since they specify only what the robot is able to do without saying how the robot will do that. Then, planners will compute how the task will be solved in the specific environment according to the capabilities of the robot that will be allocated to this task.
- *Domain-specific languages*, as for instance PROMISE (Sect. 4.2), enable operators who are not required to have expertise in programming nor robotics, to specify in an easy and correct way the mission they would like the robots to safely accomplish. Each domain-specific language will make use of the language operators that we will define. There will also be specific "dialects" for specializing the language to the various domains. In this way, healthcare operators will find in the domain-specific language for assistive healthcare concepts that are specific to the domain, expressed in terms of domain-specific declarative tasks for assistive healthcare. The language enables the description of complex and sophisticated missions, which will also take into account non-functional properties, such as timing constraints. These properties are captured by composition operators, like *sequence*, *selector* (fallback), or *event-handler*, which are inspired by behavior trees [20, 49] or by PROMISE.[43] The DSL will help healthcare operators (with a sort of wizard or recommendation system) to deal with the variability that characterizes the environments in which missions are executed. This includes "exceptional" behaviors, such as a robot running out of battery, an unforeseen obstacle hampering the mission satisfaction, an object falling down from the hand of the robot, and so on. As testified by MAR [84] and also highlighted in a recent study [39], one of the most difficult aspects in mission specification is to deal with the variability of real-world scenarios.

Example of Mission Specification During the day, "robot" welcomes newcomers when the bell "s2" of the door rings. According to the needs of the guests, "robot" will provide the needed information or ask them to enter the dining room, and if a

---

[43] The PROMISE DSL has been developed in the context of the EU H2020 project Co4Robots [35]. PROMISE webpage: https://github.com/SergioGarG/PROMISE_implementation.

human intervention is needed, "robot" informs a caregiver. When "robot" is in the dining room, it acts as a caregiver and interacts with people by calling them to drink and offering water that "tray" carries. During night, "robot" patrols for security, and if it finds humans in the environment, it calls an operator. Robots recharge autonomously while guaranteeing the welcoming and caregiving service. Notice that the example does not include quality aspects, such as timing constraints, since patterns including these aspects are not yet available, but they will be developed during the project execution.

Mission Specification A healthcare operator will specify the mission by means of the following domain-specific macros: *Welcome, Security Patrolling, Caregiver*, and *Call caregiver*. The following figure shows the mission specified foreseeing two different graphical languages, one (a) based on the *blockly*[44] approach and the other one (b) using PROMISE's style [35]. This is just to explain what we mean by graphical and easy-to-use language for mission specification.

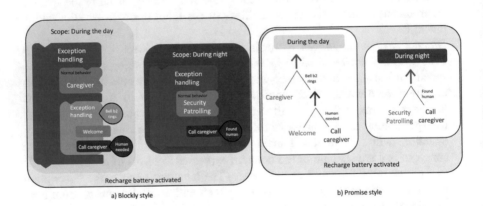

Domain-Specific Macros, Mission Specification Patterns, and Library of Tasks Behind the scene, i.e. invisible to the end-users, the macros will be built by using the mission specification patterns and the tasks stored in the library. For instance, *welcoming* might be realized by composing the *sequenced visit specification pattern*[45] to reach from the current location of the robot the door (LTL formula: $\Diamond$ (door_location)), with a *delayed action*[46] when the robot reaches the door to welcome and activate the speech recognition—LTL formula: $\Box$(door_location $\Rightarrow \Diamond$(welcome)), where "welcome" is a task in the library of tasks.

---

[44]https://developers.google.com/blockly.

[45]http://roboticpatterns.com/pattern/sequencedvisit/.

[46]http://roboticpatterns.com/pattern/delayedreaction/.

Acknowledgments The authors acknowledge the financial support from the Centre of EXcellence on Connected, Geo-Localized and Cybersecure Vehicle (EX-Emerge), funded by the Italian Government under CIPE resolution n. 70/2017 (Aug. 7, 2017). The work is also supported by the European Research Council under the European Union's Horizon 2020 research and innovation program GA No. 694277 and GA No. 731869 (Co4Robots). More support for this work was provided by the SIDA Bright 317 project.

References

1. S. Arias, F. Boudin, R. Pissard-gibollet, D. Simon, S. Arias, F. Boudin, R. Pissard-gibollet, D.S. Orccad, S. Arias, F. Boudin, R. Pissard-gibollet, D. Simon, ORCCAD, robot controller model and its support using Eclipse Modeling tools (2010)
2. R. Arkin, Missionlab: multiagent robotics meets visual programming. Working notes of Tutorial on Mobile Robot Programming Paradigms, ICRA, vol. 15 (2002)
3. M. Autili, L. Grunske, M. Lumpe, P. Pelliccione, A. Tang, Aligning qualitative, real-time, and probabilistic property specification patterns using a structured English grammar. IEEE Trans. Software Eng. **41**(7), 620–638 (2015)
4. J.C. Baillie, URBI: towards a universal robotic body interface, in 4th *IEEE/RAS International Conference on Humanoid Robots, 2004*, vol. 1 (2004), pp. 33–51. https://doi.org/10.1109/ICHR.2004.1442112
5. T. Balch, Teambots 2.0 (2000). https://www.cs.cmu.edu/~trb/TeamBots/
6. H. Båtelsson, Behavior Trees in the Unreal Engine: Function and Application (Uppsala University, Uppsala, 2016)
7. T. Berger, J.P. Steghöfer, T. Ziadi, J. Robin, J. Martinez, The state of adoption and the challenges of systematic variability management in industry. Empir. Softw. Eng. **25**, 1755–1797 (2020)
8. J. Bézivin, On the unification power of models. Softw. Syst. Modeling **4**(2), 171–188 (2005)
9. G. Biggs, B. Macdonald, A survey of robot programming systems, in *Proceedings of the Australasian Conference on Robotics and Automation, CSIRO* (2003), p. 27
10. R. Bischoff, T. Guhl, E. Prassler, W. Nowak, G. Kraetzschmar, H. Bruyninckx, P. Soetens, M. Hägele, A. Pott, P. Breedveld, J. Broenink, D. Brugali, N. Tomatis, BRICS – best practice in robotics, in *ISR 2010 (41st International Symposium on Robotics) and ROBOTIK 2010 (6th German Conference on Robotics)* (2010), pp. 1–8
11. D. Blank, D. Kumar, L. Meeden, H. Yanco, Pyro: a Python-based versatile programming environment for teaching robotics. J. Educ. Res. Comput. **3**(4) (2003)
12. J. Bohren, S. Cousins, The SMACH high-level executive [ROS news]. IEEE Robot. Autom. Mag. **17**(4), 18–20 (2010)
13. D. Bozhinoski, D. Di Ruscio, I. Malavolta, P. Pelliccione, M. Tivoli, Flyaq: enabling non-expert users to specify and generate missions of autonomous multicopters, in *2015 30th IEEE/ACM International Conference on Automated Software Engineering (ASE)* (2015), pp. 801–806
14. M. Brambilla, J. Cabot, M. Wimmer, Model-Driven Software Engineering in Practice (Morgan & Claypool, San Rafael, 2012)
15. D. Brugali, A. Agah, B. MacDonald, I.A. Nesnas, W.D. Smart, Trends in robot software domain engineering, in *Software Engineering for Experimental Robotics* (Springer, Berlin, 2007), pp. 3–8
16. W. Burnett (2018). http://www.legoengineering.com/alternative-programming-languages/
17. R.W. Button, J. Kamp, T.B. Curtin, J. Dryden, *A Survey of Missions for Unmanned Undersea Vehicles* (RAND Corporation, Santa Monica, CA, 2009). https://www.rand.org/pubs/monographs/MG808.html. Also available in print form
18. D. Caron, Competitive robotics the best brings out in students. Tech Directions, v69 n6 p21–23 Jan 2010. (2010), pp. 21–24. https://eric.ed.gov/?id=EJ894879

19. J. Cicolani, *Beginning Robotics with Raspberry Pi and Arduino: Using Python and OpenCV* (Apress, New York, 2018)
20. M. Colledanchise, Behavior trees in robotics. Ph.D. Thesis, Royal Institute of Technology, Stockholm (2017)
21. M. Colledanchise, P. Ögren, How behavior trees modularize hybrid control systems and generalize sequential behavior compositions, the subsumption architecture, and decision trees. IEEE Trans. Robotics **33**(2), 372–389 (2017). https://doi.org/10.1109/TRO.2016.2633567
22. M. Colledanchise, P. Ögren, *Behavior Trees in Robotics and AI: An Introduction* (CRC Press, Boca Raton, 2018)
23. Copella simulator (2020). https://www.coppeliarobotics.com/
24. D. Di Ruscio, I. Malavolta, P. Pelliccione, A family of domain-specific languages for specifying civilian missions of multi-robot systems. CEUR Workshop Proc. **1319**, 16–29 (2014)
25. D. Di Ruscio, I. Malavolta, P. Pelliccione, M. Tivoli, Automatic generation of detailed flight plans from high-level mission descriptions, in *International Conference on Model Driven Engineering Languages and Systems, MODELS* (ACM, New York, 2016)
26. P. Doherty, F. Heintz, J. Kvarnström, High-level mission specification and planning for collaborative unmanned aircraft systems using delegation. Unmanned Syst. **1**(01), 75–119 (2013)
27. S. Dragule, B. Meyers, P. Pelliccione, A generated property specification language for resilient multirobot missions, in *Software Engineering for Resilient Systems*, ed. by A. Romanovsky, E.A. Troubitsyna (Springer International Publishing, Cham, 2017), pp. 45–61
28. P. Estefo, J. Simmonds, R. Robbes, J. Fabry, The robot operating system: package reuse and community dynamics. J. Syst. Softw. **151**, 226–242 (2019)
29. D. Faconti, Models and Tools to design Robotic Behaviors. Tech. Rep. 732410, Eurecat Tecnológic, Barcelona (2020). https://github.com/BehaviorTree/BehaviorTree.CPP/blob/master/MOOD2Be_final_report.pdf
30. C. Finucane, G. Jing, H. Kress-Gazit, LTLMoP: experimenting with language, temporal logic and robot control, in *International Conference on Intelligent Robots and Systems (IROS)* (IEEE, Piscataway, 2010), pp. 1988–1993
31. FLYAQ (2019). http://www.flyaq.it/
32. M. Fowler, R. Parsons, *Domain-Specific Languages* (Addison-Wesley, Boston, 2011)
33. Fraunhofer IAIS (2019). https://lab.open-roberta.org/
34. S. García, C. Menghi, P. Pelliccione, T. Berger, R. Wohlrab, An architecture for decentralized, collaborative, and autonomous robots, in *2018 IEEE International Conference on Software Architecture (ICSA)* (IEEE, Piscataway, 2018), pp. 75–7509
35. S. García, P. Pelliccione, C. Menghi, T. Berger, T. Bures, High-level mission specification for multiple robots, in *Proceedings of the 12th ACM SIGPLAN International Conference on Software Language Engineering, SLE 2019* (2019)
36. S. García, D. Strüber, D. Brugali, A. Di Fava, P. Schillinger, P. Pelliccione, T. Berger, Variability modeling of service robots: experiences and challenges, in *Proceedings of the 13th International Workshop on Variability Modelling of Software-Intensive Systems* (2019), pp. 1–6
37. S. García, P. Pelliccione, C. Menghi, T. Berger, T. Bures, Promise: high-level mission specification for multiple robots, in *42nd International Conference on Software Engineering (ICSE 2020 Demos)* (2020)
38. S. García, P. Pelliccione, C. Menghi, T. Berger, T. Bures, Promise: high-level mission specification for multiple robots, in *ICSE '20: Proceedings of the ACM/IEEE 42nd International Conference on Software Engineering: Companion Proceedings* (2020)
39. S. García, D. Strüber, D. Brugali, T. Berger, P. Pelliccione, Robotics software engineering: A perspective from the service robotics domain, in *ACM Joint European Software Engineering Conference and Symposium on the Foundations of Software Engineering (ESEC/FSE 2020)* (2020)
40. R. Ghzouli, T. Berger, E.B. Johnsen, S. Dragule, A. Wasowski, Behavior trees in action: a study of robotics applications, in *13th ACM SIGPLAN International Conference on Software Language Engineering (SLE)* (2020)

41. S. Götz, M. Leuthäuser, J. Reimann, J. Schroeter, C. Wende, C. Wilke, U. Aßmann, A role-based language for collaborative robot applications, in *International Symposium On Leveraging Applications of Formal Methods, Verification and Validation* (Springer, Berlin, 2011), pp. 1–15
42. M. Guo, K.H. Johansson, D. Dimarogonas, Revising motion planning under linear temporal logic specifications in partially known workspaces, in *International Conference on Robotics and Automation* (2013)
43. S. Hart, P. Dinh, J.D. Yamokoski, B. Wightman, N. Radford, Robot task commander: a framework and ide for robot application development, in *2014 IEEE/RSJ International Conference on Intelligent Robots and Systems* (2014), pp. 1547–1554
44. A. Hentout, A. Maoudj, B. Bouzouia, A survey of development frameworks for robotics, in *2016 8th International Conference on Modelling, Identification and Control (ICMIC)* (2016), pp. 67–72
45. R.P.Y. Ho, Configuration of robotics solutions in microsoft robotics developer studio (2009). http://aunilo.uum.edu.my/Find/Record/sg-ntu-dr.10356-20828
46. G.J. Holzmann, The logic of bugs, in *Symposium on Foundations of Software Engineering, SIGSOFT '02/FSE-10* (2002)
47. L. Hugues, N. Bredeche, Simbad: an autonomous robot simulation package for education and research, in *International Conference on Simulation of Adaptive Behavior* (Springer, Berlin, 2006), pp. 831–842
48. J. Hutchinson, J. Whittle, M. Rouncefield, S. Kristoffersen, Empirical assessment of mde in industry, in *ICSE* (2011), pp. 471–480. http://doi.acm.org/10.1145/1985793.1985858
49. D. Isla, Handling complexity in the Halo 2 AI, in *In Game Developers Conference* (2005)
50. E.M. Jarvinen, A. Karsikas, J. Hintikka, Children as innovators in action – a Study of microcontrollers in finnish comprehensive schools. J. Technol. Edu. **18**, 37–52 (2007)
51. B. Jost, M. Ketterl, R. Budde, T. Leimbach, Graphical programming environments for educational robots: open roberta – yet another one? in *2014 IEEE International Symposium on Multimedia* (2014), pp. 381–386
52. H.S. Juang, K.Y. Lurrr, Design and control of a two-wheel self-balancing robot using the arduino microcontroller board, in *2013 10th IEEE International Conference on Control and Automation (ICCA)* (IEEE, Piscataway, 2013), pp. 634–639
53. M. Ketterl, T. Leimbach, R. Budde, Open Roberta (14), 1–22 (2015). https://lab.open-roberta.org/
54. A.G. Kleppe, *Software Language Engineering: Creating Domain-Specific Languages Using Metamodels* (Addison-Wesley, Boston, 2009)
55. M. Klotzbücher, H. Bruyninckx, Coordinating robotic tasks and systems with rFSM statecharts (2012)
56. N. Koenig, A. Howard, Design and use paradigms for gazebo, an open-source multi-robot simulator, in *2004 IEEE/RSJ International Conference on Intelligent Robots and Systems (IROS)(IEEE Cat. No. 04CH37566)*, vol. 3 (IEEE, Piscataway, 2004), pp. 2149–2154
57. M. Kouzehgar, Y.K. Tamilselvam, M.V. Heredia, M.R. Elara, Self-reconfigurable façade-cleaning robot equipped with deep-learning-based crack detection based on convolutional neural networks. Autom. Const. **108**, 102959 (2019)
58. J. Kramer, M. Scheutz, Development environments for autonomous mobile robots: A survey. Auton. Robot. **22**(2), 101–132 (2007)
59. B.S. Krishna, J. Oviya, S. Gowri, M. Varshini, Cloud robotics in industry using Raspberry PI, in *2016 Second International Conference on Science Technology Engineering and Management (ICONSTEM)* (IEEE, Piscataway, 2016), pp. 543–547
60. LEGO MINDSTORMS EV3 (2019). https://www.lego.com/en-us/mindstorms/downloads/download-software
61. C. Lignos, V. Raman, C. Finucane, M. Marcus, H. Kress-Gazit, Provably correct reactive control from natural language. Auton. Robot. **38**(1), 89–105 (2015)

62. S. Magnenat, P. Rétornaz, M. Bonani, V. Longchamp, F. Mondada, ASEBA: a modular architecture for event-based control of complex robots. IEEE/ASME Trans. Mechatron. **16**(2), 321–329 (2011)
63. M. Matamoros, C. Rascon, J. Hart, D. Holz, L. van Beek, Robocup@home 2018: rules and regulations (2018). http://www.robocupathome.org/rules/2018_rulebook.pdf
64. C. Menghi, S. García, P. Pelliccione, J. Tumova, Multi-robot LTL planning under uncertainty, in *International Symposium on Formal Methods* (Springer, Berlin, 2018), pp. 399–417
65. C. Menghi, C. Tsigkanos, T. Berger, P. Pelliccione, PsAlM: specification of dependable robotic missions, in *International Conference on Software Engineering (ICSE): Companion Proceedings* (2019)
66. C. Menghi, C. Tsigkanos, P. Pelliccione, C. Ghezzi, T. Berger, Specification patterns for robotic missions, in *IEEE Transactions on Software Engineering*. https://doi.org/10.1109/TSE.2019. 2945329
67. O. Michel, Cyberbotics Ltd. Webots^Û: professional mobile robot simulation. Int. J. Adv. Robot. Syst. **1**(1), 5 (2004)
68. M.A. Miskam, S. Shamsuddin, H. Yussof, A.R. Omar, M.Z. Muda, Programming platform for NAO robot in cognitive interaction applications, in *2014 IEEE International Symposium on Robotics and Manufacturing Automation (ROMA)* (IEEE, Piscataway, 2014), pp. 141–146
69. A. Nordmann, N. Hochgeschwender, D. Wigand, S. Wrede, A survey on domain-specific modeling and languages in robotics. J. Softw. Eng. Robot. **7**(1), 75–99 (2016)
70. T. Ohkawa, D. Uetake, T. Yokota, K. Ootsu, T. Baba, Reconfigurable and hardwired orb engine on FPGA by java-to-hdl synthesizer for realtime application. ACM SIGARCH Comput. Archit. News **41**(5), 77–82 (2014)
71. PICAXE (2019). http://www.picaxe.com/software
72. F. Piltan, N. Sulaiman, M. Marhaban, A. Nowzary, M. Tohidian, Design of FPGA based sliding mode controller for robot manipulator.Int. J. Robot. Autom. **2**(3), 183–204 (2011)
73. F. Piltan, M.H. Yarmahmoudi, M. Shamsodini, E. Mazlomian, A. Hosainpour, PUMA-560 robot manipulator position computed torque control methods using Matlab/Simulink and their integration into graduate nonlinear control and Matlab courses. Int. J. Robot. Autom. **3**(3), 167–191 (2012)
74. E. Pot, J. Monceaux, R. Gelin, B. Maisonnier, Choregraphe: a graphical tool for humanoid robot programming, in *RO-MAN 2009 – The 18th IEEE International Symposium on Robot and Human Interactive Communication* (2009), pp. 46–51
75. M. Quigley, K. Conley, B. Gerkey, J. Faust, T. Foote, J. Leibs, R. Wheeler, A.Y. Ng, ROS: an open-source robot operating system. ICRA Workshop Open Source Softw. **3**(3.2), 5 (2009)
76. Robot Mesh (2019). http://docs.robotmesh.com/ide-project-page
77. ROBOTC, ROBOTC's Graphical feature (2019). http://www.robotc.net/graphical/
78. T. Röfer, CABSL – C-Based agent behavior specification language. Lecture Notes in Computer Science, vol. 11175 (2018), pp. 135–142. https://doi.org/10.1007/978-3-030-00308-1_11
79. Ros development studio (2020). https://www.theconstructsim.com/rds-ros-development-studio
80. S.L. Salcedo, A.M.O. Idrobo, New tools and methodologies for programming languages learning using the SCRIBBLER robot and Alice, in *Proceedings – Frontiers in Education Conference, FIE* (2011), pp. 1–6
81. S. Schauss, R. Lämmel, J. Härtel, M. Heinz, K. Klein, L. Härtel, T. Berger, A chrestomathy of DSL implementations, in *10th ACM SIGPLAN International Conference on Software Language Engineering (SLE)* (2017)
82. B. Schwartz, L. Nägele, A. Angerer, B.A. MacDonald, Towards a graphical language for quadrotor missions. Preprint. arXiv:1412.1961 (2014)
83. B. Selic, The pragmatics of model-driven development. IEEE Softw. **20**(5), 19–25 (2003). http://csdl.computer.org/comp/mags/so/2003/05/s5019abs.htm
84. SPARC, Robotics 2020 Multi-Annual Roadmap (2016). https://eu-robotics.net/sparc/upload/about/files/H2020-Robotics-Multi-Annual-Roadmap-ICT-2016.pdf
85. T. Stahl, M. Völter, *Model-Driven Software Development* (Wiley, Hoboken, 2005)

86. D. Stampfer, A. Lotz, M. Lutz, C. Schlegel, The SmartMDSD Toolchain: an integrated MDSD workflow and integrated development environment (IDE) for robotics software. J. Softw. Eng. Robot. **7**, 3–19 (2016)
87. M. Tamre, R. Hudjakov, D. Shvarts, A. Polder, M. Hiiemaa, M. Juurma, Implementation of integrated wireless network and MatLab system to control autonomous mobile robot. Int. J. Innov. Technol. Interdis. Sci. **1**(1), 18–25 (2018)
88. U. Thomas, G. Hirzinger, B. Rumpe, C. Schulze, A. Wortmann, A new skill based robot programming language using UML/P Statecharts, in *2013 IEEE International Conference on Robotics and Automation* (IEEE, Piscataway, 2013), pp. 461–466
89. Thymio (2019). https://www.thymio.org/en:start
90. P. Ulam, Y. Endo, A. Wagner, R. Arkin, Integrated mission specification and task allocation for robot teams – Design and implementation, in *Proceedings – IEEE International Conference on Robotics and Automation* (2007), pp. 4428–4435
91. A. van Deursen, P. Klint, J. Visser, Domain-specific languages: an annotated bibliography. SIGPLAN Not. **35**(6), 26–36 (2000). https://doi.org/10.1145/352029.352035
92. VEX Robotics (2019). https://www.vexrobotics.com
93. M. Voelter, DSL Engineering. Designing, implementing and using domain specific languages (2013). http://www.dslbook.org/
94. F.Y. Wang, K.J. Kyriakopoulos, A. Tsolkas, G.N. Saridis, A petri-net coordination model for an intelligent mobile robot. IEEE Trans. Syst. Man Cybern. **21**(4), 777–789 (1991)
95. D. Weintrop, A. Afzal, J. Salac, P. Francis, B. Li, D.C. Shepherd, D. Franklin, Evaluating coblox: a comparative study of robotics programming environments for adult novices, in *Proceedings of the 2018 CHI Conference on Human Factors in Computing Systems, CHI '18* (ACM, New York, 2018), pp. 366:1–366:12
96. E.M. Wolff, U. Topcu, R.M. Murray, Automaton-guided controller synthesis for nonlinear systems with temporal logic, in *2013 IEEE/RSJ International Conference on Intelligent Robots and Systems* (IEEE, Piscataway, 2013), pp. 4332–4339
97. V.A. Ziparo, L. Iocchi, D. Nardi, P.F. Palamara, H. Costelha, Petri net plans: a formal model for representation and execution of multi-robot plans, in *Proceedings of the 7th International Joint Conference on Autonomous Agents and Multiagent Systems*, vol. 1. International Foundation for Autonomous Agents and Multiagent Systems (2008), pp. 79–86

Chapter 13
RoboStar Technology: Modelling Uncertainty in RoboChart Using Probability

Jim Woodcock, Simon Foster, Alexandre Mota, and Kangfeng Ye

Abstract RoboChart is a UML-like language designed for modelling autonomous and mobile robots. It includes timed and probabilistic primitives. In this chapter, we discuss first why we need probability by surveying how we use it in designing robots. To illustrate our approach, we focus on the verification of probabilistic robotic algorithms for pose estimation. We verify a model-fitting algorithm: random sample consensus (Ransac). This is a popular algorithm, representative of a class of particle-filter algorithms. We analyse the aspects of the algorithm and show how to model-check properties and how to get stronger guarantees using a program logic. Our contributions are a survey of probabilistic robotic applications and an approach to verifying probabilistic algorithms developed using RoboStar technology.

1 Introduction

Luckcuck et al. survey the formal specification and verification of autonomous robotic systems [59]. They describe them as complex, hybrid, often safety-critical, and operating in uncertain environments. Their formal specification and verification are challenging. Luckcuck et al. identify probabilistic models as a common way of specifying uncertainty.

An example is given by Morse et al. [65], who describe a robotic domestic assistant. They model non-robotic actors in the environment with probabilistic behaviours, representing uncertain actions. They use PRISM [49] to reason about the composition of the robot and its environment in uncertain scenarios.

Our contributions in this chapter are twofold. The first is to discuss the practical applications of using probability in robotic applications. Thrun has established the

J. Woodcock (✉) · S. Foster · K. Ye
University of York, York, UK
e-mail: jim.woodcock@york.ac.uk; simon.foster@york.ac.uk; kangfeng.ye@york.ac.uk

A. Mota
Universidade Federal de Pernambuco, Recife, Brazil
e-mail: acm@cin.ufpe.br

area of probabilistic robotics [75], but there is not a comprehensive survey of what is being done for which purpose. Our chapter is a start in this direction. The second is an approach applying both model checking and theorem proving to reasoning about probabilistic robotic algorithms. The formal verification of these algorithms is novel (to the best of our knowledge).

Our research builds on RoboChart [56, 57, 64]. This is a robot modelling language based on UML [11], but with a restricted set of constructs to simplify semantics and automate reasoning. Miyazawa et al. have implemented RoboChart in RoboTool [62], an Eclipse-based graphical editor with transformation and verification tools. RoboChart has a formal semantics in CSP [40]. It supports real-time modelling (see Ribeiro et al. [71]). It also supports probabilistic modelling (see Woodcock et al. [19, 77]).

We survey how probability is used to model robotic systems (Sect. 2). We describe our approach to probabilistic modelling in RoboChart (Sect. 3). We discuss a probabilistic pose-estimation algorithm: random sample consensus (Ransac) (Sect. 3.2). Ransac is a simple example of a robust regression algorithm, but it is a practical technique.[1] For example, it is used in the Aethon TUG robot that automates material delivery in hospitals, manufacturing units, and hotels [2, 16]. These are collaborative robots reported to be operating safely and efficiently in making over five million deliveries per year.

We start with a formalisation of a representative deterministic model-fitting method to motivate the Ransac algorithm (Sect. 4). We then present our treatment of the Ransac algorithm (Sect. 5). We analyse Ransac using probabilistic model checking in PRISM (Sect. 6). We address the stronger guarantees that theorem proving can give by verifying a simplified form of Ransac using Hehner's probabilistic predicative programming method [34–38] (Sect. 7). Finally, we draw our conclusions and set out ideas for future work (Sect. 8). Throughout the chapter, our reasoning is meticulous in preparation for mechanisation of theories for reasoning about probabilistic algorithms in robotic applications.

2   Probabilistic Robotics

Mobile autonomous robots face operational uncertainties in how they perceive and react to their environments. Luckcuck et al.'s survey shows that managing uncertainty is still a major robotics research challenge [59]. As Thrun says, it arises in practice from several sources [75], and we deal with three of these: abstractions of the real world (see Sect. 2.1), unreliable hardware (see Sect. 2.2), and approximate control algorithms (see Sect. 2.3). Despite these uncertainties, robots in important

---

[1] In software engineering, we define a robust algorithm as one that works for any input. In formal methods, we define a robust algorithm as one that works for any input within the precondition for its declared interface.

applications must still meet their mission requirements. A model-based design technique must anticipate uncertainties and provide strategies to overcome them.

Uncertainty can enter mathematical models and operational measurements in various ways (see Kennedy and O'Hagan [47] for a detailed discussion). These sources are often divided into two categories: epistemic and aleatoric uncertainty. Epistemic uncertainty describes things we could know in principle but do not know in practice and is reducible by gaining more knowledge. Aleatoric uncertainty describes the natural randomness of physical processes and is irreducible. Both categories can be modelled using probabilistic semantics. See Gretz et al. [29] and McIver and Morgan [61] for good overviews of the different kinds of semantics.

One way of modelling both kinds of uncertainty is by using model variables relating to conditional probability distributions. Epistemic uncertainty can then be modelled by computations that produce sets of distributions and may be refined by reducing the different distributions. Aleatoric uncertainty can be modelled by a computation that produces a single distribution and cannot be further refined. We use these kinds of models in RoboChart to provide probabilistic guarantees that robots satisfy their requirements. The models are then analysed with probabilistic and statistical model checkers, theorem provers, simulators, and test frameworks. Techniques using these tools provide varying degrees of confidence.

We can also use RoboChart's probabilistic models in synthesis. For example, we can synthesise a controller from a model (see Sect. 2.4) or use the model for planning (see Sect. 2.5) or self-adaptation (see Sect. 2.6).

In the rest of this section, we discuss the sources of uncertainty in robotics and the contributions made by modelling. We divide our examples into six major categories: (1) abstractions of the real world, (2) hardware failures, (3) approximate control algorithms, (4) controller synthesis, (5) motion planning, and (6) self-adaptation. We give one or more examples of each category.

2.1   Abstractions of the Real World

Sardar and Hasan consider control algorithms in cell injection [72]. This is a technique used by biologists studying individual cells. They use cell cultures to isolate cells from plant or animal tissue and then put them in an environment where they can grow. Cell injection delivers small volumes of samples into cells. The sample may be genetic material for gene therapy, sperm for in vitro fertilisation, or a drug for targeted therapy or development.

Existing manual and semi-automated cell injection systems need lengthy training and suffer from a high probability of contamination and a low success rate. Autonomous robotic cell injection systems have recently been introduced to overcome this. The force of the injection used by these robots is an important success factor: too much can destroy the membrane or tissue of the cell. A force impedance algorithm is used to control the injection and its validity is assessed by simulation. Uncertainties in the system are generally ignored in this assessment.

Sardar and Hasan present a formal analysis method based on probabilistic model checking to analyse cell injection robots using this kind of algorithm [72].[2] They use probability to model real-world disturbances and measurement noise in their models. These randomised elements have a major effect on the insertion force and thus survivability of cells. Their inclusion in the model provides useful insights into the effectiveness of a given cell injection robot. They provide these insights as quantitative properties of the robots. Their method uses PRISM [49] and has found a previously unknown discrepancy in the algorithm.[3]

There are several random factors affecting the results in these robot injection systems: internal disturbances, such as plant uncertainty, and external disturbances such as environmental effects and measurement noise of encoders, and these are what lead to uncertain states. The key property for the robot is the probability that position and force errors will exceed certain thresholds.

2.2 Hardware Failures

We divide our discussion of hardware failure into two parts: sensor and actuation noise and other kinds of failures.

2.2.1 Sensor and Actuation Noise

Lahijanian et al. [52] deploy robots automatically from their specifications by repurposing a verification algorithm. They use PCTL [32][4] to specify properties that the robot must meet on the regions of a partitioned environment. They assume that they can determine the current region occupied by the robot. They also assume that they can predict the outcome of a control action only probabilistically, due to sensor and actuation noise. They take inspiration from the PCTL model checking algorithm in generating a control strategy for an MDP that maximises the probability of satisfying the PCTL specification. They base their implementation on PRISM 3.3 [39].

---

[2]Model checking compares a temporal-logic formula p and a structure M with initial state s [10]. The model-checking problem is to decide if the model has the property: does M model p? We write this as $M, s \models p$. Probabilistic model checking uses probabilistic models and probabilistic temporal logics.

[3]PRISM is a probabilistic model checker that analyses several kinds of probabilistic models [49]. These include discrete-time and continuous-time Markov chains (DTMCs and CTMCs), Markov decision processes (MDPs), and probabilistic timed automata (PTAs).

[4]Computation tree logic (CTL) is a branching-time logic: its model of time is a tree-like structure. This means that there are different paths in the future, any one of which might be followed in the future. CTL temporal operators that describe properties constraining how these futures can unfold. Probabilistic CTL (PCTL) is an extension of CTL with probabilistic quantification of properties.

Lahijanian et al.'s verification is quantitative, giving probabilistic satisfaction guarantees due to noisy hardware. They motivate their work with an example of an emergency rescue mission. A property of interest is to assess the probability of reaching a destination by car. A dangerous route is divided into two kinds of geographic areas: either entirely safe regions or relatively safe regions. You can enter the latter only if medical assistance is available. Transition probabilities are computed given the sensor and actuator noise model or through experimental trials.

2.2.2 Hardware Resilience

Tarasyuk et al. define the *resilience* of a multi-robotic system as the ability to achieve goals despite robotic failures [74]. They show how to rigorously specify and verify the essential properties of resilience mechanisms of multi-robotic systems by refinement in Event-B [1]. To assess resilience, they annotate their formal models with statistical data on failures and rely on probabilistic verification.

They use PRISM [49] to calculate the probability of goal reachability in the presence of robot failures. They use these analyses to compare different reconfiguration strategies for various architectures. The authors apply their technique to a multi-robot cleaning system. They want their robotic system to clean the whole arena in the presence of certain failures. Their objective is to compute the probability that their goal is eventually reachable.

The key result depends on assumptions about the robots' work rates (assignment and reassignment of tasks) and hardware failure rates (both robots and base stations). To capture these properties, the authors use CTMCs.

2.3 Approximate Control Algorithms

In this section, we discuss examples of approximate control algorithms in two classes: robot swarm algorithms and bio-inspired control.

2.3.1 Robot Swarms

Konur et al. [48] and Brambilla et al. [5] have developed similar approaches to studying robot swarms using probabilistic models and using probabilistic model checking for validation. Brambilla et al. propose a property-driven design method. They model an arbitrary swarm member's interactions with the rest of the swarm and the environment as stochastic events.[5]

---

[5]By "stochastic", we mean a random variable indexed over time.

For example, a robot swarm needs to cluster in an area of the environment. The main property is that, eventually, all robots form an aggregate either on area a or area b. They set a time limit that depends on N, the size of the swarm: larger swarms need more time to aggregate. To check this property, Brambilla et al. develop a discrete-time macroscopic model, where each state tracks robot activity.[6]

Their model is non-spatial, ignores robot trajectories, and assumes that robots move instantaneously from c to a or b, and vice versa. The model could be described abstractly with nondeterministic actions for moving or staying put. A more useful model comes from using the geometric properties of the areas to compute p_{ca}: the instantaneous probability that a robot moves from c to a.

A robot in c can either go to a, go to b, or stay at c. A robot at c has a probability of going from c to a equal to $p_{ca} = A_a/A_{arena}$, where A_a is the size of area a and A_{arena} is the size of the entire environment. It has a probability of going from c to b equal to $p_{cb} = A_b/A_{arena}$, where A_b is the size of area b, and a probability of staying in c equal to $p_{cc} = A_c/A_{Arena} = 1 - (p_{ca} + p_{cb})$, where $A_c = A_{arena} - (A_a + A_b)$. The remaining probabilities depend on the behaviour of the robots.

Probability is used here to construct an approximate control algorithm arising from the need to cope with environmental abstraction.[7]

2.3.2   Bio-inspired Control

Gainer et al. conduct a formal analysis of an algorithm inspired by the foraging behaviour of ants [27]. A swarm of flying micro aerial vehicles (MAVs) searches for a target at some unknown location. It has ant-inspired behaviour based on virtual pheromones.[8] Gainer et al. demonstrate probabilistic and statistical model checking of properties to complement simulation results. Their algorithm is approximate and involves probabilistic behaviour.

Part of the swarm forms a temporary grid of nodes, while others search for the target. While in the node state each MAV broadcasts its pheromone level to other nodes within its communication range of 100 m. An MAV A in its exploring state may reach a position where there is already an MAV B. A continues moving outward, making a probabilistic choice between moving left and moving right. A decides which way to go from the levels of virtual pheromone deposited on the two directions.

---

[6]This macroscopic model is also known as a *population model*, a type of mathematical model applied in the study of population dynamics. Life scientists use these models to study ageing populations, population growth, or population decline.

[7]See Dixon et al. for a discussion of verifying robot swarm behaviour [14]. They discuss the formal verification of a robot swarm algorithm using temporal-logic model checking and the exploration and analysis of different abstractions for tractability.

[8]Trail pheromones are secreted from the body of an ant to change the behaviour of other ants that detect them. They are often used to lead their companions towards a food source. Gainer et al. [27] use this as a metaphor for their control algorithm.

Probability is used here to construct an approximate control algorithm inspired by the behaviour of ants in the real world.

2.4 Controller Synthesis

Johnson and Kress-Gazit consider the creation of a robot controller for a complex task by synthesising a hybrid controller from a high-level task specification [44, 45]. They replace the usual assumption that actuation of the robot is perfect by a probabilistic guarantee. They compose probabilistic models of the environment's behaviour and the robot actuation error with the synthesised controller. They then use PRISM [49] to find the probability that the robot satisfies the specifications. They have a preliminary approach to analysing the composite model to automatically generate revisions to improve the robot's high-level behaviour.

Probability is used here to guide controller synthesis with imperfect actuators.

2.5 Motion Planning

In this section, we discuss four examples of probabilistic motion planning, each optimising an aspect of quality of service.

2.5.1 Probabilistic Satisfaction Guarantees

Bustamante et al. [6] and Guerrero et al. [30] use probabilistic methods to decide the next action for a robotic player in a football match. Llarena and Rosenblueth work with humanoid robotic football teams [58]. They maximise or minimise scoring probability, depending on whether their team is attacking or defending.

They use MDPs to analyse the teams' playing strategies, proposing a playing model as a global strategy for attacking or defending in part of a match. They analyse the model to tune parameters depending on the robotic capabilities of both teams using PRISM [39]. They suggest adding on-board model checking to develop specific strategies during the match, such as passing the ball or the goalkeeper becoming an attacker (a robotic Peter Schmeichel).

Probability is used here to deduce plans by optimising the satisfaction of goals.

2.5.2 Cost-optimal Planning

Lacerda et al. [50, 51] present a method to calculate cost-optimal policies for co-safe linear temporal-logic task specifications over an MDP model of a stochastic system.[9]

They consider scenarios where the task may not be achievable with probability 1. They formalise a task progression metric and use multi-objective probabilistic model checking to generate policies that are formally guaranteed to maximise the probability of finishing the task, maximise progress towards completion if this is not possible, and minimise the expected time or cost required. They illustrate their approach with an example of dynamic robot task planning: visiting rooms that may become inaccessible during the robot's mission.

Probability is used here for deducing cost-optimal plans.

Fentanes et al. consider coping with change in planning problems in robotic navigation [18]. Effective planning requires assumptions about the future state of the world and the robot's corresponding chances of accomplishing its actions. The authors observe that the plan is only as good as its predictions about the world.

The authors represent changes from periodic events in the environment, such as opening or closing a door. These events may influence the probability of success of planned actions. The approach models the probability of action success as a set of supervising periodic processes. They then use this in a probabilistic planning framework to devise better navigation plans.

Probabilistic analysis is used here to evaluate planning for uncertainty.

2.5.3 Mission Goals and Expected Rewards

Chaki and Giampapa consider robotic applications in dangerous or inaccessible locations [7]. One way to maximise the chances of mission success is to deploy teams of robots. These teams operate under inherently uncertain conditions: the robots might fail and they must continuously adapt to changing environmental conditions. Constructing a mission requires specifying the team's members and how they coordinate and plan their mission. The goal is to maximise a utility function, such as the probability of mission success.

The authors propose an approach to compute the quantitative utility of robotic missions using probabilistic model checking. They show how to express a robotic mine-clearing mission as a restricted type of DTMC αPA and its utility as LTL formulas and expected rewards. They prove theorems that compute the utility of a system composed of several αPAs by combining the utilities of each αPA in isolation.

Probability is used here with Markov reward models to quantify mission goals.

---

[9]The co-safe fragment of LTL contains formulas that can be satisfied within a finite time. The satisfying infinite sequences for these formulas always have a satisfiable finite prefix.

2.6   Self-Adaptation

Ramaswamy et al. consider reactive and self-adaptive robotic architectures with the context-dependent dynamic invocation of software components [70]. They show how self-adaptation can be formally defined and modelled in an architecture-independent way. They propose a probabilistic approach to quality requirements for system design. They implement a semi-automatic transformation process to generate a subset of the solution space model that satisfies non-functional requirements. The best possible solution is then dynamically selected during runtime by maximising probabilities for quality of service goals. These techniques form the basis of their model-driven self-adaptive framework for robotic systems.

Their motivating example involves a Velodyne Lidar sensor in an autonomous vehicle's tracking system. The tracking system should detect vehicles in the environment and compute its own state (position and velocity) using a segmentation algorithm.[10]   The choice of algorithm depends on several factors: the type of sensor (lidar, time-of-flight camera), properties of the data (density, resolution, colour, intensity), and environmental features (indoor, outdoor, cluttered, flat or sloppy terrain, vegetative land). In this context, the algorithm must satisfy non-functional requirements (performance, availability, effectiveness) and quality of service parameters (the ability of a service to provide a quality level to the different demands of the clients).

Probability is used here in maximising quality of service parameters at runtime.

2.7   Summary

In this section, we have described different sources of uncertainty in robotic applications. We have given examples from the literature to illustrate each category discussing how they are modelled and analysed. In the next section, we discuss the probabilistic features of RoboChart.

3   Probability in RoboChart

In RoboChart (Miyazawa et al. [64]), the behaviour of a robot is described by several state machines that model different robotic controllers. These state machines may execute particular operations and react to events from their environment. RoboChart also includes architectural constructs for describing robotic platforms and their

---

[10]Image segmentation partitions a digital image into multiple segments. It abstracts the representation of a complex image to make it easier to analyse. Image segmentation is typically used to locate objects and boundaries, such as lines or curves.

controllers with synchronous communication between controllers and asynchronous communication between controllers and their hardware. It has constructs to specify time properties: budgets and deadlines for operations and events. Moreover, state machines can also make probabilistic choices, which allows them to model different kinds of uncertainty, such as those mentioned in Sect. 2.

In this section, we start by giving an example of a simple probabilistic algorithm in RoboChart. We then give an overview of the Ransac pose-estimation algorithm.

3.1 RoboChart with Probabilities

We start with a simple probabilistic algorithm described in RoboChart.

Example 1 (Uniform Distribution) We take as an example the state machine in Fig. 13.1 named ChooseUniform() and describe an algorithm to choose a random value in the range $1 .. N$. It has an initial node (marked with 'i'), a final node (marked with 'F'), and a single state, named 'TestLoop'. It has a fourth node inscribed 'P': a probabilistic choice between two branches. There are two state variables: integer i and Boolean c, used to record the number of iterations and whether to continue iterating, respectively.

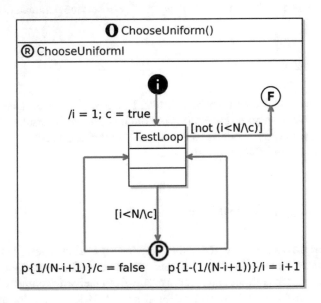

Fig. 13.1 ChooseUniform() state machine in RoboChart.

The state machine implements the following pseudocode:[11]

$i, c := 1, true;$
while $i < N \wedge c$ **do**
$\quad (c := false) \oplus_{1/(N-i+1)} (i := 1)$

In the state machine, the probabilistic choice node chooses its left-hand branch with probability $1/(N-i+1)$ and its right-hand branch with probability $1 - (1/(N-i+1)) = (N-i)/(N-i+1)$. The loop is executed at most N times. This is cut short if the value of c changes to false. When the loop terminates, i has a value between 1 and N. The precise value is chosen probabilistically: it depends on the probabilistic choice following the left-hand branch, since that's where c changes from true to false. Clearly, this path in the state chart is taken only once.

We assume that the choice to go left is made on the arbitrary kth iteration. The right-hand path must have been taken on all the previous k − 1 iterations. So the probability that we take the left-hand path on the k'th iteration is

$$
\left(\prod_{j=1}^{k-1} \frac{N-j}{N-j+1} \right) \times \frac{1}{N-k+1}
$$
$$
= \frac{N-1}{N} \times \frac{N-2}{N-1} \times \cdots \times \frac{N-k+1}{N-k+2} \times \frac{1}{N-k+1}
$$
$$
= \frac{1}{N}
$$

So the state machine terminates with $i = k$ with probability $1/N$: the state machine computes the uniform distribution over $1 \mathbin{..} N$.

The purpose of this example is to show the structure of a probabilistic algorithm in RoboChart and to give a simple correctness argument. This argument is informal but it relies on more rigorous ideas explored below.

3.2 *Example Pose-Estimation Algorithm: Ransac*

Thrun poses some key questions for probabilistic robotic applications [75]:

1. What can we do with these models?
2. How much can we recover about the true state of the world?
3. Can we still control robots to meet their requirements?

---

[11]The operator $\oplus_p$ is probabilistic choice. It chooses its left-hand argument with probability p and its right-hand argument with probability $1 - p$.

Probabilistic state estimation is an important technique to recover state from sensor data. Common state variables include:

1. **Localisation** The robot's location relative to an external coordinate frame
2. **Mapping** The location of items in the environment: walls and doors
3. **Moving mapping** Current locations of moving objects: people, doors, and robots

Recursive Bayesian estimation is a general probabilistic approach for estimating an unknown probability distribution recursively.[12] It uses a time series of measurements and a mathematical model.[13] This estimation technique uses an algorithm known as a *Bayesian filter*.[14] This gives an estimate as a probability distribution over the state x, conditioned on all available data: controls and sensor measurements [75]. It is not guessing the state. Instead, it calculates the probability that a particular state x is correct. Robotic controllers use Bayesian filters for interpreting streams of data in applications. These include robot location and mapping problems. The techniques used include Markov models, Kalman filters, dynamic Bayesian networks, and partially observable MDPs. See Doucet's overview [15].

We illustrate our engineering approach with an example of a simple filter algorithm: Ransac (random sample consensus).[15] Fischler and Bolles's algorithm is an iterative method to estimate the parameters of mathematical models from a set of observed data [20]. It does this by detecting outliers and giving a probabilistic estimate of the parameters. It assumes that the observed data contains only a minority of outliers so that an accurate model of the world exists.

Fischler and Bolles invented Ransac to solve the *location determination problem* (LDP): given an image of landmarks with known locations, determine the viewpoint. This is the point in space where an observer captured the image. The algorithm computes the minimum-landmark solutions in closed form.

Some vision-based SLAM algorithms use Ransac [53].[16] The implementation is simple and robust. Standard Ransac sometimes may not solve the problem before the algorithm terminates, but there is a relationship (explored below) between the number of iterations and the probability of there being no outliers.

---

[12] *Frequentist probability* interprets an event's probability as the limit of its relative frequency over many trials. *Bayesian probability* is an alternative where probability is interpreted as reasonable expectation representing a state of knowledge or the quantification of a personal belief.

[13] For the background to this topic, see Doucet et al.'s tutorial account of particle filters [15].

[14] A Bayesian filter is a probabilistic algorithm for estimating an unknown probability density function recursively over time using incoming measurements and a mathematical process model. The process relies on *Bayesian statistics*, the study of prior and posterior probabilities. Prior probability represents what is originally believed before new evidence is discovered, and posterior probability takes this new evidence into account.

[15] Our original interest in Ransac was in its contribution to the evolution of a practical autonomous system: the Aethon TUG robot. Ransac is used as part of the robot controller's autonomous wall-following behaviour. See Dubrawski and Thorne [16].

[16] Simultaneous localisation and mapping (SLAM) is the computational problem of constructing or updating a map of an unknown environment while keeping track of an agent's location [17].

Raguram et al. give a comparative analysis of a wide variety of Ransac techniques [68]. Although it is 40 years old, Raguram points out that Ransac remains one of the most popular tools for robust estimation [69]. For example, the Mars Exploration Rovers used Ransac in their visual odometry function because they had very limited computational resources and Ransac is an inexpensive algorithm [60]; this is typical of space-flight applications. In Sect. 1, we discussed the Aethon TUG robot [2, 16], where it is used because of its simplicity. These two examples are representative of many current robotic applications. Recently, there has been considerable research into the algorithm and its efficiency and robustness (see, e.g., Raguram et al. [69] and Ci et al. [8]).

One of our contributions is to explore why algorithms like Ransac are correct. For example, there is no existing proof of correctness of Ransac or its many optimisations using a program verification logic. We look at a naïve but familiar model-fitting algorithm that is particularly sensitive to outliers. This will motivate the technique at the centre of particle filters like Ransac.

4   Model-Fitting Methods

In this section, we look at the well-known least-squares regression algorithm. We describe a derivation of the algorithm from first principles and formulate a correctness criterion (Theorem 1). We make it clear why the algorithm is "least", why "squares", and why "regression". We give an example that shows how the algorithm is not robust against outliers. This is our motivation for Ransac. Least-squares regression can also be used as a part of the Ransac algorithm (other regression algorithms would do equally well).

The examples in this section use linear models: $y = mx + b$ for simplicity of presentation. Samples with two points are enough to find the model's parameters m and b. A line will connect any two points, so a linear equation is an exact fit providing they have distinct x coordinates. A quadratic equation will exactly fit three points. In general, an m-degree polynomial will exactly fit $m + 1$ points. Of course, if there are more than $m + 1$ points, an exact fit is not guaranteed and there may be several candidate solutions. This is what least-squares regression is doing: it decides between these solutions by comparing their deviations.

We assume that we have a model M with parameter x and a set of data D. A model-fitting method finds the best model that fits the data by optimising the parameter using an error function on D.[17]

We discuss a simple model-fitting method: *least-squares regression*. We explore its mathematical basis and show that this simple approach is not robust.

---

[17]We speak about fitting a model (a dynamic object) to a dataset (a static object). Other sources speak about fitting a dataset to a model. For example, see Wolfram www.wolframalpha.com/examples/mathematics/statistics/regression-analysis/.

We consider a straight-line model in the Cartesian plane: $y = mx + b$. Our task is to fit this model to a set of observations to discover the two model parameters m (slope) and b (y-axis intercept).

We take a pair of observed values (x_i, y_i). The error between this observation and the corresponding predicted value is the vertical residual: $y_i - (mx_i + b)$. If the observation is above the model line, then the vertical residual is positive; otherwise, it is negative. We need to normalise the residual to obtain an absolute value. A simple way to do it is to take the square: $(y_i - (mx_i + b))^2$. This has two advantages: it is a positive value and it exaggerates outliers.[18]

Our goal is to minimise the total error in the vertical residuals as we fit the model to the dataset. Call this error, the sum of normalised residuals, ε:

$$\varepsilon(m, b) = \sum_{i=1}^{N}(y_i - (mx_i + b))^2$$

Now calculate its value.

$\varepsilon(m, b)$

$=$ { by definition: $\varepsilon(m, b) = \sum_{i=1}^{N}(y_i - (mx_i + b))^2$ }

$\sum_{i=1}^{N}(y_i - (mx_i + b))^2$

$=$ { algebraic expansion (square of sum) }

$\sum_{i=1}^{N}(y_i^2 - 2y_i(mx_i + b) + (mx_i + b)^2)$

$=$ { algebraic expansion (multiplication over sum) }

$\sum_{i=1}^{N}(y_i^2 - 2my_ix_i - 2by_i + m^2x_i^2 + mb2x_i + b^2)$

$=$ { $\sum$ distributes over sums }

$\left(\sum_{i=1}^{N}y_i^2\right) - \left(\sum_{i=1}^{N}2mx_iy_i\right) - \left(\sum_{i=1}^{N}2by_i\right)$
$+ \left(\sum_{i=1}^{N}m^2x_i^2\right) + \left(\sum_{i=1}^{N}2bmx_i\right) + \left(\sum_{i=1}^{N}b^2\right)$

$=$ { $\sum$ distributes through constant factors }

$\left(\sum_{i=1}^{N}y_i^2\right) - 2m\left(\sum_{i=1}^{N}x_iy_i\right) - 2b\left(\sum_{i=1}^{N}y_i\right)$
$+ m^2\left(\sum_{i=1}^{N}x_i^2\right) + 2bm\left(\sum_{i=1}^{N}x_i\right) + b^2\left(\sum_{i=1}^{N}1\right)$

---

[18]Exaggerating outliers is an advantage because it isolates them. As we see later in the chapter, it is also a disadvantage in distorting models that we fit to datasets.

$$= \quad \{ \text{ summing a series of constants: } \left(\sum_{i=1}^{N} 1 \right) = N \; \}$$

$$\left(\sum_{i=1}^{N} y_i^2 \right) - 2m \left(\sum_{i=1}^{N} x_i y_i \right) - 2b \left(\sum_{i=1}^{N} y_i \right)$$
$$+ m^2 \left(\sum_{i=1}^{N} x_i^2 \right) + 2bm \left(\sum_{i=1}^{N} x_i \right) + Nb^2$$

We want to minimise the expression in the final line of this derivation. This is a function of two variables: m and b. All the other terms are constants, even the summations. One way to minimise this function is to use multivariate calculus. We can find the stationary point using partial derivatives for each of the variables:

$$\frac{\partial \varepsilon}{\partial m} = 0 \quad \text{and} \quad \frac{\partial \varepsilon}{\partial b} = 0$$

Of course, the stationary point might be a minimum, but it might also be a maximum or even a saddle point. The second derivatives will tell us.

There is another, simpler method to minimise ε. We begin by fixing b and minimising ε over m. We rearrange our formula to find a quadratic expression in m:

$$\varepsilon(m, b)$$
$$= \quad \{ \text{ previous derivation } \}$$

$$\left(\sum_{i=1}^{N} y_i^2 \right) - 2m \left(\sum_{i=1}^{N} x_i y_i \right) - 2b \left(\sum_{i=1}^{N} y_i \right)$$
$$+ m^2 \left(\sum_{i=1}^{N} x_i^2 \right) + 2mb \left(\sum_{i=1}^{N} x_i \right) + Nb^2$$

$$= \quad \{ \text{ rearranging as quadratic in } m \text{ and abstracting fresh variables } p, \, q, \text{ and } r \; \}$$

$$\left(\sum_{i=1}^{N} x_i^2 \right) m^2 \qquad\qquad\qquad =: pm^2$$
$$+ 2 \left(b \sum_{i=1}^{N} x_i - \sum_{i=1}^{N} x_i y_i \right) m \qquad =: qm$$
$$+ (Nb^2 - 2b \sum_{i=1}^{N} y_i + \sum_{i=1}^{N} y_i^2) \qquad =: r$$

$$= \quad \{ \text{ complete the square: } pm^2 + qm + r \;=\; p(m + q/2p)^2 + r - q^2/4p \; \}$$

$$p(m + q/2p)^2 + r - q^2/4p$$

The final manipulation, completing the square, reveals something interesting: $\varepsilon(m, b)$ describes a parabola. We review the definition we need next.

Definition 1 (Parabola) A parabola with vertex (h, k) and constant $a \in \mathbb{N}$ is one of four formulas:

Orientation	Vertex form	Orientation	Vertex form
up	$y = a(x - h)^2 + k$	down	$y = -a(x - h)^2 + k$
right	$x = a(y - k)^2 + h$	left	$x = -a(y - k)^2 + h$

The orientation of the parabolas is made clear in the following example.

Example 2 (Parabolas) There are four equations:

(1) $y = x^2/4 + 2$ (2) $y = -x^2/4 - 2$
(3) $x = y^2/4 + 2$ (4) $x = -y^2/4 - 2$

Each of these equations describes a parabola:

Equation	Vertex form	h	k	Open
(1) $y = x^2/4 + 2$	$y = \frac{1}{4}(x - 0)^2 + 2$	0	2	up
(2) $y = -x^2/4 - 2$	$y = -\frac{1}{4}(x - 0)^2 + -2$	0	-2	down
(3) $x = y^2/4 + 2$	$x = \frac{1}{4}(y - 0)^2 + 2$	2	0	right
(4) $x = -y^2/4 - 2$	$x = -\frac{1}{4}(y - 0)^2 + -2$	-2	0	left

We depict the parabolas in Fig. 13.2. Each has a different orientation.

Returning to minimising $\varepsilon(m, b)$, we discovered the parabola in m:

$$p(m + q/2p)^2 + r - q^2/4p \qquad \textbf{vertex} \quad (-q/2p, r - q^2/4p)$$

$$p = \left(\sum_{i=1}^{N} x_i^2 \right)$$
$$q = 2 \left(b \left(\sum_{i=1}^{N} x_i \right) - \left(\sum_{i=1}^{N} x_i y_i \right) \right)$$
$$r = \left(Nb^2 - 2b \left(\sum_{i=1}^{N} y_i \right) + \left(\sum_{i=1}^{N} y_i^2 \right) \right)$$

Discovering the parabola solves our problem. First, we take advantage of the orientation. The parabola is quadratic in m, a value in the x-axis, so its orientation is up-down, not left-right. Second, we need to find out if $p(m + q/2p)^2$ is positive (up-open) or negative (down-open).

$$p(m + q/2p)^2 \geq 0$$
$$\Leftarrow \quad \{ \text{ arithmetic: } \forall a, b : \mathbb{Z} \bullet a \geq 0 \Rightarrow ab^2 \geq 0 \}$$
$$p \geq 0$$

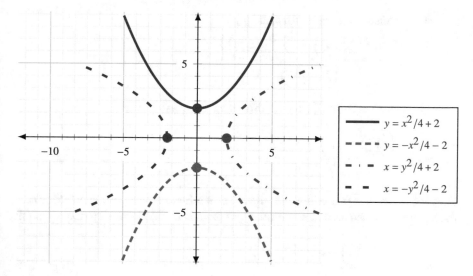

Fig. 13.2 Four parabolas with different orientations.

$=$ { assumption: $p = \left(\sum_{i=1}^{N} x_i^2\right)$ }

$\left(\sum_{i=1}^{N} x_i^2\right) \geq 0$

$\Leftarrow$ { summation algebra: $\forall s : \text{seq } \mathbb{Z};\ i : 1 .. \#s \bullet s_i \geq 0 \Rightarrow \left(\sum_{i=1}^{\#s} s_i\right) \geq 0$ }

$\forall i : 1 .. N \bullet x_i^2 \geq 0$

$=$ { arithmetic: $\forall a : \mathbb{Z} \bullet a^2 \geq 0$ }

$true$

So we have a parabola that is open upwards. This gives us the smallest value for ε: it is at the parabola's vertex. We have proved the following theorem.

Theorem 1

$\min\{\, m : \mathbb{N} \bullet \varepsilon(m)\,\} \;=\; \varepsilon(-\tfrac{q}{2p})$

Summarising:

m_{min}

$=$ { parabolic vertex }

$-q/2p$

$=$ { substituting values for p and q }

$$\frac{-2\left(b\left(\sum_{i=1}^{N} x_i\right) - \left(\sum_{i=1}^{N} x_i y_i\right)\right)}{2\left(\sum_{i=1}^{N} x_i^2\right)}$$

This gives us an equation for the minimum values of m:

$$m = \frac{\left(\left(\sum_{i=1}^{N} x_i y_i\right) - b\left(\sum_{i=1}^{N} x_i\right)\right)}{\sum_{i=1}^{N} x_i^2} \qquad \text{[Equ.1 } (m)]$$

Now we repeat this procedure and express ε as a function of b. When we complete the square, we discover that it is another parabola in b. This leads us to the equation:

$$b = \frac{\left(\sum_{i=1}^{N} y_i\right) - m\left(\sum_{i=1}^{N} x_i\right)}{N} \qquad \text{[Equ.2 } (b)]$$

Now we have two simultaneous equations. Use Equ.2 to substitute for b in Equ.1:

$$m = \frac{\left(\left(\sum_{i=1}^{N} xy\right) - \left(\left(\left(\sum_{i=1}^{N} y\right) - m\left(\sum_{i=1}^{N} x\right)\right)/N\right)\left(\sum_{i=1}^{N} x\right)\right)}{\sum_{i=1}^{N} x^2}$$

$$= m\left(\sum_{i=1}^{N} x^2\right) = \left(\sum_{i=1}^{N} xy\right) - \left(\frac{\left(\sum_{i=1}^{N} y\right) - m\left(\sum_{i=1}^{N} x\right)}{N}\right)\left(\sum_{i=1}^{N} x\right)$$

$$= Nm\left(\sum_{i=1}^{N} x^2\right) = N\left(\sum_{i=1}^{N} xy\right) - \left(\left(\sum_{i=1}^{N} y\right) - m\left(\sum_{i=1}^{N} x\right)\right)\left(\sum_{i=1}^{N} x\right)$$

$$= Nm\left(\sum_{i=1}^{N} x^2\right) = N\left(\sum_{i=1}^{N} xy\right) - \left(\left(\sum_{i=1}^{N} x\right)\left(\sum_{i=1}^{N} y\right) - m\left(\sum_{i=1}^{N} x\right)^2\right)$$

$$= Nm\left(\sum_{i=1}^{N} x^2\right) = N\left(\sum_{i=1}^{N} xy\right) - \left(\sum_{i=1}^{N} x\right)\left(\sum_{i=1}^{N} y\right) + m\left(\sum_{i=1}^{N} x\right)^2$$

$$= Nm\left(\sum_{i=1}^{N} x^2\right) - m\left(\sum_{i=1}^{N} x\right)^2 = N\left(\sum_{i=1}^{N} xy\right) - \left(\sum_{i=1}^{N} x\right)\left(\sum_{i=1}^{N} y\right)$$

$$= m\left(N\left(\sum_{i=1}^{N} x^2\right) - \left(\sum_{i=1}^{N} x\right)^2\right) = N\left(\sum_{i=1}^{N} xy\right) - \left(\sum_{i=1}^{N} x\right)\left(\sum_{i=1}^{N} y\right)$$

$$= m = \frac{N\left(\sum_{i=1}^{N} xy\right) - \left(\sum_{i=1}^{N} x\right)\left(\sum_{i=1}^{N} y\right)}{N\left(\sum_{i=1}^{N} x^2\right) - \left(\sum_{i=1}^{N} x\right)^2}$$

We now have the parameters for the best-fitting model for our data. We summarise this in the following algorithm.[19]

Definition 2 (Least-Squares Regression) Given the model parameters m (slope) and b (intercept), calculate the least-squares regression by the following steps:

Step 1:   For each (x, y) point, calculate x^2 and xy.

Step 2:   Sum all x, y, x^2, and xy.

Step 3:   Calculate slope m:

$$m = \frac{N\left(\sum xy\right) - \left(\sum x\right)\left(\sum y\right)}{N\left(\sum x^2\right) - \left(\sum x\right)^2}$$

Step 4:   Calculate intercept b:

$$b = \frac{\left(\sum y\right) - m\left(\sum x\right)}{N}$$

Figure 13.3 shows the implementation of the algorithm in RoboChart using a state machine. We use two constants dataX and dataY (sequences of real numbers) to represent the N data points in the x-axis and y-axis. The x^2 and xy values of each data point are stored in the two variables X2 and XY, and calculated by a self-transition of the state calcX2_XY. The variable i is the current index in processing and will be added by 1 to move to the next data point if the transition is taken when i is less than N. After all N data points are calculated (i is equal to N), the transition from calcX2_XY to sum is enabled. When the transition is taken, i is reset to 0. The sums of x, y, x^2, and xy are stored in the four variables sumX, sumY, sumX2, and sumXY. These summations are modelled by a step-wise addition, as shown in the action of the self-transition of the state sum. After the summations are performed, the control of the state machine is moved to the state fitModel. Upon the entry of fitModel, the model parameters m and b are calculated.

We now know how least-squares regression works. It minimises the vertical residual errors between a model and a set of observations. What could be better?

Example 3 (Least-Squares Regression with Outliers) We assume that we have a series of observations of a robot's position. The robot starts from a known point A and moves at the speed of $1\,\text{m}\cdot\text{s}^{-1}$ in a fixed direction. This is depicted in Fig. 13.4.

---

[19]The algorithm is least-squares regression. It should now be clear why "least" and why "squares". The Victorian statistician Sir Francis Galton originated the idea of "regression to the mean". In his sense, regression transforms messy, hard-to-interpret data, to a clearer, more meaningful model.

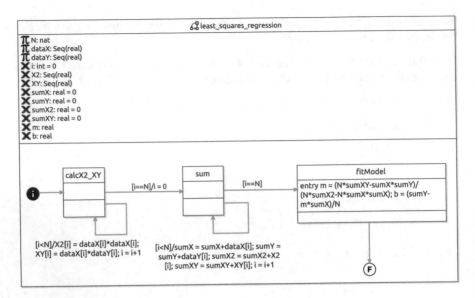

Fig. 13.3 State machine for least-squares regression algorithm.

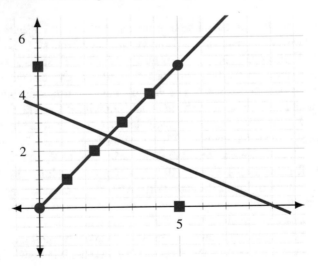

Fig. 13.4 Least-squares with two outliers.

The timed observations (*time, distance*) are:

$$\langle\, (0, 0), (1, 1), (2, 2), (3, 3), (4, 4), (5, 5) \,\rangle$$

Of course, this is just $t = s$, that is, *time = distance*. We apply least-squares regression to see what we make of this. First, a few useful calculations:

x	y	x^2	xy
0	0	0	0
1	1	1	1
2	2	4	4
3	3	9	9
4	4	16	16
5	5	25	25
$\sum x$	$\sum y$	$\sum x^2$	$\sum xy$
15	15	55	55

Now we start by calculating the slope:

$$m$$
$$= \quad \{ \text{ least-squares regression slope } \}$$
$$\frac{N\left(\sum xy\right) - \left(\sum x\right)\left(\sum y\right)}{N\left(\sum x^2\right) - \left(\sum x\right)^2}$$
$$= \quad \{ \text{ observed values } \}$$
$$\frac{6 \cdot 55 - 15 \cdot 15}{6 \cdot 55 - 15^2}$$
$$= \quad \{ \text{ arithmetic } \}$$
$$1$$

Yes, the slope is 1! Now for the intercept:

$$b$$
$$= \quad \{ \text{ least-squares regression intercept } \}$$
$$\frac{\left(\sum y\right) - m\left(\sum x\right)}{N}$$
$$= \quad \{ \text{ observed values } \}$$
$$\frac{15 - 1 \cdot 15}{6}$$
$$= \quad \{ \text{ arithmetic } \}$$
$$0$$

So there we have it: the model is

$$(y = mx + b)$$

$$= (y = 1 \cdot x + 0)$$

$$= (y = x)$$

Now we consider an example where this fails. We assume that the sensor for our observations gives default readings when the robot starts to move and when it comes to a halt. Otherwise, it is faithful. The dataset for this is as follows:

$$\langle (0, 5), (1, 1), (2, 2), (3, 3), (4, 4), (5, 0) \rangle$$

Most of the observations are correct, so where's the problem? We see what the least-squares regression says.

(x, y)	x^2	xy	$\sum x$	$\sum y$	$\sum x^2$	$\sum xy$
$(0, 5)$	0	0	15	15	55	30
$(1, 1)$	1	1				
$(2, 2)$	4	4				
$(3, 3)$	9	9				
$(4, 4)$	16	16				
$(5, 0)$	25	0				

$$\begin{aligned}
& \quad m \\
= & \quad \{ \text{ least-squares regression slope } \} \\
& \quad \frac{N \left(\sum xy \right) - \left(\sum x \right) \left(\sum y \right)}{N \left(\sum x^2 \right) - \left(\sum x \right)^2} \\
= & \quad \{ \text{ observed values } \} \\
& \quad \frac{6 \cdot 30 - 15 \cdot 15}{6 \cdot 55 - 15^2} \\
= & \quad \{ \text{ arithmetic } \} \\
& \quad \frac{180 - 225}{330 - 225} \\
= & \quad \{ \text{ arithmetic } \} \\
& \quad -\frac{45}{105} \\
\approx & \quad \{ \text{ arithmetic } \} \\
& \quad -0.43
\end{aligned}$$

The two outlying observations dominate the model and the robot appears to be getting closer to A as time goes on (the line is sloping down, not up). See Fig. 13.4.

So we see that the least-mean squares algorithm cannot adequately handle outliers. Clearly we need a better parameter estimation method, which Ransac provides.

5   Ransac Algorithm

We are ready to present the Ransac algorithm. We start by discussing the general requirements for pose-estimation algorithms (Sect. 5.1). Then we present the algorithm itself (Sect. 5.2). We conclude with an analysis of the properties of the Ransac algorithm (Sect. 5.3).

5.1   Pose Estimation

Autonomous robots process images from cameras and other devices. They determine the position and orientation (the *pose*) relative to a frame of reference so they can manipulate or avoid objects. But sensors are imperfect: they suffer from Gaussian noise and unreliable readings. Other sources of outliers include excessive noise, sensor errors, unexpected events, network hacking, and adversarial data.[20]

Given a set of observations about a robot's environment, we need an algorithm to analyse the data and fit a model to it. We want the algorithm to be fast, accurate, and robust. We need the results fast enough to be usable in a robotic control application. We need them to be accurate so that we can rely on them in critical applications. We need them to be robust to cope with the errors occurring during execution.

A pose-estimation algorithm must learn the parameters for the model. Outliers in the dataset can have a disproportionate effect on learning parameters. We saw this with the least-squares fitting method in Example 3. This distortion of the model happens partly because the algorithm uses the squares of the residuals, instead of the absolute offset values. Outliers have larger offsets and will affect the line more than the points closer to the line.

We now go into the details of the Ransac algorithm.

---

[20]Adversarial data comprises small, deliberate perturbations to make a robot find a bad model. Machine learning models are vulnerable to adversarial examples. See Wiyatno et al. for a review [76].

```
ransac_stm.rct
 1  import set_toolkit::*
 2  import sequence_toolkit::*
 3  stm ransac {
 4    const data : Set( int * int )
 5    const t : int
 6    const d : int
 7    const k : int
 8    const maxint : int
 9    var i : int = 0
10    var maybeInliers : Set( int * int )
11    var alsoInliers : Set( int * int )
12    var cands : Set( int * int )
13    var cand : int * int
14    var maybeModel : int * int
15    var betterModel : int * int
16    var bestFit : int * int
17    var bestErr : int
18    var thisErr : int
19    var err : int
20
21    initial i0
22    final f0
23    state iter_loop {
24    }
25    state calcInliers {
26    }
27    state compareModel {
28    }
29    transition t0 {
30      from i0
31      to iter_loop
32      action i = 0 ; bestErr = maxint
33    }
34    transition t1 {
35      from iter_loop
```

```
35      from iter_loop
36      to f0
37      condition i >= k
38    }
39    transition t2 {
40      from iter_loop
41      to calcInliers
42      condition i < k
43      action
44        maybeInliers = getRandPair(data);
45        maybeModel = fitModel(maybeInliers);
46        alsoInliers = { } ;
47        cands = diff(data , maybeInliers);
48        thisErr = 0
49    }
50    transition t3 {
51      from calcInliers
```

```
51      from calcInliers
52      to calcInliers
53      condition cands != { }
54      action
55        cand = chooseMinXPoint(cands);
56        cands = diff(cands , {cand});
57        err = errFit(maybeModel , cand);
58        thisErr = thisErr + err ;
59        if(err < t)then
60          alsoInliers = union(alsoInliers, {cand})
61        end
62    }
63    transition t4 {
64      to compareModel
65      from calcInliers
66      condition cands == { }
67      action
```

```
ransac_stm.rct
68        if size(alsoInliers)> d then
69          betterModel = fitModel(union(maybeInliers , alsoInliers)) ;
70          if thisErr < bestErr then
71            bestFit = betterModel ; bestErr = thisErr
72          end
73        end
74    }
75    transition t5 {
76      from compareModel
77      to iter_loop
78      action i = i + 1
79    }
80  }
81  function getRandPair(data : Set(int * int)) : Set(int * int){ }
82  function fitModel(data : Set(int * int)) : int * int { }
83  function errFit(model : int * int , cand : int * int): int { }
84  function chooseMinXPoint(data : Set(int * int)) : int * int { }
```

Fig. 13.5 Random sample consensus (Ransac).

5.2   The Algorithm

Ransac's two main parameters are the dataset D and a parametric model $M(\mathbf{x})$. Other parameters are:

- The number of points needed to estimate the model.
- The positive integer t: the threshold required for a data point to be an inlier. The threshold is part of the quality specification of a model for a model-fitting algorithm. It gives the tolerance for the residual between an observation and its corresponding model prediction.
- The quorum d: the number of data points required for a model. This is another part of the quality specification of a model.
- The number of iterations required k. This is the third quality requirement: the probability of finding a well-fitting model for the data (see Sect. 5.3 for an analysis of this parameter).

The algorithm is described in Fig. 13.5 using the textual notation of state machines in RoboChart [63].

A state machine ransac implements the algorithm. Here we assume the type of all data and parameters is integer numbers (int), and a fitted model takes two parameters and so is modelled by a pair of type int ∗ int.

The state machine is composed of several parts: variable declarations, node declarations, and transition definitions.

The variables data, maybeInliers, alsoInliers, and cands all have the same type: a set of pairs, representing a set of data points; cand is a pair, representing a data point; maybeModel, betterModel, and bestFit are tuples, representing parametric models; err, thisErr, and betsErr are integer numbers; and the constant

variable maxint is an integer number, representing a relatively large number. A
variable i is the loop index.

In this machine, there are five nodes: one initial junction i0, one final state f0, and
three other states iter_loop, calcInliers, and compareModel. Six transitions from
t0 to t5 connect these nodes and implement the actual algorithm.

The initialisation stage of the algorithm is implemented by the action of the
transition t0. An iteration loop is modelled by the transitions between the three
states. The functions that are called inside the state machine, such as diff, size, and
getRandPair, are either imported from set and sequence toolkits or declared.

5.3 Analysis of the Algorithm's Performance

We often do not know the proportion of outliers in the dataset beforehand. Instead,
it is usually estimated. Assume that the ratio of $\#outliers/\#dataset = q \in$
$[0, 1]$. Fischler and Bolles use this estimate to determine the number of iterations
needed [20].

We define $Pure(S_{i,j})$ to mean that on the ith iteration, the jth point sampled is an
inlier for model $M(\mathbf{x}_i)$. Of course, $\neg Pure(S_{i,j})$ means that the point is an outlier.

We write $P(Pure(S_{i,j}))$ to denote the probability that, on the ith iteration, the
jth point sampled is pure. Using our previous assumption, we have that

$$P(\neg Pure(S_{i,j})) = q$$

The algorithm selects on the ith iteration the points needed to find a model (the
random sample sequence S_i) and then calculates the parameters for the model (the
vector $\mathbf{x}_i$). We have run $S_i \subseteq D \wedge \#S_i = n$, the number of data points needed to
estimate the model. The sample S_i is a sequence of data points of length n.

The algorithm terminates having considered the sequence $\langle S_1, S_2, \cdots, S_k \rangle$ of
samples. Of course, the algorithm does not keep all these results, only the best model
for the dataset according to the number of inliers and gross error.

Contaminated samples will yield poor models, so we want a high probability that
at least one of these random sample records exactly n inliers. This probability is:

$$P(\exists i : 1 .. k \bullet \forall j : 1 .. n \bullet Pure(S_{i,j}))$$

That is, on some iteration the algorithm finds a completely uncontaminated sample.

$$P\left(\exists i : 1 .. k \bullet \forall j : 1 .. n \bullet Pure\left(S_{i,j}\right)\right)$$
$$= \quad \{ \text{complement: } P(x) = (1 - P(\neg x)) \}$$
$$1 - P\left(\neg \exists i : 1 .. k \bullet \forall j : 1 .. n \bullet Pure\left(S_{i,j}\right)\right)$$
$$= \quad \{ \text{predicate calculus} \}$$

$$1 - P\left(\forall i : 1 .. k \bullet \neg \; \forall j : 1 .. n \bullet Pure\left(S_{i,j}\right)\right)$$

$=$ { joint probability: $P(e_1 \wedge e_2) = P(e_1) \cdot P(e_2)$ }

$$1 - \prod_{i=1}^{k} \bullet P\left(\neg \; \forall j : 1 .. n \bullet Pure\left(S_{i,j}\right)\right)$$

$=$ { complement: $P(\neg \, x) = (1 - P(x))$ }

$$1 - \prod_{i=1}^{k} \bullet \left(1 - P\left(\forall j : 1 .. n \bullet Pure\left(S_{i,j}\right)\right)\right)$$

$=$ { jointprobability: $P(e_1 \wedge e_2) = P(e_1) \cdot P(e_2)$ }

$$1 - \prod_{i=1}^{k} \bullet \left(1 - \prod_{j=1}^{n} \bullet P\left(Pure\left(S_{i,j}\right)\right)\right)$$

$=$ { complement: $P(x) = (1 - P(\neg \, x))$ }

$$1 - \prod_{i=1}^{k} \bullet \left(1 - \prod_{j=1}^{n} \bullet \left(1 - P\left(\neg \; Pure\left(S_{i,j}\right)\right)\right)\right)$$

$=$ { assumption: $P\left(\neg \; Pure\left(S_{i,j}\right)\right) = q$ }

$$1 - \prod_{i=1}^{k} \bullet \left(1 - \prod_{j=1}^{n} \bullet (1 - q)\right)$$

$=$ { product of constants: $\left(\prod_{i=1}^{n} c\right) = c^n$ }

$$1 - \prod_{i=1}^{k} \bullet \left(1 - (1 - q)^n\right)$$

$=$ { product of constants }

$$1 - \left(1 - (1 - q)^n\right)^{k}$$

This expression is a quality parameter that we could specify as, say, p. For example, we may wish to achieve a correct model 90% of the time, in which case we set $p = 0.9$ and calculate k. Now we solve the resulting equation $p = 1 - (1 - (1 - q)^n)^k$ for the number of iterations needed to achieve this quality.[21]

$$1 - (1 - (1 - q)^n)^k = p$$

$=$ { arithmetic }

$$1 - p = (1 - (1 - q)^n)^k$$

$=$ $\left\{ \begin{array}{l} \text{log defined: } 1 - p \geq 0 \\ \text{logarithm function injective: } (x = y) = (\log(x) = \log(y)) \end{array} \right\}$

$$\log(1 - p) = \log((1 - (1 - q)^n)^k)$$

$=$ { cancelling exponential: $(b^x) = x \log(b)$ }

$$\log(1 - p) = k \cdot \log(1 - (1 - q)^n)$$

$=$ { arithmetic }

$$k = \frac{\log(1 - p)}{\log(1 - (1 - q)^n)}$$

---

[21] It does not matter which base we use when we take logarithms in this derivation since we end up with a ratio of logarithms. Recall the logarithmic identity $\log_a(x) / \log_a(y) = \log_b(x) / \log_b(y)$.

5.4 Summary

In Sect. 5.2, we presented the Ransac algorithm in RoboChart. Our analysis in Sect. 5.3 produces a well-known result about the performance of the Ransac algorithm from first principles required for our verification technique.

6 Model Checking Ransac

In this section we continue our analysis of the Ransac algorithm using the PRISM probabilistic model checker [49]. We show in some detail how the RoboChart Ransac algorithm is implemented in PRISM's input language (Alur and Henzinger's Reactive Modules formalism [3], widely used among various probabilistic model checkers). The main result in this section is confirmation that the analysis in the last section holds, for example, runs of the algorithm. We point out that we can do this only for small examples. This is an issue that we address in Sect. 7, where we give stronger guarantees about the algorithm's correctness.

6.1 The Ransac PRISM Model

We use probabilistic model checkers to analyse the different probabilistic models, including DTMCs and CTMCs, MDPs, and probabilistic timed automata (PTAs). We write verification properties in various probabilistic temporal logics.

A probabilistic model checker answers the fundamental question: what is the probability that we can reach a set of goal states? The answer to this problem is in the unique solution of a system of linear equations. Model checkers get an exact solution by using linear algebraic techniques, such as Gaussian elimination, or by using approximate iterative methods, such as power iteration.

Probabilistic model checkers include the following: PRISM [49], MRMC [46], LiQuor [9], iscasMC [31], PAT [73], and Storm [12].

We encoded the Ransac algorithm in PRISM (see Appendix), starting from our RoboChart model. To overcome the state explosion problem, we restrict our model to six observations, like in Example 3. We can analyse much larger sample sizes approximately using statistical model checking. This is a technique that checks temporal-logic formulas using Monte Carlo discrete-event simulation [54, 55, 79].

The biggest problem is that RoboChart is more abstract than PRISM. RoboChart has high-level support for algorithms: rich control flow and structured, mathematical data types. We have a prototype automated translation from RoboChart to PRISM, integrated into RoboTool [78]. As the RoboChart model for the Ransac algorithm contains mathematical types and corresponding operators, such as sets and the set difference operator, which are not currently supported in the translator, the PRISM code discussed here is not a result of the automatic translation. Instead, it results from modelling the algorithm in PRISM directly.

We present our PRISM code in Appendix. Bear in mind that PRISM code is very low level: it is the assembly code for RoboChart.

We start with several constants:

```
const int M=2; // number of points needed for
    building the model
const int N=6; // number of data points
const int t=0; // threshold value for outliers
const int d=4; // number of points needed for a valid
    model
const int k=14; // number of iterations
```

We represent the six-point dataset by the constants dataX0 .. dataX5 and dataY0 .. dataY5. To change the dataset, we must edit the code. We take the data from Example 3: $(0, 5), (1, 1), (2, 2), (3, 3), (4, 4), (5, 0)$.

The next part of the code contains a long series of formulas. A formula consists of an identifier and an expression. We use the identifier as a shorthand for the expression in the rest of the code. Formulas do not take parameters: there are no function definitions in PRISM. Unlike RoboChart, PRISM has no structured data types, only scalars. This first group of formulas simulate arrays. For example, the constants dataX0 .. dataX5 simulate the contents of a six-element array x. The formula[22]

```
formula x1 =
  (maybeInlierIndex1=0?dataX0:
    (maybeInlierIndex1=1?dataX1:
      (maybeInlierIndex1=2?dataX2:
        (maybeInlierIndex1=3?dataX3:
          (maybeInlierIndex1=4?dataX4:
            dataX5))))));
```

simulates the array expression x[maybeInlierIndex1].

The next sequence of formulas performs the computations needed for the Ransac algorithm, including what we need for the least-squares regression. There is a methodological issue here. The algorithm carries out these computations during execution. The first version of our PRISM model did exactly this, but the time taken to model-check it in PRISM was excessive. Pre-calculating these values increases the time taken to build the model, but checking the model becomes much faster. This is worthwhile if we compile the model once and then check many properties.

PRISM modules are like Back's action systems [4]: they consist of a state, an initialisation, and a set of labelled actions on the state. The actions are guarded commands (as in Dijkstra's language [13]). The state update can be probabilistic.

---

[22]In PRISM, (b?f:g) is the conditional expression **if** b **then** f **else** g.

The Ransac module contains the main code for the algorithm. The module has eight local variables:

```
pc : [pcOuterLoopTest..pcTerm] init pcOuterLoopTest;
i : [1..k+1] init 1;
maybeInlierIndex1 : [0..N-1] init 0;
maybeInlierIndex2 : [0..N-1] init 0;
bestM : [-ySize..ySize] init 0;
bestB : [-ySize..ySize] init 0;
bestErr : [0..N*ySize+1] init N*ySize+1;
bestNumInliers : [0..N-1] init 0;
```

The first variable, pc, acts as a program counter, recording the current state. We need this because the module is an action system without a richer control structure. The variable i records the current iteration. The two variables maybeInlierIndex1 and maybeInlierIndex2 are indexes into the simulated arrays holding the dataset. They constitute the current random sample. The variables bestM and bestB contain the slope and intercept for the best model so far and bestErr and bestNumInliers contain its gross error and number of inliers.

There are eight guarded commands. The first pair deals with the loop's termination:

```
[] pc=pcOuterLoopTest &i=k+1 -> (pc'=pcTerm);
[] pc=pcOuterLoopTest &i<k+1 -> (pc'=
   pcSelectMaybeInlierIndex1);
```

The next guarded command selects the next random sample:

```
[] pc=pcSelectMaybeInlierIndex1 ->
   1/N: (maybeInlierIndex1'=0) &(maybeInlierIndex2
      '=0) &
      (pc'=pcInnerLoopTest) +
   1/N: (maybeInlierIndex1'=1) &(maybeInlierIndex2
      '=1) &
      (pc'=pcInnerLoopTest) +
   1/N: (maybeInlierIndex1'=2) &(maybeInlierIndex2
      '=2) &
      (pc'=pcInnerLoopTest) +
   1/N: (maybeInlierIndex1'=3) &(maybeInlierIndex2
      '=3) &
      (pc'=pcInnerLoopTest) +
   1/N: (maybeInlierIndex1'=4) &(maybeInlierIndex2
      '=4) &
      (pc'=pcInnerLoopTest) +
   1/N: (maybeInlierIndex1'=5) &(maybeInlierIndex2
      '=5) &
      (pc'=pcInnerLoopTest) ;
```

The command has six probabilistic alternatives. N is the size of the dataset, so the command assigns one of the indexes to `maybeInlierIndex1'` with probability `1/N`.[23] The command also gives a default value to the state variable `maybeInlierIndex2'` that ensures the execution of the next guarded command occurs at least once.

The RoboChart ensures that the two sample points are distinct. We achieve this by the following two guarded commands:

```
[] pc=pcInnerLoopTest &maybeInlierIndex2=
   maybeInlierIndex1 ->
   1/N: (maybeInlierIndex2'=0) + 1/N: (
      maybeInlierIndex2'=1) +
   1/N: (maybeInlierIndex2'=2) + 1/N: (
      maybeInlierIndex2'=3) +
   1/N: (maybeInlierIndex2'=4) + 1/N: (
      maybeInlierIndex2'=5) ;
[] pc=pcInnerLoopTest &maybeInlierIndex2!=
   maybeInlierIndex1
     -> (pc'=pcTestBetterModel);
```

The next two commands compare the current model with the best model so far and update the latter if necessary.

```
[] pc=pcTestBetterModel &newNumInliers>=d &i<k+1 &
   newErr<bestErr ->
     (bestM'=newM) &(bestB'=newB) &(bestErr'=newErr)
        &
     (bestNumInliers'=newNumInliers) &(i'=i+1) &
     (pc'=pcOuterLoopTest) ;
[] pc=pcTestBetterModel &i<k+1 &!(newNumInliers>=d &
   newErr<bestErr) ->
     (i'=i+1) &(pc'=pcOuterLoopTest) ;
```

This completes the description of the loop body. The final guarded command avoids the possibility of deadlock when the loop terminates. The state update `true` does not change the state, so once the program counter reaches `pcTerm`, it stays there.[24]

```
[] pc=pcTerm -> true;
```

---

[23]The expression (`maybeInlierIndex1'=0`) is the assignment of the value 0 to the state variable `maybeInlierIndex1`.

[24]This is needed since PRISM requires that every state has at least one outgoing transition. This is a healthiness condition for the underlying probability transition matrix.

6.2 Results

We might want a 99% chance that the sample does not contain any outliers: $p = 0.99$. Since there are 2 outliers, the dataset has 6 elements ($q = 1/3$), and we need a sample size of $n = 2$, we want the following number of iterations:

$$\frac{\log(1 - p)}{\log(1 - (1 - q)^n)}$$

$$= \frac{\log(1 - 0.99)}{\log(1 - (1 - \frac{1}{3})^2)}$$

$$= 7.8$$

We check this by using the PLTL (Probabilistic LTL) formula

```
P=? [ F (pc=pcTerm&bestM=1&bestB=0) ]
```

This queries the model for the probability that in the future we reach the final state having found the right model: slope $= 1$ and intercept $= 0$. This implies the property we want, that there are no outliers in the sample. As we vary k, we get the following:

k	probability	k	probability
1	0.40000	6	0.95334
2	0.64000	7	0.97200
3	0.78400	8	0.98320
4	0.87040	9	0.98992
5	0.92224	10	0.99395

There is a difference between the theoretical prediction that eight iterations meet the 99% requirement and the empirical result that eight iterations give only 98%. This seems to be due to the imprecise arithmetic in the PRISM code. We can see evidence of this if we check two further properties with eight iterations. First,

```
P=? [ F ( pc=pcTerm &bestNumInliers=4 ) ]
```

This queries the probability that the model we discover has four inliers. The models all give a straight line through points $1 .. 4$, excluding points 0 and 5: the best model discussed above. The probability is 0.98320, exactly the entry in the table. Now check that there are no inliers:

```
P=? [ F ( pc=pcTerm &bestNumInliers=0 ) ]
```

This gives us the remaining probability of 0.01679. This model does not have any inliers, not even the sample points.

7 Stronger Guarantees

We present a proof of the correctness of a simplified form of the Ransac algorithm. In the last section, the simplification was in terms of the number of observations. The examples were still useful as a way of checking instances of the algorithm to look for bugs, and this is analogous to testing. In this section, our proof of correctness is independent of the number of observations. Our simplification this time reduces the number of proof steps to keep the presentation within the page limit for the chapter. Our future work on automated proof is intended to remove this restriction by mechanising proof steps.

We use a program verification technique where programs are predicates, following Hehner's original work [34] and its extension to probabilistic predicative programming [37]. We have also taken inspiration from McIver and Morgan [61].

7.1 RoboChart Model

We model this form of the Ransac algorithm in a state machine simpRansacSTM in RoboChart, shown in Fig. 13.6. The interface DataInf required in the machine is displayed in Fig. 13.7. In this model, we consider N equal to 6 and q equal to 1/3. In the state machine, we use a binary probabilistic junction (between the states ChooseIndex, UpdateChosenIndex, and repeat) to implement a uniform distribution described in Example 1. Here, we use a different algorithm from that in Fig. 13.1 because the model of this algorithm is currently supported by the RoboChart to PRISM translator, which allows us to analyse the model automatically. This selection algorithm will be used twice in order to choose the two random values in the range $0 .. (N - 1)$. The two values are recorded in index0 and index1 and represent the indexes of the two randomly selected data points. We use a special value -1 to represent that the corresponding index has not been selected.

After the two indexes are selected, if they represent the same data point (the guard of the transition from the state determineInliers to the state badFit), it is a bad fit and so the control moves to badFit. Otherwise, they are different data points and the one represented by index0 or index1 has probability q of being an outlier and so $1 - q$ of being an inlier. If either is an outlier, then this is a bad fit again, modelled by two probabilistic junctions around the state index0In. Otherwise, this is a good fit. For a bad fit, the diagram resets i and both indexes in order to repeat the selection.

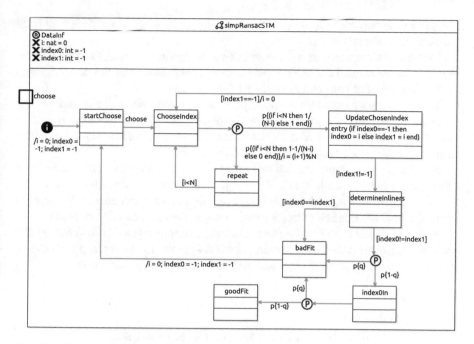

Fig. 13.6 State machine for the simplified Ransac algorithm.

Fig. 13.7 DataInf interface.

7.2   Formal Proof of Ransac

The algorithm in Fig. 13.6 is related to Hehner's probabilistic predicative programming approach [37]. We abstract the choice from the uniform distribution by using
a random number generator *rand*. We fit our model with parameters u and v to the
dataset D. We let $M(u, v)$ denote the number of inliers that results from this and we
let d be the number of inliers for the model to be valid. In case $u = v$, M returns the
value 0. We start with the following program:

$$ R \; \widehat{=} \; u, v := rand(N), rand(N) \; ; \; \textbf{if } M(u, v) \geq d \textbf{ then } t' = t \textbf{ else } (t := t + 1 \; ; \; R) $$

This is an abstraction of the Ransac algorithm. The tail-recursive program repeat-
edly chooses random values for u and v. If the resulting model fits the data with
enough inliers, the program terminates. Otherwise, the program increments the
integer variable t and then repeats. To account for execution time, we are using

Hehner's time variable: t is the time the execution starts and t' is the time it ends. In the case of nontermination, $t' = \infty$.

Our abstraction has reduced the checking of fitting the model to the data by using an uninterpreted function M and streamlining the rest of the code. A more convincing proof of correctness needs all these details.

We use a pseudo-random number generator to get values for u and v. The function *rand* returns a value whose distribution is uniform (constant) over a nonempty finite range. For $N \in \mathbb{N}$, the notation $rand(N)$ produces natural numbers uniformly distributed over the range $1 .. N$, inclusively.

Program R is a simple version of Ransac. It ignores the gross error in fitting a model to the data. Instead, it finds the first pair of parameters that makes the model fit. When it terminates, we have the time taken to find a suitable model. We want to know the final distribution of this variable: how many iterations do we need?

As Hehner points out, $rand$ is not a function, since it can produce different results from the same parameter. In reasoning about this program, we replace each use of $rand(N)$ with fresh variables $r(t)$ and $s(t)$, indexed by the current time.[25]

$$R$$
$$= \quad \{ \text{ by definition } \}$$
$$u, v := rand(N), rand(N) \ ; \ \textbf{if } M(u, v) \geq d \textbf{ then } t' = t \textbf{ else } (t := t + 1 \ ; \ R)$$
$$= \quad \{ \text{ introduce random variables } \}$$
$$\textbf{if } M(r(t), s(t)) \geq d \textbf{ then } t' = t \textbf{ else } (t := t + 1 \ ; \ R)$$

The definition of the recursive program is a fixed-point equation in R.[26] The solutions to this equation are predicates that are invariants of the program. We start by thinking about the program's postcondition.[27] R finds the model that fits the data. If it terminates, then we have $M(r(t'), s(t')) \geq d$. At all earlier times, the

---

[25]Hehner is encoding in a proof strategy a semantics for the random number generator as a sequence of values. This is a nondeterministic time-ordered implementation of a uniform distribution. His semantics for a program is then a function from this distribution to a final distribution of states.

[26]The fixed points of an equation $y = f(x)$ are all the solutions x, where $x = f(x)$. Defining a recursive program is similar: the fixed points of a recursive program $Y = F(X)$ are all the conditions X, where $X = F(X)$. An important point here is that the condition X that we are considering doesn't change: it is *invariant*. So if we have a recursive program defined as $R = \textbf{if } z \neq 0 \textbf{ then } (z := z - 1 \ ; \ R)$ then this has the fixed point $z \geq 0$. That is, we guarantee that if we start with $z \geq 0$, then when we terminate, we will also have $z \geq 0$. The condition is *invariant*. The guarantee is made by the definition of the program's semantics, and from this we can prove that the condition is a fixed point.

[27]The program's *precondition* is the assumption of what must be true for a program's correct execution. The *postcondition* is the guarantee of what must be true on termination.

model is not valid:

$$(\forall i : t .. t' \bullet M(r(i), s(i))) < d)$$

Our postcondition is the conjunction of these two properties:

$$Q \widehat{=} (\forall i : t .. t' - 1 \bullet M(r(i), s(i)) < d) \wedge M(r(t'), s(t')) \geq d$$

We show that this is a solution by substituting it for R on the right-hand side.

$$\textbf{if } M(r(t), s(t)) \geq d \textbf{ then } t' = t \textbf{ else } (t := t + 1 \; ; \; R)$$

$=$ { substitute Q for R }

    if $M(r(t), s(t)) \geq d$ **then**

       $t' = t$

    else

       $(t := t + 1 \; ; \; (\forall i : t .. t' - 1 \bullet M(r(i), s(i)) < d) \wedge M(r(t'), s(t')) \geq d)$

$=$ { assignment: $(x := e \; ; \; P) = (P[e/x])$ }

    if $M(r(t), s(t)) \geq d$ **then**

       $t' = t$

    else

       $((\forall i : t + 1 .. t' - 1 \bullet M(r(i), s(i)) < d) \wedge M(r(t'), s(t')) \geq d)$

$=$ { conditional: $(\textbf{if } b \textbf{ then } P \textbf{ else } Q) = ((b \wedge Q) \vee (\neg b \wedge Q))$ }

    $(M(r(t), s(t)) \geq d \wedge t' = t)$

    $\vee \, (M(r(t), s(t)) < d$

       $\wedge \, (\forall i : t + 1 .. t' - 1 \bullet M(r(i), s(i)) < d)$

       $\wedge \, M(r(t'), s(t')) \geq d)$

$=$ $\left\{ \begin{array}{l} \text{predicate calculus:} \\ \quad (P(i) \wedge (\forall x : i + 1 .. n \bullet P(x)) = (\forall x : i .. n \bullet P(x)) \end{array} \right\}$

    $(M(r(t), s(t)) \geq d \wedge t' = t)$

    $\vee$

    $(\forall i : t .. t' - 1 \bullet M(r(i), s(i)) < d) \wedge M(r(t'), s(t')) \geq d)$

$=$ { propositional calculus }

    $((t' = t) \vee (\forall i : t .. t' \bullet M(r(i), s(i)) < d) \wedge M(r(t'), s(t')) \geq d)$

$=$ { propositional calculus: $(p \Rightarrow q) \Rightarrow ((p \vee q) = q))$ }

    $(\forall i : t .. t' - 1 \bullet M(r(i), s(i)) < d) \wedge M(r(t'), s(t')) \geq d$

$=$ { definition of Q }

    Q

So Q is a solution for R and therefore an invariant of the program.

We treat $r(i)$ and $s(i)$ as variables in their own right. We write $Out(r(i))$ to denote that $r(i)$ is an outlier in the model with the current parameters. The probability that $Out(r(i))$ is q (since $r(i)$ is uniformly distributed over $1 .. N$). That is, the probability of selecting an outlier is q. We have

$$\sum r(i) : 1 .. N \bullet Out(r(i)) = qN$$

For inliers, we have the complementary result:

$$\sum r(i) : 1 .. N \bullet \neg\, Out(r(i)) = (1-q)N$$

For combinations, we have

$$\left(\sum r(i), s(i) : 1 .. N \bullet Out(r(i)) \wedge \neg\, Out(s(i))\right) = qN(1-q)(N-1)$$

$$\left(\sum r(i), s(i) : 1 .. N \bullet (r(i) \neq s(i)) \wedge Out(r(i)) \wedge Out(s(i))\right) = qNq(N-1)$$

We can use these results to answer the following question: how many ways can $M(r(i), s(i))$ be strictly less than d?

Hehner's probabilistic predicative semantics treats truth values as integers: *true* is 1 and *false* is 0. So our question can be formalised as the sum

$$\sum r(i), s(i) : 1 .. N \bullet M(r(i), s(i)) < d$$

The predicate $M(r(i), s(i)) < d$ is true in three cases: either of the parameters could be outliers, but also $r(i)$ and $s(i)$ could be equal. We can decompose these overlapping cases into seven disjoint cases using the exclusive-or operator ($\underline{\vee}$):

$$\sum r(i), s(i) : 1 .. N \bullet M(r(i), s(i)) < d$$

$$= \quad \left\{ \begin{array}{l} \text{property of} \\ \quad M\colon M(r(i), s(i)) < d = (r(i) = s(i)) \vee Out(r(i)) \vee Out(s(i)) \end{array} \right\}$$

$$\sum r(i), s(i) : 1 .. N \bullet (r(i) = s(i)) \vee Out(r(i)) \vee Out(s(i))$$

$$= \quad \{\ \text{proposition calculus}\ \}$$

$$\sum r(i), s(i) : 1 .. N \bullet$$

$$((r(i) = s(i)) \wedge Out(r(i)) \wedge Out(s(i)))$$
$$\underline{\vee}\ ((r(i) = s(i)) \wedge Out(r(i)) \wedge \neg\, Out(s(i)))$$
$$\underline{\vee}\ ((r(i) = s(i)) \wedge \neg\, Out(r(i)) \wedge Out(s(i)))$$
$$\underline{\vee}\ ((r(i) = s(i)) \wedge \neg\, Out(r(i)) \wedge \neg\, Out(s(i)))$$
$$\underline{\vee}\ ((r(i) \neq s(i)) \wedge Out(r(i)) \wedge Out(s(i)))$$
$$\underline{\vee}\ ((r(i) \neq s(i)) \wedge Out(r(i)) \wedge \neg\, Out(s(i)))$$

$$\underline{\vee}\ ((r(i) \neq s(i)) \wedge \neg\ Out(r(i)) \wedge Out(s(i)))$$

=   { summation distributes over disjoint predicates }

$$\sum r(i), s(i) : 1 .. N \bullet (r(i) = s(i)) \wedge Out(r(i)) \wedge Out(s(i))$$
$$+ \sum r(i), s(i) : 1 .. N \bullet (r(i) = s(i)) \wedge Out(r(i)) \wedge \neg\ Out(s(i))$$
$$+ \sum r(i), s(i) : 1 .. N \bullet (r(i) = s(i)) \wedge \neg\ Out(r(i)) \wedge Out(s(i))$$
$$+ \sum r(i), s(i) : 1 .. N \bullet (r(i) = s(i)) \wedge \neg\ Out(r(i)) \wedge \neg\ Out(s(i))$$
$$+ \sum r(i), s(i) : 1 .. N \bullet (r(i) \neq s(i)) \wedge Out(r(i)) \wedge Out(s(i))$$
$$\vdash \sum r(i), s(i) : 1 .. N \bullet (r(i) \neq s(i)) \wedge Out(r(i)) \wedge \neg\ Out(s(i))$$
$$+ \sum r(i), s(i) : 1 .. N \bullet (r(i) \neq s(i)) \wedge \neg\ Out(r(i)) \wedge Out(s(i))$$

=   { proposition calculus, summation }

$$N$$
$$+ \sum r(i), s(i) : 1 .. N \bullet (r(i) \neq s(i)) \wedge Out(r(i)) \wedge Out(s(i))$$
$$+ \sum r(i), s(i) : 1 .. N \bullet (r(i) \neq s(i)) \wedge Out(r(i)) \wedge \neg\ Out(s(i))$$
$$+ \sum r(i), s(i) : 1 .. N \bullet (r(i) \neq s(i)) \wedge \neg\ Out(r(i)) \wedge Out(s(i))$$

=   { property: $Out(r(i)) \wedge \neg\ Out(s(i)) \Rightarrow (r(i) \neq s(i))$, etc. }

$$N$$
$$+ \sum r(i), s(i) : 1 .. N \bullet (r(i) \neq s(i)) \wedge Out(r(i)) \wedge Out(s(i))$$
$$+ \sum r(i), s(i) : 1 .. N \bullet Out(r(i)) \wedge \neg\ Out(s(i))$$
$$+ \sum r(i), s(i) : 1 .. N \bullet \neg\ Out(r(i)) \wedge Out(s(i))$$

=   { symmetry }

$$N$$
$$+ \sum r(i), s(i) : 1 .. N \bullet (r(i) \neq s(i)) \wedge Out(r(i)) \wedge Out(s(i))$$
$$+ 2 \left(\sum r(i), s(i) : 1 .. N \bullet Out(r(i)) \wedge \neg\ Out(s(i)) \right)$$

= $\left\{ \begin{array}{l} \text{previous results:} \\ (\sum r(i), s(i) : 1 .. N \bullet Out(r(i)) \wedge \neg\ Out(s(i))) = qN(1 - q)(N - 1) \\ (\sum r(i), s(i) : 1 .. N \bullet (r(i) \neq s(i)) \wedge Out(r(i)) \wedge Out(s(i))) = qNq(N - 1) \end{array} \right\}$

$$N + q^2 N(N - 1) + 2q(1 - q)N(N - 1)$$

=   { arithmetic }

$$(1 - q)^2 N + q(2 - q)N^2$$

=   { introduce definition }

$$W_1$$

We calculate the complement:

$$\sum r(i), s(i) : 1 .. N \bullet M(r(i), s(i)) \geq d$$

=   { negation }

$$\sum r(i), s(i) : 1 .. N \bullet \neg\, M(r(i), s(i)) < d$$

$=$ { summation complement }

$$\left(\sum r(i), s(i) : 1 .. N \bullet \textbf{true}\right) - \left(\sum r(i), s(i) : 1 .. N \bullet \neg\, M(r(i), s(i)) < d\right)$$

$=$ { summation }

$$N^2 - \left(\sum r(i), s(i) : 1 .. N \bullet \neg\, M(r(i), s(i)) < d\right)$$

$=$ { summation over constant, previous result }

$$N^2 - (1-q)^2 N + q(2-q)N^2$$

$=$ { arithmetic }

$$(1-q)^2 N(N-1)$$

$=$ { introduce definition }

$$W_2$$

These results allow us to answer the following important question: How are $M(r(i), s(i)) < d$ and $M(r(i), s(i)) < d$ distributed? The distribution of final states satisfying postcondition S is given by

$$\sum X, Y \bullet S * X * Y$$

where $X * Y$ describes the distribution of the initial states. So the distribution of $M(r(i), s(i)) < d$ is given by $W_1 * (1/N)$ and the distribution of $M(r(i), s(i)) \geq d$ is given by $W_2 * (1/N)$. We calculate:

$$W_1/N^2$$

$=$ { previous result }

$$((1-q)^2 N + q(2-q)N^2)/N^2$$

$=$ { arithmetic }

$$((1-q)^2 + q(2-q)N)/N$$

$=$ { introduce definition }

$$D_1$$

$$W_2/N^2$$

$=$ { previous result }

$$((1-q)^2 N(N-1))/N^2$$

$=$ { arithmetic }

$$(1-q)^2 (N-1)/N$$

$=$ { introduce definition }

$$D_2$$

Of course, $D_1 + D_2 = 1$. We combine our results about these two distributions in the hypothesis that the final time t' has the distribution

$$E = (t' \geq t) * D_1^{t'-t} * D_2$$

The distribution of the implementation (right-hand side) is

$$\sum r(t), s(t) \bullet (\textbf{if } M(r(t), s(t)) \geq d \textbf{ then } (t' = t) \textbf{ else } (t := t + 1 \; ; \; E))$$

= { conditional: $(\textbf{if } b \textbf{ then } P \textbf{ else } Q) = (b * P + (1 - b) * Q)$ }

$$\left(\sum r(t), s(t) \bullet M(r(t), s(t)) \geq d \wedge (t' = t)\right)$$
$$+ \left(\sum r(t), s(t) \bullet M(r(t), s(t)) < d \wedge (t := t + 1 \; ; \; E)\right)$$

= { definition: $D_2 = (\sum r(t), s(t) \bullet M(r(t), s(t)) \geq d)$ }

$$D_2 * (t' = t) + (\sum r(t), s(t) \bullet M(r(t), s(t)) < d \wedge (t := t + 1 \; ; \; E))$$

= { definition: $D_1 = (\sum r(t), s(t) \bullet M(r(t), s(t)) < d)$ }

$$D_2 * (t' = t) + D_1 * (t := t + 1 \; ; \; E)$$

= { definition: $E = (t' \geq t) * D_1^{t'-t} * D_2$ }

$$D_2 * (t' = t) + D_1 * (t := t + 1 \; ; \; (t' \geq t) * D_1^{t'-t} * D_2)$$

= { assignment: $(x := e \; ; \; P) = P[e/x]$ }

$$D_2 * (t' = t) + D_1 * (t' \geq t + 1) * D_1^{t'-t-1} * D_2$$

= { arithmetic }

$$D_2 * (t' = t) + (t' \geq t + 1) * D_1^{t'-t} * D_2$$

= { arithmetic }

$$(t' = t) * D_1^{t'-t} * D_2 + (t' > t) * D_1^{t'-t} * D_2$$

= { arithmetic }

$$(t' \geq t) * D_1^{t'-t} * D_2$$

= { definition }

$$E$$

Now that we know the distribution computed by the Ransac algorithm, we can calculate its average termination time:

$$\sum t' \bullet t' * E$$

= { definition }

$$\sum t' \bullet t' * (t' \geq t) * D_1^{t'-t} * D_2$$

= {summation algebra }

$$\left(\sum t' \bullet t' * (t' \geq t) * D_1^{t'-t}\right) * D_2$$

In the simple case where the starting time is $t = 0$, this reduces to

$$\left(\sum t' \bullet t' * D_1^{t'}\right) * D_2/N$$

$$= \quad \{ \text{ power series identity: } \sum_{k=0}^{\infty} kz^k = z/(1-z)^2 \}$$

$$\left(D_1/(1-D_1)^2\right) * D_2$$

$$= \quad \{ \text{ previous result: } D_2 = 1 - D_1 \}$$

$$\left(D_1/D_2^2\right) * D_2$$

$$= \quad \{ \text{ arithmetic } \}$$

$$D_1/D_2$$

This is a simple and appealing result. We use it to evaluate the average termination time for our example with $q = 1/3$ and $N = 6$, and starting time $t = 0$. We start with D_2, the distribution of final states that represent valid models.

$$D_2$$

$$= \quad (1-q)^2(N-1)/N$$

$$= \quad (1 - (1/3))^2 * (6-1)/6$$

$$= \quad 10/27$$

Consequently, D_1 (the distribution of final states that represent invalid models) $= 1 - (10/27) = 17/27$. For this example, the average termination time is $D_1/D_2 = \dfrac{17/27}{10/27} = 1.7$. On average, it takes just two attempts for the simplified Ransac algorithm to find a sample with no outliers for this dataset. This is not finding the best fit, just the first fit.

As a final sanity check to make sure that our sums are correct, we translate the simplified algorithm from RoboChart into PRISM and check some properties.[28]

Figure 13.8 contains a transcript of the properties specified in the RoboChart assertion language. The first assertion requires that the algorithm is free from deadlock. As we noted in footnote 24, this is a well-formedness condition for a PRISM model. PRISM can automatically add self-loop transition to any deadlocked states, but here we prefer to make the check explicit.

We translate the RoboChart to a DTMC. We specify the reward to be the number of iterations of the algorithm.[29] This is achieved by specifying in nrtries that the reward is accumulated while the algorithm has chosen a bad fit for the model.

---

[28] Our sanity check is an important part of model verification, which is the useful checking of conformance of redundant descriptions.

[29] Technically, this property defines a co-safe LTL reward (see footnote 9 for an explanation of co-safety). We can also specify cumulative and total reward properties.

```
prob assertion P_deadlock_free:
    !Exists [Finally "deadlock"]

rewards nrtries =
    (ransacMOD::ransacCTRL::stm_ref0 is in ransacMOD::ransacCTRL::stm_ref0::badFit) : 1;
endrewards

prob assertion P_nr_of_tries:
    Reward {nrtries} =? [
        Reachable ransacMOD::ransacCTRL::stm_ref0 is in ransacMOD::ransacCTRL::stm_ref0::goodFit
    ]

rewards nrchoices =
    [ransacMOD::ransacCTRL::stm_ref0::choose] true : 1;
endrewards

prob assertion P_nr_of_choices:
    Reward {nrchoices} =? [
        Reachable ransacMOD::ransacCTRL::stm_ref0 is in ransacMOD::ransacCTRL::stm_ref0::goodFit
    ]
```

Fig. 13.8 Assertions for Ransac algorithm.

Assertion: P_deadlock_free

Assertion	States	Transitions	Time	Result
P_deadlock_free	744	839	0.004 seconds	true

Assertion: P_nr_of_tries

Assertion	States	Transitions	Time	Result
P_nr_of_tries	744	839	0.031 seconds	1.6998350819503547

Assertion: P_nr_of_choices

Assertion	States	Transitions	Time	Result
P_nr_of_choices	744	839	0.035 seconds	2.6998316695824656

Fig. 13.9 Report on assertions for Ransac algorithm.

The following assertion `P_nr_of_tries` calculates the reward accumulated along a path up until some point. Here this is specified to be the point at which a good fit for the model is produced.

Figure 13.9 contains the results of running the model and checking the properties that we specified. It states that the model is deadlock-free and that the expected reward (the number of iterations) is 1.6998250819503547, which confirms our theoretical prediction. The second reward-assertion pair in Fig. 13.8 describes the number of choices of data points.

We now have consistent results with complementary methods for this little example: proof that the average number of iterations is just 1.7 and a probabilistic model check that confirms the result. The advantage of the former is that it applies to datasets of all sizes, while the latter is restricted to this small example.

7.3   *Engineering Method*

In this chapter, we discussed the use of probabilistic modelling as a way of coping with uncertainty in robotic systems and increasing their level of robustness in real-world situations and applications. The work is the start of our investigation of a practical engineering method. We envisage using RoboTool to create probabilistic RoboChart models. These models are then automatically translated for analysis with the PRISM model checker and correctness criteria suggested as temporal-logic formulas. It will be necessary to restrict the size of the model for probabilistic model checking, but this will still be useful for debugging. Larger versions of the model can be explored using the approximate technique of statistical model checking. When confidence in the model is strong enough, Isabelle/UTP will be used to verify that the properties hold for any instance of its parameters.

We intend this engineering method to be used by practising roboticists. This will require that we provide practical modelling guidelines, a usable property language, friendly interfaces, and a very high level of automation in all tools: translators, model checkers, and theorem provers. Our vision is that we provide strong model engineering support without requiring undue expertise in using our tools. The skill that is required must be amply repaid by the results that can be achieved.

7.4   *Mechanisation*

As we saw in Sect. 6, model checking can verify correctness properties for a bounded instance of a RoboChart model. The advantage is that checking some properties becomes automatic. Of course, you need to find the right model to check, but model checking trades proof skills for modelling skills, which might be easier for hard-pressed engineers in industry.

There are disadvantages too. The model has to have a small number of states so that the model checking does not take an unreasonable amount of time or space. It is disappointing to check a linear version of Ransac for only six data points. What about larger datasets? What about non-linear models? Second, we have to limit the kinds of formulas we check to those expressed in simple logics.

A solution to this problem lies in mechanisation of our probabilistic theories in a theorem prover. A theorem prover can model the state space symbolically rather than explicitly, which allows it to represent an unbounded, even infinite, state space. This is particularly important for systems using real-valued variables, which are uncountably infinite. In a generic proof assistant, like Isabelle [67], we are not limited to a single logic, like PCTL, but can express properties in a variety of potentially more powerful logics. However, theorem proving also requires considerably more time and expertise than model checking, and so we see these two techniques as complementary.

We have a long-term research agenda: *the mechanised verification of RoboChart models*. Our research goal is to build:

A sound, automated theorem prover for diagrammatic descriptions of reactive, timed, probabilistic controllers for robotics and autonomous systems.

The development of our RoboChart theorem prover is being supported by Isabelle/UTP [24], a formal semantics and verification framework based on Hoare and He's Unifying Theories of Programming (UTP) [41] built on top of Isabelle [67]. UTP follows Hehner [34] in modelling programs as predicates, and therefore is highly amenable to proof automation. We are specialising the tool to prove theorems about RoboChart models [22, 26] using mechanised theories of reactive [23, 25] and hybrid programming [21, 66]. We have developed UTP theories for probabilistic programs [77] based on the "weakest completion" approach of He, Morgan, and McIver [33]. We are now mechanising these in Isabelle/UTP, utilising the sophisticated Isabelle libraries for probability and MDPs [42]. With our mechanisation, we have been able to prove some small examples, including that the algorithm in Fig. 13.1 gives rise to a uniform distribution. Our work will, in the future, allow us to support verification of probabilistic programs using theorem proving.

8   Related Work and Conclusion

Cavalcanti et al. describe the RoboStar technology based on RoboCalc and RoboTool in full detail in Chap. 9. Dennis and Fisher discuss probabilistic temporal logic in verifying autonomy and responsible robotics in Chap. 7. Gotlieb et al. outline the issues in generating test oracles for AI and machine learning (ML) systems in Chap. 4. This is intrinsically complex because ML introduces probabilistic reasoning. Gómez-Abajo et al. introduce the Wodel tools that they use to seed faults with RoboTool and FDR [28]. They have applied this to probabilistic automata. Ingrand discusses the use of probabilistic and statistical model checking in verifying robotic behaviours, but warns about the knowledge and skills required for roboticists to use them effectively [43]. Dragule et al. propose the use of patterns to overcome this skills gap (Chap. 12).

Our take-home message for roboticists is that uncertainty is an essential feature of robotics and autonomous systems, but using probability is an approach to modelling uncertainty. RoboChart supports probability. We can translate from RoboChart to PRISM and model-check properties. In the future, we will use Isabelle/UTP for theorem proving and analysis. Our long-term goal is for probabilistic guarantees of behaviour in the face of uncertainty.

Acknowledgements The ideas in this chapter benefited from presentations at several meetings, and we would like to thank the organisers and participants for their valuable feedback. The meetings include the symposium in honour of Professor Marie-Claude Gaudel at the University of Paris-Saclay in September 2019; the UTP symposium in Porto in October 2019; the RoboSoft meeting at the Royal Academy of Engineering in London in November 2019; the workshop on Formal Methods for AI at Southwest University in Chongqing in December 2019; and the DIGIT workshop on digital twins at Aarhus University in January 2020. Finally, we would like to thank the other authors and the reviewers for their comments that improved the chapter.

Ransac in PRISM

```
dtmc

const int M = 2;
const int N = 6;
const int t = 0;
const int d = 4;
const int k = 14;
const int xSize = 5;
const int ySize = 5;
const int dataX0 = 0;
const int dataX1 = 1;
const int dataX2 = 2;
const int dataX3 = 3;
const int dataX4 = 4;
const int dataX5 = 5;
const int dataY0 = 5;
const int dataY1 = 1;
const int dataY2 = 2;
const int dataY3 = 3;
const int dataY4 = 4;
const int dataY5 = 0;

formula x1 =
  (maybeInlierIndex1=0?dataX0:
    (maybeInlierIndex1=1?dataX1:
      (maybeInlierIndex1=2?dataX2:
        (maybeInlierIndex1=3?dataX3:
          (maybeInlierIndex1=4?dataX4:
      dataX5)))));

formula y1 =
  (maybeInlierIndex1=0?dataY0:
    (maybeInlierIndex1=1?dataY1:
      (maybeInlierIndex1=2?dataY2:
        (maybeInlierIndex1=3?dataY3:
```

```
          (maybeInlierIndex1=4?dataY4:
       dataY5)))));

formula x2 =
  (maybeInlierIndex2=0?dataX0:
    (maybeInlierIndex2=1?dataX1:
      (maybeInlierIndex2=2?dataX2:
        (maybeInlierIndex2=3?dataX3:
          (maybeInlierIndex2=4?dataX4:
      dataX5)))));

formula y2 =
  (maybeInlierIndex2=0?dataY0:
    (maybeInlierIndex2=1?dataY1:
      (maybeInlierIndex2=2?dataY2:
        (maybeInlierIndex2=3?dataY3:
          (maybeInlierIndex2=4?dataY4:
      dataY5)))));

formula sumX = x1+x2;
formula sumY = y1+y2;
formula sumXY = x1*y1+x2*y2;
formula sumXSq = pow(x1,2)+pow(x2,2);
formula sumXAllSq = pow(sumX,2);
formula maybeM = ceil((N*sumXY-sumX*sumY)/(N*sumXSq-
   sumXAllSq)-1/2);
formula maybeB = ceil((sumY-maybeM*sumX)/N-1/2);

formula possInlierFit0 = maybeM*dataX0+maybeB-dataY0;
formula possInlierFit1 = maybeM*dataX1+maybeB-dataY1;
formula possInlierFit2 = maybeM*dataX2+maybeB-dataY2;
formula possInlierFit3 = maybeM*dataX3+maybeB-dataY3;
formula possInlierFit4 = maybeM*dataX4+maybeB-dataY4;
formula possInlierFit5 = maybeM*dataX5+maybeB-dataY5;

formula possInlierFitNorm0 = (possInlierFit0<0?-
   possInlierFit0:possInlierFit0);
formula possInlierFitNorm1 = (possInlierFit1<0?-
   possInlierFit1:possInlierFit1);
formula possInlierFitNorm2 = (possInlierFit2<0?-
   possInlierFit2:possInlierFit2);
formula possInlierFitNorm3 = (possInlierFit3<0?-
   possInlierFit3:possInlierFit3);
```

```
formula possInlierFitNorm4 = (possInlierFit4<0?-
   possInlierFit4:possInlierFit4);
formula possInlierFitNorm5 = (possInlierFit5<0?-
   possInlierFit5:possInlierFit5);

formula possInlier0 = ( possInlierFitNorm0<=t ? true
   : false );
formula possInlier1 = ( possInlierFitNorm1<=t ? true
   : false );
formula possInlier2 = ( possInlierFitNorm2<=t ? true
   : false );
formula possInlier3 = ( possInlierFitNorm3<=t ? true
   : false );
formula possInlier4 = ( possInlierFitNorm4<=t ? true
   : false );
formula possInlier5 = ( possInlierFitNorm5<=t ? true
   : false );

formula newSumX =
   (possInlier0?dataX0:0) + (possInlier1?dataX1:0) +
   (possInlier2?dataX2:0) + (possInlier3?dataX3:0) +
   (possInlier4?dataX4:0) + (possInlier5?dataX5:0) ;

formula newSumY =
   (possInlier0?dataY0:0) + (possInlier1?dataY1:0) +
   (possInlier2?dataY2:0) + (possInlier3?dataY3:0) +
   (possInlier4?dataY4:0) + (possInlier5?dataY5:0) ;

formula newSumXY =
   (possInlier0?dataX0*dataY0:0) + (possInlier1?dataX1
      *dataY1:0) +
   (possInlier2?dataX2*dataY2:0) + (possInlier3?dataX3
      *dataY3:0) +
   (possInlier4?dataX4*dataY4:0) + (possInlier5?dataX5
      *dataY5:0) ;

formula newSumXSq =
   (possInlier0?dataX0*dataX0:0) + (possInlier1?dataX1
      *dataX1:0) +
   (possInlier2?dataX2*dataX2:0) + (possInlier3?dataX3
      *dataX3:0) +
   (possInlier4?dataX4*dataX4:0) + (possInlier5?dataX5
      *dataX5:0) ;
```

```
formula newSumXAllSq = pow(newSumX,2);

formula newM =
  (N*newSumXSq=newSumXAllSq?1:
   ceil((N*newSumXY-newSumX*newSumY)/(N*newSumXSq-
      newSumXAllSq)-1/2));
formula newB =
  (N*newSumXSq=newSumXAllSq?1:
   ceil((newSumY-newM*newSumX)/N-1/2));

formula possInlierErr0 =
  ( possInlier0 ? newM*dataX0+newB-dataY0 : 0 );
formula possInlierErr1 =
  ( possInlier1 ? newM*dataX1+newB-dataY1 : 0 );
formula possInlierErr2 =
  ( possInlier2 ? newM*dataX2+newB-dataY2 : 0 );
formula possInlierErr3 =
  ( possInlier3 ? newM*dataX3+newB-dataY3 : 0 );
formula possInlierErr4 =
  ( possInlier4 ? newM*dataX4+newB-dataY4 : 0 );
formula possInlierErr5 =
  ( possInlier5 ? newM*dataX5+newB-dataY5 : 0 );

formula newErr =
  possInlierErr0 + possInlierErr1 + possInlierErr2 +
  possInlierErr3 + possInlierErr4 + possInlierErr5 ;

formula newNumInliers =
  (possInlier0?1:0) + (possInlier1?1:0) + (
     possInlier2?1:0) +
  (possInlier3?1:0) + (possInlier4?1:0) + (
     possInlier5?1:0) ;

const int pcOuterLoopTest = 0;
const int pcSelectMaybeInlierIndex1 = 1;
const int pcInnerLoopTest = 2;
const int pcTestBetterModel = 3;
const int pcTerm = 4;

module Ransac
  pc : [pcOuterLoopTest..pcTerm] init pcOuterLoopTest
     ;
  i : [1..k+1] init 1;
  maybeInlierIndex1 : [0..N-1] init 0;
```

```
maybeInlierIndex2 : [0..N-1] init 0;
bestM : [-ySize..ySize] init 0;
bestB : [-ySize..ySize] init 0;
bestErr : [0..N*ySize+1] init N*ySize+1;
bestNumInliers : [0..N-1] init 0;
[] pc=pcOuterLoopTest &i=k+1 -> (pc'=pcTerm);
[] pc=pcOuterLoopTest &i<k+1 -> (pc'=
   pcSelectMaybeInlierIndex1);
[] pc=pcSelectMaybeInlierIndex1 ->
   1/N: (maybeInlierIndex1'=0) &(maybeInlierIndex2
      '=0) &(pc'=pcInnerLoopTest) +
   1/N: (maybeInlierIndex1'=1) &(maybeInlierIndex2
      '=1) &(pc'=pcInnerLoopTest) +
   1/N: (maybeInlierIndex1'=2) &(maybeInlierIndex2
      '=2) &(pc'=pcInnerLoopTest) +
   1/N: (maybeInlierIndex1'=3) &(maybeInlierIndex2
      '=3) &(pc'=pcInnerLoopTest) +
   1/N: (maybeInlierIndex1'=4) &(maybeInlierIndex2
      '=4) &(pc'=pcInnerLoopTest) +
   1/N: (maybeInlierIndex1'=5) &(maybeInlierIndex2
      '=5) &(pc'=pcInnerLoopTest) ;
[] pc=pcInnerLoopTest &maybeInlierIndex2=
   maybeInlierIndex1 ->
   1/N: (maybeInlierIndex2'=0) + 1/N: (
      maybeInlierIndex2'=1) + 1/N: (
      maybeInlierIndex2'=2) +
   1/N: (maybeInlierIndex2'=3) + 1/N: (
      maybeInlierIndex2'=4) + 1/N: (
      maybeInlierIndex2'=5) ;
[] pc=pcInnerLoopTest &maybeInlierIndex2!=
   maybeInlierIndex1 ->
   (pc'=pcTestBetterModel);
[] pc=pcTestBetterModel &newNumInliers>=d &i<k+1 &
   newErr<bestErr ->
   (bestM'=newM) &(bestB'=newB) &(bestErr'=newErr) &
   (bestNumInliers'=newNumInliers) &(i'=i+1) &(pc'=
      pcOuterLoopTest);
[] pc=pcTestBetterModel &i<k+1 &!(newNumInliers>=d
   &newErr<bestErr) ->
   (i'=i+1) &(pc'=pcOuterLoopTest);
[] pc=pcTerm -> true;
endmodule // Ransac
```

References

1. J.-R. Abrial, *Modeling in Event-B – System and Software Engineering* (Cambridge University Press, Cambridge, 2010)
2. Aethon. Tug website. aethon.com/why-mobile-robots-from-aethon/ (2020)
3. R. Alur, T.A. Henzinger, Reactive modules. Formal Methods Syst. Des. **15**(1), 7–48 (1999)
4. R.-J. Back, R. Kurki-Suonio, Decentralization of process nets with centralized control. Distrib. Comput. **3**(2), 73–87 (1989)
5. M. Brambilla, A. Brutschy, M. Dorigo, M. Birattari, Property-driven design for robot swarms: A design method based on prescriptive modeling and model checking. ACM Trans. Auton. Adapt. Syst. **9**(4), 17:1–17:28 (2015)
6. C. Bustamante, L. Garrido, R. Soto, Comparing fuzzy naive Bayes and Gaussian naive Bayes for decision making in RoboCup 3D, in *Proceedings of the MICAI 2006: Advances in Artificial Intelligence, 5th Mexican International Conference on Artificial Intelligence, Apizaco, 13–17 November 2006* ed. by A.F. Gelbukh, C.A.R. García, vol. 4293. Lecture Notes in Computer Science (Springer, Berlin, 2006), pp. 237–247
7. S. Chaki, J.A. Giampapa, Probabilistic verification of coordinated multi-robot missions, in *Proceedings of the SPIN 2013: 20th International Symposium on Model Checking Software, Stony Brook, 8–9 July 2013*, ed. by E. Bartocci, C.R. Ramakrishnan, vol. 7976. Lecture Notes in Computer Science (Springer, Berlin, 2013), pp. 135–153
8. W. Ci, J. Wang, M. Zhu, A. Dou, Extensions to the standard Ransac algorithm for efficiency and robustness. Int. J. Recent Sci. Res. **9**(12), 29842–29846 (2018)
9. F. Ciesinski, C. Baier, LiQuor: a tool for qualitative and quantitative linear time analysis of reactive systems, in *QEST 2006: Third International Conference on the Quantitative Evaluation of Systems, 11–14 September 2006, Riverside* (IEEE Computer Society, Washington, 2006), pp. 131–132
10. E.M. Clarke, O. Grumberg, D.A. Peled, *Model Checking* (MIT Press, Cambridge, 1999)
11. S. Cook, C. Bock, P. Rivett, T. Rutt, E. Seidewitz, B. Selic, D. Tolbert, Unified modeling language (UML) version 2.5.1. Standard, Object Management Group (OMG) (2017)
12. C. Dehnert, S. Junges, J.-P. Katoen, M. Volk, A storm is coming: a modern probabilistic model checker, in *Proceedings of the Part II CAV 2017: 29th International Conference on Computer Aided Verification, Heidelberg, 24–28 July 2017*, ed. by R. Majumdar, V. Kuncak, vol. 10427. Lecture Notes in Computer Science (Springer, Berlin, 2017), pp. 592–600
13. E.W. Dijkstra, Guarded commands, nondeterminacy and formal derivation of programs. Commun. ACM **18**(8), 453–457 (1975)
14. C. Dixon, A.F.T. Winfield, M. Fisher, C. Zeng, Towards temporal verification of swarm robotic systems. Robotics Auton. Syst. **60**(11), 1429–1441 (2012)
15. A. Doucet, A. Johansen, A tutorial on particle filtering and smoothing: fifteen years later. Technical report, University of British Columbia (2008)
16. A. Dubrawski, H. Thorne, Evolution of a useful autonomous system, in *Robot Motion and Control 2009*, vol. 396. Part of the Lecture Notes in Control and Information Sciences (Springer, Berlin, 2009), pp. 453–462
17. H.F. Durrant-Whyte, T. Bailey, Simultaneous localization and mapping: part I. IEEE Robot. Automat. Mag. **13**(2), 99–110 (2006)
18. J.P. Fentanes, B. Lacerda, T. Krajník, N. Hawes, M. Hanheide, Now or later? Predicting and maximising success of navigation actions from long-term experience, in *IEEE International Conference on Robotics and Automation, ICRA 2015, Seattle, 26–30 May 2015* (IEEE, Piscataway, 2015), pp. 1112–1117
19. M.S.C. Filho, R. Marinho, A. Mota, J. Woodcock, Analysing RoboChart with probabilities, in *Proceedings of the SBMF 2018: 21st Brazilian Symposium on Formal Methods: Foundations and Applications, Salvador, 26–30 November 2018* ed. by T. Massoni, M.R. Mousavi, vol. 11254. Lecture Notes in Computer Science (Springer, Berlin, 2018), pp. 198–214

20. M.A. Fischler, R.C. Bolles, Random sample consensus: a paradigm for model fitting with applications to image analysis and automated cartography. Commun. ACM **24**(6), 381–395 (1981)
21. S. Foster, Hybrid relations in Isabelle/UTP, in *UTP*, vol. 11885. LNCS (Springer, Berlin, 2019), pp. 130–153
22. S. Foster, J. Baxter, A. Cavalcanti, A. Miyazawa, J. Woodcock, Automating verification of state machines with reactive designs and Isabelle/UTP, in *Proceedings of the FACS 2018: 15th International Conference on Formal Aspects of Component Software, Pohang, 10–12 October 2018*, ed. by K. Bae, P.C. Ölveczky, vol. 11222. Lecture Notes in Computer Science (Springer, Berlin, 2018), pp. 137–155
23. S. Foster, K. Ye, A. Cavalcanti, J. Woodcock, Calculational verification of reactive programs with reactive relations and Kleene algebra, in *Proceedings of the RAMiCS 2018: 17th International Conference on Relational and Algebraic Methods in Computer Science, Groningen, 29 October–1 November 2018*, ed. by J. Desharnais, W. Guttmann, S. Joosten, vol. 11194. Lecture Notes in Computer Science (Springer, Berlin, 2018), pp. 205–224
24. S. Foster, J. Baxter, A. Cavalcanti, J. Woodcock, F. Zeyda, Unifying semantic foundations for automated verification tools in Isabelle/UTP. Sci. Comput. Program. **197**, 102510 (2020)
25. S. Foster, A. Cavalcanti, S. Canham, J. Woodcock, F. Zeyda, Unifying theories of reactive design contracts. Theor. Comput. Sci. **802**, 105–140 (2020)
26. S. Foster, Y. Nemouchi, C. O'Halloran, N. Tudor, K. Stephenson, Formal model-based assurance cases in Isabelle/SACM: an autonomous underwater vehicle case study, in *Proceedings of the 8th International Conference on Formal Methods in Software Engineering (FormaliSE)* (ACM, New York, 2020)
27. P. Gainer, C. Dixon, U. Hustadt, Probabilistic model checking of ant-based positionless swarming, in *Proceedings of the TAROS 2016: 17th Annual Conference – Towards Autonomous Robotic Systems, Sheffield, 26 June–1 July 2016*, ed. by L. Alboul, D.D. Damian, J.M. Aitken, vol. 9716. Lecture Notes in Computer Science (Springer, Berlin, 2016), pp. 127–138
28. P. Gómez-Abajo, R.M. Hierons, R. Lefticaru, M.G. Merayo, Mutation testing for RoboChart, in *RoboSoft: Software Engineering for Robotics* (Springer, Berlin, 2020)
29. F. Gretz, J.-P. Katoen, A. McIver, Operational versus weakest pre-expectation semantics for the probabilistic guarded command language. Perform. Eval. **73**, 110–132 (2014)
30. P. Guerrero, J. Ruiz-del-Solar, G. Díaz, Probabilistic decision making in robot soccer, in *RoboCup 2007: Robot Soccer World Cup XI, 9–10 July 2007, Atlanta*, ed. by U. Visser, F. Ribeiro, T. Ohashi, F. Dellaert, vol. 5001. Lecture Notes in Computer Science (Springer, Berlin, 2008), pp. 29–40
31. E.M. Hahn, Y. Li, S. Schewe, A. Turrini, L. Zhang, iscasMC: a web-based probabilistic model checker, in *Proceedings of the FM 2014: 19th International Symposium on Formal Methods, Singapore, 12–16 May 2014*, ed. by C.B. Jones, P. Pihlajasaari, J. Sun, vol. 8442. Lecture Notes in Computer Science (Springer, Berlin, 2014), pp. 312–317
32. H. Hansson, B. Jonsson, A logic for reasoning about time and reliability. Formal Asp. Comput. **6**(5), 512–535 (1994)
33. J. He, C. Morgan, A. McIver, Deriving probabilistic semantics via the 'weakest completion', in 6th International Conference on Formal Engineering Methods, ICFEM 2004, Seattle, 8–12 November, 2004, ed. by J. Davies, W. Schulte, M. Barnett, vol. 3308. Lecture Notes in Computer Science (Springer, Berlin, 2004), pp. 131–145
34. E.C.R. Hehner, Predicative programming, part I. Commun. ACM **27**(2), 134–143 (1984)
35. E.C.R. Hehner, Predicative programming, part II. Commun. ACM **27**(2), 144–151 (1984)
36. E.C.R. Hehner, *A Practical Theory of Programming*. Texts and Monographs in Computer Science (Springer, Berlin, 1993)
37. E.C.R. Hehner, Probabilistic predicative programming, in *Proceedings of the MPC 2004: 7th International Conference on Mathematics of Program Construction, Stirling, 12–14 July 2004*, ed. by D. Kozen, C. Shankland, vol. 3125. Lecture Notes in Computer Science (Springer, Berlin, 2004), pp. 169–185

38. E.C.R. Hehner, L.E. Gupta, A.J. Malton, Predicative methodology. Acta Inf. **23**(5), 487–505 (1986)
39. A. Hinton, M.Z. Kwiatkowska, G. Norman, D. Parker, PRISM: a tool for automatic verification of probabilistic systems, in *Proceedings of the TACAS 2006: 12th International Conference on Tools and Algorithms for the Construction and Analysis of Systems, Held as Part of the Joint European Conferences on Theory and Practice of Software, ETAPS 2006, Vienna, 25 March–2 April 2006*, ed. by H. Hermanns, J. Palsberg, vol. 3920. Lecture Notes in Computer Science (Springer, Berlin, 2006), pp. 441–444
40. C.A.R. Hoare, *Communicating Sequential Processes* (Prentice-Hall, Upper Saddle River, 1985)
41. C.A.R. Hoare, J. He, *Unifying Theories of Programming* (Prentice-Hall, Upper Saddle River, 1998)
42. J. Holzl, Markov chains and Markov decision processes in Isabelle/HOL. J. Autom. Reasoning **59**, 345–387 (2017)
43. F. Ingrand, A roboticist's bottom-up approach? in *RoboSoft: Software Engineering for Robotics* (Springer, Berlin, 2020)
44. B. Johnson, H. Kress-Gazit, Analyzing and revising high-level robot behaviors under actuator error, in *2013 IEEE/RSJ International Conference on Intelligent Robots and Systems, Tokyo, 3–7 November 2013* (IEEE, Piscataway, 2013), pp. 741–748
45. B. Johnson, H. Kress-Gazit, Analyzing and revising synthesized controllers for robots with sensing and actuation errors. I. J. Robotics Res. **34**(6), 816–832 (2015)
46. J.-P. Katoen, I.S. Zapreev, E.M. Hahn, H. Hermanns, D.N. Jansen, The ins and outs of the probabilistic model checker MRMC. Perform. Eval. **68**(2), 90–104 (2011)
47. M.C. Kennedy, A. O'Hagan, Bayesian calibration of computer models. J. R. Stat. Soc. Ser. B (Stat. Methodol.) **63**(3), 425–464 (2001)
48. S. Konur, C. Dixon, M. Fisher, Analysing robot swarm behaviour via probabilistic model checking. Robotics Auton. Syst. **60**(2), 199–213 (2012)
49. M.Z. Kwiatkowska, G. Norman, D. Parker, PRISM 4.0: verification of probabilistic real-time systems, in *Proceedings of the CAV 2011: 23rd International Conference on Computer Aided Verification, Snowbird, 14–20 July 2011*, ed. by G. Gopalakrishnan, S. Qadeer, vol. 6806. Lecture Notes in Computer Science (Springer, Berlin, 2011), pp. 585–591
50. B. Lacerda, D. Parker, N. Hawes, Optimal and dynamic planning for Markov decision processes with co-safe LTL specifications, in *2014 IEEE/RSJ International Conference on Intelligent Robots and Systems, Chicago, 14–18 September 2014* (IEEE, Piscataway, 2014), pp. 1511–1516
51. B. Lacerda, D. Parker, N. Hawes, Optimal policy generation for partially satisfiable co-safe LTL specifications, in *Proceedings of the Twenty-Fourth International Joint Conference on Artificial Intelligence, IJCAI 2015, Buenos Aires, 25–31 July 2015*, ed. by Q. Yang, M.J. Wooldridge (AAAI Press, Palo Alto, 2015), pp. 1587–1593
52. M. Lahijanian, J. Wasniewski, S.B. Andersson, C. Belta, Motion planning and control from temporal logic specifications with probabilistic satisfaction guarantees, in *IEEE International Conference on Robotics and Automation, ICRA 2010, Anchorage, 3–7 May 2010* (IEEE, Piscataway, 2010), pp. 3227–3232
53. G.H. Lee, F. Fraundorfer, M. Pollefeys, RS-SLAM: RANSAC sampling for visual FastSLAM, in *2011 IEEE/RSJ: International Conference on Intelligent Robots and Systems, IROS 2011, San Francisco, 25–30 September 2011* (IEEE, Piscataway, 2011), pp. 1655–1660
54. A. Legay, B. Delahaye, S. Bensalem, Statistical model checking: an overview, in *Proceedings of the RV 2010: First International Conference on Runtime Verification, St. Julians, 1–4 November 2010*, ed. by H. Barringer, Y. Falcone, B. Finkbeiner, K. Havelund, I. Lee, G.J. Pace, G. Rosu, O. Sokolsky, N. Tillmann, vol. 6418. Lecture Notes in Computer Science (Springer, Berlin, 2010), pp. 122–135

55. A. Legay, A. Lukina, L.-M. Traonouez, J. Yang, S.A. Smolka, R. Grosu, Statistical model checking, in *Computing and Software Science: State of the Art and Perspectives*, ed. by B. Steffen, G.J. Woeginger, vol. 10000. Lecture Notes in Computer Science (Springer, Berlin, 2019), pp. 478–504

56. W. Li, A. Miyazawa, P. Ribeiro, A. Cavalcanti, J. Woodcock, J. Timmis, From formalised state machines to implementations of robotic controllers. CoRR. abs/1702.01783 (2017)

57. W. Li, A. Miyazawa, P. Ribeiro, A. Cavalcanti, J. Woodcock, J. Timmis, From formalised state machines to implementations of robotic controllers, in *DARS 2016: 13th International Symposium on Distributed Autonomous Robotic Systems, Natural History Museum, London, 7–9 November 2016*, ed. by R. Groß, A. Kolling, S. Berman, E. Frazzoli, A. Martinoli, F. Matsuno, M. Gauci, vol. 6. Springer Proceedings in Advanced Robotics (Springer, Berlin, 2018), pp. 517–529

58. A. Llarena, D.A. Rosenblueth, Model checking applied to humanoid robotic soccer, in *Joint Proceedings of the 13th Annual TAROS Conference and the 15th Annual FIRA RoboWorld Congress: Advances in Autonomous Robotics, Bristol, 20–23 August 2012*, ed. by G. Herrmann, M. Studley, M.J. Pearson, A.T. Conn, C. Melhuish, M. Witkowski, J.-H. Kim, P. Vadakkepat, vol. 7429. Lecture Notes in Computer Science (Springer, Berlin, 2012), pp. 256–269

59. M. Luckcuck, M. Farrell, L.A. Dennis, C. Dixon, M. Fisher, Formal specification and verification of autonomous robotic systems: a survey. ACM Comput. Surv. **52**(5), 100:1–100:41 (2019)

60. M.W. Maimone, Y. Cheng, L.H. Matthies, Two years of visual odometry on the mars exploration rovers. J. Field Robotics **24**(3), 169–186 (2007)

61. A. McIver, C. Morgan, *Abstraction, Refinement and Proof for Probabilistic Systems*. Monographs in Computer Science (Springer, Berlin, 2005)

62. A. Miyazawa, RoboTool: RoboChart tool manual (2019). www.cs.york.ac.uk/circus/publications/techreports/reports/robotool-manual.pdf

63. A. Miyazawa, P. Ribeiro, A. Cavalcanti, W. Li, J. Woodcock, J. Timmis, RoboChart reference manual (2019). www.cs.york.ac.uk/circus/publications/techreports/reports/robochart-reference.pdf

64. A. Miyazawa, P. Ribeiro, W. Li, A. Cavalcanti, J. Timmis, J. Woodcock, Robochart: modelling and verification of the functional behaviour of robotic applications. Softw. Syst. Modeling **18**(5), 3097–3149 (2019)

65. J. Morse, D. Araiza-Illan, K. Eder, J. Lawry, A. Richards, A fuzzy approach to qualification in design exploration for autonomous robots and systems, in *IEEE: International Conference on Fuzzy Systems, (FUZZ-IEEE), Italy, 9–12 July 2017* (2017), pp. 1–6

66. J.H.Y. Munive, G. Struth, S. Foster, Differential Hoare logics and refinement calculi for hybrid systems with Isabelle/HOL, in *RAMiCS*, vol. 12062. LNCS (Springer, Berlin, 2020)

67. T. Nipkow, M. Wenzel, L. Paulson, *Isabelle/HOL: A Proof Assistant for Higher-Order Logic*, vol. 2283. LNCS (Springer, Berlin, 2002)

68. R. Raguram, J.-M. Frahm, M. Pollefeys, A comparative analysis of RANSAC techniques leading to adaptive real-time random sample consensus, in *Proceedings of the ECCV 2008: 10th European Conference on Computer Vision, Marseille, 12–18 October 2008, Part II*, ed. by D.A. Forsyth, P.H.S. Torr, A. Zisserman, vol. 5303. Lecture Notes in Computer Science (Springer, Berlin, 2008), pp. 500–513

69. R. Raguram, O. Chum, M. Pollefeys, J. Matas, J.-M. Frahm, USAC: a universal framework for random sample consensus. IEEE Trans. Pattern Anal. Mach. Intell. **35**(8), 2022–2038 (2013)

70. A. Ramaswamy, B. Monsuez, A. Tapus, Model-driven self-adaptation of robotics software using probabilistic approach, in *ECMR 2015: European Conference on Mobile Robots, Lincoln, 2–4 September 2015* (IEEE, Piscataway, 2015), pp. 1–6

71. P. Ribeiro, A. Miyazawa, W. Li, A. Cavalcanti, J. Timmis, Modelling and verification of timed robotic controllers, in *Proceedings of the IFM 2017: 13th International Conference on Integrated Formal Methods, Turin, 20–22 September 2017*, ed. by N. Polikarpova, S. Schneider, vol. 10510. Lecture Notes in Computer Science (Springer, Berlin, 2017), pp. 18–33

72. M.U. Sardar, O. Hasan, Towards probabilistic formal modeling of robotic cell injection systems, in *MARS@ETAPS 2017: Proceedings of the 2nd Workshop on Models for Formal Analysis of Real Systems, Uppsala, 29th April 2017*, ed. by H. Hermanns, P. Höfner, vol. 244. EPTCS (2017), pp. 271–282

73. S. Song, J. Sun, Y. Liu, J.S. Dong, A model checker for hierarchical probabilistic real-time systems, in *Proceedings of the CAV 2012: 24th International Conference on Computer Aided Verification, Berkeley, 7–13 July 2012*, ed. by P. Madhusudan, S.A. Seshia, vol. 7358. Lecture Notes in Computer Science (Springer, Berlin, 2012), pp. 705–711

74. A. Tarasyuk, I. Pereverzeva, E. Troubitsyna, L. Laibinis, Formal development and quantitative assessment of a resilient multi-robotic system, in *Proceedings of the SERENE 2013: 5th International Workshop on Software Engineering for Resilient Systems, Kiev, 3–4 October 2013*, ed. by A. Gorbenko, A.B. Romanovsky, V.S. Kharchenko, vol. 8166. Lecture Notes in Computer Science (Springer, Berlin, 2013), pp. 109–124

75. S. Thrun, W. Burgard, D. Fox, *Probabilistic Robotics* (MIT Press, Cambridge, 2005)

76. R.R. Wiyatno, A. Xu, O. Dia, A. de Berker, Adversarial examples in modern machine learning: a review. CoRR. abs/1911.05268 (2019)

77. J. Woodcock, A. Cavalcanti, S. Foster, A. Mota, K. Ye, Probabilistic semantics for RoboChart – a weakest completion approach, in *Proceedings of the UTP 2019: 7th International Symposium on Unifying Theories of Programming, Dedicated to Tony Hoare on the Occasion of His 85th Birthday, Porto, 8 October 2019*, ed. by P. Ribeiro, A. Sampaio, vol. 11885. Lecture Notes in Computer Science (Springer, Berlin, 2019), pp. 80–105

78. K. Ye, A. Cavalcanti, S. Foster, A. Miyazawa, J. Woodcock, RoboChart: formal modelling and verification of probabilistic behaviour of robotic applications. Technical report, University of York (2020)

79. H.L.S. Younes, R.G. Simmons, Statistical probabilistic model checking with a focus on time-bounded properties. Inf. Comput. **204**(9), 1368–1409 (2006)

Chapter 14
Panel Discussion: Regulation and Ethics of Robotics and Autonomous Systems

Brijesh Dongol, Ron Bell, Ibrahim Habli, Mark Lawford, Pippa Moore, and Zeyn Saigol

Abstract This chapter summarises a panel discussion on the topics of Regulation and Ethics in Software Engineering for Robotics at RoboSoft, chaired by Jon Timmis (University of Sunderland). The panel members were Ron Bell (Engineering Safety Consultants), Mark Lawford (McMaster University), Ibrahim Habli (University of York), and Pippa Moore (Civil Aviation Authority), whose views are summarised below. The chapter also integrates the issues raised in a talk on "Extending automotive certification processes to handle autonomous vehicles" by Zeyn Saigol (Connected Places Catapult), which provides a case study for the issues being discussed.

1 Introduction

This introduction compiles the introductory remarks made by each of the panel members, which summarises ethics and regulation in the safety of autonomous systems. A case study in autonomous driving is provided as an example in Sect. 2, and a summary of the Q&A session with panel members is given in Sect. 3.

B. Dongol (✉)
University of Surrey, Guildford, UK
e-mail: b.dongol@surrey.ac.uk

R. Bell
Engineering Safety Consultants, London, UK

I. Habli
University of York, York, UK

M. Lawford
McMaster University, Hamilton, ON, Canada

P. Moore
Civil Aviation Authority, London, UK

Z. Saigol
Connected Places Catapult, Milton Keynes, UK

© Springer Nature Switzerland AG 2021
A. Cavalcanti et al. (eds.), *Software Engineering for Robotics*,
https://doi.org/10.1007/978-3-030-66494-7_14

467

1.1   The Value of Autonomous Systems

To understand the value of robotics and AI, one must not only consider what they can do, but also remember that robots do not get tired or bored and can work in environments that human beings cannot work in. This resolves many possible sources of human error. In addition, the ways of working with and thinking about developing AI and robotic systems have evolved the way we solve engineering problems and how we think about engineering problems. The ability and potential for this to spread beyond robotics and AI into wider engineering is phenomenal.

However, there is a range of regulatory and ethical considerations for robotics and autonomous systems: "Do we know what we are getting into?" One way to think about how to answer this question is to think about a robot as an *embodied AI* and to consider the regulatory and ethical questions around this definition. Taking it one step further, robotics and AI are tools. They are either a system in their own right or part of a system. This means one needs to know what this tool does and, equally importantly, what the tool does not do. One also needs an understanding of how the tool works, how any faults will manifest themselves, and how one can maintain the tool. Without this understanding, one can end up spending a lot of time and money on unanticipated tasks, fixing unexpected problems, and doing tasks that the *tool* was supposed to do.

Equally, as a developer, one has to understand the amount of work that the tool enables and the additional work required to operate the tool, particularly for a safety domain and for something that has ethical considerations. More work is also needed in *verifying* the work done and *validating* the conclusions, and actually why the verification and validation are worth anything at all. Why should regulators believe any of the conclusions drawn about the system by a developer? One may also have to produce information for the end users. Aircraft, for example, often stay in service for 20–30 years. One has to provide information to the people using them about how they will be maintained and how people should use them during this time. The last thing one wants is to take something developed over several decades, do something bizarre with it, and then come back to claim that the tool does not work. While all of the above is common sense, it is really difficult to add afterwards, and hence, one needs to understand the capabilities of the tools from the outset.

Organisations have been known to motivate the inclusion of an advanced tool into the workflow as a mechanism for attracting talented graduates or increasing the levels of investment from their stakeholders. While this can keep both graduates and stakeholders busy and happy, it should never be the primary motivating factor for choosing advanced technology. Organisations can run out of money if they have not fully understood the implications of the technology they have used.

1.2 Functional Safety in Critical Systems

Functional safety is part of the overall safety of a system and depends upon the system operating correctly in response to its inputs. In practical terms, this means the achievement of safety through the application of control systems and requires identifying what has to be done (functionality) and how well it should be done (safety integrity). The concept of functional safety, in the general industrial sector, was developed through international standardisation in the mid-1980s to facilitate the application of complex technologies to achieve a defined risk target.

A cornerstone of functional safety is the concept of a "safety function". Achievement of functional safety necessitates the undertaking of hazard analysis to define the functionality of a safety function and a risk assessment to define the safety integrity (Safety Integrity Level) of the safety function.

Two key standards on functional safety are published by the International Electrotechnical Commission (IEC 61508) and the International Organization for Standardization (ISO 26262).

IEC 61508 adopts a safety lifecycle approach to all phases from initial concept to final decommissioning. Functional safety is not solely concerned with achievement through adoption of the technical requirements. A key aspect of the achievement of very few dangerous failures for complex systems is the need to have very stringent assurance measures (e.g. functional safety assessment, functional safety audit, verification, validation). Not only do the technical requirements have to be addressed, there are strong requirements relating to the management of functional safety and competence of all those having responsibilities for one or more safety lifecycle phases. Safety culture is increasingly an issue that has to be addressed. Key drivers for a company involved in functional safety include business reputation.

International standards are becoming the norm for what constitutes good practice. Where international standards are lagging are in the development of critical systems employing new approaches and techniques (e.g. AI or other new approaches in software development or verification) and for applications such as autonomous robots. Safety regulators would expect a comprehensive justification of the achievement of functional safety and are likely to apply the "precautionary principle". It is beneficial to have a viable working relationship between those developing new systems and those regulating new systems.

Tools, techniques, and measures that are applied to current non-AI systems are not likely to be sufficient for AI-based systems.

Legal requirements in industry do not always mandate adoption of a specific IEC or ISO standard for any target risk level. However, a safety regulator may use a particular IEC or ISO standard as a basis of good practice. It is not enough to claim that a good practice standard has been applied, it is also necessary to justify that all the relevant requirements have been met and that the standard covers the technology to be applied. In the context of autonomous systems, such justifications may be difficult to make. Some key issues to consider are:

- What specific legal requirements apply, e.g., Health and Safety at Work Act 1974 and Machinery Directive?
- What is deemed a tolerable risk and what is the methodology for showing it has been achieved? Such a risk is often quantified, but if using AI, this would be a challenging issue.
- The environment and sector in which the risk is present. This is important since there are different cultures within each sector, and this should be considered.
- Collaboration with other research organisations, which requires consensus since an industry typically needs to move forward as a whole. Moreover, there is strength in numbers in building a consensus, with different points of view.
- Understanding the market, which requires good communications with all those in the supply chain, including end users.
- Whether guidance has been developed as a precursor to standardisation at ISO or IEC international levels.
- The state of play for standardisation in this area. For example, at the present time, there is work on autonomous vehicles, but there is a gap between the development work and the standardisation work, which currently does not provide comprehensive requirements for autonomous systems.

1.3 Cultural Differences in Regulation

There are specific challenges in establishing ethical and regulatory positions for mobile and autonomous robots as opposed to autonomous systems in general. In the virtual world, there can sometimes be life-changing mistakes (missing out on a loan, being incorrectly filtered for a job), but there is a way to roll this back. In a Tesla accident in China, a car rear-ended a street sweeper and someone was killed.

This gets to the issue of *informed consent*. The users have the right to know what they are getting into. What are the risks, and how will a company use your interactions for future training of the system? Within autonomous driving, for example, there is currently no good consensus on what informed consent is and how you should go about getting that from the driver. For the public, i.e. those interacting with these systems, there is a tension between the autonomous driving utility of getting from A to B, smoothly with drivability, and safety (not hitting things). But setting the balance can be a business decision that is also influenced by the safety culture within an organisation. For instance, General Motors has supercruise, which uses LIDAR integrated with high-definition 3D maps and driver gaze detection. On the other hand, the Tesla autopilot sensing system is much more limited, using only cameras, radar, and ultrasonic sensors rather than adding LIDAR, and it only uses a torque sensor in a steering wheel to measure driver attention. But, of course, traditional car makers are not under quite the same financial pressure as Tesla, and they have been hurt by litigation due to mistakes in the past.

The regulatory position in the USA, e.g. in terms of automotive and medical devices, is to not stifle innovation with overburdensome regulation as a result of directives from Congress. We recall the Toyota unintended acceleration issue, where the company president had to testify to Congress and the source code was accessed and examined by the regulator. In the USA, there is in part a reliance on lawsuits to create pressure on companies to produce acceptably safe systems. As a result, this means that actuarial science and lawyers will likely determine the safety of future vehicles in the USA. Contrast this with the European style, which is based on legislated safety regulations. Would Tesla's development of autopilot have been possible if it started in Germany with a heavier regulatory burden? It is unclear where regulations on autonomous vehicles and robots will end up, but finding that balance between innovation and safety is going to be important. Some would argue that Tesla has been allowed to stray too far from acceptable safety norms, but they are also driving the innovation.

1.4 Developing Safety Standards

An interesting inclusion in DO-178C (a key guidance document in the civil aerospace domain) is the statement that it was written by consensus of the community. This statement reflects the fact that many experts discussed their positions and existing evidence and came to a consensus after many arguments and compromises. This brings two issues into focus:

- First, participation in standardisation committees is voluntary. On the one hand, committees are driven by the top experts in the field, but on the other hand, these experts must have adequate resources (time and money) to attend. They might also not be incentivised to discuss details of the latest systems they are developing.
- Second, the document is produced by consensus, where compromises are not necessarily based on any evidence, e.g. trials, systematic reviews, meta-analysis, or theoretical evidence.

When it comes to robotics and autonomous systems, we cannot afford to continue with the same model for developing standards, since the stakes and uncertainty are so high. One cannot appeal to the experience of experts because the systems have not yet been developed. Therefore, judgement (sometimes ideological) will inevitably be made, which means there needs to be greater transparency in the process of developing standards. This can mean the development of domain-specific case studies that explain why an artefact is safe. Such safety cases should evolve with the domain, and this is what forms good practice. This also informs engineers who can use the case studies to explain the risks to the wider public, and the public can decide what trade-offs they are making when using a robotic or AI-based system. There are also security, safety, and privacy trade-offs that one has to make. For example, AI can be highly reliant on personal data to improve safety and security.

But there is a clash between safety and privacy, and this has to be made explicit to the end user.

Before rushing into generating new standards, new principles, and new techniques, one must first explain why the existing ones do not work and how any new standards build on existing work rather than starting from scratch.

2 Case Study: Certification in Autonomous Vehicles

All major car manufacturers are now, somewhat to their surprise, actually multibillion-dollar robotics startups. This creates a safety challenge: (1) OEMs (car manufacturers) have limited experience of verifying complex robotics systems; (2) they have a lot of experience of verifying complex mechanical systems, but this experience doesn't directly translate; and (3) the verification of autonomous vehicles is a hard problem.

From a technical perspective, autonomous vehicles are robots. They take inputs from percepts and make decisions based on pre-programmed algorithms. However, they are much bigger, operate at a larger scale, are faster, and operate alongside the general public. The three major challenges for developing autonomous vehicles are:

- Challenge 1: Operation in a complex, diverse, and changeable environment
- Challenge 2: Complexity of road rules and interaction with humans
- Challenge 3: Perception challenges, e.g. due to weather conditions or other road obstacles

Traditional automotive-industry safety processes are highly effective (see, e.g. Fig. 14.1). The processes are well established and very prescriptive. A well-used standard is ISO 26262, which defines a risk-based functional safety methodology that is designed to apply to all electronic and software systems on a vehicle. This includes driver assistant systems (e.g. lane-keeping assist) but also electronic stability control, braking systems, and even fuel injection systems. The standard defines processes to be followed at all stages of the V-cycle (see Fig. 14.2). For functional safety, ISO 26262 considers all possible failures and the likely severity of the consequences, which is used to assign an Automotive Safety Integrity Level (ASIL) to the failure. Higher ASILs require more robust processes for specification, development, verification, and validation. ISO 26262 also enables traceability of requirements, specification, and implementation, use of change control, and use of safe coding standards such as MISRA C.

Section 2.1 outlines some of the issues in regulating autonomous vehicles. Then Sects. 2.2 and 2.3 describe two simulation-based testing systems for autonomous vehicles developed by Connected Places Catapult.

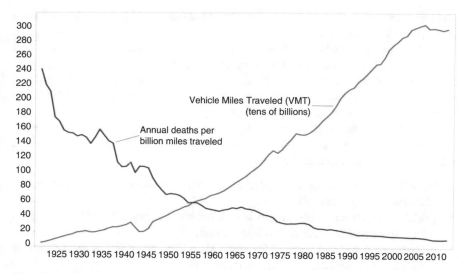

Fig. 14.1 US fatality rate per 100 million vehicle miles travelled. Copyright Dennis Bratland (2011) released under CC BY-SA 3.0 (https://creativecommons.org/licenses/by-sa/3.0/deed.en).

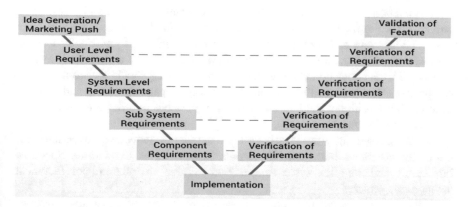

Fig. 14.2 V-Model for systems engineering. Copyright 2019 Connected Places Catapult, Creative Commons Attribution-Share Alike 3.0.

2.1   Regulating Autonomous Vehicles

ISO 26262 only considers failures of electrical and software systems. For autonomous systems, manufacturers aim to establish SOTIF (safety of the intended functionality) via standards such as ISO/PAS 21448. SOTIF fills in the gaps by focusing on complex systems that use sensors to build up situational awareness. Like ISO 26262, SOTIF (e.g. ISO/PAS 21448) is a hazard-focused, process-based standard. However, unlike ISO 26262, the standard enables one to distinguish faults such as "functional insufficiencies of the intended functionality", which is a failure in a system's specification.

Testing in the automotive industry is an exhaustive and manual process. It proceeds through simulation, hardware-in-the-loop, vehicle-in-the-loop, private track tests, and public road tests. Final testing takes place with multiple vehicles over multiple continents (to ensure coverage of all weather conditions). Test drivers typically work in shifts and can take several months.

These testing techniques do not easily map to autonomous vehicles. To see why, one can consider the processes for ensuring vehicles are driven safely on roads in the UK, which can be categorised as follows:

- *Vehicles are safe*, which is ensured by checks such as vehicle safety tests (i.e. MOT checks), vehicle recalls, etc.
- *Vehicles are driven safely*, which is ensured by ensuring all legal drivers pass driving test checked against the highway code.
- *Infrastructure and roads are safe*, which is covered by adhering to road design standards and management of the infrastructure.

The introduction of autonomous vehicles means that one can no longer ensure that the vehicles are driven safely, requiring a completely new type of testing to be introduced.

Certification of automated driving systems cannot be achieved by testing on public roads. From a recent study, "To demonstrate that fully autonomous vehicles have a fatality rate of 1.09 fatalities per 100 million miles [...] with a fleet of 100 autonomous vehicles being test-driven 24 h a day, 365 days a year at an average speed of 25 miles per hour, this would take about 12.5 years" [2]. From a testing perspective, the dynamic driving task has an input space too large and complex to test using traditional methods, and it is not possible to write a comprehensive specification for the task.

ISO 26262 and the V-cycle only apply to simpler systems. Moreover, the ASIL categories assume a human driver is present to mitigate any failure. Specification errors and complex system interaction failures are other major factors in autonomous vehicles.

There are several requirements for the effective regulation of such systems. From the perspective of the underlying architecture and competitive fairness, test processes must work with any ADS (Automated Driving System) architecture and must be seen as fair, i.e. should not favour any specific developer or technology. Test processes must also not constrain innovation. The testing framework must itself be comprehensive and rigorous, e.g. manufacturers should not be able to design to test and thus should incorporate some form of randomisation. At the same time, tests should be repeatable. Finally, tests must work internationally and fit within the existing regulatory regime.

The design of regulation should promote independent certification and testing. For example, in Europe, regulators strive to provide independent assurance of the safety of products, which implies certification tests should be conducted by an impartial organisation. Autonomous systems are most likely to be treated as black boxes, given independent testing, architecture neutrality, and (current) reluctance of manufacturers to provide access within their systems. However, this prevents testing of individual components, e.g. regulators might not be able to test the perceptions separately. Treating systems as black boxes also prevents application of verification and validation methodologies, such as model checking.

Overall, the concepts surrounding autonomous systems regulations are novel and different from existing regulations. Even concepts that apply to the regulation of software systems cannot be easily applied. The next two sections present simulation-based testing techniques developed by Connected Places Catapult for autonomous vehicles. These have been developed to be compatible with upcoming changes to regulations while simultaneously informing the design of the regulations themselves. The projects are motivated by two guidelines.

Observation 1: The use of simulation as a testing mechanism for automotive safety. Given the number of different scenarios that are possible, real-world testing cannot adequately provide coverage [1]. Simulation avoids this by (cheaply) running many tests in parallel and potentially running tests faster than real time. Moreover, this avoids danger to participants and enables test parameters to be precisely controlled. A difficulty here is that the simulation must take into account the whole environment. This is much more difficult than previous simulations used in the automotive domain. The modelling challenges include the physical environment, ideally in sub-millimetre detail; sensors, corresponding exactly to sensor models used on the autonomous vehicle; weather; and actions of other road users.

Observation 2: The use of scenarios to boost test coverage. Testing is often uninformative and unlikely to find failure cases. Instead, testing should be against *scenarios*, which define more edge cases than would be encountered in everyday driving. Scenarios are better aligned with how regulations are defined. Traditionally, regulations specify concise, explicit performance standards. Most stakeholders agree that scenarios represent the most effective way of specifying the test cases for certification. Given enough scenarios, at the right level of abstraction, almost all cases can be captured. The certification process would then be driven from a shared international database of scenarios.

2.2 MUSICC: Multi-User Scenario Catalogue for CAVs (Connected Autonomous Vehicles)

MUSICC (Multi-User Scenario Catalogue for CAVs) is a project funded by the UK's Department of Transport, and it focuses on regulatory testing of autonomous vehicles. It has implemented a language for describing scenarios (test cases) aligned with industry standards, and an open and expandable library for CAV certification scenarios represented in this language. The project involves collaboration with vehicle manufacturers, Automated Driving System developers, organisations with

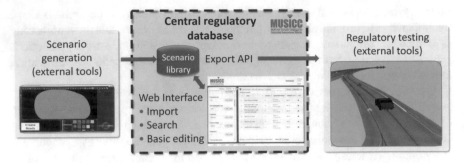

Fig. 14.3 Overview of MUSICC, including screenshots of VectorZero RoadRunner for scenario generation and es-mini for simulation testing.

expertise in CAV validation, and regulators. An overview of the framework of MUSICC is given in Fig. 14.3. Both the scenario language and the system software have been released as open source.[1]

MUSICC is intended for use in certification testing, which means it will have to store scenarios that cover AD systems from all manufacturers. However, the technical challenges of ADS development mean early systems will be restricted to an Operational Design Domain (ODD), which will be different for each manufacturer. An ODD may include factors such as weather, time of day, location type (e.g. cities only, motorways only), road-type restrictions, explicit geofences, traffic levels, allowable vehicle manoeuvres, and others. It is critical to test a specific ADS against all scenarios applicable for its ODD and only such scenarios. MUSICC supports this by tagging scenarios with metadata that indicate which ODDs they relate to and allowing ODD-aligned queries of the library.

The current release of MUSICC has relatively simple metadata, but the increasing maturity of AD systems will require a more sophisticated language to describe their ODD, backed by an ontology (see Fig. 14.4 for an example). Learnings from MUSICC [3] have contributed to ongoing international standardisation projects for ODD languages.

MUSICC also enables representation of the required performance standard. This spans beyond collisions (since collisions may be unavoidable in certain scenarios). Performance standards include considerations such as rule compliance, safety margins, confusing behaviour, and making progress. Within the scenario-specific language, scoring on a particular metric depends on inputs such as position within the lane, lane and road departures, speeds and accelerations, and minimum distances to other actors, providing a rich and powerful language for expressing many different types of scenarios [4].

The framework consists of a Python core together with a set of parameterised variables that can be used in pass/fail criteria, a standard way of reporting failures

---

[1] https://gitlab.com/connected-places-catapult/musicc.

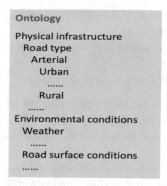

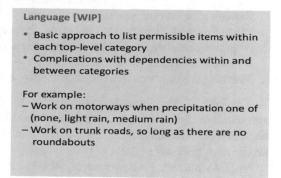

Ontology

Physical infrastructure
    Road type
        Arterial
            Urban
            ......
        Rural
        ......
Environmental conditions
    Weather
    ......
    Road surface conditions
    ......

Language [WIP]

- Basic approach to list permissible items within each top-level category
- Complications with dependencies within and between categories

For example:
– Work on motorways when precipitation one of (none, light rain, medium rain)
– Work on trunk roads, so long as there are no roundabouts

Fig. 14.4 Example of ODD.

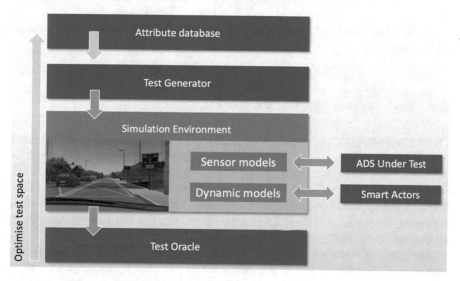

Fig. 14.5 Overview of the VeriCAV framework.

or scores, and a library of common functions (e.g. assert-vehicle-did-not-collide). This will be developed further in the follow-on project, CertiCAV.

2.3 VeriCAV

VeriCAV is a 2-year collaborative research project led by HORIBA MIRA and involving Connected Places Catapult, University of Leeds, and Aimsun. The project is developing a test framework to allow safe and efficient testing of automated vehicles using simulation (see Fig. 14.5 for an overview). Specifically, three challenging areas associated with testing in simulation are addressed:

1. Reducing the level of human effort in setting up tests and then analysing results—currently, this is considerable and slows the pace of testing.
2. Improving the behaviour of the other actors in the simulation—currently, this limits the realism of the test scenario.
3. Maturing the interfaces and dataflow between ADS and test framework tools.

The core of the framework is enabled by a test oracle, which outputs an automated analysis of the performance of the ADS under test. Test results are fed back into the randomisation engine, which uses these results to focus the next tests on the most informative areas of the test space.

VeriCAV also aims to develop smart actors, i.e. algorithms to produce human-like behaviour in the other drivers, pedestrians, and cyclists in the scene. Incorporating this behaviour is vital to replicating real-world conditions in simulation tests and making the test cases more challenging and realistic. These actor models will be developed from cognitive and AI models of human decision-making.

VeriCAV is supported by the Centre for Connected and Autonomous Vehicles and Innovate UK.[2]

2.4 Conclusions and Future Work

Overall, the MUSICC and VeriCAV projects have enabled regulatory dialogue with UNECE WP.29,[3] in particular the GRVA working party, which is focused on autonomous vehicles. A key sub-group within this party is the Verification Methods for Automated Driving (VMAD), which addresses closed-road tests, real-world test drives, auditing, and simulation. Internationally, there is ongoing regulatory work including the USA (NHTSA, SAE, and UL 4600), the EU Commission (Joint Research Centre (JRC)), and Singapore (CETRAN programme).

The projects have also identified several key challenges that demand further exploration. From a technical perspective, there is still a challenge in finding "good" scenarios, ensuring test space coverage, fault injection, etc. Fidelity of autonomous simulation environments (sensor models, maps, and 3D world models) also requires further improvement. Work is also needed in areas such as explainable AI and AI verification, modelling, and formal methods. In terms of certification, there are questions such as identification of the acceptable level of safety-relevant performance and how the scenario-based test processes (simulation versus physical) interface to the ADS. Simulation of scenarios such as the safety of semi-autonomous vehicles and their handover between humans and AI is also challenging. How can one assure safety by following systems engineering processes, and how can one verify this? Other day-to-day considerations include non-functional requirements

---

[2]More information is available here: https://vericav-project.co.uk/.

[3]https://www.unece.org/trans/main/wp29/introduction.html.

such as component redundancy, product lifecycle, verification of software installed as an over-the-air update, MOT-type testing, handling damage and dirt, etc. Continuous in-service performance monitoring, data recording, accident investigation, and sharing of new safety-relevant scenarios are all likely to require advances. Collaboration is critical to making progress on such a wide range of issues. This includes collaboration between regulators, industry, and academia, who must come together to build an international ecosystem of projects and initiatives to address these aspects.

In the near term, a workable certification methodology requires growing the amount of activity in verification and certification aspects. Regulators will rely on advice and research from the whole community—industry, consultancies, and academia. Longer-term research must work towards advancements in coverage with existing test methods, improved speed and fidelity in simulation tools, and improved search optimisation. Application of formal verification methods to real-world systems and application in verified AI systems will also be increasingly important. To speed up development, there needs to be a cultural shift towards openness and collaboration on the part of the manufacturers, who will ultimately benefit from any advances in verification and validation methodologies.

3   Q&A Session

What Is the Difference Between Safety of Autonomous Systems in General and for Autonomous Robotics in Particular?

Ron Bell One issue is the level of complexity, which from a standards perspective can cause difficulty in gaining consensus on any particular issue. It is not unknown for a group to spend 18 months on the wording of a particular clause within a standards document, so the added complexity of autonomous robotics means that it is much harder to reach consensus and hence much harder to produce standards.

To give an example, there was a debate on whether cybersecurity should be included as a normative requirement in the functional safety standard—not on whether cybersecurity should be included or not—but where the best place was to do it. In the end, cybersecurity was taken out of the task group, split into two maintenance teams who put in three proposals with requests for comment. Not only that, standards like ISO 26262 and IEC 61508 also go back to all the national committees, who provide additional comments. In one case, there were over 100 comments; for each of these, one might have to go back to the points that have not been agreed, working back to the cause of the disagreement.

Pippa Moore To distinguish specifically between general automation and robotics, one has to bear in mind the area of public perception. So when you're working as a scientist, engineer, or regulator, you could have a very purist, evidence-based view of what safety is, and you may have data to demonstrate that something is acceptably safe or is subject to acceptable controls. But when something goes into service, you have to accept that public opinion will be based on what the public perceives, which is usually heavily influenced by what they see in the media or have heard from friends.

So when you start bringing in driverless trains, for instance, the public may form very strong opinions based on what they see in the media or believe to be true. The media and human interactions being what they are, issues that are perceived to be contentious will be promulgated, whether they are completely accurate or not. This means that engineering and regulatory decisions can be directly affected by public perception. This can sometimes be for the better and makes us do a better job, and sometimes it actually stops us doing the things that we need to do, because it may be a step too far and a step too complicated for people to easily grasp or accept.

Are There Different Challenges in Different Sectors?

Mark Lawford It comes back to the risk/benefit and the market. In the USA, when they do pre-market assessments of medical devices, it often is for people who are terminal, with the device being their last hope. So, if there is a chance to help patients and save their lives, then the risks can be worth it. But in aircraft, public perception is that there is too great a risk.

Autonomous driving is a very funny one. I think what's going to happen is that the evidence is going to come back that the autonomous driving systems will soon, and under most conditions, be better than humans. But even in those cases, when people are killed—if you lose your child to an autonomous vehicle, then what's going to happen?

Ibrahim Habli I think there is something we are all uncomfortable about—autonomy versus automation. It's important to analyse such systems with lawyers and regulators and think about the *semantic gap*. Consider a deceleration system for an aircraft or an implantable medical device. They do something very, very specific, and we understand the rules. Compare this with an application for medical diagnosis, where you give a scan to a machine, it gives you the answer. Instead of getting an answer from a professional, we're delegating this responsibility to a machine without necessarily having sufficient control. We become uncomfortable with the knowledge and that's what creates a *responsibility gap*. So these two issues, the *epistemic knowledge* about what's explainable and *control*, have to be explicit in our requirements and explicit in the safety cases.

Ron Bell In principle, there's no difference between now and what happened in the 1980s—many different methods were being used for process control. In time, this led to simple hardwired systems with little quantification of the performance measure and a standard which says, "we can't modify the software". Although potentially sub-optimal, that facilitates the application of the technology because it means that manufacturers can use a standard with some degree of guarantee almost that they will not be prosecuted if something happens within that standard.

There are also differences between the American and European models for adopting standards. The European model is to do what is good practice at this point. The American model will aim for a particular level in practice but also want to make sure that when the standard comes out, they're already beating it. As an example, I did some work 10–12 years ago for an American company, one of the biggest manufacturers of implantable medical devices, that wanted to use ISO 26262. To my surprise, they would not discuss the devices with the Food and Drug Administration (FDA). Compare this with the UK, where Zeyn is in discussion with the Department of Transport (even at this early stage). I would expect anybody who's developing robots to do the same. We are still in contact with the Health and Safety Executive (HSE), and they are represented in our standards group.

Is There a Belief that Regulation Is a Barrier to Innovation?

Pippa Moore It does not have to be. But, developers must speak to regulators early on. This is true for innovation and anything that requires a certificate or relevant training. Regulators can add value to the process since they can see organisations across the whole industry, from small to large, and all different types of use. Regulators are aware that from a practical point of view, improvements require resources, and safety cannot be improved unless companies innovate, and hence, they work hard to avoid limiting innovation.

What can happen frequently is that a product may be in development for several years before the developer approaches a regulator. In some cases, this forces the regulator to impose changes on the product because the law itself requires that the problems identified must be addressed. So regulation does not stop the innovation, but regulation can only help innovation if developers open up a dialogue with regulators, and early on if possible.

Ron Bell In the UK, the guidance used is: "as far as is reasonably practical", and that wording has been around since the 1900s. For example, certain regulations that were made in the early 1900s have not changed, even though technology has moved forward, since "what is reasonable in practice" has moved with time. The proportionality in regulation in the UK for general industrial systems, including robots, is built into the law, not in inspection. This is unique to the UK across Europe. Contrast this with the rest of Europe, the proportionality is absolute in the regulations and proportional at the point of inspection.

Concepts such as Safety Integrity Levels (SILs) provide massive transparency, but sometimes this leads to problems. For instance, before the publication of IEC 61508, the medical sector worldwide voted to *not* use IEC 61508 saying that it was the "wrong standard". What a manufacturer would prefer is a system that complies with the standard but with a law that defines the level of integrity or other performance measures.

Mark Lawford In terms of the US, the view that regulations stifle innovations is a view that comes from lobbyists and Congress. We had attempted to form a software certification consortium for medical devices with manufacturers and regulators, but we had one medical device manufacturer trying to shut down the consortium altogether because they did not want the restrictions imposed by the FDA regulators. So, there are bad actors that sometimes take the view that regulation is the problem, so the bar should be set as low as possible. Although this is speaking in broad brushstrokes, there is less inclination to regulate in North America. Part of it is motivated by the desire to innovate, but it is also because of the way Congress works with paid lobbyists.

What About Societal Acceptance of Autonomous Systems?

Pippa Moore Within human factors, we draw a distinction between actual knowledge and perceived knowledge. One of the things we fight against, when setting new standards and driving new technology, is differentiating between different levels of expertise. A lot of that comes down to the fact that there are special interest groups, e.g. politicians, unions, and the media, that can (negatively) impact societal expectations, creating difficulty.

Ron Bell Some of the standards interfaced with the public directly in the same way that standards interface with manufacturers and other organisations. Adhering to such standards is difficult when dealing with products that are produced by the million or for a product that has to last a long time, e.g. a dryer that has to last 15 years.

Ibrahim Habli Engaging with the public in safety cases can be beneficial and help raise awareness. Having taught safety and security to a range of practitioners from engineers to 15–16-year-old children has led to many surprises. There is always a range of different perspectives of what safety and security are, which shows that there is a lot to be learnt from the wider public, regardless of their background.

Mark Lawford Considering autonomous driving in the USA, the fatalities have been tech-savvy engineers who were early adopters. For example, an Apple engineer died on autopilot in San Francisco. If tech-savvy people do not understand the underlying systems, then what is the general public going to do? How do we make autonomous systems understandable to the general public so that they are informed well enough to be able to provide input?

In the 1980s, there was an enquiry into a train crash resulting in a fatality, where the signals were controlled by a computerised system. Although the system was designed with an incredibly low probability of failure, when asked if the system could fail the next day, the designers were forced to answer "Yes" (even though the probability was a very small number). This is the difficulty in explaining safety and integrity to the public—they might not always understand the numbers.

References

1. A. Dosovitskiy, G. Ros, F. Codevilla, A. Lopez, V. Koltun, CARLA: an open urban driving simulator, in *Proceedings of the 1st Annual Conference on Robot Learning* (2017), pp 1–16
2. N. Kalra, S.M. Paddock, Driving to safety: how many miles of driving would it take to demonstrate autonomous vehicle reliability? Technical Report, RAND Corporation (2016). https://www.rand.org/pubs/research_reports/RR1478.html
3. R. Myers, Z. Saigol, Design considerations for ODD ontology. Technical Report, Connected Places Catapult (2020). https://cp.catapult.org.uk/case-studies/musicc/#Publications
4. R. Myers, Z. Saigol, Pass-fail criteria for scenario-based testing of automated driving systems. Technical Report, Connected Places Catapult. (2020). arXiv:2005.09417. https://cp.catapult.org.uk/case-studies/musicc/#Publications

Printed in the United States
by Baker & Taylor Publisher Services